Quantum Computing

Fundamental Principles of Quantum Computing Systems

Richard Skiba

AFTER MIDNIGHT PUBLISHING

Skiba, Richard (author)

Quantum Computing: Fundamental Principles of Quantum Computing Systems

ISBN 978-1-7638811-7-4 (Paperback) 978-1-7638811-8-1 (eBook) 978-1-7638811-9-8 (Hardcover)

Non-fiction

Contents

Introduction

Overview of Quantum Computing

Imagine you have a regular computer—like your laptop or smartphone. These devices process information using bits, which can be either 0 or 1 (think of a light switch that is either on or off). Everything a computer does—whether it's playing a video, solving math problems, or running an app—is based on these binary bits. Traditional computers follow a step-by-step process to compute data, meaning they solve problems one calculation at a time.

Now, imagine if a computer didn't have to choose between 0 and 1 but could be both at the same time—this is what a quantum computer does. It uses quantum bits (qubits), which can exist in multiple states at once due to a strange rule in physics called superposition. This unique property allows quantum computers to process vast amounts of information simultaneously, making them incredibly powerful for certain tasks.

One of the reasons quantum computing is so powerful is because of three key quantum principles: superposition, entanglement, and quantum interference. Superposition allows qubits to represent multiple possibilities at once, so a quantum computer can explore many solutions in parallel rather than trying one at a time like a classical computer. Entanglement is another fascinating property, where two or more qubits become linked, meaning that changing the state of one qubit instantly affects the state of another—no matter how far apart they are. This connection enables faster and more complex calculations that classical computers cannot achieve. Quantum interference further enhances this process by amplifying the probability of correct answers while reducing the likelihood of incorrect ones, ensuring more accurate results.

Quantum computers are still in development, but they could revolutionize several fields. They have the potential to break modern encryption, which is currently used to secure online banking, messaging, and sensitive data. They could also accelerate drug discovery by simulating molecules at the atomic level, helping scientists develop new medicines much faster than conventional methods. In logistics and transportation, quantum computers could optimize airline scheduling, supply chains, and traffic management, leading to more efficient and cost-effective solutions. Additionally, they could help solve complex problems in physics, chemistry, and artificial intelligence, tackling challenges that even the most powerful supercomputers would take thousands of years to compute.

Despite their enormous potential, quantum computers are not meant to replace regular computers. Instead, they will be used for specialized tasks where quantum mechanics provides a significant advantage. Your laptop will still be perfect for watching YouTube or sending emails, but for highly complex calculations, quantum computers could be game-changers. They will work alongside classical computers, handling problems that require immense computational power while leaving everyday tasks to conventional machines.

However, there are still significant challenges in making quantum computers practical. Right now, they are highly experimental, requiring extremely cold temperatures near absolute zero to function. They also suffer from instability and errors, making calculations difficult to sustain over time. Additionally, building quantum hardware is incredibly expensive, and large-scale, error-free quantum computing is still years—if not decades—away. Scientists and engineers are actively working to overcome these issues by developing more stable, scalable, and efficient quantum systems.

Think of quantum computers as super-powered calculators that can solve problems much faster than any regular computer—but only for certain types of problems. While they are not yet ready for mainstream use, the possibilities they offer are truly groundbreaking. If researchers can solve the current technical challenges, quantum computing could change the world, transforming industries and solving problems that are currently beyond human capability. For now, however, they remain a fascinating and promising area of research, with much more development needed before they become part of our everyday lives.

Quantum computing represents a significant advancement in computational technology, merging principles from computer science, physics, and mathematics to harness the peculiarities of quantum mechanics for solving complex computational problems. Unlike classical computers that utilize binary bits (0s and 1s) for data processing, quantum computers operate using quantum bits or qubits. These qubits can exist in multiple states simultaneously due to the principles of superposition and quantum entanglement, which fundamentally alter the landscape of computation [1, 2].

The multidisciplinary nature of quantum computing is essential for its development and application. As highlighted by Juárez-Ramírez et al. [1], the field necessitates collaboration across various disciplines, including physics, mathematics, and computer science, to address the challenges posed by quantum computation. This interdisciplinary approach is echoed in the work of Schuld et al. [3], who emphasize the potential of quantum machine learning as an emerging area that combines these fields to optimize learning processes through quantum algorithms. Furthermore, Swan et al [4]. discuss how quantum information science can provide solutions that extend beyond traditional computing paradigms, particularly in the context of energy efficiency and problem modelling.

The computational advantages of quantum computers are particularly evident in specific problem domains. For instance, Shor's algorithm demonstrates that quantum computers can factor large integers in polynomial time, a task that is believed to be infeasible for classical computers [5]. This capability has profound implications for fields such as cryptography, where the security of many systems relies on the difficulty of factoring large numbers. Additionally, quantum computing has shown promise in areas such as artificial intelligence and complex system simulations, where the ability to process vast amounts of data simultaneously can lead to breakthroughs in efficiency and effectiveness [3, 6].

Moreover, the principles of quantum mechanics, such as entanglement, allow for new forms of data processing and storage that are not possible with classical systems. For example, the work of Wang [7] emphasizes the intersection of quantum computation and data science, suggesting that the stochastic nature of quantum mechanics can enhance data processing capabilities. This interplay between quantum mechanics and computational theory not only broadens the scope of potential applications but also fosters innovation in algorithm development and problem-solving techniques [6].

Quantum computers are a completely different type of machine from the computers we use every day. Instead of using regular bits (which can be either 0 or 1), they use qubits—special quantum bits that can be both 0 and 1 at the same time. This ability allows quantum computers to process information much more efficiently for certain types of problems.

However, qubits are extremely fragile and need special conditions to work properly. To keep them stable and free from outside disturbances, scientists have developed different methods to create and maintain qubits. Some of the main approaches include:

- Superconducting Qubits – These qubits, used by IBM, Google, and Rigetti, are made using electrical circuits that have no resistance when cooled to extremely low temperatures (almost absolute zero). This helps them maintain their quantum state for longer periods.
- Trapped Ion Qubits – Used by IonQ and Honeywell, this method traps charged atoms (ions) inside an electromagnetic field, keeping them suspended in a controlled environment. Lasers are then used to manipulate them.
- Photonic Qubits – These qubits, developed by Xanadu, use light particles (photons) to carry quantum information through optical circuits, which could lead to faster and more energy-efficient quantum computers.

- Topological Qubits – Being researched by Microsoft, these qubits rely on exotic quantum states that help reduce errors in calculations, potentially making quantum computers more stable and scalable.

To actually use qubits to perform calculations, quantum computers rely on quantum gates. These gates are similar to the logic gates in normal computers, but they work differently because they manipulate qubits using quantum mechanics. Some of the most common quantum gates include:

- Hadamard Gate (H) – This gate puts a qubit into superposition, meaning it can be both 0 and 1 at the same time, allowing quantum computers to explore multiple possibilities simultaneously.
- Pauli Gates (X, Y, Z) – These gates flip or rotate a qubit's state, similar to how a light switch can be turned on or off, but in more complex ways.
- Controlled-NOT Gate (CNOT) – This gate connects two qubits so that changing one automatically affects the other, taking advantage of quantum entanglement for more powerful computations.

By combining these quantum gates, scientists can create quantum circuits that execute complex calculations. These circuits act like the "software" of a quantum computer, guiding how the qubits interact and process information.

Why Quantum Computing Matters

Quantum computing is a game-changer because it can solve problems that are too complex for even the most powerful supercomputers today. While traditional computers are great at handling everyday tasks like browsing the internet, running apps, and playing games, they struggle with highly advanced calculations that involve enormous amounts of data and intricate relationships between variables. Quantum computers have the potential to revolutionize industries by making these impossible problems solvable.

One of the biggest reasons quantum computing matters is its potential to break modern encryption. Most of today's online security, including banking transactions, secure messaging, and passwords, relies on encryption methods that take classical computers thousands or even millions of years to crack. A sufficiently advanced quantum computer could solve these encryption puzzles in minutes, forcing the world to rethink cybersecurity and develop new, quantum-resistant encryption techniques.

Another area where quantum computing could make a huge impact is drug discovery and medical research. Simulating molecules and chemical reactions is incredibly difficult for classical computers because atoms behave according to quantum mechanics. Quantum computers, which follow the same rules as nature, can accurately model these interactions, leading to faster drug development, better treatments, and even the discovery of new materials for medical applications.

Quantum computing also has the power to optimize complex systems in ways that classical computers cannot. Take logistics and transportation as an example. A delivery company like FedEx or Amazon needs to find the fastest and most efficient routes for thousands of packages across the world, considering factors like traffic, weather, fuel costs, and shipping delays. A quantum computer could analyse all these factors simultaneously, finding the best solution in seconds, saving billions of dollars and reducing environmental impact.

In the field of artificial intelligence (AI) and machine learning, quantum computing could greatly enhance the way machines process and understand information. AI models require massive amounts of data to learn and make predictions, but training them can take weeks or months using classical computers. Quantum computers could speed up this process dramatically, leading to more advanced AI systems that improve everything from healthcare diagnostics to financial market predictions.

Even in climate science, quantum computing could play a major role in understanding and addressing global challenges. Simulating climate patterns and predicting extreme weather events is one of the most difficult tasks in computing because there are so many interacting variables. Quantum computers could process these calculations far more accurately and quickly, helping scientists develop better climate models and solutions for tackling global warming.

Quantum computing isn't about replacing regular computers but expanding the possibilities of what we can compute. While still in its early stages, quantum technology is expected to revolutionize industries like cybersecurity, healthcare, finance, logistics, AI, and environmental science. By solving problems that were once thought impossible, quantum computing has the potential to change the world in ways we are only beginning to understand.

Quantum computing is currently in a transitional phase characterized by significant advancements and ongoing challenges. The field is moving from theoretical research to practical applications, with notable developments in quantum hardware, achievements in quantum supremacy, and the emergence of cloud-based quantum computing services.

The most advanced quantum computers today operate with hundreds of qubits, yet they continue to face issues related to error rates and instability. For instance, IBM and Google have developed quantum processors that exceed 100 qubits, while companies like IonQ and Rigetti are also contributing to the progress in this area [8]. However, achieving fully functional quantum computing will require millions of qubits, as current systems are hindered by quantum decoherence, which results from environmental noise causing qubits to lose their quantum state [9]. This challenge is compounded by the limitations of existing hardware, which often necessitates extreme conditions, such as cryogenic temperatures, to maintain qubit stability [10].

In 2019, Google announced it had achieved quantum supremacy, demonstrating that its quantum computer could solve a specific problem in seconds—a task that would take classical supercomputers thousands of years [11]. However, this milestone does not imply that quantum computers are ready to replace classical systems; rather, it highlights the need for further advancements to reach quantum advantage, where quantum systems can consistently outperform classical computers in practical applications [12]. The distinction between quantum supremacy and quantum advantage is crucial, as the former pertains to isolated tasks while the latter focuses on real-world utility [13].

The accessibility of quantum computing has been greatly enhanced through cloud services offered by major tech companies such as IBM, Google, Microsoft, and Amazon. These platforms allow researchers and businesses to experiment with quantum algorithms without the need for extensive hardware investments [14]. This cloud-based approach is facilitating the development of new applications in various fields, including cryptography, artificial intelligence, and materials science, by enabling users to leverage quantum computing capabilities remotely [15].

Although quantum computers are not yet ready for widespread commercial use, several early-stage applications are being explored. These include quantum-resistant encryption methods in cryptography, drug discovery through molecular simulations, optimization problems in logistics and finance, and enhancements in artificial intelligence [16]. Each of these areas demonstrates the potential of quantum computing to revolutionize traditional processes, although practical implementations remain in the developmental stage [17].

Despite the progress, quantum computing faces several significant challenges that must be addressed before it can achieve mainstream adoption. Key obstacles include the need for improved quantum error correction methods, the scalability of current systems, hardware limitations that require extreme operational conditions, and the high costs associated with developing and maintaining quantum technology [18]. These challenges highlight the complexity of transitioning from theoretical models to practical, reliable quantum computing solutions.

Looking ahead, experts predict that practical quantum advantage may be realized within the next 5 to 10 years, with fault-tolerant quantum computers potentially emerging in the next two to three decades [19]. This evolution is expected to transform various industries by enabling the resolution of complex problems that are currently intractable for classical computers. The rapid pace of advancements in quantum technology suggests that significant breakthroughs could occur in the near future, bringing the full potential of quantum computing closer to reality [20].

Classical vs. Quantum Computing

Computers have revolutionized the way we live, work, and solve problems. From smartphones and supercomputers to artificial intelligence and cloud computing, classical computers power much of the modern world. However, there are some problems that classical computers struggle with—problems that could take even the fastest supercomputers thousands or millions of years to solve. This is where quantum computing comes in. Unlike classical computers, which rely on binary bits (0s and 1s) to process information, quantum computers use quantum bits (qubits), which follow the rules of quantum mechanics to solve complex problems much faster than classical systems.

Classical computers process information using bits, which are the smallest unit of data. A bit can be either 0 or 1, similar to a light switch that is either on or off. Every task a classical computer performs—whether it's opening an app, playing a video, or solving a math problem—is done through combinations of millions or billions of bits. These bits are processed using logic gates, which perform operations like AND, OR, and NOT to manipulate data.

Classical computers work in a sequential manner, meaning they process one calculation at a time or use parallel processing (such as in modern multi-core processors) to speed things up. However, for extremely complex problems—such as simulating molecules, optimizing massive supply chains, or breaking encryption—the sheer number of possible combinations grows exponentially, making even the most powerful classical computers inefficient.

Quantum computers, on the other hand, operate based on qubits, which are fundamentally different from classical bits. A qubit is not just 0 or 1; it can exist in both states simultaneously due to a phenomenon called superposition. Imagine flipping a coin—it can be either heads or tails in a classical system, but in a quantum system, it's like the coin is spinning in the air, existing as both heads and tails at the same time until measured.

Another important quantum phenomenon is entanglement, where two qubits become linked regardless of distance. If one entangled qubit is measured, the state of the other qubit is instantly determined, no matter how far apart they are. This interconnectedness allows quantum computers to perform calculations in parallel, solving multiple possibilities at once, unlike classical computers that process one solution at a time.

Quantum computers also leverage quantum interference, which helps them amplify correct solutions and suppress incorrect ones. This means they can arrive at the right answer more efficiently than classical systems.

Key Differences Between Classical and Quantum Computers:

1. Information Processing: Bits vs. Qubits

Classical computers use bits, which can only be 0 or 1 at any given time. They process information in a step-by-step manner, which can be slow for certain types of problems.

Quantum computers use qubits, which can be in a superposition of both 0 and 1 simultaneously, allowing them to process multiple possibilities at once. This makes them exponentially faster for certain types of calculations.

2. Computational Power

A classical computer with N bits can represent only one state at a time out of 2^N possible states. For example, a 10-bit classical computer can store only one of 1,024 possible states at a time.

A quantum computer with N qubits, however, can represent all 2^N states simultaneously due to superposition. A 10-qubit quantum computer can exist in 1,024 states at once, and as the number of qubits increases, the computational power grows exponentially.

3. Speed and Efficiency

Classical computers are efficient for everyday tasks like browsing the web, running applications, and handling structured data. They follow deterministic algorithms, meaning they execute commands in a predictable way. However, they struggle with problems involving massive calculations or combinatorial complexity, such as optimizing flight routes for thousands of airplanes or cracking complex encryption algorithms.

Quantum computers, on the other hand, can explore multiple solutions at the same time, making them incredibly fast for problems that require probability-based solutions, optimization, and simulations.

4. Use of Logic Gates

Classical computers rely on Boolean logic gates (AND, OR, NOT, XOR, etc.) to process data through circuits.

Quantum computers use quantum gates, such as the Hadamard gate, Pauli gates, and CNOT gate, which manipulate qubits in complex ways that classical logic gates cannot.

5. Error Handling and Stability

Classical computers are relatively stable, with well-developed error correction techniques. Even if there are errors, they can often be easily detected and corrected.

Quantum computers, however, are extremely sensitive to their environment. Decoherence—the loss of quantum state due to external interference—can cause errors in calculations. Scientists are still developing quantum error correction techniques to make quantum computers more reliable.

In the realm of cryptography and cybersecurity, quantum computers have the potential to disrupt traditional encryption methods, such as RSA, which classical computers find challenging to break due to their computational limitations. Quantum algorithms, particularly Shor's algorithm, can factor large integers exponentially faster than the best-known classical algorithms, posing a significant threat to current cryptographic practices [21]. This necessitates the development of quantum-resistant security systems to safeguard sensitive information against potential quantum attacks [2].

Drug discovery and healthcare also stand to benefit immensely from quantum computing. Classical computers struggle with simulating molecular interactions due to the complexity of quantum states involved in chemical reactions. Quantum computers can model these interactions more accurately and efficiently, potentially leading to faster drug development and personalized medicine solutions [22]. For instance, quantum annealers have been employed to solve optimization problems related to molecular configurations, showcasing their ability to handle complex chemical simulations that are infeasible for classical systems [23].

In the field of artificial intelligence and machine learning, quantum computing offers the promise of accelerating the training of AI models. Quantum algorithms can exploit quantum superposition and entanglement to process vast amounts of data simultaneously, enhancing pattern recognition and decision-making capabilities [24]. This could lead to significant improvements in the efficiency of machine learning tasks, allowing for more sophisticated AI applications that classical systems cannot achieve within a reasonable timeframe [25].

Logistics and optimization are other areas where quantum computing can provide substantial advantages. Quantum algorithms can optimize complex systems such as supply chains and air traffic control by evaluating numerous potential solutions simultaneously, a task that classical computers find time-consuming and often intractable [23, 26]. The ability to solve combinatorial optimization problems more efficiently could revolutionize industries reliant on logistics and operational efficiency [27].

Lastly, in climate science and material design, quantum simulations can aid in the development of new materials and technologies aimed at combating climate change. For example, quantum computing can facilitate the design of better batteries and superconductors, as well as enhance carbon capture technologies [22]. The ability to simulate quantum systems accurately allows researchers to explore new materials that could lead to breakthroughs in energy storage and environmental sustainability [28].

Quantum computers are not replacing classical computers; rather, they are expected to coexist with classical systems, each serving distinct roles based on their unique capabilities. Classical computers remain essential for everyday tasks such as web browsing, spreadsheets, and gaming, while quantum computers are poised to tackle specialized problems where quantum mechanics offers a significant advantage. This division of labour is crucial because classical computers excel in general-purpose computing, utilizing binary bits and processing tasks sequentially, whereas quantum computers leverage qubits, superposition, and entanglement to potentially solve complex problems much faster in specific domains like cryptography and optimization [21, 29, 30].

The fundamental differences between classical and quantum computing architectures underscore their complementary nature. Classical computers rely on bits that can exist in one of two states (0 or 1), making them well-suited for traditional computing tasks. In contrast, quantum computers utilize qubits, which can represent multiple states simultaneously due to superposition. This property allows quantum computers to perform numerous calculations in parallel, significantly enhancing their computational power for particular applications [21, 28]. However, the current state of quantum computing is characterized by the Noisy Intermediate-Scale Quantum (NISQ) era, where quantum systems are still developing and face challenges such as error correction and hardware scalability [31, 32].

As researchers continue to advance quantum technologies, the potential applications of quantum computing are expanding. For instance, hybrid quantum-classical algorithms are being explored to optimize the performance of quantum processors while still relying on classical systems for pre-processing and post-processing tasks [33, 34]. This hybrid approach is essential as it allows for the integration of quantum computing's strengths with the robustness of classical computing, ensuring that both types of systems can be utilized effectively in various scientific and industrial applications [27, 35].

While quantum computers hold the promise of revolutionizing specific fields by outperforming classical systems in certain tasks, they are not set to replace classical computers. Instead, they will enhance computational capabilities by working alongside classical systems, each fulfilling roles that leverage their respective strengths [31, 33, 34, 36].

A Brief History of Quantum Theory

Quantum theory, also known as quantum mechanics, has undergone significant evolution since its inception in the early 20th century. The initial breakthroughs in quantum theory emerged in the late 1800s and early 1900s, primarily driven by the need to explain phenomena that classical physics could not account for. Max Planck's introduction of the concept of quantization of energy in 1900, aimed at resolving the blackbody radiation problem, marked the formal birth of quantum theory [37]. Albert Einstein further advanced this field in 1905 by explaining the photoelectric effect, proposing that light consists of discrete packets of energy called photons, thus establishing the dual wave-particle nature of light [37].

The 1920s and 1930s witnessed the formal development of quantum mechanics as a comprehensive theory. Key contributions came from physicists such as Louis de Broglie, who introduced the wave-particle duality concept, and Werner Heisenberg, who formulated matrix mechanics. Erwin Schrödinger's wave mechanics, encapsulated in the Schrödinger equation, provided a mathematical framework that was later shown to be equivalent to Heisenberg's approach [37]. The introduction of the Uncertainty Principle by Heisenberg in 1927 fundamentally challenged classical determinism by asserting that certain pairs of physical properties cannot be simultaneously known with arbitrary precision [37]. Niels Bohr's Copenhagen interpretation further emphasized the probabilistic nature of quantum systems, suggesting that particles exist in multiple states until measured [37].

The application of quantum mechanics expanded significantly during and after World War II, particularly in the context of nuclear physics. The Manhattan Project, which led to the development of atomic bombs, relied heavily on quantum principles to understand nuclear reactions [37]. Post-war, quantum mechanics laid the groundwork for the development of modern electronics, including the invention of transistors, which became pivotal in the evolution of computers and digital technology [37].

The 1950s to 1970s marked a period of refinement and expansion of quantum theory into areas such as Quantum Electrodynamics (QED) and Quantum Chromodynamics (QCD). QED, developed by Richard Feynman, Julian Schwinger, and Sin-Itiro Tomonaga, describes how light interacts with charged particles and earned them the Nobel Prize in Physics in 1965 [37]. The emergence of the Standard Model of particle physics during this time further solidified quantum mechanics as the foundational framework for understanding fundamental forces and particles [37].

The convergence of quantum mechanics and computer science began in the 1980s, spurred by Richard Feynman's proposal that quantum systems could be utilized to simulate complex physical processes more efficiently than classical computers [37]. This led to the development of quantum algorithms, such as Shor's algorithm for factoring large numbers and Grover's algorithm for searching unsorted databases, which demonstrated the potential for quantum computing to outperform classical methods in specific tasks [37]. The experimental realization of quantum computers began with small-scale systems using trapped ions and superconductors, culminating in Google's announcement of achieving quantum supremacy in 2019 with a 54-qubit machine [37].

Timeline of Quantum Theory and Quantum Computing

Late 1800s – Early 1900s: Birth of Quantum Ideas

- **1900** – Max Planck introduces the idea of quanta to explain blackbody radiation, marking the birth of quantum theory.
- **1905** – Albert Einstein explains the photoelectric effect, proving that light behaves as both a wave and a particle (introducing the concept of photons).

- **1921** – Einstein is awarded the Nobel Prize in Physics for his work on the photoelectric effect.

1920s – 1930s: The Foundation of Quantum Mechanics

- **1924** – Louis de Broglie proposes that particles (like electrons) also exhibit wave-like behaviour (wave-particle duality).
- **1925** – Werner Heisenberg develops matrix mechanics, one of the first mathematical models of quantum mechanics.
- **1926** – Erwin Schrödinger formulates wave mechanics and introduces the famous Schrödinger equation.
- **1927** – Heisenberg's Uncertainty Principle states that the position and momentum of a particle cannot be simultaneously measured with precision.
- **1927** – Niels Bohr introduces the Copenhagen Interpretation, suggesting that quantum systems exist in multiple states until measured.

1930s – 1950s: Quantum Mechanics and the Atomic Age

- **1930s** – Quantum mechanics becomes fundamental to atomic physics, explaining the structure of atoms.
- **1945** – The Manhattan Project develops the first nuclear weapons, applying quantum physics to nuclear fission.
- **1947** – Transistors are invented, leading to the rise of modern computing and digital electronics based on quantum principles.

1950s – 1970s: Quantum Field Theory and Advanced Quantum Mechanics

- **1950s** – Development of Quantum Electrodynamics (QED) to describe interactions between light (photons) and charged particles (electrons).
- **1965** – Richard Feynman, Julian Schwinger, and Sin-Itiro Tomonaga receive the Nobel Prize for contributions to QED.
- **1970s** – Expansion of quantum mechanics into Quantum Chromodynamics (QCD), explaining the strong nuclear force that holds atomic nuclei together.
- **1970s** – The Standard Model of Particle Physics emerges, describing all known fundamental particles (except gravity).

1980s – Present: The Birth of Quantum Computing

Quantum Computing

- **1980** – Paul Benioff introduces the quantum Turing machine, applying quantum principles to computation.
- **1981–1982** – Richard Feynman suggests using quantum computers to simulate physical systems more efficiently than classical computers.
- **1984** – Bennett and Brassard propose Quantum Key Distribution (QKD), applying quantum mechanics to cryptography.
- **1985** – Deutsch's Algorithm introduces quantum parallelism, proving that quantum computers can process multiple calculations at once.
- **1993–1994** – Quantum algorithms like the Bernstein–Vazirani Algorithm (1993), Simon's Algorithm (1994), and Shor's Algorithm (1994) demonstrate quantum computing's potential.
- **1996** – Grover's Algorithm shows that quantum computers can speed up search problems.
- **1998** – The first two-qubit quantum computer is successfully built, demonstrating feasibility.

2000s – 2010s: Early-Stage Quantum Computing

- **2000s** – Companies like IBM, Google, Microsoft, and D-Wave begin developing early quantum computers.
- **2010s** – Improvements in trapped ions, superconducting qubits, and photonic qubits help quantum computers scale up.
- **2019** – Google and NASA claim to have achieved quantum supremacy, performing a calculation faster than a classical supercomputer could.
- **2020s** – Ongoing developments in fault-tolerant quantum computing, error correction, and quantum networks bring us closer to practical quantum computers.

As quantum theory continues to evolve, it remains one of the most successful yet enigmatic frameworks in modern science. Its implications extend beyond theoretical physics into practical applications, including quantum cryptography, quantum teleportation, and the potential for fault-tolerant quantum computing [37]. The ongoing research in quantum mechanics not only promises to revolutionize technology but also aims to unify quantum mechanics with general relativity, potentially leading to a deeper understanding of the universe [37].

The Future of Quantum Mechanics and Computing

- **Near Future (2030s – 2040s)** – Scientists aim to achieve fault-tolerant quantum computers that can solve real-world problems more efficiently than classical computers.
- **Long-Term Future** – Quantum research will continue to explore quantum cryptography, quantum networking, and unification with general relativity, leading to a deeper understanding of the universe.

Structure of This Book

This book is structured to provide a comprehensive understanding of quantum computing, beginning with the fundamental principles of quantum mechanics and gradually introducing key concepts such as quantum algorithms, hardware, applications, and future challenges. Each chapter builds upon the previous one, ensuring that readers with varying levels of experience, from beginners to those with a background in quantum physics or computer science, can follow along seamlessly.

The book opens with an introduction to quantum computing, covering its significance, how it differs from classical computing, and a brief history of quantum theory. This sets the stage for the next section, which delves into the foundations of quantum mechanics. Topics in this section include the origins of quantum mechanics, key principles such as superposition, entanglement, and measurement, as well as fundamental concepts like quantum states, wave functions, the uncertainty principle, and thought experiments such as Schrödinger's cat.

Following this, the discussion shifts to quantum bits (qubits) and quantum systems, explaining the nature of qubits, how they compare to classical bits, and their physical implementations. This section also introduces the Bloch sphere representation and explores challenges related to decoherence and error correction. The book then moves into quantum logic gates and circuit design, detailing different types of quantum gates, their role in quantum computing, and the principles of reversible computing.

A significant portion of the book is dedicated to quantum algorithms, starting with an overview of their importance. It then explores key algorithms such as the Deutsch-Jozsa algorithm, Grover's search algorithm, Shor's algorithm for factoring large numbers, and the quantum Fourier transform, as well as variational quantum algorithms. Since quantum systems are prone to errors, the next section focuses on quantum error correction and noise mitigation. It explains the problem of decoherence, introduces quantum error correction codes, discusses fault-tolerant quantum computing, and presents strategies for mitigating noise in quantum systems.

Quantum Computing

The book then shifts to an in-depth discussion of quantum computing hardware, covering superconducting qubits developed by companies like IBM, Google, and Rigetti, as well as trapped ion qubits used by IonQ and Honeywell. It also introduces topological qubits, photonic quantum computing, and quantum annealers used for optimization problems. Quantum cryptography and secure communications follow, explaining quantum key distribution (QKD), post-quantum cryptography, quantum random number generation, and the security implications of quantum computing.

An exploration of quantum machine learning and AI follows, introducing concepts such as quantum data encoding, quantum neural networks, variational quantum algorithms for machine learning, and applications in big data and optimization. This leads naturally into a discussion of real-world applications of quantum computing, including its impact on chemistry and material science, financial modeling and risk analysis, healthcare and drug discovery, climate modeling, and logistics and supply chain optimization.

With the increasing availability of quantum computing through cloud-based services, the book also examines cloud-based quantum computing platforms such as IBM Quantum Experience, Google Quantum AI, and Amazon Braket. It explains the concept of Quantum as a Service (QaaS), how users can access and run quantum algorithms online, and the challenges of scaling quantum cloud services.

The business and industry landscape of quantum computing is another critical aspect covered in the book. This section explores the current state of the quantum computing industry, investment trends, the involvement of governments and corporations, the path to commercial quantum advantage, and the ethical and economic implications of quantum technologies.

The final major section discusses the challenges and future of quantum computing. It outlines the technical barriers that must be overcome, theoretical and physical constraints, the potential for a quantum internet and distributed quantum computing, and predictions for the field over the next decade or two. It also considers the role of quantum computing in a post-Moore's Law era.

The book concludes with a summary of key takeaways, guidance for researchers and developers interested in getting involved in quantum computing, and final thoughts on its future. Several appendices provide additional resources, including an overview of the mathematical foundations of quantum computing, programming quantum computers with tools like Qiskit, Cirq, Braket, and PennyLane, a glossary of key terms, and recommended readings. A comprehensive list of references is also included to guide readers toward further study.

Chapter 1

Foundations of Quantum Mechanics

Before we can fully appreciate the power of quantum computing, we must first understand the fundamental principles that make it possible. Unlike classical computing, which is built on well-established mathematical and logical structures, quantum computing operates within the strange and counterintuitive framework of quantum mechanics. This foundational science describes the behaviour of particles at the smallest scales, where the rules of classical physics no longer apply. The ability of quantum systems to exist in multiple states simultaneously (superposition), to be intrinsically linked across vast distances (entanglement), and to behave probabilistically rather than deterministically are what enable quantum computers to surpass their classical counterparts in certain computational tasks.

This chapter explores the historical development of quantum mechanics, beginning with the challenges faced by classical physics and the revolutionary discoveries that led to the birth of quantum theory. It introduces key principles such as wave-particle duality, the uncertainty principle, and quantum state measurement, all of which play crucial roles in how quantum computers process information. By understanding the theoretical foundations of quantum mechanics, we can better grasp how qubits function, how quantum gates operate, and why quantum algorithms hold such transformative potential. This chapter serves as a crucial starting point, laying the groundwork for the more advanced topics covered in later chapters, including quantum circuits, quantum error correction, and real-world applications of quantum computing.

Origins of Quantum Mechanics

The origins of quantum mechanics can be traced back to the early 20th century, when classical physics faced challenges in explaining various phenomena. The need for a new theoretical framework emerged from several pivotal discoveries, including the problem of blackbody radiation, the photoelectric effect, atomic structure, and the development of mathematical formulations that would ultimately redefine our understanding of the physical world.

One of the first significant challenges was the problem of blackbody radiation. Classical physics predicted that an ideal blackbody would emit infinite energy at short wavelengths, leading to what is known as the ultraviolet catastrophe. In 1900, Max Planck proposed a revolutionary hypothesis, suggesting that energy is quantized and emitted in discrete packets called "quanta" [38]. This concept laid the groundwork for quantum theory, as it provided a solution to the discrepancies observed in blackbody radiation spectra, which could be accurately described by Planck's law [39]. Planck's work was pivotal in establishing the idea of quantized energy levels, which became a cornerstone of quantum mechanics.

Following Planck's hypothesis, Albert Einstein's work on the photoelectric effect in 1905 further advanced quantum theory. Einstein proposed that light consists of discrete packets of energy, or photons, and that the energy of these photons is proportional to their frequency [40]. This contradicted classical wave theory, which suggested that increasing light intensity should increase the energy of emitted electrons. Einstein's explanation of the photoelectric effect not only validated Planck's quantization of energy but also introduced the concept of wave-particle duality, which became fundamental to quantum mechanics. His contributions earned him the Nobel Prize in Physics in 1921, solidifying the importance of quantum ideas in explaining physical phenomena.

The structure of the atom posed another significant challenge to classical physics. In 1911, Ernest Rutherford's discovery of the atomic nucleus led to questions about the stability of atoms, as classical mechanics suggested that orbiting electrons should radiate energy and spiral into the nucleus. Niels Bohr addressed this issue in 1913 by proposing a model in which electrons occupy fixed, quantized orbits around the nucleus [41]. Bohr's model successfully explained the discrete spectral lines observed in atomic emissions, reinforcing the concept of quantization in atomic structure and paving the way for future developments in quantum mechanics.

The 1920s saw the emergence of wave-particle duality as a fundamental principle of quantum mechanics, primarily through the work of Louis de Broglie, who suggested that particles like electrons exhibit wave-like behaviour [42]. This idea was experimentally confirmed through the double-slit experiment, demonstrating that particles can interfere with themselves, a hallmark of wave behaviour. The synthesis of wave and particle descriptions of matter became a key aspect of quantum theory, influencing subsequent theoretical advancements.

The mathematical foundations of quantum mechanics were established through the competing formulations of Werner Heisenberg and Erwin Schrödinger in the mid-1920s. Heisenberg's matrix mechanics and Schrödinger's wave mechanics provided two equivalent frameworks for describing quantum systems [43]. These formulations unified earlier discoveries and allowed for a comprehensive understanding of quantum behaviour, leading to the development of the Schrödinger equation, which describes the evolution of quantum states over time.

In 1927, Heisenberg introduced the Uncertainty Principle, which posits that it is impossible to simultaneously know both the position and momentum of a particle with arbitrary precision [44]. This principle challenged classical determinism and introduced a probabilistic interpretation of quantum mechanics, suggesting that particles exist in probability distributions rather than fixed orbits. The philosophical implications of this principle prompted further exploration into the nature of reality and measurement in quantum systems.

The Copenhagen Interpretation, developed by Bohr and Heisenberg, became the dominant framework for understanding quantum mechanics. It posited that quantum particles exist in multiple states until measured, at which point the act of measurement collapses the system into a single state [45]. This interpretation led to thought experiments, such as Schrödinger's Cat, which illustrated the counterintuitive nature of quantum mechanics and sparked debates about the implications of quantum measurement.

By the 1930s, quantum mechanics had successfully explained a range of atomic behaviours and laid the groundwork for numerous technological advancements, including semiconductors and lasers. The theory continued to evolve, leading to the development of Quantum Electrodynamics (QED) and Quantum Chromodynamics (QCD), which describe fundamental interactions in particle physics. Today, quantum mechanics remains a cornerstone of modern physics, underpinning advancements in quantum computing, cryptography, and other fields.

Relevance of Quantum Mechanics to Quantum Computing

Quantum computing would not exist without quantum mechanics. The fundamental principles of quantum mechanics—superposition, entanglement, and quantum interference—are the core concepts that allow quantum computers to perform calculations in ways that classical computers cannot. The development of quantum mechanics provided the theoretical foundation that later led to the concept of quantum computing, as scientists realized that quantum systems could be used for computational purposes.

One of the most important discoveries in quantum mechanics was wave-particle duality, introduced by Louis de Broglie in 1924 and later mathematically described by Schrödinger's Wave Equation in 1926. This idea laid the groundwork for superposition, a fundamental feature of quantum computing. In classical computing, bits can only be 0 or 1 at any given time. In quantum computing, qubits (quantum bits) can exist in both 0 and 1 at the same time due to superposition. This allows quantum computers to process multiple calculations simultaneously, making them exponentially faster for certain problems. Without the development of quantum wave mechanics, we would not have understood how quantum states can exist in multiple configurations at once, a concept that is directly applied in quantum computing algorithms.

The Copenhagen Interpretation, formulated by Niels Bohr and Werner Heisenberg in 1927, introduced the idea that particles do not have definite states until they are measured. This led to the discovery of quantum entanglement, where two particles remain instantaneously connected, regardless of distance. In quantum computing, entangled qubits can communicate instantaneously, allowing for ultra-fast parallel computations. Entanglement is essential for quantum teleportation, quantum cryptography, and distributed quantum computing networks. Without entanglement, quantum speedup and secure quantum communication would not be possible, making it a crucial feature of quantum computing.

Heisenberg's Uncertainty Principle, also introduced in 1927, established that measuring a quantum system inherently disturbs it, introducing the idea that quantum states must be manipulated carefully. In quantum computing, this principle affects quantum measurement—when we measure a quantum state, it collapses to a single value. This means that quantum algorithms must be designed to extract information efficiently before measurement disrupts the quantum state. Quantum systems require error correction techniques, based on our understanding of quantum uncertainty, to preserve quantum information.

By the 1980s, physicists such as Richard Feynman realized that classical computers could not efficiently simulate quantum systems because quantum mechanics followed exponential complexity rather than linear calculations. Feynman proposed in 1982 that if we wanted to simulate quantum behaviour, we should build a computer that operates on quantum principles, marking the birth of quantum computing. Shor's Algorithm in 1994 demonstrated that quantum computers could factor large numbers exponentially faster than classical computers, showing real-world implications for cryptography and cybersecurity. Grover's Algorithm, developed in 1996, showed that quantum computers could search unsorted databases quadratically faster than classical computers, proving that quantum mechanics could speed up problem-solving.

The principles of quantum wave functions, energy quantization, and electron states—discovered by Planck, Einstein, and Schrödinger—are directly applied in designing quantum hardware. Superconducting qubits, used by companies like IBM and Google, rely on quantum tunnelling and superconducting circuits to maintain quantum coherence. Trapped ion qubits, developed by IonQ and Honeywell, use electromagnetic fields to trap individual quantum particles for computations. Photonic qubits, explored by Xanadu, use light-based quantum states to perform calculations. Without quantum mechanics, there would be no qubit designs, no understanding of how to control and manipulate quantum states, and no way to build a working quantum processor.

Quantum mechanics is not just the foundation of quantum computing—it is also driving breakthroughs in other fields. Quantum cryptography uses quantum mechanics for ultra-secure encryption that classical computers cannot break. Quantum networks leverage quantum entanglement to create unhackable communication systems. Quantum sensors offer ultra-sensitive detection capabilities for gravity, magnetic fields, and even brain activity with extreme precision.

Quantum computing is a direct application of quantum mechanics. Every breakthrough in quantum computing—from qubits and entanglement to quantum algorithms and error correction—is rooted in principles discovered by early quantum physicists. The origins of quantum mechanics explain why quantum computing is possible, how it works, and where it is headed. As researchers continue to explore quantum theory, we can expect even more powerful quantum computing advancements that will revolutionize science, cryptography, artificial intelligence, and beyond.

Key Principles: Superposition, Entanglement, and Measurement

Quantum computing is fundamentally different from classical computing because it relies on the principles of superposition, entanglement, and measurement. These principles, rooted in quantum mechanics, allow quantum computers to perform computations that would be impractical or impossible for classical computers. Understanding these concepts is essential to grasp how quantum computers achieve their superior processing power.

Imagine flipping a coin. In classical physics, once the coin lands, it is either heads or tails—one definite outcome. However, in quantum mechanics, a qubit (the quantum version of a computer bit) is like a coin that is still spinning in the air. Instead of having a single outcome, it exists in a state where it is both heads and tails at the same time. This concept is called superposition.

A simple way to understand superposition is to think about a light switch. A regular light switch is either on or off, just like a classical computer bit being 0 or 1. But in a quantum world, the switch can be both on and off at the same time until you look at it. This means that quantum computers, which use qubits instead of classical bits, can process multiple possibilities simultaneously, making them incredibly powerful for solving certain complex problems.

For example, imagine you are trying to find the best route on a road map. A classical computer would check one route at a time until it finds the shortest one. A quantum computer, using superposition, could consider all possible routes at once, drastically reducing the time needed to solve the problem.

Quantum Computing

Entanglement is one of the strangest and most fascinating aspects of quantum mechanics. It means that two particles, even if they are light-years apart, can be instantly connected in a way that classical physics cannot explain. If one particle changes, the other changes immediately, no matter how far apart they are.

A good analogy is a pair of magic dice. Imagine you and a friend each take one die, and you travel to opposite sides of the planet. When you roll your die, you instantly know what number your friend rolled, even though you are thousands of miles away. In a normal world, this would be impossible because the dice are separate objects. But in the quantum world, entangled particles share the same fate, just like our magic dice.

Entanglement is a key resource in quantum cryptography, where it is used to create ultra-secure communication channels. If an eavesdropper tries to intercept an entangled message, the act of spying will disturb the quantum state, making the intrusion detectable. This feature makes quantum encryption potentially unbreakable.

One of the most mind-boggling ideas in quantum mechanics is that simply observing a quantum system changes it. Before we measure a qubit, it exists in a superposition of multiple possibilities. But the moment we measure it, the superposition collapses into a single definite state—just like a coin landing as either heads or tails.

To illustrate this, imagine a mystery box containing a cat, inspired by Schrödinger's famous thought experiment. The cat is in a weird quantum state where it is both alive and dead at the same time. However, the moment you open the box to check, you force the universe to make a decision—the cat is now either alive or dead, but no longer both.

In practical terms, this means that quantum computers must carefully avoid measuring qubits too soon. If they do, they will lose the advantage of superposition, making them behave like regular classical computers. Scientists use special techniques such as quantum interference and error correction to keep qubits in their quantum state as long as possible before extracting results.

Superposition, entanglement, and measurement work together to make quantum computing revolutionary. Superposition allows a quantum system to be in many states at once, entanglement connects qubits in a way that defies classical intuition, and measurement forces quantum uncertainty into a single, concrete result. These principles unlock computing power that could transform fields like medicine, artificial intelligence, and cryptography, solving problems that even the most advanced classical supercomputers struggle with.

The quantum world is strange, but its potential is enormous. Just as early classical computers reshaped the modern world, quantum computers could revolutionize science and technology in ways we can barely imagine today.

Superposition: Processing Multiple States at Once

Superposition is one of the most fundamental and counterintuitive principles of quantum mechanics. In classical computing, bits represent information as either 0 or 1. However, in quantum computing, qubits (quantum bits) can exist in a combination of both 0 and 1 simultaneously.

To illustrate, imagine a classical light switch that is either on or off. A classical bit works the same way, being either 0 or 1 at any given time. A qubit, however, is like a dimmer switch that can be in multiple brightness levels at once, meaning it is in a superposition of 0 and 1 until it is measured. This allows quantum computers to explore multiple possibilities simultaneously rather than sequentially, as classical computers do.

Mathematically, a qubit in superposition is represented as:

$$|\psi\rangle = \alpha|0\rangle + \beta|1\rangle$$

where α and β are probability amplitudes that determine the likelihood of measuring the qubit as either 0 or 1.

In practical terms, this means that a quantum computer with n qubits can represent 2^n different states simultaneously, exponentially increasing the computational power for certain types of problems. For example, with just 10 qubits, a quantum computer can process 1,024 states at once, whereas a classical computer would need to process each state one at a time.

Superposition is a unique property of quantum mechanics that allows a quantum system, such as an electron or a photon, to be in multiple states at once until it is measured. In classical physics, a coin toss results in either heads or tails. However, in the quantum realm, a coin could be both heads and tails at the same time until someone observes it. This is the essence of quantum superposition.

Mathematically, superposition states that if a system has two or more possible states, it can exist in a combination of those states simultaneously. For example, in quantum computing, a qubit (quantum bit) can be in a superposition of 0 and 1, rather than being restricted to just one value at a time. This ability is crucial because it allows quantum computers to process vast amounts of information in parallel, unlike classical computers, which process information sequentially.

When a quantum system is measured, the superposition collapses, and the system settles into one definite state. The probability of measuring a particular state is determined by the quantum wave function, which follows the Schrödinger Equation—a mathematical framework that describes how quantum states evolve over time.

Quantum superposition is not just limited to theoretical physics; it plays a crucial role in nature and chemistry. The behaviour of atoms, molecules, and chemical bonds is governed by quantum mechanics. Electrons in an atom do not exist in fixed orbits, as classical physics might suggest, but rather in a superposition of different energy states.

For example, in photosynthesis, plants use quantum superposition to optimize energy transfer during light absorption. This efficiency, which classical physics struggles to explain, is an example of nature's quantum computations in action. Similarly, quantum superposition underlies biological systems, including enzyme reactions and genetic information processing in DNA.

Quantum superposition is not just limited to theoretical physics; it plays a crucial role in nature and chemistry. The behaviour of atoms, molecules, and chemical bonds is governed by quantum mechanics. Electrons in an atom do not exist in fixed orbits, as classical physics might suggest, but rather in a superposition of different energy states.

For example, in photosynthesis, plants use quantum superposition to optimize energy transfer during light absorption. This efficiency, which classical physics struggles to explain, is an example of nature's quantum computations in action. Similarly, quantum superposition underlies biological systems, including enzyme reactions and genetic information processing in DNA.

One of the most famous thought experiments that illustrates quantum superposition is Schrödinger's Cat, proposed by physicist Erwin Schrödinger in the 1930s. The paradox involves a cat placed inside a sealed box with a vial of poison, a radioactive atom, and a Geiger counter.

According to quantum mechanics, the radioactive atom has a 50% chance of decaying and releasing the poison, killing the cat, and a 50% chance of remaining intact, keeping the cat alive. Until the box is opened and an observation is made, the cat exists in a superposition of both alive and dead states. The moment we observe the cat, the quantum state collapses, and it becomes either alive or dead.

While Schrödinger originally proposed this as a critique of the Copenhagen interpretation of quantum mechanics, the experiment provides a powerful analogy for how quantum systems remain in superposition until measured.

In quantum computing, qubits function much like Schrödinger's cat. They exist in a superposition of 0 and 1, allowing quantum computers to perform multiple calculations simultaneously. The collapse of the quantum state upon measurement is a fundamental challenge in quantum computing, as it means quantum information can be lost if not carefully manipulated.

The behaviour of quantum superposition is mathematically described by Schrödinger's Equation, one of the most important equations in quantum mechanics. Schrödinger's Equation determines how quantum states evolve over time and why particles can exist in multiple states at once.

This equation is linear, meaning that if two quantum states ψ_1 and ψ_2 are valid solutions, then any combination $a\psi_1 + b\psi_2$ is also a valid solution, where a and b are complex numbers. This property of linearity enables superposition to exist, making it not just an abstract concept but a fundamental mathematical reality of quantum mechanics.

By solving Schrödinger's Equation, physicists can describe how a quantum system behaves and predict the probability of measuring different outcomes. This concept is what enables quantum algorithms to leverage superposition for high-speed problem-solving in quantum computing.

Quantum mechanics is deeply connected to probability theory. In classical computing, the probability of an event (such as flipping a coin) is determined by classical statistics. In quantum mechanics, probability is determined by the wave function of a quantum system.

Unlike classical probability, which follows linear logic, quantum probability follows complex geometric principles. Before measurement, a quantum system does not have a single definite outcome but exists in a superposition of multiple outcomes. When a measurement occurs, the wave function collapses, and the system settles into one of its possible states, much like a roulette wheel landing on a single number after spinning.

This quantum probability theory allows quantum computers to process many possible outcomes at once, making them vastly superior for solving optimization problems, cryptography, and artificial intelligence applications.

Superposition is the key to quantum computing's power. In a classical computer, information is processed using bits, which can be either 0 or 1. A classical computer must process each possible solution one at a time.

In contrast, quantum computers use qubits, which can be in superposition of both 0 and 1 simultaneously. This means that a quantum computer can explore many different solutions at once, vastly increasing its computational efficiency.

For example, in Shor's Algorithm (used for factoring large numbers), a quantum computer can check many possible factors simultaneously using superposition. In Grover's Algorithm, quantum superposition helps search through unsorted databases quadratically faster than classical computers.

By leveraging superposition, quantum computers can perform computations that would take millions of years on classical supercomputers. This makes them particularly useful for encryption, drug discovery, financial modelling, and artificial intelligence.

Superposition is not just a theoretical concept; it can be experimented with in quantum computing platforms. Cloud-based quantum computing platforms, like IBM's Qiskit, Google Cirq, and BlueQubit, allow users to create quantum circuits that manipulate qubits into superposition.

These platforms enable researchers, students, and quantum enthusiasts to experience how quantum gates control superposition and how quantum algorithms leverage this principle to solve real-world problems. By experimenting with superposition, users can develop quantum computing applications without needing physical quantum hardware.

Entanglement: Instantaneous Quantum Connection

Quantum entanglement is one of the most fascinating and counterintuitive concepts in quantum mechanics, providing a foundation for quantum computing, cryptography, and next-generation communication networks. It describes a unique connection between two or more quantum systems, such that their states become inseparably correlated, even when separated by vast distances. This instantaneous connection enables quantum computers to perform highly complex computations faster than classical computers.

Understanding Quantum Entanglement

At its core, quantum entanglement is a phenomenon where two or more quantum particles (such as electrons or photons) become so deeply linked that the quantum state of one particle is directly correlated with the state of the other, regardless of the distance between them. This correlation means that if one particle's state is measured, the state of its entangled partner is instantly determined, even if they are light-years apart.

To visualize this, imagine two identical dice that are quantum entangled. If you roll one die and it lands on a 6, the other die will always land on a 6, no matter where it is in the universe. However, before rolling, both dice exist in a superposition of all possible outcomes (1, 2, 3, 4, 5, and 6), and only when one is measured does the other instantly "decide" its outcome. This instantaneous connection between entangled particles was famously referred to by Albert Einstein as “spooky action at a distance.”

In classical physics, information cannot travel faster than the speed of light. However, quantum entanglement challenges this idea, as entangled particles instantly influence each other, no matter the distance. While entanglement does not allow for faster-than-light communication, it does enable parallel processing in quantum computing and secure communication in quantum cryptography.

Quantum systems are described using wavefunctions, mathematical objects that define the probabilities of different outcomes when a quantum measurement is performed. If two particles are not entangled, their combined wavefunction can be expressed as a simple product of individual wavefunctions:

$$\Psi = \Psi_1 \times \Psi_2$$

However, when two quantum particles become entangled, their wavefunctions can no longer be written separately. Instead, they form a single wavefunction, which represents the combined system:

$$\Psi = \frac{1}{\sqrt{2}}(|0\rangle_A|1\rangle_B + |1\rangle_A|0\rangle_B)$$

This equation means that neither particle has an independent state—they exist in a shared quantum state until one is measured. If one particle collapses into state 0, the other immediately collapses into state 1, and vice versa. This mathematical representation shows that measuring one qubit directly affects the other, regardless of the distance between them.

In quantum mechanics, wavefunctions are mathematical objects that describe the state of a quantum system. A wavefunction, denoted as Ψ, contains all the possible information about a quantum system, including the probabilities of different measurement outcomes. The fundamental idea behind quantum mechanics is that particles do not have definite properties until they are measured; instead, they exist in a superposition of multiple possible states, each associated with a certain probability.

If two quantum particles (such as electrons or photons) are not entangled, their quantum states can be described independently. In this case, their combined wavefunction is simply the product of their individual wavefunctions:

$$\Psi = \Psi_1 \times \Psi_2$$

This means that the particles do not influence each other. Each particle behaves independently, and measuring one particle does not affect the state of the other. This is similar to flipping two separate coins—each flip is independent, and the result of one does not determine the result of the other.

When two quantum particles become entangled, their individual wavefunctions can no longer be separated. Instead, they form a single wavefunction that describes the entire system, rather than individual particles. This entangled wavefunction is written as:

$$\Psi = \frac{1}{\sqrt{2}}(|0\rangle_A|1\rangle_B + |1\rangle_A|0\rangle_B)$$

This equation represents a maximally entangled state of two qubits (quantum bits), often referred to as a Bell state. Here's what it means:

- The system is in a superposition of two possibilities:
    - Qubit A is in state 0 while qubit B is in state 1

- Qubit A is in state 1 while qubit B is in state 0

- The key point is that neither qubit has a definite state on its own. The two qubits share a single quantum state, meaning their outcomes are always correlated.

Before measurement, both qubits exist in a superposition of the two possible states. However, the moment one qubit is measured, the entangled system "collapses" into a definite state.

For example:

- If qubit A is measured and found to be 0, then qubit B must be 1.
- If qubit A is measured and found to be 1, then qubit B must be 0.

This correlation holds no matter how far apart the two qubits are, which is what makes entanglement so extraordinary. Even if the qubits are separated by light-years, measuring one instantaneously determines the state of the other. This is why Albert Einstein famously called entanglement "spooky action at a distance."

Entanglement is one of the key principles that make quantum computing more powerful than classical computing. Because entangled qubits share information instantaneously, quantum computers can perform parallel computations and solve complex problems much faster than classical computers.

In practical applications, entanglement enables:

- Quantum Teleportation – The transfer of quantum states over long distances using entangled pairs.
- Quantum Cryptography – Secure communication using entangled particles to detect eavesdroppers.
- Quantum Parallelism – Performing multiple calculations at once, increasing computational efficiency.
- Quantum Error Correction – Using entangled states to detect and correct errors in quantum processors.

In quantum computing, entanglement plays a crucial role in enabling parallel computation and increasing processing efficiency. Classical computers process bits independently, whereas quantum computers leverage entanglement to process multiple qubits simultaneously, improving computational power exponentially.

Entanglement allows quantum computers to:

- Store and process exponentially large datasets efficiently by encoding complex information in multiple qubits at once.

- Execute algorithms faster than classical computers, such as Shor's Algorithm for breaking encryption and Grover's Algorithm for database searching.
- Enhance error correction methods, as entangled qubits help detect and correct quantum errors more effectively than classical systems.

For instance, if two qubits are entangled, changing the state of one immediately changes the state of the other, regardless of how far apart they are. This enables quantum parallelism, where computations occur in multiple quantum states simultaneously, making quantum computers vastly more powerful than traditional computers.

One of the most promising applications of quantum entanglement is quantum cryptography, which enables ultra-secure communication. Traditional encryption methods rely on mathematical complexity, making them vulnerable to future quantum attacks. Quantum cryptography, on the other hand, leverages the properties of entanglement to create tamper-proof communication channels.

The most well-known quantum cryptographic protocol is Quantum Key Distribution (QKD), specifically the BB84 and E91 protocols. In QKD, two parties (Alice and Bob) share a secret key using entangled qubits. If an eavesdropper (Eve) attempts to intercept the key, the entangled state collapses, alerting Alice and Bob to the intrusion. This ensures unbreakable encryption, as any attempt to measure or intercept the quantum key alters the system.

Entanglement is also the foundation of quantum teleportation, a process where quantum states (rather than physical particles) are transferred instantaneously between distant locations. This technique has been experimentally demonstrated over distances of hundreds of kilometres using entangled photons.

Quantum entanglement is actively being researched and applied in various fields, including:

- Quantum Networks – Creating entangled-based networks for secure quantum communication.
- Quantum Sensors and Metrology – Using entangled states to achieve unprecedented precision in measurement devices, such as gravitational wave detectors.
- Quantum Repeaters – Extending the range of entangled quantum networks, enabling the development of the quantum internet.
- Quantum Teleportation – Transferring quantum information across long distances using entangled states.

Challenges in Quantum Entanglement

Despite its potential, quantum entanglement faces significant challenges:

- Decoherence – Entangled states are fragile and can be easily disrupted by their environment, causing loss of quantum information.
- Error Correction – Quantum computers require advanced error correction techniques to maintain entanglement over time.
- Scalability – Generating and maintaining entanglement in large-scale quantum computers is a major technical hurdle.

To overcome these challenges, researchers are exploring methods such as quantum error correction codes, entanglement purification, and topological quantum computing to enhance the stability of entangled states.

Understanding Quantum Teleportation, Superdense Coding, and the CHSH Game

Quantum information science explores the use of quantum mechanics to encode, transmit, and process information in ways that classical systems cannot. Three key protocols demonstrate the power of quantum entanglement and information processing:

1. **Quantum Teleportation** – A method for transmitting quantum information using entanglement and classical communication.
2. **Superdense Coding** – A technique that allows the transmission of two classical bits using one quantum bit (qubit) through entanglement.
3. **The CHSH Game** – An experimental setup that illustrates quantum nonlocality, proving that quantum mechanics violates classical expectations about correlations between distant systems.

These protocols are not just theoretical exercises but form the foundation of quantum information science. They rely on quantum entanglement, which allows particles to share information instantaneously, making these methods possible.

In quantum information science, Alice and Bob are frequently used as hypothetical characters to represent two parties exchanging information. This naming convention simplifies explanations of complex quantum protocols by assigning familiar names to abstract entities involved in cryptographic or quantum communication systems. Alice and Bob serve as placeholders for the sender and receiver of information, making it easier to describe and analyse various quantum protocols without delving into unnecessary specifics about the participants.

Alice is typically the sender of information. In quantum communication scenarios, she might prepare a quantum state, perform certain operations on it, and then send it to Bob. Her role varies depending on the protocol in question. For instance, in quantum teleportation, Alice wants to send a quantum state to Bob without physically transferring the particle itself. To achieve this, she must perform specific quantum operations and send classical bits to Bob, who then reconstructs the original state. In quantum cryptography, Alice might generate a secure key and transmit quantum-encoded information to Bob, ensuring that their communication remains private.

Bob is usually the receiver of information. He takes the quantum state or classical message sent by Alice and processes it according to the specific protocol. In superdense coding, Bob decodes a message that Alice has compressed into a quantum system, extracting two classical bits from just one qubit. In quantum key distribution (QKD), Bob receives quantum-encoded bits from Alice and measures them to establish a shared secret key that can be used for secure communication. His role is often reactive—waiting for Alice's transmission and performing necessary operations to extract, interpret, or reconstruct the information.

Depending on the scenario, additional participants may be introduced. A third party, often named Charlie, may facilitate certain protocols or mediate interactions between Alice and Bob. For example, in quantum secret sharing, Charlie could be responsible for distributing quantum-encoded information among multiple parties, ensuring that no single participant can access the complete information without cooperation from the others. Another commonly used character is Eve, who represents an eavesdropper attempting to intercept or tamper with Alice and Bob's communication. In quantum cryptographic protocols like BB84, Alice and Bob design their strategy specifically to detect Eve's presence, ensuring the security of their exchanged information.

The Alice and Bob framework is a fundamental tool in quantum information science because it provides a structured way to describe complex communication protocols. By assigning these roles to generic participants, researchers and students can focus on understanding how information flows through a system rather than being distracted by unnecessary details. This abstraction is particularly useful when discussing entanglement, teleportation, and quantum cryptography, as it allows for a clear representation of how different parties interact in quantum systems.

While Alice and Bob are not real people, their presence in quantum mechanics discussions has become a standardized way to illustrate and explain information exchange. This convention is rooted in cryptographic research from the 1970s, where these names were first introduced to describe secure communication protocols. Since then, the tradition has expanded into quantum computing, where it remains a crucial framework for understanding how information is transmitted, manipulated, and secured at the quantum level.

Entanglement is one of the most fundamental and mysterious phenomena in quantum mechanics. It describes a situation where two or more quantum particles become so deeply connected that their states can no longer be described independently, regardless of the distance between them. If one particle is measured, its entangled partner's state is instantly determined, even if the two are separated by vast distances. This non-local connection, which Einstein famously referred to as "spooky action at a distance," defies classical intuition and is a cornerstone of quantum computing, quantum cryptography, and quantum communication.

A maximally entangled state of two qubits is often represented mathematically using what is known as a Bell state. One of the simplest Bell states is written as:

$$|\phi^+\rangle = \frac{1}{\sqrt{2}}(|00\rangle + |11\rangle)$$

This equation signifies that the two qubits exist in a superposition where they are both in the state $|00\rangle$ and $|11\rangle$ at the same time. However, as soon as one qubit is measured, the state of the other is instantly determined. If the first qubit collapses to $|0\rangle$, the second must also be $|0\rangle$. If the first qubit is found in state $|1\rangle$, the second must also be $|1\rangle$. This shared quantum state is what makes entanglement fundamentally different from classical correlations, where two objects might share a pattern but do not instantaneously influence each other across space.

Entanglement is often described as a resource because it enables quantum technologies to perform tasks that classical systems cannot achieve. In quantum computing, entanglement allows multiple qubits to be linked together in a way that exponentially increases computational power. This is the basis of quantum parallelism, which enables quantum computers to process vast amounts of data simultaneously. Unlike classical bits, which operate independently as either 0 or 1, entangled qubits exist in a correlated state that allows computations to scale in ways that classical computers cannot replicate.

Beyond computation, entanglement plays a crucial role in quantum communication. In quantum teleportation, a quantum state can be transferred from one location to another without physically moving the particle itself. This is achieved by using an entangled pair as a medium for the transfer, where one qubit remains with the sender (Alice) and the other with the receiver (Bob). By performing specific quantum operations and sending classical information, Alice can effectively "teleport" the quantum state of her qubit onto Bob's qubit, preserving all quantum information without directly transmitting the particle.

Entanglement also enhances data transmission efficiency through superdense coding, which allows for the transmission of two classical bits of information using only one qubit. This is another example of how entanglement can be used as a computational resource, enabling more information to be packed into fewer quantum particles. In secure communication, entanglement forms the foundation of quantum key distribution (QKD), where entangled pairs ensure that any attempt at eavesdropping alters the system and is immediately detectable. This property allows for theoretically unbreakable encryption methods that classical cryptographic techniques cannot match.

The idea of entanglement as a resource has broader implications in fundamental physics. It plays a crucial role in understanding quantum mechanics at the deepest level, particularly in how quantum information is stored, manipulated, and transferred. Some physicists even propose that spacetime itself may be linked to the structure of quantum entanglement, suggesting that the fabric of the universe could be fundamentally quantum in nature.

In practical applications, entanglement is one of the driving forces behind the development of a quantum internet, where entangled qubits distributed across large distances enable ultra-secure, instantaneous communication. Researchers are working to develop quantum networks that use entanglement to synchronize and transfer information between quantum processors, leading to more powerful and interconnected quantum systems.

Because entanglement is essential to many aspects of quantum computing and quantum communication, it is often compared to energy in classical computing. Just as classical computers require energy to perform calculations and process information, quantum computers and quantum networks require entanglement as a fundamental building block to enable advanced quantum operations. In many ways, entanglement serves as the fuel that powers quantum technologies, allowing them to outperform their classical counterparts in ways that were once thought to be impossible.

Understanding entanglement as a resource is critical to appreciating the vast potential of quantum computing and quantum communication. Whether it is used for faster computations, secure communications, teleportation of quantum states, or fundamental physics experiments, entanglement continues to be one of the most valuable and intriguing phenomena in modern science. Researchers are still exploring new ways to harness its power, making it one of the most exciting frontiers in quantum information science.

Quantum teleportation is a fundamental process in quantum information science that enables the transfer of a quantum state from one location to another without physically moving the particle itself. Unlike the teleportation depicted in science fiction, this phenomenon does not involve the instant transport of matter. Instead, it relies on the principles of quantum entanglement and classical communication to transmit quantum information securely and accurately between two distant parties. This process is crucial for quantum networks, distributed quantum computing, and the development of the quantum internet.

The teleportation protocol begins with two parties, Alice and Bob, who wish to communicate quantum information. They must first share a pair of entangled qubits, which serve as the fundamental resource enabling the teleportation process. Alice possesses one qubit, labelled as A, while Bob holds the other, labelled as B. These qubits exist in an entangled state, meaning that their quantum properties are interdependent, even when separated by vast distances. This entanglement ensures that any operation performed on one qubit will instantaneously influence the other, regardless of the spatial separation.

At some point, Alice receives an additional qubit, Q, which contains the quantum state she wants to send to Bob. The critical challenge she faces is that quantum mechanics prohibits the direct measurement of an unknown quantum state without disturbing it. Measuring Q would cause its quantum information to collapse into a classical value, destroying the original quantum state and preventing faithful transmission. Because of this fundamental limitation, Alice cannot simply measure Q and send the result to Bob. Instead, she must use a quantum operation to transfer the information while preserving its quantum nature.

To initiate the teleportation, Alice performs a controlled-NOT (CNOT) operation on her two qubits, Q and A. This operation entangles them, linking their states together in a way that facilitates information transfer. Following this step, Alice applies a Hadamard gate to Q, which creates a superposition and prepares the system for measurement. These quantum gates are essential because they ensure that the quantum state of Q is encoded in a way that can be effectively communicated using classical information.

After applying these quantum operations, Alice proceeds to measure both Q and A. The measurement results yield two classical bits, which correspond to one of four possible outcomes: 00, 01, 10, or 11. These classical bits are then transmitted to Bob over a traditional communication channel. This step is crucial because, although the quantum state itself is not being transmitted physically, the classical information helps Bob reconstruct the original state. The transmission of these two bits ensures that Bob knows precisely how to manipulate his entangled qubit, B, to complete the teleportation process.

Once Bob receives the two classical bits from Alice, he applies the appropriate quantum operation to his qubit, B. The specific transformation Bob must perform depends on the classical bits he receives. If the classical bits are 00, Bob does nothing because his qubit is already in the correct state. If he receives 01, he applies a phase flip using a Z gate. If he receives 10, he applies a bit flip using an X gate. If he receives 11, he applies both a bit flip and a phase flip, effectively performing a ZX operation. These quantum corrections ensure that his qubit, B, is now in the exact quantum state that Alice's qubit Q was in before the measurement.

At the end of the teleportation process, Bob's qubit B contains the quantum information that was originally in Alice's qubit Q. However, Alice's original qubit is no longer in the same state because the act of measurement has collapsed its quantum information. This aspect of quantum teleportation aligns with the no-cloning theorem, which states that an arbitrary unknown quantum state cannot be perfectly copied. Instead of creating a duplicate of Q, the teleportation process effectively relocates the quantum information to Bob's qubit while erasing it from Alice's.

One of the most remarkable aspects of quantum teleportation is that it enables the transfer of quantum information without any physical transport of qubits. The quantum state appears to "jump" from Alice to Bob, but in reality, this is a consequence of quantum entanglement and classical information transfer. The process still respects the limitations of relativity because it requires classical communication, which cannot exceed the speed of light. This means that while entanglement creates an instantaneous correlation between qubits, the actual transfer of usable quantum information remains bound by classical limits.

Quantum teleportation is a critical tool for the development of quantum networks and the quantum internet. It enables the secure transfer of quantum states over long distances, which is essential for quantum cryptography and distributed quantum computing. Future advancements in quantum teleportation will allow for more sophisticated quantum communication protocols, where entangled qubits are distributed across vast distances, enabling secure, ultra-fast data transfer that classical systems cannot achieve. As quantum technology advances, teleportation will play a foundational role in connecting quantum processors and establishing robust quantum communication channels worldwide.

Superdense coding is a quantum communication protocol that takes advantage of entanglement to transmit two classical bits of information using only one qubit. This capability is impossible in classical systems, where a single unit of information (such as a classical bit) can only carry one bit of information. By utilizing quantum mechanics, specifically the properties of entanglement and quantum gates, superdense coding allows for more efficient data transmission, making it an essential tool for high-speed quantum communication networks.

The process of superdense coding begins with Alice and Bob sharing an entangled qubit pair. They start by preparing a Bell state, which is a maximally entangled state of two qubits. A common choice is the Φ+ Bell state, which is mathematically represented as:

$$|\phi^+\rangle = \frac{1}{\sqrt{2}}(|00\rangle + |11\rangle)$$

In this state, the two qubits are perfectly correlated, meaning that their quantum properties are interdependent. Alice holds one qubit (A), and Bob holds the other (B). Since these qubits are entangled, any operation performed on Alice's qubit will influence the entire system, which is a crucial aspect of superdense coding.

To send her message, Alice encodes two classical bits into her qubit by applying specific quantum operations. The message she wants to send can be one of four possible values: 00, 01, 10, or 11. Since her qubit is entangled with Bob's, she can manipulate the entangled state in a way that encodes two classical bits. Depending on the message she wants to send, Alice applies the following quantum gates:

- If she wants to send 00, she does nothing.
- If she wants to send 01, she applies a bit flip (X gate), which swaps the states $|0\rangle$ and $|1\rangle$.
- If she wants to send 10, she applies a phase flip (Z gate), which changes the phase of the qubit but does not alter the probability of measuring 0 or 1.

- If she wants to send 11, she applies both a bit flip and a phase flip (ZX gates), completely transforming the state.

After encoding the message, Alice physically sends her qubit to Bob. This is a key difference from quantum teleportation, where quantum information is transmitted without sending the qubit itself. In superdense coding, Alice must physically transfer the qubit to Bob through a quantum communication channel. Once Bob receives the qubit, he now possesses both qubits that were originally entangled.

With both qubits in his possession, Bob decodes the message by performing a Bell state measurement. This is a type of quantum measurement designed to distinguish between the four possible Bell states. By performing this measurement, Bob can determine which quantum operations Alice applied and thus extract the two classical bits of information she encoded. Even though Alice only sent a single qubit, the entanglement between them allowed her to transmit two classical bits.

Superdense coding is remarkable because it doubles the classical capacity of a quantum communication channel. Normally, a single qubit carries at most one bit of classical information when measured. However, by leveraging entanglement, superdense coding enables Alice to send two bits of information instead of one, making it a highly efficient form of communication. This efficiency is why the protocol is considered fundamental in quantum networking and quantum communication.

Another key insight from superdense coding is that entanglement is a necessary resource for the protocol to function. Without an initial entangled pair, Alice and Bob would not be able to achieve this information compression. If Alice simply sent an unentangled qubit, Bob would only be able to extract one classical bit upon measurement. The shared entanglement allows Bob to correctly interpret the two-bit message by measuring the relationship between both qubits rather than just measuring a single qubit in isolation.

The ability to efficiently encode and transmit information makes superdense coding valuable for various applications. It can be used in quantum data compression, allowing large amounts of information to be transmitted using fewer quantum resources. It is also a crucial tool in quantum networks, where high-speed quantum communication is required for secure and reliable data transfer. By reducing the number of qubits needed to transmit classical data, superdense coding plays a vital role in optimizing quantum communication systems.

In the context of future quantum internet and distributed quantum computing, superdense coding is expected to play a crucial role in optimizing bandwidth and reducing transmission delays. Quantum communication networks will benefit from this protocol by maximizing the efficiency of quantum channels, enabling faster and more secure data exchange between quantum nodes. As researchers develop more advanced quantum technologies, superdense coding will likely be integrated into real-world quantum communication infrastructure, enhancing the speed and capacity of quantum information transfer.

By making use of quantum entanglement and quantum gates, superdense coding demonstrates a clear advantage over classical communication methods. The protocol highlights the unique capabilities of quantum mechanics and its potential to revolutionize information transfer, making it a cornerstone of quantum communication and networking.

The CHSH game is a theoretical experiment used to test one of the most intriguing principles of quantum mechanics: nonlocality. This game is a quantum version of a classical test based on Bell's theorem, which challenges the idea that reality behaves according to classical physics. The game provides a concrete way to demonstrate that quantum entanglement allows for stronger correlations between distant particles than classical physics predicts. By playing the CHSH game, scientists and researchers can experimentally confirm that quantum mechanics fundamentally differs from classical intuition.

CHSH stands for Clauser-Horne-Shimony-Holt, named after the four physicists John Clauser, Michael Horne, Abner Shimony, and Richard Holt, who formulated the CHSH inequality in 1969.

The CHSH inequality is a mathematical expression used to test Bell's theorem, which explores whether quantum mechanics violates classical physics in terms of correlations between distant particles. The CHSH game is a reformulation of this inequality into a game-like setting, demonstrating quantum nonlocality by showing that quantum systems can achieve correlations stronger than what is possible with classical physics.

In the CHSH game, two players—Alice and Bob—are separated so that they cannot communicate. A referee gives each of them a random bit, which is either 0 or 1. Their task is to produce an output bit—again either 0 or 1—without sharing any information between them. The referee sets a rule that determines when Alice and Bob win the game: their outputs must satisfy a certain mathematical condition that depends on the input bits given to them. Since Alice and Bob cannot communicate once they receive their inputs, they must devise a pre-determined strategy to maximize their chances of winning.

If Alice and Bob were using classical strategies, they would have to rely on pre-agreed rules without any form of communication. Mathematically, the best they can do is correctly satisfy the game's condition 75% of the time. This limitation arises because classical physics follows local realism, meaning that their outputs must be predetermined based on their independent inputs and shared initial strategy.

However, if Alice and Bob share a pair of entangled qubits, they can use quantum mechanics to outperform classical strategies. Instead of following a deterministic rule, they apply quantum measurements to their entangled qubits. Because of the way entanglement works, their measurement outcomes exhibit correlations that are stronger than any classical system could achieve. Using this quantum strategy, they can win the CHSH game 85% of the time, which is significantly better than the classical maximum of 75%. This increased success rate provides experimental proof that quantum mechanics enables stronger-than-classical correlations.

The key takeaway from the CHSH game is that quantum entanglement allows for correlations that defy classical expectations. This result supports Bell's theorem, which states that any theory based on local hidden variables—the idea that particles carry pre-existing, deterministic properties—cannot explain the behaviour of entangled quantum particles. The fact that quantum mechanics allows Alice and Bob to win at a higher probability than classical physics predicts demonstrates that entanglement is a real, non-classical phenomenon.

Beyond being a theoretical curiosity, the CHSH game has profound implications for quantum cryptography and quantum security. Since quantum entanglement produces stronger-than-classical correlations, it enables ultra-secure communication methods that classical physics cannot replicate. Quantum key distribution (QKD) protocols, such as BB84 and E91, rely on this principle to ensure that an eavesdropper cannot intercept quantum-encrypted messages without being detected. The CHSH game serves as a practical test to confirm that a quantum system is truly behaving non-classically, making it a crucial tool in designing secure quantum networks.

Furthermore, the CHSH game is an experimental benchmark for testing quantum devices. If a quantum system can achieve the 85% success rate, it confirms that entanglement is present and functioning correctly. This makes the CHSH game a valuable method for validating quantum computers, quantum sensors, and other emerging quantum technologies. Researchers use variations of the game to test the limits of quantum mechanics, exploring how entanglement behaves under different conditions and refining our understanding of the quantum world.

At its core, the CHSH game provides direct experimental evidence that the universe does not follow purely classical rules. The ability to outperform classical systems by leveraging entanglement proves that quantum mechanics is fundamentally different from any classical description of reality. This insight not only deepens our understanding of quantum physics but also serves as the foundation for next-generation quantum technologies, including quantum computing, secure communication, and advanced sensing applications.

The CHSH game is more than just a theoretical experiment—it is proof that entanglement offers advantages that classical physics cannot replicate. It highlights the power of quantum mechanics and paves the way for quantum technologies that will revolutionize information science.

Quantum teleportation, superdense coding, and the CHSH game are cornerstones of quantum information science, showing how entanglement can be harnessed for communication, computation, and security.

- Quantum teleportation enables the transfer of quantum states without physically sending qubits.
- Superdense coding allows for efficient data transmission, using fewer quantum resources.
- The CHSH game provides experimental evidence for quantum nonlocality, proving that quantum mechanics defies classical expectations.

These protocols demonstrate why quantum computing is so powerful and why entanglement is a key resource for the future of quantum technologies.

Measurement: The Collapse of Quantum States

A key feature of quantum mechanics is that measurement affects the system being observed. Quantum measurement is the process of observing a qubit's state, but unlike classical bits, once a qubit is measured, it collapses into a definite value—either 0 or 1. Before measurement, the qubit exists in a superposition of states, described mathematically as a wavefunction that includes probabilities for both 0 and 1. The act of measurement forces the wavefunction to collapse, meaning that the superposition is lost and the qubit becomes indistinguishable from a classical bit.

This collapse introduces a major challenge for quantum computing. Because measuring a qubit disrupts its quantum state, quantum algorithms must be carefully designed to avoid premature measurement. If a qubit is observed too early in a computation, the rich quantum information encoded in its superposition is destroyed, and the computation loses its quantum advantage. This problem makes it necessary to develop techniques that allow quantum systems to perform complex operations before measurement occurs.

One such technique is quantum interference, which allows different possible quantum states to interfere with each other, amplifying correct solutions and cancelling out incorrect ones. By carefully manipulating qubits through quantum gates, quantum algorithms can perform operations while preserving quantum coherence. Quantum error correction is another essential strategy that helps maintain quantum states. Since quantum systems are highly sensitive to disturbances from their environment, error correction methods work by distributing quantum information across multiple entangled qubits, making it more resilient to decoherence and external noise.

In quantum computing, the challenge is not just in maintaining qubits in superposition but in extracting useful results without collapsing the entire system too soon. Quantum algorithms such as Shor's algorithm for factoring large numbers and Grover's search algorithm take advantage of superposition and interference to perform computations efficiently. The success of these algorithms depends on controlling quantum states without measuring them directly until the final step of computation.

The irreversible nature of quantum measurement also has profound implications for quantum cryptography and quantum security. Because observation fundamentally changes quantum states, quantum encryption techniques, such as Quantum Key Distribution (QKD), rely on the fact that any eavesdropper attempting to measure quantum information will necessarily alter it. This property enables ultra-secure communication methods that classical cryptography cannot achieve.

The measurement problem in quantum computing highlights one of the most counterintuitive aspects of quantum mechanics—the idea that reality is not fixed until it is observed. Unlike classical systems, where information can be stored, retrieved, and copied without disruption, quantum information is fragile and must be carefully manipulated to ensure meaningful computation. Future advancements in quantum hardware, such as topological qubits and fault-tolerant quantum computing, aim to address these challenges by making quantum states more stable and resistant to measurement-induced errors.

Understanding quantum measurement is crucial for the development of practical quantum computers. The ability to preserve quantum coherence, delay measurement, and extract results efficiently will determine the success of quantum technologies in fields like cryptography, drug discovery, artificial intelligence, and materials science. As researchers continue to refine quantum computing techniques, managing the effects of measurement remains one of the most fundamental hurdles to unlocking the full potential of quantum computing.

In summary, superposition allows quantum computers to represent multiple states at once, exponentially increasing computational power. Entanglement creates deep interconnections between qubits, enabling rapid communication and parallel processing. Measurement plays a crucial role in extracting results but also collapses the quantum state, making it necessary to carefully design quantum algorithms. Together, these principles define how quantum computers function and why they have the potential to revolutionize computing, cryptography, artificial intelligence, and materials science.

Quantum States and Wave Functions

In quantum mechanics, the concept of quantum states and wave functions is essential to understanding how particles behave at the microscopic level. Unlike classical physics, where objects have well-defined properties such as position and velocity, quantum mechanics describes particles using probability distributions governed by a wave function.

Quantum States

A quantum state is the complete description of a quantum system. It contains all the information about a particle, such as its position, momentum, spin, and energy levels. However, unlike classical objects, a quantum system does not have a definite state until it is measured. Instead, it exists in a superposition of multiple possible states, and measurement collapses the system into one of these possible outcomes.

Imagine a spinning top that can spin in any direction. In classical physics, we can precisely measure its orientation and speed at any given moment. In quantum mechanics, however, the spinning top exists in multiple orientations at once until it is measured. Only then does it "choose" a single direction.

Mathematically, a quantum state is represented by a wave function, usually denoted by Ψ (psi). The wave function describes the probability of finding the particle in different locations or states when measured.

Quantum states are fundamental to quantum mechanics and play a crucial role in describing how quantum systems evolve, interact, and are measured. They differ significantly from classical states, which are used in Newtonian physics to describe macroscopic objects like cannonballs, planets, or cars. Understanding quantum states requires an exploration of how they evolved from classical mechanics and how they differ in terms of determinism, probability, and measurement.

In classical mechanics, a state is a complete set of values describing a system at a given moment in time. For example, to fully describe the motion of a cannonball, we need to know its position and velocity at a specific instant. Once these values are known, Newton's laws allow us to predict exactly where the cannonball will be at any future moment. The equations of motion in classical physics are deterministic, meaning that if we start with precise initial conditions, we will always obtain a specific outcome.

Quantum mechanics, however, does not work this way. The state of a quantum system is represented by a wave function (denoted as Ψ) rather than a set of exact position and velocity values. The wave function does not provide definite outcomes but instead gives a probability distribution of possible results. Unlike classical mechanics, where measurements do not affect the state, quantum measurements fundamentally alter the system, introducing an element of randomness.

For instance, in the double-slit experiment, an electron's quantum state is spread across multiple paths, meaning it is not simply following a single trajectory like a classical cannonball. Instead, quantum states exist as superpositions of multiple possibilities until they are observed, at which point they collapse into a single outcome.

The double-slit experiment is a fundamental demonstration in quantum mechanics that illustrates the wave-particle duality of matter. Originally conducted by Thomas Young in 1801, the experiment involves directing a coherent light source (or particles such as electrons) at a barrier with two closely spaced slits. When light or particles pass through the slits, they create an interference pattern on a screen behind the barrier, indicative of wave behaviour. This phenomenon occurs because the waves emanating from the two slits overlap and interfere with each other, producing regions of constructive and destructive interference [46, 47].

The significance of the double-slit experiment extends beyond classical optics; it has profound implications in quantum mechanics. When particles such as electrons are sent through the slits one at a time, they still produce an interference pattern over time, suggesting that each particle behaves as a wave that passes through both slits simultaneously. This observation challenges classical intuitions about particles and raises questions about the nature of measurement and observation in quantum mechanics [48, 49]. Notably, if an attempt is made to measure which slit a particle passes through, the interference pattern disappears, demonstrating the principle of complementarity proposed by Niels Bohr [49].

Recent advancements in technology have allowed for more sophisticated versions of the double-slit experiment. For instance, experiments utilizing electron beams and advanced detection techniques have confirmed the wave-like behaviour of electrons, reinforcing the concept of wave-particle duality [50, 51]. Additionally, variations such as the cross-double-slit experiment have been proposed to further explore the nuances of quantum behaviour and the implications of measurement [52]. These experiments continue to challenge our understanding of quantum mechanics and the fundamental nature of reality.

In quantum mechanics, a system's state is fully described by a quantum state vector, which contains all the information about the system. Unlike classical states, which involve fixed numerical values, quantum state vectors are complex-valued functions governed by Schrödinger's equation. These states evolve deterministically according to mathematical rules but yield probabilistic results upon measurement.

For example, if an electron is in a quantum state that represents both spin-up and spin-down simultaneously, we cannot predict its exact spin before measurement. Instead, we can only assign a probability to each possible outcome. When we perform a measurement, the state "collapses" into one of the possible outcomes with a probability dictated by the wave function.

This is fundamentally different from classical physics. If we throw a die in classical physics and know all the forces acting on it, we could, in principle, predict exactly which number it will land on. In quantum physics, however, even if we have complete knowledge of the system, we can only predict probabilities—never a guaranteed outcome.

Measurement plays a completely different role in quantum mechanics than in classical mechanics. In classical physics, measuring a system does not change its state. If we measure the position of a moving car, we can check it again later without altering the car's motion.

In quantum mechanics, however, measurement fundamentally alters the system. Before measurement, a quantum state exists in superposition, meaning it holds multiple possible values at the same time. But as soon as a measurement is made, the state collapses into a single definite value.

For example, if we measure the position of an electron in an atom, its wave function, which was previously spread over multiple locations, collapses into one specific point. However, if we later try to measure its momentum, we find that our original measurement of position has destroyed any precise knowledge of momentum. This is an example of Heisenberg's uncertainty principle, which states that certain pairs of properties, like position and momentum, cannot be precisely known at the same time.

Quantum measurements fall into two categories:

- Compatible measurements, where repeated measurements do not alter the state. For example, measuring an electron's spin along one axis multiple times will always give the same result.
- Incompatible (or complementary) measurements, where measuring one property disturbs another. For instance, if we measure an electron's momentum, we lose information about its position, and vice versa.

Eigenstates and Pure States

When a quantum measurement is performed, the system collapses into an eigenstate corresponding to the measurement result. In simple terms, an eigenstate is a quantum state that remains unchanged under a particular measurement. If a system is already in an eigenstate of a certain observable, repeated measurements will always give the same value.

For example, if an electron is in an eigenstate of spin-up, measuring its spin multiple times will always yield the same result. However, if it is in a superposition of spin-up and spin-down, measuring it once will force it into either spin-up or spin-down, destroying the original superposition.

- A pure state is a quantum state that is completely known and not mixed with any statistical uncertainty. It corresponds to a single vector in a quantum system.
- A mixed state is a statistical mixture of multiple possible quantum states. In practical experiments, pure states are rarely achieved, and instead, we deal with probabilistic ensembles of quantum states.

Different Representations of Quantum States

Quantum states can be represented in different mathematical forms, depending on what is most useful for solving a problem.

- The wave function representation describes a quantum state as a function of position or momentum. This is commonly used in Schrödinger's equation.
- The matrix representation expresses quantum states and operations using linear algebra, making it useful for quantum computing and quantum mechanics in higher dimensions.
- The bra-ket notation, developed by Paul Dirac, provides a convenient abstract representation of quantum states in terms of vectors in a Hilbert space, which is useful for general quantum theories.

The wave function is a fundamental way to describe a quantum state in terms of position (x) or momentum (p). It is a complex-valued function denoted as Ψ(x, t) that gives the probability amplitude of finding a particle at a particular position.

Example: Particle in a Box (Position Representation)

For a quantum particle confined in a 1D box of length **L**, the wave function is given by:

$$\Psi_n(x) = \sqrt{\frac{2}{L}} \sin\left(\frac{n\pi x}{L}\right)$$

where:

- **n** is the quantum number (1, 2, 3, ...)
- **L** is the box length
- **x** is the position

This wave function describes a standing wave pattern where the probability of finding the particle is determined by **Ψ^2(x)**. The higher the value of Ψ^2(x), the more likely the particle is to be found there.

Example: Free Particle (Momentum Representation)

In momentum space, a free particle's wave function is given as a plane wave:

$$\Psi(p) = \frac{1}{\sqrt{2\pi\hbar}} e^{ipx/\hbar}$$

where:

- **p** is the momentum

- **ħ** is the reduced Planck's constant

This representation is useful when working with momentum-space operators or analysing quantum scattering problems.

Quantum states can also be represented as column vectors, while quantum operations (like measurements and transformations) are represented as matrices. This is widely used in quantum computing, where qubits are manipulated using matrix operations.

Example: Qubit Representation

A single qubit can be expressed as a vector in the computational basis:

$$|0\rangle = \begin{bmatrix} 1 \\ 0 \end{bmatrix}, \quad |1\rangle = \begin{bmatrix} 0 \\ 1 \end{bmatrix}$$

A general qubit state (a superposition of |0⟩ and |1⟩) is given by:

$$|\psi\rangle = \alpha\,|0\rangle + \beta\,|1\rangle = \begin{bmatrix} \alpha \\ \beta \end{bmatrix}$$

where **α** and **β** are complex numbers satisfying $|\alpha|^2 + |\beta|^2 = 1$.

Example: Quantum Gates as Matrices

Quantum gates manipulate qubits and are represented by matrices. Some common gates include:

- **Hadamard Gate (H):** Creates superposition

$$H = \frac{1}{\sqrt{2}} \begin{bmatrix} 1 & 1 \\ 1 & -1 \end{bmatrix}$$

Acting on **|0⟩**, the Hadamard gate produces:

$$H\,|0\rangle = \frac{1}{\sqrt{2}}\,(|0\rangle + |1\rangle)$$

Pauli-X Gate: Equivalent to the classical NOT gate

$$X = \begin{bmatrix} 0 & 1 \\ 1 & 0 \end{bmatrix}$$

Acting on **|0⟩**, the Pauli-X gate flips the qubit to **|1⟩**.

Quantum Computing

The bra-ket notation is a powerful abstract representation of quantum states, introduced by Paul Dirac, and is widely used in quantum mechanics and quantum information theory.

Example: General Quantum State Representation

A quantum state **$|\psi\rangle$** (ket) is written as:

$$|\psi\rangle = \alpha |0\rangle + \beta |1\rangle$$

The corresponding dual state (bra) is written as:

$$\langle\psi| = \alpha^* \langle 0| + \beta^* \langle 1|$$

where **α^*** and **β^*** are the complex conjugates of **α** and **β**.

Example: Inner Product in Bra-Ket Notation

The probability of measuring a state **$|\psi\rangle$** in the **$|0\rangle$** basis is given by the inner product:

$$\langle 0|\psi\rangle = \alpha$$

which represents the probability amplitude of obtaining **$|0\rangle$** as the outcome.

Example: Outer Product for Projectors

A projector operator that keeps a qubit in the **$|0\rangle$** state is written as:

$$|0\rangle \langle 0| = \begin{bmatrix} 1 & 0 \\ 0 & 0 \end{bmatrix}$$

This is useful in quantum computing for measuring and projecting quantum states onto specific bases.

Table 1: Comparison and Applications of Representations.

Representation	Example Use Cases	Mathematical Form
Wave Function Representation	Describes quantum states in position and momentum space, used in Schrödinger's equation	Ψ(x) or Ψ(p) (continuous function)
Matrix Representation	Used in quantum computing and quantum mechanics of discrete systems	Column vectors and matrices

Bra-Ket Notation	Abstract framework for quantum mechanics and quantum information theory	**

Each representation has its own advantages depending on the problem being solved. The wave function representation is essential for studying quantum mechanics and particle physics, the matrix representation is widely used in quantum computing, and the bra-ket notation provides a flexible way to express quantum states in theoretical physics.

By understanding and switching between these representations, physicists and engineers can develop quantum algorithms, analyse wave functions, and explore quantum systems across various fields of research.

Different representations are like choosing between different coordinate systems in geometry. Some representations make certain calculations easier while making others more complicated.

Quantum states are the fundamental building blocks of quantum mechanics, replacing the deterministic states of classical physics with probability distributions described by wave functions. Unlike classical systems, quantum states do not have definite values until they are measured, and measurement itself alters the state.

The probabilistic nature of quantum states, governed by Schrödinger's equation, allows quantum mechanics to explain phenomena that classical physics cannot, such as superposition, entanglement, and quantum tunnelling. This understanding forms the backbone of quantum computing, quantum cryptography, and quantum communication, leading to revolutionary technologies that harness these strange but powerful principles.

Wave Functions

A wave function (Ψ) is a mathematical function that encodes all possible states of a quantum system. It does not provide a single outcome but instead gives the probability of different outcomes when measured. The wave function evolves over time according to Schrödinger's Equation, which describes how quantum states change dynamically.

For a single particle, the wave function $\Psi(x, t)$ depends on position (x) and time (t). The square of the wave function's absolute value, $|\Psi(x, t)|^2$, gives the probability of finding the particle at a particular location.

For example, if we shoot an electron toward a wall with two slits, the wave function of the electron will pass through both slits simultaneously, creating an interference pattern on the other side, much like ripples in water. This is a fundamental difference from classical physics, where the electron would have to go through just one slit.

Before measurement, a quantum system exists in a superposition of states. However, the moment it is measured, the wave function collapses into a single definite outcome.

For example, if we measure an electron's spin, which can be either up (↑) or down (↓), before measurement, the electron is in a superposition of both states. The moment we observe it, however, it "chooses" one outcome. The other possibilities vanish, and the electron no longer remains in superposition.

This process, known as wave function collapse, is one of the most puzzling and debated aspects of quantum mechanics. It raises deep philosophical questions about the nature of reality and observation.

Wave functions also play a crucial role in quantum entanglement, where two or more particles share a single, unified quantum state. When two particles are entangled, their wave function cannot be described separately; instead, it forms a single system. Measuring one entangled particle instantaneously determines the state of the other, even if they are light-years apart.

For example, if two entangled electrons are in a superposition of spin states, measuring one to be spin up (↑) immediately collapses the other into spin down (↓), no matter how far apart they are. This is what Einstein famously called "spooky action at a distance."

Wave Function Representations in Quantum Mechanics

Wave function representations provide a fundamental way to describe quantum states, particularly in terms of position or momentum. Unlike classical states, which are well-defined in terms of specific values like position and velocity, quantum states are probabilistic in nature and evolve according to the Schrödinger equation. The wave function, denoted as $\boldsymbol{\Psi(x, t)}$ in position representation, encodes the probability amplitude of a particle's location at a given time. Squaring the wave function, $\boldsymbol{|\Psi(x, t)|^2}$, provides the probability density of finding the particle at a particular position. This is a key departure from classical mechanics, where a particle's position and momentum are well-defined at all times.

Pure quantum states describe systems in definite, well-characterized conditions. They are solutions to the Schrödinger equation and are often represented as eigenstates, labelled by quantum numbers corresponding to physical observables. For example, in the hydrogen atom, the pure states of an electron are characterized by the principal quantum number (**n**), angular momentum quantum number ($\boldsymbol{\ell}$), magnetic quantum number (**m**), and spin component ($\mathbf{s_x}$). These quantum numbers define the shape, orientation, and energy levels of the electron's wave function.

A common example of a pure state is an electron in a Stern-Gerlach experiment, where its spin state can only be measured as "up" or "down." In mathematical terms, this is represented as a two-dimensional complex vector:

$$|\psi\rangle = \alpha|\uparrow\rangle + \beta|\downarrow\rangle$$

where **α** and **β** are complex numbers that satisfy $|\alpha|^2 + |\beta|^2 = 1$, ensuring the total probability remains 1. These coefficients describe the probability of measuring the electron's spin in either the "up" or "down" state.

In quantum mechanics, every measurable quantity, such as energy or momentum, is associated with a mathematical operator. The possible values that can be observed experimentally are given by the eigenvalues of these operators. If a quantum state is in an eigenstate of an operator, measurement of that observable will always yield a specific value without uncertainty. However, if the state is a superposition of different eigenstates, measurement introduces quantum uncertainty, leading to a probabilistic outcome.

A quantum state evolving over time is represented as a linear combination of eigenstates:

$$|\Psi(t)\rangle = \sum_n C_n(t)|\Phi_n\rangle$$

where $C_n(t)$ are time-dependent coefficients that describe how the quantum state evolves dynamically. This evolution is governed by the time evolution operator, which determines how quantum states change under external influences.

Not all quantum states are pure. Mixed states describe statistical ensembles of quantum states where the system is not in a well-defined, single eigenstate but instead exists as a probabilistic mixture of several possible states. Unlike pure states, mixed states cannot be represented by a single wave function. Instead, they require a density matrix formalism, which allows for a more general description of quantum states.

For example, in a two-state system such as an electron's spin, a mixed state can be represented using a 2×2 density matrix:

$$\rho = \begin{bmatrix} p_1 & 0 \\ 0 & p_2 \end{bmatrix}$$

where p_1 and p_2 represent the probabilities of the system being in the spin-up and spin-down states, respectively. The trace of the density matrix must always equal 1, ensuring a valid probability distribution.

One notable example of a mixed state is the singlet state, which arises in quantum entanglement:

Quantum Computing

$$|\psi\rangle = \frac{1}{\sqrt{2}}(|\uparrow\downarrow\rangle - |\downarrow\uparrow\rangle)$$

In this state, the spins of two entangled particles are perfectly correlated, meaning that measuring the spin of one particle instantaneously determines the spin of the other, regardless of distance. This is a fundamental aspect of quantum mechanics that has been experimentally verified in Bell's inequality tests.

Mathematically, a mixed state is represented as a statistical mixture of pure states:

$$\rho = \sum_s p_s |\psi_s\rangle\langle\psi_s|$$

where $\mathbf{p_s}$ represents the probability of the system being in the pure state $|\psi_s\rangle$. Unlike a pure superposition, which involves complex coefficients allowing for interference effects, a mixed state is merely a probability-weighted sum of different states, lacking quantum coherence.

Measurement and the Heisenberg Uncertainty Principle

In classical physics, measuring an object does not change its state. For example, if you check the speed of a moving car, the car continues moving as before. However, in quantum mechanics, measurement is different—it actually changes the state of the system. This happens because quantum particles, like electrons or photons, do not have definite properties until they are measured. Before measurement, a quantum system exists in a superposition of multiple possible states, meaning it can be in a combination of different outcomes at the same time. But once it is observed, it instantly collapses into one specific state, losing all other possibilities.

This collapse is crucial in quantum computing because quantum computers rely on qubits—the quantum equivalent of classical bits—which can exist in superposition. However, when a qubit is measured, it behaves like a classical bit, taking a definite value of either 0 or 1. This means that while quantum computers can perform powerful parallel computations using superposition, extracting useful information requires careful handling. If qubits collapse too early, valuable quantum information is lost, reducing the advantage over classical computers.

One of the key consequences of quantum measurement is the Heisenberg uncertainty principle, which states that certain properties of a quantum system, like position and momentum, cannot be precisely known at the same time. If you measure the exact position of an electron, for example, its momentum becomes highly uncertain, and vice versa. The same applies to other quantum properties, like spin direction—measuring spin along one axis disrupts the spin along another axis. This uncertainty is not due to flaws in measurement instruments but is a fundamental property of nature.

For quantum computing, this principle means that measurements must be performed strategically. Quantum algorithms are designed to delay measurement until the computation is complete, ensuring that quantum states remain in superposition for as long as possible. If measured too soon, the quantum computer loses its ability to perform multiple calculations simultaneously, making it no better than a classical computer.

Moreover, quantum mechanics allows for entanglement, where two qubits become linked, meaning measuring one qubit instantly determines the state of the other, no matter how far apart they are. While entanglement is a powerful resource for quantum computing, it also presents challenges, as unintended measurements or interactions with the environment can disrupt entangled states, a problem known as quantum decoherence.

To overcome these challenges, quantum computers use techniques like error correction codes and quantum interference to maintain delicate quantum states. Researchers are also exploring fault-tolerant quantum computing, which ensures that useful computations can still be performed despite the fragile nature of qubits.

Unlike classical physics, quantum measurements inherently alter the state of a system. When an observable **A** is measured, the quantum state collapses into one of its eigenstates corresponding to the measured value. If the system was not initially in an eigenstate of **A**, the measurement outcome is probabilistic. After the first measurement, repeating the measurement will always yield the same result, as the system is now in a definite eigenstate.

This measurement process introduces a fundamental limit to the precision with which certain pairs of observables, such as position and momentum, can be simultaneously known. This is formalized in the Heisenberg uncertainty principle:

$$\Delta x \cdot \Delta p \geq \frac{\hbar}{2}$$

where **Δx** and **Δp** represent the standard deviations of position and momentum, and **ħ** is the reduced Planck's constant. This principle implies that preparing a system in a state with a precise position inevitably increases uncertainty in its momentum, and vice versa.

This principle also extends to incompatible observables, such as spin measurements along different axes. If an electron's spin is measured along the x-axis and then along the z-axis, the second measurement will disturb the first, resulting in a statistical spread of outcomes rather than a definite value.

the unique way quantum measurements work—causing state collapse, introducing uncertainty, and disrupting entangled qubits—makes quantum computing both powerful and challenging. Quantum computers must be carefully designed to maximize the advantages of quantum superposition and entanglement while minimizing the impact of measurement-induced disturbances. This is what makes quantum computing fundamentally different from classical computing and why developing stable quantum systems remains one of the biggest challenges in the field.

Quantum Entanglement and Bell's Theorem

A particularly striking consequence of wave function representations is quantum entanglement, where two particles share a joint quantum state. Entangled states cannot be factorized into independent wave functions for each particle. Instead, they must be described by a single wave function for the entire system.

Entanglement is the basis for many quantum technologies, including quantum teleportation and superdense coding. The experimental confirmation of entanglement was demonstrated through Bell's theorem, which showed that entangled particles exhibit correlations that cannot be explained by classical physics. These non-local correlations violate Bell's inequalities, providing strong evidence that quantum mechanics is fundamentally different from classical probability theory.

Schrödinger vs. Heisenberg Pictures

Quantum mechanics offers two equivalent formalisms for describing the evolution of quantum states. The Schrödinger picture treats quantum states as evolving over time while keeping observables fixed. The Heisenberg picture, in contrast, keeps states fixed while allowing observables to evolve dynamically. Both pictures yield the same physical predictions but provide different conceptual approaches to quantum mechanics.

In the Schrödinger picture, the state **$|\Psi(t)\rangle$** evolves according to the time-dependent Schrödinger equation:

$$i\hbar\frac{d}{dt}|\Psi(t)\rangle = H|\Psi(t)\rangle$$

where **H** is the Hamiltonian operator governing the system's energy. In the Heisenberg picture, operators such as position and momentum evolve in time according to the Heisenberg equation of motion:

$$\frac{dA}{dt} = \frac{i}{\hbar}[H, A] + \frac{\partial A}{\partial t}$$

where **[H, A]** is the quantum commutator between the Hamiltonian and observable **A**. The Heisenberg picture is especially useful in quantum field theory and high-energy physics.

The Schrödinger Picture: Evolution of Quantum States

In the Schrödinger picture, the quantum state |Ψ(t)⟩ changes over time, while observables (such as position, momentum, and energy) remain fixed. The time evolution of the quantum state is governed by the time-dependent Schrödinger equation:

$$i\hbar\frac{d}{dt}|\Psi(t)\rangle = H|\Psi(t)\rangle$$

where:

- i is the imaginary unit $(\sqrt{-1})$
- *ħ* (h-bar) is the reduced Planck's constant
- *H* is the Hamiltonian operator, which represents the total energy of the system (including kinetic and potential energy)
- |Ψ(t)⟩ is the quantum state vector that evolves over time

This equation dictates how a quantum system changes from one state to another. The solution to this equation is:

$$|\Psi(t)\rangle = e^{-iHt/\hbar}|\Psi(0)\rangle$$

which shows that the quantum state evolves according to a unitary operator $U(t) = e^{-iHt/\hbar}$, often called the time evolution operator.

Example: Free Particle Evolution

For a free particle with mass mmm and Hamiltonian:

$$H = \frac{p^2}{2m}$$

where *p* is the momentum operator, the wave function evolves as:

$$\Psi(x,t) = e^{-ip^2t/2m\hbar}\Psi(x,0)$$

This equation describes how a quantum wave function spreads over time, which is fundamental in understanding phenomena such as wave packet dispersion.

The Heisenberg Picture: Evolution of Observables

In the Heisenberg picture, the state remains fixed, and instead, the observables (operators) evolve over time. The equation governing the time evolution of an observable A is given by the Heisenberg equation of motion:

$$\frac{dA}{dt} = \frac{i}{\hbar}[H, A] + \frac{\partial A}{\partial t}$$

where:

- *[H,A]=HA−AH* is the **commutator** between the Hamiltonian and the observable A
- $\frac{\partial A}{\partial t}$ accounts for explicit time dependence of A

This equation shows that if an observable does not explicitly depend on time, its time evolution is fully determined by its commutator with the Hamiltonian.

Example: Evolution of Momentum and Position Operators

For a quantum particle, the position xxx and momentum p obey the canonical commutation relation:

$$[x, p] = i\hbar$$

Using the Heisenberg equation, we compute the time evolution of p:

$$\frac{dp}{dt} = \frac{i}{\hbar}[H, p] + \frac{\partial p}{\partial t}$$

For a simple Hamiltonian $H = \frac{p^2}{2m} + V(x)$, we calculate:

$$[H, p] = \left[\frac{p^2}{2m} + V(x), p\right] = -i\hbar\frac{dV}{dx}$$

which gives Newton's second law in quantum form:

$$\frac{dp}{dt} = -\frac{dV}{dx}$$

Similarly, the position operator evolves as:

$$\frac{dx}{dt} = \frac{i}{\hbar}[H, x]$$

Using $[p^2, x] = -2i\hbar p$, we get:

$$\frac{dx}{dt} = \frac{p}{m}$$

which is the quantum analogue of classical velocity.

Comparison and Relevance to Quantum Computing

The Schrödinger and Heisenberg pictures provide different perspectives on quantum evolution, but they are mathematically equivalent. The Schrödinger picture is typically used in quantum mechanics for solving problems involving wave functions and probability amplitudes, while the Heisenberg picture is more convenient in quantum field theory and quantum computing.

Relevance to Quantum Computing:

Schrödinger Picture in Quantum Computing:

- Quantum circuits manipulate quantum states, evolving them step-by-step.
- Quantum gates, such as Hadamard (*H*) and CNOT, act as unitary transformations, similar to the time evolution operator $e^{-iHt/\hbar}$.
- Algorithms like Shor's factoring algorithm and Grover's search algorithm use transformations in the Schrödinger picture to process quantum information.

Heisenberg Picture in Quantum Computing:

- Useful in error correction and quantum control theory, where observables like qubit states evolve over time.
- In quantum optics and quantum field theory, the Heisenberg picture describes interactions in quantum processors more naturally.
- Quantum hardware engineers use Heisenberg dynamics to model the evolution of physical qubits.

Both frameworks are essential in different aspects of quantum mechanics and computing, allowing researchers to model and control quantum systems effectively.

Why Do Quantum States and Wave Functions Matter?

Quantum states and wave functions are pivotal in the realms of quantum computing, quantum cryptography, and quantum communication, serving as the foundational elements that enable the unique capabilities of quantum technologies.

In quantum computing, qubits, which are the basic units of quantum information, can exist in a superposition of states. This property allows quantum computers to process multiple possibilities simultaneously, vastly increasing their computational power compared to classical computers. For instance, superposition enables quantum algorithms to explore many solutions at once, which is a significant advantage in tasks such as optimization and search problems [28]. Quantum states evolve through quantum gates, which manipulate these qubits to perform complex computations. The manipulation of entangled states is also crucial, as it allows for operations that are fundamentally impossible in classical computing [53]. The robustness of quantum states against decoherence is essential for maintaining the integrity of quantum computations, as errors can accumulate during processing [54].

Quantum cryptography leverages the unique properties of quantum states to achieve secure communication. The no-cloning theorem states that quantum states cannot be copied perfectly, which underpins the security of quantum key distribution (QKD) protocols. In QKD, entangled states are used to generate keys that are secure against eavesdropping, as any attempt to measure the quantum states will disturb them, alerting the communicating parties to potential interception [55]. This principle of entanglement not only enhances security but also enables protocols like superdense coding, which allows for the transmission of more information than classical methods would permit [56].

Quantum states also play a critical role in quantum sensing, where they enable high-precision measurements that surpass classical limits. The sensitivity of quantum sensors is often enhanced through the use of entangled states, which can provide better measurement outcomes due to their non-classical correlations [57]. Furthermore, quantum teleportation, which relies on entangled states and wave functions, allows for the transfer of quantum information from one location to another without physically moving the quantum state itself. This process is essential for quantum communication and information processing, as it enables the transmission of quantum states over long distances while preserving their integrity [58].

Quantum states and wave functions are the mathematical backbone of quantum mechanics. They describe how particles behave at the microscopic level, enabling powerful phenomena like superposition, entanglement, and wave function collapse. These principles not only challenge our classical understanding of reality but also open the door to revolutionary quantum technologies that will shape the future of computing, cryptography, and communication.

The Uncertainty Principle

The Heisenberg Uncertainty Principle is one of the most profound and counterintuitive concepts in quantum mechanics. It states that certain pairs of physical properties, such as position and momentum, cannot be measured with absolute precision at the same time. The more precisely we determine one property, the less precisely we can know the other.

At its core, it tells us that there is a fundamental limit to how precisely we can measure certain pairs of properties of a particle, such as position and momentum. The more accurately we measure one, the less accurately we can know the other. This is not due to any limitations in our measurement tools but is a fundamental feature of nature itself.

Imagine you are trying to take a photo of a moving car at night. If you use a very short exposure time, the image is sharp, and you can clearly see the car's position. However, because you only saw it for a brief moment, you can't accurately determine its speed. On the other hand, if you use a long exposure, the car appears blurry, making it easier to estimate its speed but much harder to pinpoint its exact position.

This is similar to what happens at the quantum level. If we try to measure the exact position of a tiny particle, such as an electron, we disturb its motion, making its momentum more uncertain. If we try to measure its momentum precisely, its position becomes more uncertain.

Unlike classical objects, quantum particles do not behave like tiny billiard balls moving along a predictable path. Instead, they exist in a wave-like state, described by a mathematical function called a wavefunction. These wave-like properties mean that a particle's position and momentum are not fixed until we measure them.

This principle is mathematically expressed as:

$$\Delta x \cdot \Delta p \geq \frac{\hbar}{2}$$

where:

- Δx is the uncertainty in position
- Δp is the uncertainty in momentum
- $\hbar$ (h-bar) is the reduced Planck's constant, approximately 1.054×10^{-34} Joule-seconds

This equation means that if we try to measure a particle's exact position with high precision (making Δx very small), the uncertainty in its momentum Δp must increase, making it impossible to determine both values simultaneously with infinite accuracy, or that the more precisely we know one variable (like position **x**), the more uncertainty there must be in the other variable (momentum **p**).

The Uncertainty Principle, a cornerstone of quantum mechanics, fundamentally alters our understanding of measurement and the behaviour of subatomic particles. It is not merely a limitation of measurement tools but a reflection of the intrinsic properties of quantum systems, particularly the wave-particle duality that governs their behaviour. This principle manifests in various phenomena, including the behaviour of electrons in atoms, the results of the double-slit experiment, and the implications of particle confinement.

Electrons do not orbit the nucleus in fixed paths akin to planets around the Sun; rather, they exist in probabilistic distributions known as wavefunctions. These wavefunctions describe the likelihood of finding an electron in a particular region of space, illustrating the uncertainty in both position and momentum. If an electron were to possess both a precise position and momentum, it would contravene the Uncertainty Principle, which states that the product of the uncertainties in position (Δx) and momentum (Δp) must satisfy the relation $\Delta x \Delta p \geq \hbar/2$, where $\hbar$ is the reduced Planck constant [59]. This fundamental aspect of quantum mechanics emphasizes that the more accurately we know an electron's position, the less accurately we can know its momentum, and vice versa [60].

The double-slit experiment vividly illustrates the Uncertainty Principle in action. When electrons are fired at a barrier with two slits, they exhibit wave-like behaviour, creating an interference pattern on a detection screen. This pattern emerges because the electrons are in a superposition of states, passing through both slits simultaneously. However, if a measurement is made to determine which slit an electron passes through, the act of measurement collapses the wavefunction, resulting in the loss of the interference pattern. This phenomenon highlights the trade-off between knowledge of position and momentum: measuring the position (which slit the electron goes through) increases the uncertainty in momentum, thus altering the outcome of the experiment [61, 62].

In scenarios where particles are confined to small regions, such as within quantum dots or atomic nuclei, the uncertainty in position (Δx) is significantly reduced. According to the Uncertainty Principle, this reduction leads to an increase in the uncertainty of momentum (Δp), which in turn raises the kinetic energy of the particles. This relationship is critical in understanding the behaviour of particles in confined systems, as tighter confinement results in higher energy states due to the increased momentum uncertainty [59, 63]. The implications of this principle extend to various fields, including quantum computing and nanotechnology, where controlling particle confinement is essential for device performance [63].

Beyond position and momentum, the uncertainty principle applies to other pairs of incompatible observables (operators that do not commute), such as:

$$\Delta E \cdot \Delta t \geq \frac{\hbar}{2}$$

where:

- ΔE is the uncertainty in energy
- Δt is the uncertainty in time

This relation implies that the energy of a quantum system cannot be precisely defined over an arbitrarily short time. This principle explains why virtual particles can momentarily appear in empty space due to quantum fluctuations.

The uncertainty principle plays a fundamental role in quantum technologies, particularly in quantum computing and quantum cryptography. Unlike classical computing, where information is stored in definite states of 0s and 1s, quantum computing leverages the probabilistic nature of quantum mechanics. The uncertainty principle directly impacts how quantum information is processed, measured, and secured.

In quantum computing, superposition and measurement are deeply tied to the uncertainty principle. Quantum bits, or qubits, can exist in a superposition of both 0 and 1 simultaneously. This ability allows quantum computers to process multiple possibilities at once, offering a significant computational advantage. However, measurement in quantum mechanics forces a qubit into one definite state, either 0 or 1, collapsing its superposition. This process is analogous to how an electron's wavefunction collapses when its position is measured. The challenge in quantum computing is to extract useful information without prematurely collapsing quantum states, ensuring that the quantum coherence necessary for computation is maintained as long as possible.

Another critical aspect of quantum computing influenced by the uncertainty principle is quantum error correction. Unlike classical bits, qubits are highly susceptible to noise and decoherence due to their fragile quantum states. The uncertainty principle implies that any attempt to precisely measure a qubit's state will disturb it, making direct error detection difficult. To overcome this, quantum error correction relies on entanglement, where multiple qubits are linked in a way that allows errors to be identified and corrected without direct measurement. Techniques such as the surface code and Shor's error correction code utilize redundancy across several qubits, preserving quantum information even in the presence of noise.

In quantum cryptography and security, the uncertainty principle ensures that information remains protected against eavesdropping. One of the most significant applications is Quantum Key Distribution (QKD), which allows two parties to exchange encryption keys securely. If an eavesdropper attempts to intercept a quantum key, the act of measurement will inevitably disturb the system, introducing detectable errors in the transmission. This ensures that any unauthorized attempt to access the data is revealed, making QKD theoretically unbreakable under the laws of quantum mechanics.

The uncertainty principle also plays a role in quantum computing algorithms, particularly those that take advantage of quantum interference. Shor's Algorithm, used for factoring large numbers exponentially faster than classical methods, relies on quantum superposition and entanglement. Similarly, Grover's Algorithm, which enables faster searching in unstructured databases, leverages the principles of quantum probability and interference. Both of these algorithms would not be possible without the uncertainty principle governing the behaviour of quantum states.

Overall, the uncertainty principle is not just a theoretical constraint but a foundational aspect that enables the power of quantum computing. It influences how quantum states evolve, how measurements are performed, how errors are corrected, and how security is maintained in quantum communication. By harnessing the fundamental limits of measurement and uncertainty, quantum computing is able to achieve computational advantages that classical systems cannot replicate.

One of the biggest implications of the Heisenberg Uncertainty Principle is that particles never have a definite position and momentum at the same time. This is very different from classical physics, where we can precisely determine both.

Another consequence is that absolute stillness is impossible. In classical physics, if an object has no movement, it has zero momentum. However, in quantum mechanics, since we can never precisely measure both position and momentum at the same time, a particle can never have exactly zero motion. This is why even in the coldest temperatures, atoms and molecules still exhibit some movement, known as zero-point energy.

Quantum computing relies on the strange properties of quantum mechanics, including superposition and entanglement. The uncertainty principle plays a key role in determining how quantum information is processed and measured. Since measurement collapses quantum states, quantum algorithms must be designed to extract useful results without destroying the delicate quantum states too early.

Additionally, quantum cryptography uses the uncertainty principle to ensure security. If an eavesdropper tries to measure a quantum key being transmitted, the act of measurement disturbs the system, making it immediately detectable.

The Heisenberg Uncertainty Principle is a fundamental aspect of quantum mechanics that limits the precision with which certain properties of a particle can be known simultaneously. It is not a flaw of our instruments but an inherent property of nature. This principle has profound implications for atomic physics, quantum computing, and even our understanding of the universe. In quantum technologies, it plays a crucial role in quantum cryptography, computing, and error correction, making it one of the most significant principles in modern physics.

Schrödinger's Cat and Other Thought Experiments

Quantum mechanics is renowned for its strange and counterintuitive phenomena that challenge our conventional understanding of reality. These phenomena are often illustrated through thought experiments, which are hypothetical scenarios designed to highlight the peculiar nature of quantum behaviour. Among the most famous of these thought experiments is Schrödinger's Cat, which exemplifies the concept of superposition, where a system can exist in multiple states simultaneously until measured. However, there are several other notable thought experiments, including Wigner's Friend, the EPR Paradox, and the Quantum Zeno Effect, each of which provides insight into fundamental quantum principles such as entanglement and measurement.

Wigner's Friend is a particularly intriguing thought experiment that raises questions about the nature of observation and reality in quantum mechanics. In this scenario, Wigner's friend performs a measurement on a quantum system inside a sealed laboratory, while Wigner himself remains outside. The implications of this setup challenge the notion of objective reality, as the friend's measurement outcome may differ from Wigner's perspective, leading to discussions about the relational nature of quantum states and the observer's role in defining reality [64, 65]. This thought experiment has been analysed through various interpretations of quantum mechanics, such as Relational Quantum Mechanics, which posits that physical quantities only have definite values relative to an observer [64, 66].

The EPR Paradox, proposed by Einstein, Podolsky, and Rosen, further illustrates the peculiarities of quantum entanglement, where two particles can become correlated in such a way that the state of one instantly influences the state of the other, regardless of the distance separating them. This phenomenon raises fundamental questions about locality and the completeness of quantum mechanics [67]. The paradox challenges the classical notion of separability and suggests that quantum mechanics may require a re-evaluation of our understanding of reality itself.

Another significant thought experiment is the Quantum Zeno Effect, which describes how frequent measurements can inhibit the evolution of a quantum system. This effect is analogous to Zeno's paradox, where a moving arrow is said to never reach its destination if observed continuously. In quantum mechanics, repeated measurements can effectively "freeze" the dynamics of a system, preventing it from transitioning to other states [68, 69]. This phenomenon has profound implications for quantum computing, as it suggests that the timing and frequency of measurements can be manipulated to control quantum states, potentially enhancing computational capabilities [70, 71].

Together, these thought experiments underscore the bizarre nature of quantum mechanics and its implications for our understanding of reality. They reveal that the act of measurement is not merely a passive observation but an active process that influences the state of a quantum system. As researchers continue to explore these concepts, the interplay between quantum mechanics and information theory becomes increasingly relevant, particularly in the context of developing quantum technologies and computing systems [72].

Schrödinger's Cat: A Paradox of Superposition and Measurement

In 1935, Austrian physicist Erwin Schrödinger proposed a thought experiment known as Schrödinger's Cat to illustrate the paradoxes inherent in quantum mechanics, particularly superposition and measurement. The experiment was designed to challenge the way quantum mechanics describes the behaviour of particles at a fundamental level and to highlight the issues in applying quantum principles to macroscopic systems.

The setup of the experiment is as follows: A cat is placed inside a sealed box with a radioactive atom, a Geiger counter, a vial of poison, and a hammer. The radioactive atom has a 50% probability of decaying within an hour. If the atom decays, the Geiger counter detects the radiation, triggering the hammer to break the vial of poison, which would kill the cat. If the atom does not decay, the cat remains alive. According to quantum mechanics, the state of the atom is in a superposition of decayed and undecayed states until it is observed. Because the cat's fate is directly linked to the state of the atom, it is argued that the cat itself must also exist in a superposition of being both alive and dead until someone opens the box and observes it.

This paradox highlights one of the fundamental dilemmas in quantum mechanics—the measurement problem. In quantum physics, particles exist in multiple states simultaneously, described by a mathematical function called the wavefunction. However, the moment a measurement is made, the wavefunction "collapses," forcing the system to take on a definite state. Schrödinger's experiment raises the question of when exactly this collapse occurs and whether the process of measurement itself determines reality.

Schrödinger's Cat is more than just a philosophical puzzle—it has profound implications for quantum computing, as it directly relates to the principles governing qubits and quantum information processing.

The experiment serves as an analogy for superposition in quantum computing. In classical computing, bits are always in a definite state of either 0 or 1. However, in a quantum computer, qubits can exist in a superposition of both 0 and 1 at the same time, much like Schrödinger's cat being both dead and alive before observation. This unique property allows quantum computers to perform multiple calculations simultaneously, offering an exponential speedup for certain types of problems, such as factorization, search algorithms, and cryptography.

One of the key challenges in quantum computing is quantum measurement. Just as opening the box collapses the cat's state into either dead or alive, measuring a qubit collapses its superposition into a single state (either 0 or 1). This collapse destroys the quantum information stored in the superposition, limiting the computational process. Therefore, quantum algorithms must be carefully designed to minimize unnecessary measurements until the desired computation is complete. Quantum error correction techniques also play a vital role in managing this issue by preserving quantum states as long as possible.

Another important concept related to Schrödinger's Cat is quantum coherence—the ability of a qubit to maintain its superposition without being disturbed. The longer a qubit remains in superposition, the more useful it is for quantum computation. However, quantum decoherence occurs when environmental interactions cause the qubit to lose its quantum properties and collapse into a definite state prematurely. This is similar to the cat's state being influenced by external factors before the box is opened.

Quantum computers must be operated in extremely controlled environments, often at temperatures close to absolute zero, to reduce decoherence. Scientists are actively researching ways to extend quantum coherence times using techniques such as topological qubits, error correction codes, and quantum entanglement.

Schrödinger's Cat remains one of the most influential thought experiments in quantum mechanics, illustrating the strange and non-intuitive nature of quantum superposition and measurement. Its implications extend directly into quantum computing, where qubits rely on the same principles to perform computations exponentially faster than classical computers. However, challenges such as measurement collapse, quantum decoherence, and environmental disturbances must be carefully managed to harness the full power of quantum technology. By better understanding these fundamental quantum principles, researchers continue to make progress in building more stable and scalable quantum computers.

Wigner's Friend: Extending the Schrödinger's Cat Paradox

In 1961, physicist Eugene Wigner proposed a thought experiment known as Wigner's Friend, which builds on Schrödinger's Cat to explore the role of observers and consciousness in quantum measurement. The paradox extends Schrödinger's original scenario by introducing a human observer inside the box instead of a cat. This thought experiment raises profound questions about the nature of measurement, consciousness, and reality in quantum mechanics.

Imagine a sealed laboratory in which Wigner's friend is conducting the Schrödinger's Cat experiment. Inside the lab, the friend observes whether the cat is alive or dead after the quantum measurement. However, Wigner, who remains outside the lab, has no knowledge of what his friend has observed. From Wigner's perspective, the entire system—the friend and the cat—remains in a quantum superposition of all possible outcomes until he opens the lab door and interacts with his friend. According to quantum mechanics, until Wigner directly measures the system, his friend and the cat's fate should exist in a superposition of all possible outcomes.

This raises a fundamental philosophical question: Is quantum measurement absolute, or is it observer-dependent? If Wigner's friend already observed the cat's state, does that collapse the wavefunction, or does Wigner's measurement collapse an even larger wavefunction that includes his friend? The paradox suggests that different observers could have different realities based on their respective measurements, leading to deep discussions about whether consciousness plays a role in quantum measurement.

The Wigner's Friend experiment challenges our understanding of how quantum mechanics applies to larger systems, including conscious observers. In classical physics, reality exists independently of observation. However, in quantum mechanics, observation appears to affect the state of a system. If a human observer (Wigner's friend) already knows the result, but another observer (Wigner) does not, does the quantum system still remain in superposition until both have measured it?

Several interpretations of quantum mechanics offer different explanations for this paradox.

- The Copenhagen Interpretation suggests that the measurement collapses the wavefunction, meaning that Wigner's friend has already observed a definite reality. However, Wigner himself does not experience this reality until he interacts with his friend.

- The Many-Worlds Interpretation offers an alternative view, suggesting that the wavefunction never collapses. Instead, every possible outcome branches into a new parallel universe, meaning that in some universes, Wigner's friend observed a dead cat, while in others, they observed a living cat.

- The Relational Interpretation proposes that quantum states are observer-dependent, meaning Wigner's friend and Wigner could each have a different version of reality based on their respective observations.

These interpretations highlight the philosophical and practical challenges of understanding measurement in quantum mechanics.

Wigner's Friend has direct implications for quantum computing and quantum information theory, particularly regarding measurement, observer effects, and quantum interpretations.

The paradox highlights the role of measurement in collapsing quantum states, a crucial factor in quantum computing. In a quantum computer, qubits exist in a superposition of states until they are measured. If a measurement occurs too early, it collapses the qubit into a definite state, disrupting the quantum computation. This means that quantum algorithms must be carefully designed to delay measurement until the final step, ensuring that quantum states remain coherent throughout the computation.

For example, quantum error correction techniques are essential to prevent environmental interactions from unintentionally collapsing qubits, similar to how Wigner's measurement affects his friend's observation.

The Many-Worlds Interpretation of quantum mechanics, which suggests that different outcomes exist in parallel realities, aligns with how quantum computing performs parallel processing. Quantum computers take advantage of superposition and entanglement to explore multiple possible solutions at the same time, akin to Wigner's Friend existing in multiple possible realities until observed.

In quantum cryptography, the uncertainty in quantum measurement ensures secure communication. Quantum key distribution (QKD) relies on the fact that any measurement by an eavesdropper will collapse the quantum state, alerting legitimate users to the intrusion. The Wigner's Friend paradox reinforces the idea that observation changes the state of a system, which is fundamental to cryptographic security.

Wigner's Friend extends Schrödinger's Cat to the human level, raising fundamental questions about whether reality is absolute or observer-dependent. The thought experiment challenges the way we interpret quantum measurement and wavefunction collapse, leading to debates over the role of consciousness in quantum mechanics.

For quantum computing, the paradox reinforces the importance of measurement control, quantum coherence, and interpretations of quantum mechanics that influence how quantum algorithms and cryptographic systems function. As researchers continue to explore the foundations of quantum mechanics, thought experiments like Wigner's Friend remain crucial in shaping our understanding of quantum reality and computation.

The EPR Paradox: Quantum Entanglement and Nonlocality

In 1935, Albert Einstein, Boris Podolsky, and Nathan Rosen proposed a thought experiment that would become known as the EPR Paradox. This paradox challenged the completeness of quantum mechanics, questioning whether the theory could fully describe physical reality. At the heart of the paradox is quantum entanglement, a phenomenon where two or more particles share a quantum state, meaning the measurement of one instantly determines the state of the other, regardless of the distance between them. Einstein, who was sceptical of this concept, referred to it as "spooky action at a distance," as it seemed to imply that information could travel faster than the speed of light, violating the principles of special relativity.

To understand the paradox, imagine that two particles are entangled and then separated by a vast distance—perhaps even light-years apart. According to quantum mechanics, if an observer measures a property (such as spin or polarization) of one of the entangled particles, the state of the second particle will be instantly determined, even though no classical information could have travelled between them. This instantaneous correlation suggests that either information is traveling faster than light or that quantum mechanics is incomplete and must rely on some form of hidden variables that predefine the states of the particles before measurement.

Einstein and his colleagues argued that if quantum mechanics were correct, then two entangled particles must be influencing each other in some nonlocal way, violating classical notions of causality. This led them to propose that quantum mechanics was an incomplete theory, suggesting that hidden variables must exist to account for the seemingly instantaneous correlation between entangled particles.

However, later experiments—most notably those based on Bell's Theorem—confirmed that no local hidden variables could explain quantum entanglement. John Bell mathematically demonstrated that quantum mechanics makes predictions about correlations between entangled particles that are fundamentally different from those allowed by classical physics. Experiments conducted by physicists such as Alain Aspect in the 1980s showed that these quantum correlations are real, reinforcing the idea that nature itself is fundamentally nonlocal.

The confirmation of the EPR Paradox through experimental tests of Bell's Inequality revolutionized our understanding of reality at the quantum level. It established that quantum mechanics is not just a mathematical tool but a description of how nature fundamentally operates. This has led to significant developments in quantum technologies, particularly in quantum computing and quantum communication.

One of the most important conclusions from the EPR Paradox is that measurement collapses the wavefunction not just locally, but across any distance where entangled particles exist. This means that while quantum mechanics does not allow for classical faster-than-light communication, it does enable instantaneous correlations, which can be used for advanced technologies such as quantum cryptography and distributed quantum computing.

The EPR Paradox is not just a theoretical concept; it has direct applications in modern quantum computing and quantum information science.

Quantum computers rely on qubits, which, unlike classical bits, can exist in superposition and be entangled with one another. When qubits are entangled, changes made to one qubit instantly affect the others, even if they are physically separated. This allows quantum computers to process information in parallel, greatly enhancing their computational power. Quantum algorithms like Shor's algorithm (for factoring large numbers) and Grover's algorithm (for searching databases) leverage entanglement to perform calculations exponentially faster than classical computers.

One of the most promising applications of quantum entanglement is in quantum cryptography. Quantum Key Distribution (QKD) protocols, such as BB84 and E91, use entangled particles to create encryption keys that are impossible to intercept without detection. Because any attempt to measure an entangled qubit disrupts the entire entangled system, QKD provides unbreakable encryption for secure communications. This is particularly important for governments, financial institutions, and other organizations that require highly secure data transmission.

Entanglement also plays a key role in distributed quantum computing, where multiple quantum computers located in different parts of the world could work together as a unified system. By entangling qubits across large distances, quantum computers can share information instantly, allowing for more efficient computations and reducing the need for massive quantum processors. Scientists are currently researching quantum teleportation techniques that would allow entangled states to be transmitted between quantum processors, paving the way for the quantum internet.

The EPR Paradox remains one of the most important thought experiments in quantum mechanics, challenging classical notions of locality and realism. Although originally designed to highlight the apparent contradictions in quantum mechanics, it instead led to the experimental verification of quantum entanglement and the development of new technologies.

In quantum computing, entanglement is a fundamental resource that enables faster computation, secure communication, and distributed quantum processing. The ability of entangled qubits to instantaneously share information makes quantum computing superior to classical computing for certain problems. Furthermore, the use of entanglement in quantum cryptography ensures a new era of secure communications, where encryption cannot be broken by classical means.

The EPR Paradox thus not only reshaped our understanding of the quantum world but also laid the foundation for the quantum technologies of the future.

The Quantum Zeno Effect: Preventing Change by Observation

The Quantum Zeno Effect is a counterintuitive phenomenon in quantum mechanics where frequent measurements prevent a quantum system from evolving over time. The name is derived from Zeno's paradox, a philosophical argument by the ancient Greek philosopher Zeno of Elea, which suggests that constant observation of motion prevents change. In quantum mechanics, this idea is not just a paradox—it is a real, experimentally verified effect.

In quantum systems, particles exist in superposition, meaning they can be in multiple states simultaneously. However, when a measurement is performed, the system collapses into one of these states, losing its superposition. If measurements are performed rapidly and repeatedly, the wavefunction collapses so frequently that the system is effectively "frozen" in its initial state. This prevents the natural quantum evolution that would otherwise occur according to Schrödinger's equation.

Quantum Computing

To grasp the concept, consider an analogy with a boiling pot of water. Suppose you keep opening the lid every second to check whether the water has started boiling. Although in classical physics this does not affect the boiling time, in quantum mechanics, repeated observations actually prevent the system from evolving. The boiling process would be continuously interrupted by these frequent checks, stopping the system from reaching its final state.

A more precise quantum example involves an unstable atomic nucleus undergoing radioactive decay. If an observer constantly measures the state of the nucleus at very short time intervals, the decay process is significantly slowed down or even completely halted. The reason is that each measurement forces the quantum state to "reset" itself, preventing the system from transitioning into the decayed state.

This effect was theoretically proposed in interpretations of quantum mechanics and later confirmed through experimental tests. Scientists observed that quantum particles, such as trapped ions or atoms in optical lattices, could indeed be prevented from evolving if they were frequently measured.

The Quantum Zeno Effect is not just a curiosity—it plays a fundamental role in quantum computing and other advanced quantum technologies. Since quantum computers rely on fragile quantum states, controlling and stabilizing these states is essential for their operation.

Quantum computers are highly susceptible to errors due to decoherence, which causes qubits to lose their quantum properties over time. The Quantum Zeno Effect can be used to prevent qubits from decohering by frequently measuring them in a controlled manner. This allows error correction protocols to detect and suppress errors before they spread and disrupt the computation.

One of the biggest challenges in quantum computing is maintaining quantum coherence—the ability of qubits to remain in superposition and entanglement. Controlled measurement strategies that leverage the Quantum Zeno Effect can help extend the coherence time of qubits, making them more stable for longer periods. This is especially important for running complex quantum algorithms that require sustained coherence.

To build a practical quantum computer, qubits must remain stable without unwanted state transitions. The Quantum Zeno Effect can be integrated into quantum circuits to lock qubits into specific states, preventing them from undergoing unintended changes. This is a crucial step toward fault-tolerant quantum computing, where errors are minimized, and quantum gates operate with high reliability.

The Quantum Zeno Effect is a profound example of how quantum measurement can influence the behaviour of a quantum system. While classical physics assumes that observing a system does not change it, quantum mechanics shows that frequent observation can actually prevent change from occurring.

In quantum computing, this effect is particularly valuable for error correction, qubit stabilization, and fault-tolerant computing. As quantum technologies advance, the Quantum Zeno Effect may become an essential tool for protecting quantum information, ensuring that quantum computers can perform calculations reliably without being disrupted by environmental noise or decoherence.

Thought Experiments and Their Impact on Quantum Computing

These thought experiments are more than just abstract philosophical exercises. They serve as essential conceptual tools that help illustrate the fundamental principles of quantum mechanics, principles that directly shape the design, functionality, and challenges of quantum computing. By exploring these paradoxes and their implications, scientists and engineers gain a deeper understanding of how quantum systems behave, enabling the development of advanced quantum technologies.

Schrödinger's Cat is perhaps the most well-known thought experiment in quantum mechanics. It demonstrates the principle of superposition, where a system can exist in multiple states simultaneously until measured. This directly relates to how qubits function in quantum computing—a qubit can be in a state of 0, 1, or a superposition of both. Additionally, this experiment highlights the measurement problem, which poses a significant challenge in quantum computation. The act of measuring a quantum state collapses its superposition, forcing it into a definite state and potentially disrupting calculations. Understanding and mitigating this issue is crucial in designing quantum algorithms and quantum error correction techniques.

Wigner's Friend extends the measurement problem by introducing the role of an observer in quantum mechanics. This thought experiment suggests that the outcome of a quantum measurement might be observer-dependent, meaning that quantum states could exist differently depending on who is measuring them. This raises questions about whether quantum mechanics applies universally to all systems, including conscious observers. In the context of quantum computing, this debate influences different interpretations of quantum mechanics, particularly when considering how quantum information is processed and whether quantum computers operate within a single deterministic reality or across multiple parallel universes, as suggested by the Many-Worlds Interpretation.

The EPR Paradox, proposed by Einstein, Podolsky, and Rosen, provides one of the strongest proofs of quantum entanglement—a phenomenon where two or more particles become so deeply correlated that measuring the state of one instantaneously determines the state of the other, no matter how far apart they are. This paradox was initially designed to challenge quantum mechanics, but later experiments confirmed entanglement as a fundamental reality. Quantum computing harnesses this property for quantum teleportation, quantum networking, and exponentially faster computations. By entangling qubits, quantum computers can perform operations in parallel, vastly increasing computational power. Furthermore, quantum cryptography leverages entanglement to develop ultra-secure communication methods, as any attempt to measure an entangled particle disturbs the system, making eavesdropping detectable.

The Quantum Zeno Effect offers insights into how frequent measurement can stabilize quantum systems. In quantum computing, one of the greatest challenges is maintaining qubit coherence—preventing quantum information from being lost due to environmental disturbances. The Quantum Zeno Effect suggests that by carefully measuring a quantum system at rapid intervals, it is possible to prevent its quantum state from changing, thereby improving fault tolerance and quantum error correction. This principle is being explored as a way to extend the lifespan of qubits and increase the reliability of quantum computations.

These thought experiments not only help explain the strange behaviours of quantum mechanics but also drive the development of quantum technologies. The insights gained from these paradoxes influence the engineering of quantum processors, the construction of quantum communication networks, and the refinement of quantum algorithms. As quantum computing progresses, researchers continue to investigate these fundamental quantum effects, searching for new ways to harness quantum mechanics for practical applications in computing, cryptography, and beyond.

Practical Applications of Quantum Mechanics Concepts

The concepts in this chapter describe fundamental principles of quantum mechanics and their direct applications to quantum computing. The following outlines how these concepts are practically implemented in quantum computing and how they would be used by developers of quantum computers.

1. Superposition and Its Implementation in Quantum Computing

Superposition is a core principle of quantum mechanics that allows a quantum system, such as a qubit, to exist in multiple states simultaneously. In quantum computing, this means that qubits can represent both 0 and 1 at the same time instead of being limited to a single value like classical bits.

Practical Implementation:

- **Quantum Gates:** Developers use Hadamard gates (H-gates) to place qubits into superposition. The H-gate transforms a qubit from a definite 0 or 1 state into a superposition of both.

- **Parallel Processing:** Superposition allows quantum computers to process multiple computations simultaneously, leading to an exponential increase in computing power for specific algorithms.

Usage by Developers:

- **Quantum Algorithm Design:** Algorithms such as Grover's algorithm and Shor's algorithm leverage superposition to explore multiple solutions at once.
- **Quantum Circuit Simulation:** Developers test quantum programs using software like IBM's Qiskit or Google's Cirq to ensure qubits maintain superposition until measurement.
- **Quantum Machine Learning:** Superposition enables quantum neural networks to process multiple states simultaneously, making them more efficient than classical models.

Practical Implementation Example: Hadamard Gate Creating Superposition

One of the most fundamental quantum operations is applying the Hadamard gate (H-gate) to a qubit to put it into superposition.

Let's consider a single qubit initially in the state $|0\rangle$. Applying the Hadamard gate transforms it into an equal superposition of both $|0\rangle$ and $|1\rangle$:

$$H|0\rangle = \frac{|0\rangle + |1\rangle}{\sqrt{2}}$$

Similarly, if the initial qubit state was $|1\rangle$:

$$H|1\rangle = \frac{|0\rangle - |1\rangle}{\sqrt{2}}$$

This means that after applying the Hadamard gate, the qubit has an equal probability (50%) of being measured as either 0 or 1.

In a real quantum computer, this operation is written in Python using IBM's Qiskit as:

```
from qiskit import QuantumCircuit, Aer, execute

# Create a quantum circuit with 1 qubit
qc = QuantumCircuit(1, 1)
```

```
# Apply Hadamard gate to the qubit
qc.h(0)

# Measure the qubit
qc.measure(0, 0)

# Simulate the circuit
simulator = Aer.get_backend('qasm_simulator')
result = execute(qc, simulator, shots=1024).result()
counts = result.get_counts()

print("Measurement outcomes:", counts)
```

Parallel Processing Example: Superposition in Grover's Algorithm

Superposition allows quantum computers to explore multiple possibilities simultaneously, unlike classical computers that check solutions one at a time. Grover's Algorithm is an example of how superposition helps speed up searching an unsorted database.

Consider a problem where we want to find a particular item in a list of N possible solutions. A classical search algorithm would take O(N) time, meaning we have to check N elements one by one. However, Grover's algorithm leverages superposition to check all elements at once, reducing the time complexity to $O(\sqrt{N})$.

In a quantum circuit, Grover's algorithm works as follows:

1. Apply Hadamard Gates to all qubits to create superposition, putting the system into a state where all possible solutions exist simultaneously.
2. Apply an Oracle function that marks the correct solution by flipping its amplitude.
3. Apply an Amplitude Amplification step, which increases the probability of measuring the correct answer.
4. Measure the qubits, revealing the solution with high probability.

This means that if we have a list of 1,000,000 items, instead of checking each one sequentially, a quantum computer can find the correct solution in only 1,000 operations—a massive speedup.

A simple Qiskit implementation of Grover's algorithm would look like this:

```
from qiskit import QuantumCircuit, Aer, execute
from qiskit.circuit.library import MCXGate

# Create a 2-qubit quantum circuit
qc = QuantumCircuit(2)
```

```
# Step 1: Apply Hadamard gates to put qubits in superposition
qc.h([0, 1])

# Step 2: Oracle (marking the target state |11⟩)
qc.x([0, 1])
qc.h(1)
qc.mcx([0], 1)  # Multi-Controlled X (Toffoli) gate
qc.h(1)
qc.x([0, 1])

# Step 3: Apply the Grover diffusion operator
qc.h([0, 1])
qc.x([0, 1])
qc.h(1)
qc.mcx([0], 1)
qc.h(1)
qc.x([0, 1])
qc.h([0, 1])

# Measure the qubits
qc.measure_all()

# Simulate
simulator = Aer.get_backend('qasm_simulator')
result = execute(qc, simulator, shots=1024).result()
counts = result.get_counts()
print("Grover's Algorithm Output:", counts)
```

This circuit would return '11' with a high probability, demonstrating how superposition speeds up computation.

Usage in Quantum Machine Learning

In classical machine learning, a neural network processes data sequentially or in parallel on multiple processors. Quantum machine learning (QML) uses superposition to process all training data at the same time, leading to an exponential speedup.

For example, in a quantum support vector machine (QSVM), quantum computers encode a dataset into quantum states using superposition and entanglement. This allows them to analyse all data points simultaneously and classify patterns much faster than classical methods.

A real-world application of this is the quantum kernel method used by IBM in quantum-enhanced machine learning. The Hadamard gate (H-gate) and other quantum gates are applied to encode the training data into a superposition, allowing quantum parallelism.

A simple Qiskit-based quantum classifier that uses superposition might look like this:

```
from qiskit import QuantumCircuit, Aer, execute

# Quantum circuit for a basic quantum classifier
qc = QuantumCircuit(2)

# Apply Hadamard to put qubits in superposition
qc.h(0)
qc.h(1)

# Apply quantum transformations (representing classification)
qc.cx(0, 1)

# Measure
qc.measure_all()

# Simulate
simulator = Aer.get_backend('qasm_simulator')
result = execute(qc, simulator, shots=1024).result()
counts = result.get_counts()
print("Quantum Classifier Output:", counts)
```

This demonstrates a simplified quantum classification model where qubits represent multiple possible categories at once, leveraging superposition to analyse different cases simultaneously.

2. Quantum Entanglement and Its Role in Quantum Computing

Quantum entanglement links two or more qubits so that their states become interdependent, no matter how far apart they are. Measuring one entangled qubit instantaneously determines the state of its partner.

Practical Implementation:

- **Entangled Qubit Pairs:** Quantum hardware platforms like IBM Q and Google's Sycamore use entanglement to connect qubits in a way that improves computational speed.
- **Quantum Teleportation:** Used in quantum networks to transfer qubit states without physically moving the qubit itself.

- **Quantum Cryptography:** Protocols like Quantum Key Distribution (QKD) use entanglement to ensure that any interception of an encrypted quantum communication introduces detectable changes.

Usage by Developers:

- **Developing Quantum Networking:** Entanglement enables secure quantum networks and distributed quantum computing.
- **Building Fault-Tolerant Systems:** Developers use entanglement in error correction protocols to detect and fix errors without collapsing quantum states.
- **Quantum Speedup:** Entanglement improves the performance of quantum algorithms, such as variational quantum eigensolvers used in materials science.

Examples of Quantum Entanglement in Quantum Computing

A. Entangled Qubit Pairs in Quantum Hardware

Example: Entanglement in IBM Q and Google Sycamore

In IBM's quantum processors and Google's Sycamore quantum chip, entanglement is implemented through controlled quantum gates such as the CNOT (Controlled-NOT) gate. If two qubits start in the state $|00\rangle$, applying a Hadamard gate to the first qubit and then a CNOT gate creates an entangled Bell state:

$$\frac{1}{\sqrt{2}}(|00\rangle + |11\rangle)$$

This means that measuring the first qubit immediately determines the state of the second qubit, regardless of their physical distance.

Quantum hardware developers use entanglement to increase computational power. For example, IBM Qiskit's `QuantumCircuit` module allows users to implement entanglement:

```
from qiskit import QuantumCircuit

qc = QuantumCircuit(2)  # Create a 2-qubit quantum circuit
qc.h(0)  # Apply Hadamard gate to qubit 0
qc.cx(0, 1)  # Apply CNOT gate, entangling qubit 0 with qubit 1
qc.measure_all()  # Measure both qubits

print(qc.draw())  # Print the circuit diagram
```

By executing this circuit on IBM's quantum cloud, developers can test entanglement and verify that the two qubits always have correlated states.

B. Quantum Teleportation: Transferring Qubit States

Example: Teleporting a Qubit's State

Quantum teleportation allows the transfer of qubit states between two distant locations using entanglement and classical communication. Suppose Alice and Bob share an entangled qubit pair:

1. Alice has a qubit whose state she wants to teleport.
2. She entangles her qubit with her share of the entangled pair using a CNOT gate.
3. She then applies a Hadamard gate and measures her two qubits.
4. The classical measurement results are sent to Bob.
5. Bob applies a quantum gate based on Alice's results to reconstruct the original state.

Quantum Circuit for Teleportation

Using Qiskit, developers can simulate quantum teleportation:

```
from qiskit import QuantumCircuit

qc = QuantumCircuit(3, 2)  # 3 qubits: Alice's qubit (0),
Entangled pair (1,2)

# Step 1: Create Entanglement
qc.h(1)
qc.cx(1, 2)

# Step 2: Bell Measurement on Alice's qubit and her share of
entangled qubit
qc.cx(0, 1)
qc.h(0)
qc.measure([0, 1], [0, 1])

# Step 3: Bob applies corrections
qc.cx(1, 2)
qc.cz(0, 2)

print(qc.draw())  # Display circuit
```

Quantum teleportation is crucial for quantum networking, where qubit states need to be reliably transferred between nodes in a quantum network.

C. Quantum Cryptography: Securing Communications

Example: Quantum Key Distribution (QKD)

Quantum Key Distribution (QKD) protocols, such as BB84 and E91, use entangled qubits to establish encryption keys securely.

- In the E91 protocol, two parties (Alice and Bob) share entangled qubits.
- If a third party (Eve) attempts to intercept a qubit, the entanglement is disturbed.
- This disturbance can be detected, ensuring secure communication.

Developers working on quantum-secure communications implement QKD using entanglement. Quantum SDKs like Project Q and Qiskit Aqua allow simulation of QKD. The real-world applications of QKD include China's Micius satellite and DARPA-funded quantum networks.

D. Fault-Tolerant Quantum Computing: Using Entanglement in Error Correction

Example: Quantum Error Correction with Entanglement

Quantum error correction (QEC) relies on entanglement to detect and fix quantum errors without collapsing superpositions. One widely used QEC code is the Shor Code, which encodes a single logical qubit into nine entangled physical qubits to detect and correct bit-flip and phase-flip errors.

Quantum Circuit for Shor Code Encoding

```
from qiskit import QuantumCircuit

qc = QuantumCircuit(9)

# Encode logical |0> using entanglement
qc.h(0)
qc.cx(0, 1)
qc.cx(0, 2)

qc.cx(0, 3)
qc.cx(1, 4)
qc.cx(2, 5)
```

```
qc.cx(3, 6)
qc.cx(4, 7)
qc.cx(5, 8)

print(qc.draw())   # Show the entangled error correction circuit
```

By encoding information in entangled qubit clusters, error correction prevents quantum states from collapsing due to decoherence.

E. Quantum Speedup: Entanglement in Quantum Algorithms

Example: Entanglement in Variational Quantum Eigensolvers (VQE)

Quantum algorithms like VQE and Quantum Approximate Optimization Algorithm (QAOA) rely on entanglement to efficiently solve problems in materials science and chemistry.

For example, simulating a molecule like H_2 requires entangled qubits to represent electronic interactions:

```
from qiskit.chemistry import FermionicOperator
from qiskit.aqua.algorithms import VQE
from qiskit.aqua.components.optimizers import COBYLA

# Define Hamiltonian for H2
h2_hamiltonian = FermionicOperator.from_hdf5("H2_molecule.hdf5")

# Run VQE with entanglement-enabled ansatz
vqe = VQE(h2_hamiltonian, optimizer=COBYLA())
result = vqe.run()
print("Energy of H2:", result['energy'])
```

Using entangled qubits, quantum computers outperform classical systems in solving complex chemical problems.

Quantum entanglement is the foundation of many powerful quantum computing applications. It enables faster computation, secure communication, error correction, and quantum networking. Developers leverage entanglement through quantum gates, teleportation circuits, cryptographic protocols, and error-correcting codes to build the next generation of quantum technology. As quantum hardware improves, entanglement will play an even greater role in unlocking computational capabilities beyond classical limits.

3. Quantum Measurement and the Collapse of Superposition

Measurement in quantum mechanics forces a quantum state into a definite outcome, which removes its superposition properties.

Practical Implementation:

- **Controlled Measurement Timing:** Quantum algorithms are designed to delay measurement until the end of computation to prevent information loss.
- **Error Mitigation Techniques:** Quantum developers use techniques like quantum error correction to counteract unintended measurements or decoherence.
- **Quantum Sensors:** Measurement principles are used in high-precision quantum sensors that rely on superposition and collapse dynamics.

Usage by Developers:

- **Quantum Algorithm Optimization:** Developers create algorithms that minimize unnecessary measurements to maintain quantum coherence.
- **Quantum Error Correction Codes:** Developers design redundant qubits (logical qubits) to protect against measurement errors.
- **Quantum Simulation:** Understanding measurement effects helps developers simulate quantum materials and chemical reactions.

Examples of Quantum Measurement and the Collapse of Superposition in Quantum Computing

A. Controlled Measurement Timing in Quantum Algorithms

Quantum measurement collapses a qubit's superposition state into either 0 or 1, meaning it no longer holds multiple possibilities. To prevent premature collapse, quantum algorithms are designed to delay measurement until the end of computation.

Example: Delaying Measurement in Grover's Algorithm

Grover's search algorithm, which finds the correct solution in $O(\sqrt{N})$ time instead of $O(N)$ in classical computing, relies on quantum superposition. If qubits were measured too early, the quantum advantage would be lost.

Below is a Qiskit implementation that delays measurement until the final step:

```
from qiskit import QuantumCircuit, Aer, execute

n = 3  # Number of qubits
qc = QuantumCircuit(n, n)
```

```
# Step 1: Put all qubits in superposition
qc.h(range(n))

# Step 2: Apply Grover's diffusion operator (Oracle and
Amplification step)
qc.cz(0, 1)  # Example Oracle (customized for problem)
qc.h(range(n))
qc.x(range(n))
qc.h(n-1)
qc.mcx(list(range(n-1)), n-1)  # Multi-Control X
qc.h(n-1)
qc.x(range(n))
qc.h(range(n))

# Step 3: Measure at the end
qc.measure(range(n), range(n))

print(qc.draw())  # Show the circuit diagram
```

By postponing measurement, the algorithm maintains superposition long enough to enhance computation.

B. Error Mitigation Techniques: Preventing Unintended Measurement

Quantum states are fragile, and unintended measurements can collapse superpositions prematurely due to decoherence (interaction with the environment). Developers use quantum error correction (QEC) to protect qubits.

Example: Using Redundant Qubits for Quantum Error Correction

The Shor Code protects a single logical qubit by encoding it into nine physical qubits to correct errors from unwanted measurements.

```
from qiskit import QuantumCircuit

qc = QuantumCircuit(9)

# Encode logical qubit |0> with redundancy
qc.h(0)  # Create superposition
qc.cx(0, 1)
qc.cx(0, 2)

qc.cx(0, 3)
qc.cx(1, 4)
```

```
qc.cx(2, 5)

qc.cx(3, 6)
qc.cx(4, 7)
qc.cx(5, 8)

print(qc.draw())  # Show the entanglement structure
```

This approach allows developers to identify and correct measurement errors without losing quantum information.

C. Quantum Sensors and Measurement Principles

Quantum measurement principles are applied in high-precision quantum sensors that exploit superposition and measurement collapse for increased sensitivity.

Example: Quantum Magnetometers

Quantum magnetometers, such as Nitrogen-Vacancy (NV) centers in diamonds, measure weak magnetic fields by utilizing qubit states. The measurement collapses the state into spin-up or spin-down, providing precise field data.

- Used in geophysical surveys, medical imaging, and navigation systems where high sensitivity is needed.

Developers in quantum metrology optimize sensor performance using quantum measurement protocols.

D. Quantum Algorithm Optimization: Minimizing Unnecessary Measurements

Quantum algorithms must minimize direct measurement interactions to prevent unnecessary state collapse. Instead of measuring at every step, developers encode results indirectly.

Example: Variational Quantum Eigensolver (VQE)

VQE is a hybrid quantum-classical algorithm used to find the ground state energy of molecules. To avoid excessive measurement, it samples observables without fully collapsing the quantum state.

```
from qiskit.aqua.algorithms import VQE
from qiskit.chemistry import FermionicOperator

# Define Hamiltonian for H2 molecule
h2_hamiltonian = FermionicOperator.from_hdf5("H2_molecule.hdf5")
```

```
# Run VQE without collapsing intermediate qubits
vqe = VQE(h2_hamiltonian)
result = vqe.run()

print("Estimated ground state energy:", result['energy'])
```

E. Quantum Simulation: Understanding Measurement Effects

Quantum simulation involves modeling quantum systems such as superconductors, molecules, and exotic materials. Measurement effects must be carefully handled to preserve the accuracy of the simulation.

Example: Simulating a Quantum Phase Transition

Quantum phase transitions involve changes in a material's state due to quantum fluctuations. If measured too early, the phase transition cannot be observed properly.

Using quantum simulators like Google's Sycamore or IBM's Qiskit, developers simulate such transitions by controlling when and how measurements occur.

```
from qiskit.providers.aer import AerSimulator

# Quantum phase transition model
simulator = AerSimulator()
phase_circuit = QuantumCircuit(3)
phase_circuit.h(0)
phase_circuit.cx(0, 1)
phase_circuit.cx(1, 2)
phase_circuit.measure_all()

# Run simulation
result = simulator.run(phase_circuit).result()
print(result.get_counts())  # Output quantum measurement results
```

By adjusting measurement timing, realistic simulations of quantum effects become possible.

Quantum measurement plays a fundamental role in quantum computing and must be carefully controlled to maintain superposition. Developers use controlled measurement timing, quantum error correction, and optimized quantum algorithms to prevent premature state collapse. Measurement effects are also leveraged in quantum sensors, cryptography, and simulations to enhance the performance of quantum technologies. Understanding and applying these techniques is essential for advancing practical quantum computing.

4. The Heisenberg Uncertainty Principle and Its Effects

The uncertainty principle states that certain properties, like position and momentum, cannot be precisely measured at the same time.

Practical Implementation:

- **Quantum Random Number Generators (QRNGs):** Quantum computers use uncertainty to generate truly random numbers.
- **Quantum Sensing Applications:** Quantum sensors leverage uncertainty to measure variables like gravitational waves or magnetic fields with high precision.

Usage by Developers:

- **Noise Reduction in Quantum Systems:** Developers apply uncertainty principles to optimize quantum coherence time and reduce decoherence.
- **Developing Quantum Metrology Tools:** The uncertainty principle helps improve quantum-enhanced measurement techniques.

Examples of the Heisenberg Uncertainty Principle and Its Effects in Quantum Computing

A. Quantum Random Number Generators (QRNGs)

The Heisenberg Uncertainty Principle states that certain pairs of properties—like position and momentum—cannot be precisely measured at the same time. This inherent randomness in quantum mechanics is harnessed in Quantum Random Number Generators (QRNGs) to create true random numbers, which are essential for cryptography and secure communication.

Example: Implementing a Quantum Random Number Generator (QRNG)

Quantum computers generate random numbers using superposition and measurement uncertainty. The following Qiskit code simulates a QRNG using a Hadamard gate, which puts a qubit into an equal superposition of 0 and 1, and then measures it.

```
from qiskit import QuantumCircuit, Aer, execute

# Create a quantum circuit with one qubit and one classical bit
qrng_circuit = QuantumCircuit(1, 1)

# Apply a Hadamard gate to put the qubit in superposition
qrng_circuit.h(0)

# Measure the qubit
```

```
qrng_circuit.measure(0, 0)

# Execute the circuit
simulator = Aer.get_backend('qasm_simulator')
result = execute(qrng_circuit, simulator, shots=1000).result()

# Get random bit results
random_bits = result.get_counts()
print("Random Bit Distribution:", random_bits)
```

Each time this circuit runs, the measurement collapses the qubit into either 0 or 1, with an equal probability of 50%. The uncertainty principle ensures true randomness, unlike pseudo-random generators used in classical computing.

B. Quantum Sensing Applications

Quantum sensors leverage the uncertainty principle to achieve ultra-precise measurements in fields like gravitational wave detection, medical imaging, and atomic clocks. Since quantum states cannot be measured without some level of uncertainty, researchers optimize quantum states to minimize uncertainty in one parameter while accepting a larger uncertainty in another.

Example: Using Quantum Sensors to Measure Magnetic Fields

Quantum magnetometers use Nitrogen-Vacancy (NV) centers in diamonds to measure weak magnetic fields. The Heisenberg Uncertainty Principle ensures that by reducing the uncertainty in one measurement (magnetic field strength), the system introduces uncertainty in another (spin state).

- In medical imaging, quantum magnetometers help improve MRI scan precision.
- In geophysics, quantum sensors measure Earth's magnetic field to detect underground structures.

Developers working on quantum sensing devices optimize quantum coherence to maximize sensitivity while keeping the uncertainty principle in check.

C. Noise Reduction in Quantum Systems

Quantum computers suffer from decoherence, where environmental noise disrupts delicate quantum states. The uncertainty principle plays a role in managing noise, as it limits how precisely certain quantum properties can be maintained over time. Quantum developers apply quantum error correction (QEC) techniques to counteract noise.

Example: Reducing Noise in Quantum Circuits

Quantum developers use Dynamical Decoupling (DD), a technique that extends quantum coherence by applying rapid sequences of pulsed operations to qubits.

```
from qiskit import QuantumCircuit
from qiskit.circuit.library import XGate

# Create a quantum circuit with one qubit
qc = QuantumCircuit(1)

# Apply a sequence of X gates (Pauli X pulses) to counteract
decoherence
qc.append(XGate(), [0])
qc.append(XGate(), [0])
qc.append(XGate(), [0])

print(qc.draw())  # Show the circuit
```

This approach helps reduce uncertainty-induced decoherence, allowing quantum computations to last longer before measurement collapses the state.

D. Developing Quantum Metrology Tools

Quantum metrology is the science of making ultra-precise measurements using quantum principles. The uncertainty principle plays a key role in quantum-enhanced measurement techniques, such as atomic clocks and gravitational wave detectors.

Example: Improving Atomic Clocks

Atomic clocks rely on quantum transitions in atoms to keep time more accurately than classical clocks. The Heisenberg Uncertainty Principle sets a fundamental limit on how precisely time and energy can be measured simultaneously.

- Optical lattice clocks, which use laser-cooled atoms, apply quantum techniques to minimize uncertainty in time measurement.
- Quantum-enhanced atomic clocks are 100x more precise than standard atomic clocks, improving GPS accuracy and deep-space navigation.

Developers working on quantum timing technologies apply uncertainty-limited optimizations to ensure longer stability and higher precision.

The Heisenberg Uncertainty Principle is a fundamental limitation in quantum mechanics, but quantum developers have found ways to leverage its effects for practical applications. Quantum computing benefits from the uncertainty principle in random number generation, sensing technologies, noise reduction, and metrology improvements. These applications are crucial for secure communication, precision measurements, and developing fault-tolerant quantum computers. Understanding and applying the uncertainty principle allows developers to push the boundaries of quantum technology.

5. Quantum Wave Functions and Their Role in Quantum Computing

Wave functions describe the probability distributions of quantum systems.

Practical Implementation:

- **Quantum State Manipulation:** Developers use wave function equations to predict how qubits will behave.
- **Quantum Computing Hardware Design:** Superconducting qubits and trapped ion qubits are engineered based on wave function principles.

Usage by Developers:

- **Simulating Molecules for Drug Discovery:** Quantum computers model molecular interactions more accurately using quantum wave function simulations.
- **Quantum Circuit Development:** Wave functions guide the design of quantum circuits and gate operations.

Examples of Quantum Wave Functions and Their Role in Quantum Computing

A. Quantum State Manipulation in Quantum Algorithms

Wave functions describe the probability distribution of quantum states and determine how qubits evolve over time. In quantum computing, developers use wave function equations to predict and control qubit behaviour.

Example: Simulating a Qubit's Evolution Using the Schrödinger Equation

Quantum algorithms rely on controlled wave function evolution to compute results efficiently. The Schrödinger equation governs how a quantum system changes over time:

$$i\hbar \frac{d}{dt} |\Psi(t)\rangle = H |\Psi(t)\rangle$$

where $|\Psi(t)\rangle$ is the quantum state, and ***H*** is the Hamiltonian operator.

Developers use quantum circuit simulators to test and validate how qubits behave under different operations. Below is an example of simulating wave function evolution using Qiskit:

```
from qiskit import QuantumCircuit, Aer, transpile, execute

# Create a quantum circuit with one qubit
qc = QuantumCircuit(1)

# Apply Hadamard gate to create superposition
qc.h(0)

# Apply a phase shift (simulating wave function evolution)
qc.p(3.14/2, 0)

# Display the quantum circuit
print(qc.draw())

# Simulate the quantum state
simulator = Aer.get_backend('statevector_simulator')
result = execute(qc, simulator).result()
statevector = result.get_statevector()

# Print the wave function
print("Wave Function:", statevector)
```

In this example, the qubit starts in the |0⟩|0\rangle|0⟩ state. The Hadamard gate creates a superposition, and the phase gate applies a phase shift to manipulate the wave function.

B. Quantum Computing Hardware Design

Quantum hardware platforms rely on wave function principles to engineer stable and controllable qubits. Different quantum technologies, such as superconducting qubits and trapped-ion qubits, are designed by manipulating wave functions.

Example: Superconducting Qubit Design

- Superconducting qubits in IBM's quantum computers rely on Josephson junctions, which behave like quantum harmonic oscillators.
- The wave function of a superconducting qubit exists in a double-well potential, allowing the superposition of current states.
- Engineers optimize wave functions to reduce decoherence and improve qubit stability.

Example: Trapped-Ion Qubits

- Trapped-ion qubits, used in IonQ's quantum computers, confine individual atoms using electromagnetic fields.
- The wave function of an ionized atom is manipulated using laser pulses to perform quantum gate operations.
- Precise wave function control ensures high-fidelity quantum operations.

Developers working on hardware improvements use wave function principles to optimize qubit coherence, gate fidelity, and error rates.

C. Simulating Molecules for Drug Discovery

One of the most practical applications of wave functions in quantum computing is molecular simulation, where quantum computers model how electrons behave in atoms and molecules. The electronic wave function governs molecular interactions, and classical computers struggle to simulate these complex interactions accurately.

Example: Simulating Molecular Wave Functions Using Quantum Computers

Quantum computers use Variational Quantum Eigensolver (VQE) and Quantum Phase Estimation (QPE) to calculate the ground state energy of molecules.

Below is an example using Qiskit's VQE algorithm to simulate the Hydrogen (H_2) molecule:

```
from qiskit import Aer
from qiskit.circuit.library import TwoLocal
from qiskit.algorithms import VQE
from qiskit.algorithms.optimizers import SLSQP
from qiskit_nature.drivers import PySCFDriver
from qiskit_nature.converters.second_quantization import
QubitConverter
from qiskit_nature.mappers.second_quantization import
JordanWignerMapper

# Define molecule (H₂)
driver = PySCFDriver(atom="H 0 0 0; H 0 0 0.735")
molecule = driver.run()

# Convert molecule Hamiltonian to qubits
converter = QubitConverter(mapper=JordanWignerMapper())
qubit_op = converter.convert(molecule.hamiltonian)

# Define quantum circuit ansatz
```

```
ansatz = TwoLocal(rotation_blocks='ry',
entanglement_blocks='cz', reps=3)

# Choose simulator
backend = Aer.get_backend('statevector_simulator')

# Set up and run VQE
vqe = VQE(ansatz, optimizer=SLSQP(), quantum_instance=backend)
result = vqe.compute_minimum_eigenvalue(qubit_op)

# Output ground-state energy
print("Estimated ground-state energy:", result.eigenvalue.real)
```

This simulates the wave function of hydrogen molecules, helping drug discovery and materials science by providing highly accurate quantum simulations that classical computers cannot achieve.

D. Quantum Circuit Development

Wave functions provide the foundation for designing quantum gates and circuits. Developers use wave function manipulation to create efficient quantum circuits for solving optimization and machine learning problems.

Example: Quantum Fourier Transform (QFT)

The Quantum Fourier Transform (QFT) is a quantum algorithm that efficiently transforms wave functions, used in:

- Shor's algorithm for factoring large numbers (important for cryptography).
- Quantum simulations that require Fourier analysis.

Below is an example of a QFT implementation in Qiskit:

```
from qiskit import QuantumCircuit
import numpy as np

def qft(n):
    """Creates an n-qubit QFT circuit."""
    qc = QuantumCircuit(n)
    for i in range(n):
        for j in range(i):
            qc.cp(np.pi / 2 ** (i - j), j, i)
        qc.h(i)
    return qc
```

```
# Example: 3-qubit QFT
qft_circuit = qft(3)
print(qft_circuit.draw())
```

The QFT transforms wave function representations into the frequency domain, which is useful for signal processing and solving linear equations on quantum computers.

Quantum wave functions play a fundamental role in quantum computing. Developers manipulate wave functions to design quantum algorithms, simulate molecular structures, optimize hardware architectures, and improve computational accuracy. Whether simulating molecules for drug discovery, designing superconducting qubits, or creating efficient quantum circuits, wave function principles guide the development of powerful quantum technologies.

6. Thought Experiments and Their Impact on Quantum Computing

The chapter references several famous quantum thought experiments, such as Schrödinger's Cat, Wigner's Friend, and the EPR Paradox, which have direct implications in quantum computing.

Schrödinger's Cat and Quantum Superposition

- Demonstrates that qubits remain in superposition until measured.
- Helps developers understand quantum state persistence.

Wigner's Friend and Observer Effect

- Raises questions about measurement in quantum computing.
- Developers explore delayed-choice quantum computing to leverage uncollapsed states.

EPR Paradox and Quantum Entanglement

- Validates entanglement as a real phenomenon used in quantum teleportation.
- Guides the development of quantum networking protocols.

7. The Quantum Zeno Effect and Its Application

The Quantum Zeno Effect states that frequent measurement prevents a quantum state from evolving.

Practical Implementation:

- **Quantum Error Suppression:** Using frequent quantum operations to freeze quantum states and prevent decoherence.
- **Fault-Tolerant Quantum Computing:** Developing techniques to prolong qubit coherence times.

Usage by Developers:

- **Quantum Stability Enhancement:** Engineers use the effect to protect delicate quantum states.
- **Quantum Control Systems:** Designing more stable quantum hardware by controlling measurement frequencies.

8. Quantum Algorithms That Utilize These Principles

Quantum computing developers leverage these quantum mechanics principles in specific algorithms:

- **Shor's Algorithm (Prime Factorization):** Uses superposition and entanglement to factor large numbers exponentially faster than classical computers.
- **Grover's Algorithm (Database Search):** Exploits quantum interference to search databases in $\sqrt{N}$ time.
- **Quantum Fourier Transform (QFT):** A key component of many quantum algorithms, based on wave function principles.

The principles outlined in this chapter form the backbone of quantum computing. Quantum developers implement these principles through quantum gates, quantum circuits, and advanced error correction techniques. They are used to build fault-tolerant quantum processors, improve secure communication via quantum cryptography, and enable next-generation applications in artificial intelligence, drug discovery, and materials science.

Chapter 2

Quantum Bits (Qubits) and Quantum Systems

In the previous chapter, we explored the foundations of quantum mechanics, laying the groundwork for understanding the principles that make quantum computing possible. Concepts such as superposition, entanglement, and measurement were introduced as fundamental properties that distinguish quantum systems from classical ones. These ideas, initially developed to explain the behavior of subatomic particles, are now being harnessed to build a new class of computational machines capable of solving problems beyond the reach of conventional computers.

With this foundation in place, we now turn to quantum bits (qubits)—the fundamental units of quantum information. While classical computers rely on bits that exist in one of two definite states (0 or 1), qubits can exist in a superposition of both states simultaneously, vastly increasing computational possibilities. This chapter will examine what makes qubits unique, how they compare to classical bits, and the physical systems used to create and manipulate them.

We will also introduce the Bloch sphere, a visual representation of qubit states, which serves as an essential tool for understanding quantum operations. Additionally, we will discuss the challenges of decoherence and error correction, two critical obstacles in the development of stable quantum computers. As we progress through this chapter, the transition from quantum theory to quantum computation will become increasingly clear, setting the stage for deeper discussions on quantum gates, algorithms, and real-world applications in later chapters.

What Are Qubits?

A qubit (short for quantum bit) is the fundamental unit of information in quantum computing, analogous to a classical bit in traditional computing. However, unlike classical bits, which can only be in one of two definite states (0 or 1), a qubit can exist in a superposition of both states simultaneously, thanks to the principles of quantum mechanics.

In classical computing, a bit is the smallest unit of information, represented as either 0 or 1. Modern computers use electrical signals to represent bits, where the presence of a voltage or current corresponds to 1, and the absence corresponds to 0. These bits form the foundation of all digital information, from simple binary arithmetic to complex multimedia processing.

Qubits, however, can exist in a third state, known as superposition, which means that they do not just represent a single 0 or 1, but rather a probability distribution of both states. This means a qubit can represent 0, 1, and all the possible combinations in between at the same time. The ability to store and process multiple values simultaneously makes quantum computers significantly more powerful than classical computers for solving certain types of problems.

Even though qubits exhibit quantum behaviour, they are still used within a binary system. The term bit in quantum computing can refer either to the material or process used to represent quantum information or to the measured outcome of a qubit (which collapses into either 0 or 1 upon measurement).

Qubits unlock the power of quantum parallelism, where computations happen simultaneously instead of sequentially.

To understand this, imagine solving a complex maze. A classical computer would have to test each possible path one at a time, leading to a slow and exhaustive search. A quantum computer, using qubits in superposition, could analyse multiple paths at once and determine the correct solution much faster.

Quantum interference also plays a critical role in enhancing computational power. The probability amplitudes of different qubit states can constructively or destructively interfere, leading to an efficient elimination of incorrect solutions while amplifying the correct answers.

Superposition and Quantum Probability

A key difference between qubits and classical bits is superposition, which allows a qubit to exist as a blend of both 0 and 1 states simultaneously. When a qubit is in superposition, its probability of being measured as 0 or 1 is determined by its wave function. The act of measuring a qubit collapses it into a definite state—either 0 or 1 [73].

For example, if a quantum system is used to factor a large prime number, classical bits must store and process each possible factor one at a time, while qubits in superposition can process many possible factors simultaneously, leading to a much faster computation process.

In a classical computer, a bit must be either 0 or 1. In a quantum computer, a qubit exists in a superposition of both states:

$$|\Psi\rangle = \alpha|0\rangle + \beta|1\rangle$$

where **α** and **β** are complex numbers representing probabilities, and **|0⟩** and **|1⟩** are the classical states.

For example, if a qubit is in an equal superposition, it has a 50% probability of being 0 and 50% probability of being 1 when measured.

Example: Using a Hadamard Gate to Create Superposition

In quantum programming, applying a Hadamard gate (H-gate) to a qubit creates superposition:

```
from qiskit import QuantumCircuit

qc = QuantumCircuit(1)
qc.h(0)  # Apply Hadamard gate to qubit 0
print(qc.draw())  # Display the quantum circuit
```

This operation transforms a qubit from **|0⟩** to an equal superposition of **|0⟩** and **|1⟩**.

Entanglement

Another important property of qubits is entanglement, which describes the strong correlation between two qubits that have interacted. Once entangled, qubits remain connected no matter how far apart they are. If one qubit is measured, the state of the entangled qubit is instantly determined. This property was described by Einstein as "spooky action at a distance" and is crucial for applications like quantum cryptography and quantum teleportation [73].

If one half of an entangled qubit pair is measured as 1, the other qubit in the pair will immediately collapse into 0, and vice versa. Unlike classical systems, where data is transmitted through physical signals, entanglement enables an instantaneous correlation between qubits, which can enhance quantum circuit performance.

When two or more qubits become entangled, their states become correlated, meaning measuring one qubit instantly determines the state of the other, regardless of distance. This property is crucial for quantum communication and computation.

Example: Creating an Entangled Pair

```
from qiskit import QuantumCircuit

qc = QuantumCircuit(2)
qc.h(0)  # Put first qubit in superposition
qc.cx(0, 1)  # Apply a CNOT gate to entangle qubits
print(qc.draw())  # Display circuit
```

This entangled state is fundamental in quantum teleportation and secure quantum networks.

Measurement and Quantum Collapse

Unlike classical bits, which retain their value when read, measuring a qubit collapses its wave function into either |0⟩ or |1⟩|. This means that after measurement, the qubit loses its superposition and behaves like a classical bit [73].

Example: Measuring a Qubit

```
from qiskit import Aer, execute

simulator = Aer.get_backend('qasm_simulator')
qc.measure_all()  # Measure the qubit
job = execute(qc, simulator, shots=1000)
result = job.result()
counts = result.get_counts()
print(counts)  # Output the measured results
```

Here, after measurement, we get either 0 or 1, but never both.

Qubits enable quantum computers to perform certain calculations exponentially faster than classical computers. The unique properties of qubits allow them to:

- Process multiple possibilities simultaneously (Superposition)
- Perform complex calculations efficiently (Parallelism)
- Enable ultra-secure communication (Entanglement in Quantum Cryptography)
- Solve problems that classical computers struggle with (e.g., simulating molecules, optimizing logistics, and breaking cryptographic codes)

Qubits are the building blocks of quantum computing, offering capabilities that surpass classical computing in power, efficiency, and speed for specific types of problems.

Different Types of Qubits and Their Advantages

Various physical systems can be used to create and manipulate qubits. Each type of qubit has different advantages and challenges, making them suitable for different applications in quantum computing [73].

Superconducting Qubits: These qubits are created using superconducting circuits that operate at extremely low temperatures. They are manipulated using microwave pulses and are widely used in quantum computers developed by companies such as IBM and Google. Superconducting qubits are favoured for their high-speed operation and compatibility with existing semiconductor technology, but they require extreme cooling to avoid decoherence.

Trapped Ion Qubits: In this approach, ions (charged atoms) are suspended in electromagnetic fields and controlled using lasers. Trapped ions exhibit long coherence times, meaning they can maintain their quantum states longer than many other qubit types. Their high-fidelity operations make them useful for precise quantum simulations and quantum networks.

Quantum Dot Qubits: A quantum dot is a tiny semiconductor nanostructure capable of trapping a single electron, which acts as a qubit. Quantum dots can be manipulated using magnetic and electric fields. These qubits are promising for their potential to scale up quantum computing systems, as they can be integrated into existing semiconductor fabrication methods.

Photon Qubits: Using individual light particles (photons), researchers encode quantum information based on the polarization of photons. Photon qubits are used in quantum cryptography and quantum communication, as they can travel long distances through fibre-optic cables with minimal loss.

Neutral Atom Qubits: Neutral atoms, which have a balanced charge, can be trapped and manipulated using laser beams. By exciting atoms into different energy levels, scientists can encode qubit states. These qubits are promising for scalability and high-fidelity operations in quantum computing.

Challenges in Qubit Implementation

Despite their incredible potential, qubits are extremely fragile and prone to errors. Several challenges must be overcome to build stable and reliable quantum computers [73].

Most qubits, especially superconducting qubits, require temperatures near absolute zero (close to -273°C) to minimize thermal noise and decoherence. These ultra-cold conditions are achieved using dilution refrigerators, but maintaining such an environment is expensive and complex [73].

Qubits can lose their quantum state due to interactions with their environment, a process known as decoherence. This effect limits the duration for which a quantum computation can run. Quantum error correction (QEC) techniques, such as redundant qubit encoding and quantum stabilizer codes, help mitigate errors by using multiple qubits to protect information [73].

To build practical quantum computers, thousands or even millions of qubits are needed to perform meaningful computations. However, current quantum computers only have dozens to a few hundred qubits. Scaling up quantum hardware while maintaining low error rates remains a major challenge.

As research advances, new methods for stabilizing qubits, improving quantum coherence times, and reducing hardware requirements will pave the way for powerful quantum processors. Emerging quantum technologies will enhance applications such as [73]:

- Drug discovery by simulating molecular interactions with extreme precision
- Secure communications using quantum cryptographic methods like Quantum Key Distribution (QKD)
- Optimization problems in logistics, finance, and artificial intelligence
- Quantum machine learning, where qubits accelerate data processing and pattern recognition

While quantum computing is still in its early stages, advancements in qubit technology are rapidly bringing us closer to unlocking its full potential. Future developments in fault-tolerant quantum computing and hybrid quantum-classical systems will define the next era of computational power.

Quantum vs. Classical Bits

Classical Bits: The Foundation of Traditional Computing

A classical bit is the smallest unit of data in a traditional computer. It is based on binary logic, meaning it can only exist in one of two definite states: 0 or 1. In modern computers, these bits are physically represented using:

- Electrical signals (voltage levels in transistors)
- Magnetic storage (hard drives)
- Light pulses (fibre-optic communication)
- Physical markers (punch cards or optical disks)

Quantum Computing

Each bit acts as a simple switch that is either "on" (1) or "off" (0). Classical computers process information by manipulating large sequences of bits using logic gates (AND, OR, NOT, XOR, etc.). These computations follow deterministic rules, meaning that the outcome of a calculation is predictable and repeatable.

For example, a classical computer processing a 4-bit system can only store one of 16 possible states at a time:

0000,0001,0010,...,1111

If it needs to process all 16 states, it must go through them one by one.

Quantum Bits (Qubits): The Core of Quantum Computing

A qubit is the quantum analogue of a classical bit, but it follows the laws of quantum mechanics, allowing for superposition, entanglement, and interference. These properties enable quantum computers to perform computations in ways that classical computers cannot.

1. Superposition: Qubits Can Be 0, 1, or Both

Unlike classical bits, which can only be 0 or 1, a qubit can exist in a superposition of both states at the same time:

$$|\psi\rangle = \alpha|0\rangle + \beta|1\rangle$$

where α and β are probability amplitudes that determine the likelihood of measuring 0 or 1. This means that a single qubit can encode much more information than a classical bit.

For example, if we have two classical bits, they can store one of four possible states at a time:

00,01,10,11

But if we have two qubits in superposition, they can exist in all four states simultaneously, allowing quantum computers to process multiple possibilities at once. This enables parallelism in computation, making quantum computers highly efficient for certain tasks.

2. Entanglement: Qubits Can Be Strongly Correlated

Entanglement is a uniquely quantum phenomenon where two or more qubits become linked in such a way that the state of one qubit is instantly correlated with the state of another, regardless of distance. If two qubits are entangled, measuring one qubit immediately determines the state of the other.

For example, in an entangled system:

$$|\psi\rangle = \frac{1}{\sqrt{2}}(|00\rangle + |11\rangle)$$

If we measure the first qubit and find it in state |0⟩|0\rangle|0⟩, the second qubit will also be |0⟩. If we measure the first qubit as |1⟩|1\rangle|1⟩, the second qubit will also be |1⟩—no matter how far apart they are.

Entanglement is crucial in quantum networking, quantum teleportation, and quantum cryptography.

3. Quantum Interference: Enhancing Computation

Quantum states behave like waves, meaning they can interfere with each other. This property allows quantum algorithms to amplify correct answers and cancel out incorrect ones through constructive and destructive interference.

For example, Shor's algorithm, which is used for factoring large numbers, relies on quantum interference to find factors exponentially faster than classical algorithms.

Table 2: Key Differences Between Classical Bits and Qubits.

Feature	Classical Bits	Qubits
State	Can be 0 or 1	Can be 0, 1, or a superposition of both
Processing	Processes one state at a time	Can process multiple states simultaneously
Storage	Needs *N* bits to store *N* pieces of information	Needs *N* qubits to store 2^N pieces of information
Entanglement	No entanglement; bits are independent	Entangled qubits exhibit instantaneous correlations
Interference	No wave interference	Quantum interference can enhance computations
Error Correction	Simple error correction using redundancy	Requires quantum error correction due to decoherence
Measurement	Does not alter bit values	Measurement collapses quantum states

Quantum Computing

Quantum bits, or qubits, play a crucial role in enabling quantum computers to solve problems that are infeasible for classical computers. This is primarily due to three fundamental quantum properties: superposition, entanglement, and interference. These properties allow quantum computers to perform complex calculations at speeds that are unattainable by traditional computers, unlocking new possibilities in computing and problem-solving.

One of the most significant advantages of quantum computing is the speedup achieved in quantum algorithms. Certain quantum algorithms leverage the power of qubits to provide exponential improvements over classical methods. Shor's Algorithm, for example, is used for factoring large numbers, making it highly relevant for cryptography and breaking encryption schemes that rely on the difficulty of prime factorization. Grover's Algorithm enables quantum computers to search unsorted databases with a speedup, reducing the time complexity from $O(N)$ to $O(\sqrt{N})$, which is significantly faster than classical approaches. Additionally, the Variational Quantum Eigensolver (VQE) is an algorithm designed for optimizing molecular structures in chemistry, making it useful in material science and drug discovery.

Quantum simulation is another area where qubits provide a substantial advantage. Quantum systems are inherently difficult to simulate on classical computers due to their exponential complexity, but quantum computers can directly simulate quantum mechanical systems. This capability is particularly valuable in drug discovery, where simulating molecular interactions helps researchers design more effective pharmaceuticals. Similarly, in materials science, quantum computers can help discover new superconductors and optimize material properties for various industrial applications.

Quantum cryptography is a field that benefits greatly from the unique properties of qubits. Quantum Key Distribution (QKD) ensures ultra-secure communication by utilizing entangled qubits to encode information. If an eavesdropper attempts to intercept a quantum-encoded message, the quantum state collapses, immediately revealing the intrusion. This makes quantum cryptography an essential tool for secure data transmission, offering protection against hacking and cyber threats.

Quantum computing also has the potential to revolutionize machine learning. By processing high-dimensional data more efficiently, quantum-enhanced neural networks could significantly improve pattern recognition and optimization problems. This could lead to advancements in fields such as artificial intelligence, finance, healthcare, and cybersecurity, where large-scale data processing is crucial. Quantum machine learning could enable faster training of AI models, improved data clustering, and better decision-making processes.

Overall, quantum bits empower quantum computers to perform tasks that were once thought impossible. Whether through enhanced algorithms, quantum simulation, cryptographic security, or machine learning applications, qubits provide unprecedented computational power. As quantum technology advances, its impact on various industries will continue to grow, unlocking new possibilities for solving some of the most complex challenges in science and technology.

Single-Qubit Computational Basis States

In quantum computing, a qubit is the fundamental unit of quantum information, and its states are represented in a computational basis. The computational basis states for a single qubit are typically defined in the z-basis, which consists of the two orthogonal states:

$$|0\rangle$$

$$|1\rangle$$

These states correspond to the two fundamental states of a qubit and serve as the foundation for quantum operations and measurements. When a qubit is measured in the z-basis, it collapses into either |0⟩ or |1⟩ with a probability determined by its quantum state before measurement.

In quantum mechanics, a qubit's general quantum state is represented as a linear superposition of two orthonormal basis states. These basis states, commonly denoted as |0⟩ and |1⟩, serve as the fundamental building blocks of quantum computation. The standard notation for these states follows Dirac's "bra-ket" notation, which is widely used in quantum mechanics to describe quantum states in a vector space. Mathematically, these basis states are written as:

$$|0\rangle = \begin{bmatrix} 1 \\ 0 \end{bmatrix}, \quad |1\rangle = \begin{bmatrix} 0 \\ 1 \end{bmatrix}$$

These two states form the computational basis of a qubit, meaning any arbitrary quantum state of a single qubit can be expressed as a linear combination of these two states. Because quantum states exist in a two-dimensional Hilbert space, they can be described using complex coefficients, which introduce the concepts of probability amplitudes and quantum interference.

When multiple qubits are considered, they collectively form a quantum register, and their combined states expand into higher-dimensional Hilbert spaces. For instance, a system of two qubits requires a four-dimensional vector space to represent all possible states. The product basis states of a two-qubit system are:

$$|00\rangle = \begin{bmatrix} 1 \\ 0 \\ 0 \\ 0 \end{bmatrix}, \quad |01\rangle = \begin{bmatrix} 0 \\ 1 \\ 0 \\ 0 \end{bmatrix}, \quad |10\rangle = \begin{bmatrix} 0 \\ 0 \\ 1 \\ 0 \end{bmatrix}, \quad |11\rangle = \begin{bmatrix} 0 \\ 0 \\ 0 \\ 1 \end{bmatrix}$$

These basis states define the possible measurement outcomes for a two-qubit system. However, due to quantum superposition, the system can exist in a state that is a combination of these basis states, allowing for exponentially greater computational possibilities compared to classical binary bits.

As the number of qubits increases, the dimensionality of the quantum state space grows exponentially. An n-qubit system is represented in a 2^n-dimensional Hilbert space. This means that while a single qubit exists in a two-dimensional space, a three-qubit system requires an eight-dimensional space, and a four-qubit system requires a sixteen-dimensional space, and so on. This exponential growth is the key reason why quantum computers have the potential to outperform classical computers in solving certain complex problems.

The ability to manipulate qubits in this high-dimensional space allows for advanced quantum algorithms, such as Shor's algorithm for factoring large numbers and Grover's algorithm for searching databases. Additionally, entanglement and quantum gates leverage these high-dimensional states to perform operations that classical systems cannot efficiently replicate. Quantum registers, composed of multiple qubits, form the core of quantum computation, allowing for parallel processing, entanglement-based computations, and enhanced cryptographic security.

The standard representation of qubits and their computational basis is fundamental to understanding quantum computing. Whether dealing with a single qubit or a multi-qubit system, the mathematical framework of superposition, tensor products, and Hilbert spaces provides the foundation for the extraordinary power of quantum computation.

Alternative Basis Representations of a Qubit

While the z-basis is the standard computational basis, qubits can also be described in alternative bases, such as the x-basis and y-basis. These bases are essential for quantum algorithms and quantum gate operations.

X-Basis States: The x-basis states are defined as:

$$|+\rangle = \frac{|0\rangle + |1\rangle}{\sqrt{2}}$$

$$|-\rangle = \frac{|0\rangle - |1\rangle}{\sqrt{2}}$$

These states are obtained by applying a Hadamard gate (H-gate) to the computational basis states. The Hadamard transformation maps $|0\rangle$ **to** $|+\rangle$ **and** $|1\rangle$ **to** $|-\rangle$, creating a superposition of the z-basis states. The x-basis is important because certain quantum gates, such as the Hadamard and Pauli-X gates, act naturally in this basis.

Y-Basis States

The **y-basis** states are:

$$|+i\rangle = \frac{|0\rangle + i|1\rangle}{\sqrt{2}}$$

$$|-i\rangle = \frac{|0\rangle - i|1\rangle}{\sqrt{2}}$$

Here, *i* represents the imaginary unit. The y-basis states are crucial for describing phase relationships in quantum circuits. These states can be created using the Pauli-Y gate, which introduces a 90° phase shift between the computational basis states.

A pure qubit state is a fundamental concept in quantum mechanics, representing a coherent superposition of two basis states. Unlike classical bits, which can only be in a definite state of either 0 or 1, a qubit exists in a linear combination of both states. Mathematically, a single qubit state is represented as:

$$|\psi\rangle = \alpha|0\rangle + \beta|1\rangle$$

where α\alphaα and β\betaβ are complex probability amplitudes. These amplitudes determine the likelihood of measuring the qubit in either the |0⟩ or |1⟩ state. When a measurement is performed in the standard computational basis, the probability of obtaining the outcome |0⟩ is given by $|\alpha|^2$, and the probability of obtaining the outcome |1⟩ is given by $|\beta|^2$. Since these probabilities must sum to one, the condition:

$$|\alpha|^2 + |\beta|^2 = 1$$

ensures that the qubit remains in a valid quantum state. This normalization constraint is a direct consequence of probability theory and is essential in quantum mechanics.

Beyond just representing probabilities, the probability amplitudes also contain phase information. The relative phase between α\alphaα and β\betaβ plays a crucial role in quantum interference effects, which can be observed in experiments such as the double-slit experiment. This phase difference is what allows quantum algorithms to take advantage of interference patterns to enhance computation speed and accuracy.

The Bloch Sphere Representation

The Bloch sphere provides a geometric visualization of qubit states. Every pure qubit state can be mapped to a point on the surface of a sphere. The north pole corresponds to the |0⟩|0\rangle|0⟩ state, and the south pole corresponds to the |1⟩|1\rangle|1⟩ state. Any other point on the surface of the sphere represents a superposition of these two states.

Using spherical coordinates, any qubit state can be written as:

$$\alpha = \cos\left(\frac{\theta}{2}\right), \quad \beta = e^{i\phi} \sin\left(\frac{\theta}{2}\right)$$

where θ is the polar angle, and ϕ is the azimuthal angle. These two angles uniquely define a point on the Bloch sphere. The phase factor $e^{i\phi}$ does not affect measurement probabilities but influences quantum interference, making it a key element in quantum algorithms.

At first glance, it may seem that a qubit should have four degrees of freedom since both α and β are complex numbers. However, due to the normalization constraint and the fact that an overall global phase does not affect measurement outcomes, the number of independent degrees of freedom is reduced to two. This is why a qubit's state can be fully represented using only two parameters, θ and ϕ, which define its position on the Bloch sphere.

Pure vs. Mixed States

A pure state, as described above, is a fully coherent quantum state, meaning it maintains quantum superposition and interference. However, real-world quantum systems are subject to noise and decoherence, which can cause a qubit to lose its coherence. When this happens, the qubit enters a mixed state, which can no longer be represented by a single ket vector. Instead, mixed states are described using density matrices and are represented as points inside the Bloch sphere, rather than on its surface.

Mathematically, a mixed state can be thought of as a statistical combination of different pure states rather than a coherent superposition. This means that rather than existing in a definite quantum state, the qubit exhibits probabilistic behaviour due to external interactions with its environment. The degree of mixture is represented by a parameter r, which determines how "classical" or "quantum" the state behaves. A fully mixed state sits at the centre of the Bloch sphere and represents complete uncertainty.

Quantum Error Correction and Maintaining Purity

Since decoherence leads to a loss of quantum information, quantum error correction techniques are employed to preserve the purity of qubits. These techniques involve encoding quantum information across multiple qubits in such a way that errors can be detected and corrected without disturbing the quantum state. Quantum error correction is essential for building reliable quantum computers, as uncontrolled decoherence can severely limit the ability of a quantum system to perform meaningful computations.

The mathematical representation of a qubit in terms of probability amplitudes and the Bloch sphere provides an intuitive understanding of quantum states. The concepts of superposition, probability amplitudes, and phase play a crucial role in quantum computing and quantum algorithms. The distinction between pure and mixed states highlights the challenges of maintaining coherence in real-world quantum systems, making quantum error correction a fundamental aspect of quantum computation.

Applications of Different Qubit Basis States

Quantum computation and communication are fundamentally reliant on the manipulation of qubits across different bases, which is essential for executing various tasks. This manipulation is crucial in several domains, including quantum algorithms, error correction, cryptography, and networking.

Quantum Algorithms: Quantum algorithms such as Grover's search algorithm and Shor's factoring algorithm leverage the principles of superposition and interference, which are pivotal for achieving computational speedup. The x-basis and y-basis states are integral to these computations, as they allow for the representation and manipulation of qubits in a manner that enhances the efficiency of quantum operations. For instance, Shor's algorithm utilizes the Quantum Fourier Transform (QFT), a key component that benefits from the coherent manipulation of qubits in different bases to factor large numbers efficiently [74]. The ability to operate in multiple bases is what enables quantum algorithms to outperform their classical counterparts [75].

Quantum Error Correction: Qubits are susceptible to decoherence and errors induced by environmental noise, making quantum error correction (QEC) essential for the reliability of quantum computations. QEC codes, such as those based on entangled states like Bell states, allow for the detection and correction of errors without directly measuring the quantum information, thus preserving the integrity of the qubits [76, 77]. The threshold theorem indicates that if the error rate is below a certain threshold, it is possible to perform arbitrary quantum computations reliably using QEC [78]. Furthermore, various strategies, including the use of decoherence-free subspaces, have been proposed to enhance error resilience in quantum systems [79].

Quantum Cryptography: Quantum Key Distribution (QKD), particularly the BB84 protocol, exemplifies how quantum mechanics can be employed for secure communication. This protocol relies on the random selection of different qubit bases to exchange cryptographic keys securely. The measurement of qubits in an incorrect basis can reveal the presence of an eavesdropper, thereby ensuring the security of the communication channel [80]. The inherent properties of quantum mechanics, such as entanglement and superposition, are leveraged to create a secure framework for key distribution that classical systems cannot replicate [81].

Quantum Networking: Quantum entanglement facilitates the teleportation of qubits across significant distances, which is a cornerstone of quantum networking. The establishment and verification of entanglement between remote qubits often involve measurements in the x-basis and y-basis, which are essential for confirming the integrity of the quantum states being transmitted [81, 82]. This capability is crucial for the development of quantum networks that can support secure communication and distributed quantum computing [81].

Physical Implementations of Qubits (Superconducting, Trapped Ions, Photonic Qubits)

A physical qubit serves as the fundamental building block of quantum computing, analogous to a classical transistor in traditional computing. Classical bits operate as binary units of information, representing either 0 or 1. In contrast, qubits exploit quantum mechanical principles such as superposition and entanglement, allowing them to store and process information in ways that classical bits cannot [83]. The manipulation of qubits differs significantly from classical transistors; while transistors change state through voltage variations, qubits undergo state changes via interactions with laser or microwave pulses, depending on their specific implementation [83, 84].

The nature of physical qubits presents several challenges, particularly their susceptibility to environmental noise and errors. Even minor interactions with external factors, such as temperature fluctuations or electromagnetic fields, can disrupt the fragile quantum state of a qubit, leading to decoherence—the loss of quantum information over time [85, 86]. This inherent vulnerability has driven the development of Quantum Error Correction (QEC) techniques and the concept of logical qubits, which are designed to safeguard and extend the operational lifetime of physical qubits [87, 88]. In the current Noisy Intermediate-Scale Quantum (NISQ) era, the term "qubit" is often used interchangeably with physical qubits, as demonstrated by systems like a 256-qubit neutral atom quantum computer, where each atom represents a single physical qubit [89, 90].

Various implementations of physical qubits exist, each with unique strengths and weaknesses, making them suitable for different quantum computing applications. For instance, neutral atoms offer scalability and long coherence times, while electron spins can be trapped in quantum dots or defects in crystal lattices, providing a solid-state alternative [83, 84]. Trapped ions, manipulated with laser pulses, exhibit high fidelity operations and long coherence times, making them a popular choice for quantum computing [83]. Other modalities include nitrogen vacancy centres in diamonds, which are promising for quantum sensing, and superconducting qubits, which are currently favoured by companies like IBM and Google for their potential in large-scale quantum computing [83, 91]. Additionally, topological qubits, though still theoretical, are considered highly resistant to decoherence and could pave the way for fault-tolerant quantum computing [83, 88].

The distinction between physical and logical qubits is crucial in quantum computing. Logical qubits are constructed from multiple physical qubits using QEC techniques, allowing them to function as error-protected units of quantum information [87]. This construction is vital for overcoming the high error rates associated with physical qubits, enabling reliable storage and processing of quantum data over extended periods [92, 93]. Recent advancements have demonstrated that increasing the number of physical qubits can significantly enhance the performance of logical qubits, as evidenced by research showing lower error rates with larger logical qubit configurations [88, 94]. Furthermore, innovative approaches have emerged, such as using fewer physical qubits to create multiple logical qubits through advanced error correction techniques [95].

Qubits can be physically implemented using various quantum systems. These implementations take advantage of different quantum properties, such as spin, polarization, or energy levels. Some of the most common physical qubit technologies include:

- Superconducting Qubits
- Trapped Ion Qubits
- Quantum Dots
- Photonic Qubits
- Neutral Atom Qubits

Each qubit technology has unique advantages and challenges. Some offer faster gate operations, while others provide better coherence or scalability.

Superconducting Qubits

Quantum Computing

Superconducting qubits are one of the most widely used and extensively researched types of qubits in quantum computing. These qubits are built using Josephson junctions, which are superconducting circuits that exhibit quantum behaviour when cooled to extremely low temperatures. Unlike traditional transistors in classical computing, superconducting qubits rely on the principles of quantum mechanics, such as superposition and entanglement, to perform computations at an exponentially faster rate than classical systems.

Superconducting qubits are created using superconducting circuits, typically made of niobium or aluminium, that are cooled to temperatures close to absolute zero (near 10-15 millikelvin) using dilution refrigerators. At such low temperatures, these materials lose all electrical resistance and exhibit macroscopic quantum effects.

A Josephson junction is a critical component of superconducting qubits. It consists of two superconductors separated by a thin insulating barrier. This structure allows for the creation of quantum states that can be controlled and manipulated using external signals. By applying microwave pulses, researchers can control the state of the qubit and induce quantum operations necessary for computation.

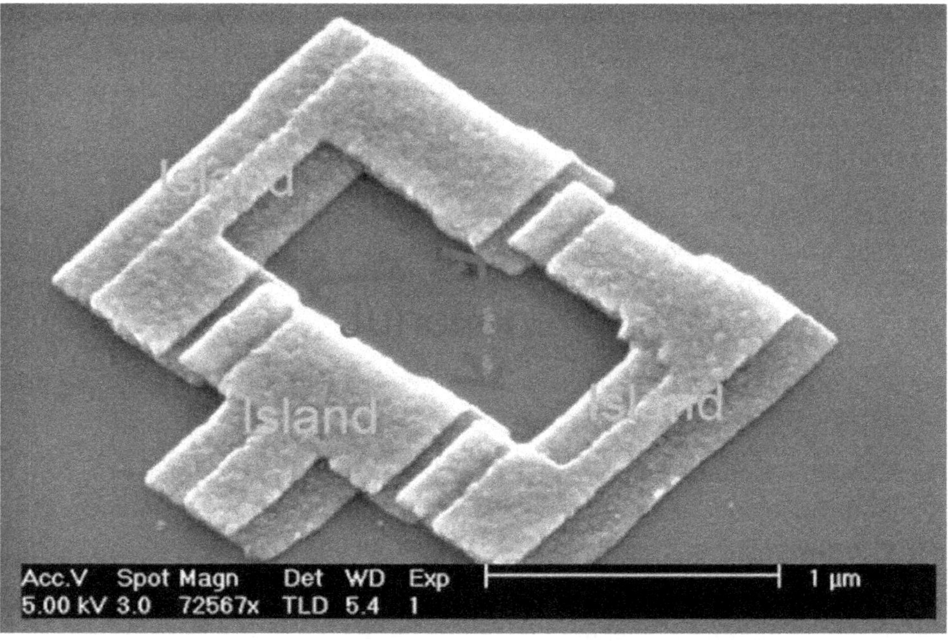

Figure 1: Solid-state superconducting quantum computing with Josephson junctions (SQUID). Cohærence, CC BY-SA 2.0, via Flickr.

The architecture of superconducting qubits, particularly the transmon design, has been pivotal in their adoption. Transmon qubits, which utilize Josephson junctions shunted by capacitors, have demonstrated reduced sensitivity to charge noise, making them suitable for scalable quantum processors [83, 96]. The rapid advancements in superconducting quantum circuits (SQC) over the past two decades have further solidified their status as a leading technology in quantum computing [8, 83]. These advancements have been accompanied by improvements in qubit quality and the ability to integrate larger numbers of qubits into coherent systems, as evidenced by the increasing number of qubits in operational quantum processors [96, 97].

Moreover, superconducting qubits have shown remarkable performance in terms of coherence times, which are critical for executing complex quantum algorithms. Recent studies highlight that superconducting qubits can achieve coherence times reaching milliseconds, which is essential for maintaining quantum information over longer periods [97, 98]. This capability is complemented by their ability to facilitate rapid measurements and gate operations, which are crucial for practical quantum computing applications [98, 99].

The commercial interest in superconducting qubits is also reflected in the strategic investments made by major quantum computing companies. IBM and Google, for example, have heavily invested in developing superconducting qubit technology, leading to significant breakthroughs in quantum supremacy and error correction [8, 98]. Rigetti Computing has similarly focused on superconducting qubits, emphasizing their potential for scalable quantum processors [8, 96]. The collective efforts of these companies underscore the consensus within the quantum computing community regarding the viability of superconducting qubits as a foundational technology for future quantum systems.

Each superconducting qubit is essentially a nonlinear LC circuit (a circuit with an inductor and capacitor) that supports two distinct energy states, which are used to represent the computational states $|0\rangle$ and $|1\rangle$. Unlike classical bits, a superconducting qubit can exist in a superposition of both states simultaneously, enabling quantum parallelism.

Superconducting qubits have become the preferred choice for many quantum computing companies because they offer several advantages:

1. **Scalability** – Superconducting qubits can be fabricated using standard lithography techniques, similar to how classical semiconductor chips are produced. This makes them easier to scale up compared to some other qubit types.
2. **Fast Gate Operations** – Quantum gate operations (basic quantum logic operations) on superconducting qubits are faster than in many other qubit technologies. Gates operate in the nanosecond range, allowing for rapid computation.
3. **Integration with Existing Technology** – Because superconducting qubits are made using semiconductor fabrication techniques, they are compatible with existing microelectronics, making them easier to integrate into larger systems.

Despite their advantages, superconducting qubits also face several challenges:

1. **Decoherence and Noise** – Superconducting qubits are highly susceptible to environmental noise, which can cause them to lose their quantum state (decoherence) quickly. This limits their coherence time, which is typically in the microsecond range.

2. **Extreme Cooling Requirements** – Superconducting qubits require dilution refrigerators to maintain the necessary ultra-cold temperatures. This makes the hardware expensive and complex to operate.

3. **Error Rates** – Quantum errors occur frequently in superconducting qubits, requiring quantum error correction (QEC) techniques to ensure reliable computations.

Many of the world's leading quantum computing companies rely on superconducting qubits to build their quantum processors:

- **IBM Quantum** – IBM has developed superconducting qubit processors, including the Eagle (127 qubits) and Osprey (433 qubits). Their roadmap includes plans to build a 1,121-qubit processor (Condor) by 2025.

- **Google Quantum AI** – Google used a superconducting quantum processor called Sycamore to demonstrate quantum supremacy in 2019 by performing a task in 200 seconds that would take classical supercomputers thousands of years.

- **Rigetti Computing** – Rigetti focuses on developing superconducting quantum processors for cloud-based quantum computing services. They provide access to quantum hardware through their Quantum Cloud Services (QCS) platform.

Superconducting qubits continue to be one of the most promising platforms for quantum computing. Researchers are actively working on improving coherence times, reducing error rates, and scaling up the number of qubits to build fault-tolerant quantum computers. Innovations such as superconducting qubits with longer lifetimes, better quantum error correction techniques, and more efficient microwave control methods are paving the way for more powerful quantum systems.

Despite challenges, superconducting qubits remain the dominant qubit technology for near-term quantum computing applications, with ongoing advancements expected to bring practical quantum computing closer to reality.

Trapped Ion Qubits

Trapped ion qubits are one of the most promising and well-developed approaches to quantum computing. They utilize ions, which are charged atoms, as qubits and are manipulated using electromagnetic fields and laser pulses to perform quantum operations. Trapped ion qubits are known for their exceptionally long coherence times and high-fidelity gate operations, making them a strong candidate for building fault-tolerant quantum computers.

In a trapped ion quantum computer, individual ions are confined using electromagnetic fields inside an ultra-high vacuum chamber. The most commonly used ions include ytterbium (Yb^+) and barium (Ba^+) because their internal electronic states are highly stable and well-suited for quantum operations.

Each trapped ion represents a single qubit, where its internal energy levels serve as the computational states |0⟩ and |1⟩. Unlike superconducting qubits, which rely on microwave circuits, trapped ion qubits are controlled using laser pulses, which drive quantum state transitions between different energy levels.

Quantum operations in trapped ion systems are performed using laser-induced interactions that manipulate the internal states of the ions. These operations include:

- **Single-Qubit Gates** – Applying a laser pulse to an individual ion to rotate its quantum state.
- **Multi-Qubit Gates** – Using a shared vibrational mode (phonons) of the trapped ions to create entanglement between qubits.
- **Quantum Measurement** – Using fluorescence detection, where a laser excites the ion and the emitted light determines whether it is in |0⟩ or |1⟩.

Trapped ion quantum computers offer several advantages that set them apart from other qubit technologies:

1. **Long Coherence Times** – One of the most significant advantages of trapped ion qubits is their exceptionally long coherence times, often measured in seconds to minutes, compared to the microseconds seen in superconducting qubits. This extended coherence time reduces the likelihood of quantum errors, making trapped ion systems highly reliable.
2. **High-Fidelity Operations** – Because laser-controlled quantum operations are highly precise, trapped ion systems achieve high gate fidelities exceeding 99.9%, meaning they have significantly lower error rates than many other qubit implementations.
3. **All-to-All Connectivity** – In a trapped ion system, every qubit in the chain is naturally connected to every other qubit via shared motion modes. This all-to-all connectivity simplifies the implementation of quantum algorithms and enhances the efficiency of multi-qubit operations.
4. **Scalability Using Modular Approaches** – Unlike superconducting qubits, which require complex wiring and cooling for scaling, trapped ion qubits can be scaled using

modular architectures, where different ion traps are interconnected using photonic links.

Despite their advantages, trapped ion quantum computers face certain challenges that impact their scalability and efficiency. One of the primary limitations is the speed of gate operations. Unlike superconducting qubits, which can execute quantum operations in nanoseconds, trapped ion qubits perform operations in the microsecond range. This difference in speed makes trapped ion systems relatively slower, particularly when implementing large-scale quantum algorithms that require numerous operations.

Another challenge is the complexity of laser control. Trapped ion qubits rely on precisely tuned lasers to manipulate their quantum states and perform computations. Each ion in the system requires multiple finely calibrated laser beams for tasks such as initialization, state manipulation, and readout. Managing and maintaining these laser systems is technically demanding, requiring high precision and stability. As the number of qubits in a system increases, the complexity of laser control grows, making scalability more difficult.

Additionally, trapped ion quantum computers require ultra-high vacuum environments and precise electromagnetic field control to maintain ion stability. While these systems do not need the extreme cryogenic temperatures required for superconducting qubits, they must operate in vacuum chambers to prevent unwanted interactions with air molecules, which can disrupt quantum coherence. The necessity for sophisticated vacuum and electromagnetic field systems adds to the overall complexity and cost of building and maintaining trapped ion quantum processors.

Several leading companies and research institutions are at the forefront of developing and commercializing trapped ion quantum computers, with notable contributions from IonQ, Honeywell Quantum Solutions, and Quantinuum.

IonQ is recognized as a pioneer in the field of trapped ion quantum computing. The company has successfully developed commercial quantum processors featuring over 20 high-fidelity qubits, which are fully coupled, allowing for robust quantum operations. IonQ's systems are designed to scale using modular architectures, which is essential for enhancing computational power and efficiency in quantum applications [100, 101]. The low error rates associated with IonQ's qubits further bolster the reliability of their quantum processors, making them suitable for complex quantum algorithms and machine learning applications [100, 102].

Honeywell Quantum Solutions has also made significant strides in trapped ion quantum computing, achieving some of the highest fidelity measurements in the industry. Their innovations include advancements in error correction techniques specifically tailored for trapped ion systems, which are crucial for maintaining the integrity of quantum computations over extended periods [101, 103]. The company's commitment to enhancing the performance of trapped ion qubits has positioned them as a key player in the quantum computing landscape [104].

The merger of Honeywell Quantum Solutions and Cambridge Quantum has led to the formation of Quantinuum, which focuses on advancing trapped ion quantum technologies. Quantinuum has developed a unique quantum charge-coupled device (QCCD) architecture that allows for low crosstalk and high fidelity in quantum operations by separating the ion trap into multiple zones. This architecture facilitates efficient ion transport and arbitrary qubit connectivity, which are critical for scaling quantum systems [104, 105]. Their work emphasizes the importance of integrating photonics with trapped ion systems to enhance communication and processing capabilities [104].

Trapped ion qubits are highly valuable for various quantum computing applications due to their long coherence times and exceptional operational fidelity. One of the key areas where they are applied is quantum cryptography. Secure communication protocols, such as entanglement-based quantum key distribution (QKD), leverage the stability and precision of trapped ion qubits to create unbreakable encryption methods. By utilizing quantum entanglement, these systems ensure that any attempt at eavesdropping is immediately detectable, making them ideal for secure data transmission in fields such as government communications and financial transactions.

Another significant application of trapped ion qubits is in quantum chemistry simulations. Accurately modelling complex molecular interactions is an enormous challenge for classical computers due to the exponential growth of quantum states in molecular systems. Trapped ion quantum computers provide a powerful tool for simulating chemical reactions, aiding in drug discovery, materials science, and the design of novel catalysts. Their ability to process quantum interactions naturally allows researchers to explore molecular structures with unparalleled accuracy.

Trapped ion qubits also play a crucial role in solving optimization problems across various industries. From logistics and supply chain management to financial modelling and machine learning, optimization problems require immense computational power. Trapped ion quantum algorithms help improve efficiency in areas such as route planning, portfolio optimization, and data clustering. Their long coherence times allow for more precise and reliable calculations, making them particularly effective in finding optimal solutions.

Quantum artificial intelligence and machine learning benefit significantly from the robustness of trapped ion qubits. Enhancing quantum-assisted machine learning models requires qubits that can maintain coherence and interact reliably over extended periods. Trapped ion quantum computers provide the stability needed for tasks such as pattern recognition, natural language processing, and predictive modelling. Their high-fidelity operations enable the development of advanced AI systems that leverage quantum principles for improved computational speed and accuracy.

Trapped ion quantum computers are expected to play a major role in the future of quantum computing, especially in the development of fault-tolerant quantum computers. Advances in error correction, modular architectures, and photonically linked quantum networks could make trapped ion systems one of the leading candidates for scalable quantum computation.

As technology continues to improve, trapped ion qubits may become one of the primary platforms for building large-scale, practical quantum computers that outperform classical systems in real-world applications.

Quantum Dots

Quantum dots (QDs) are semiconductor nanostructures that have garnered significant attention for their potential applications in quantum computing, particularly as qubits. These structures are designed to confine single electrons within a potential well, resulting in discrete energy levels that can be utilized to encode quantum information. The confinement of electrons in quantum dots is a direct consequence of quantum mechanics, where the spatial dimensions of the dot are comparable to the de Broglie wavelength of the electrons, leading to quantized energy states [106, 107]. This phenomenon is crucial for the operation of quantum dots as qubits, as it allows for the precise manipulation of electron states necessary for quantum computation [94].

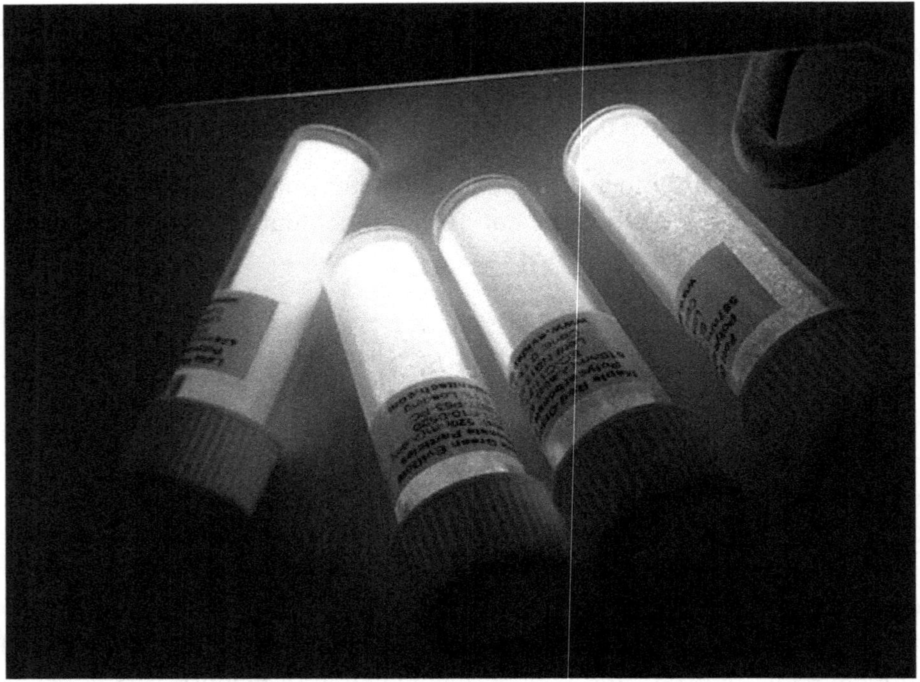

Figure 2: Quantum dots shining under UV light. Marc Vidal, CC BY-SA 4.0, via Wikimedia Commons.

One of the primary advantages of quantum dots is their ability to be manipulated using both electric and magnetic fields, which provides researchers with a high degree of control over qubit states. This manipulation enables operations such as superposition and entanglement, essential for quantum computing. For instance, studies have demonstrated that the spin entanglement of electrons in quantum dots can be effectively controlled, allowing for the generation of entangled photon pairs, which are vital for measurement-based quantum computing [108]. Furthermore, the application of external fields can tune the energy levels of quantum dots, facilitating the execution of quantum gate operations that are fundamental to quantum computation [109].

Another significant feature of quantum dots is their compatibility with traditional silicon-based semiconductor technology. Unlike many other qubit modalities that necessitate entirely new hardware and infrastructure, quantum dots can be integrated into existing semiconductor fabrication processes. This compatibility is particularly advantageous as it allows for the utilization of well-established manufacturing techniques in modern electronics, paving the way for scalable quantum computing solutions [94, 107]. The ability to operate within standard semiconductor frameworks not only enhances the feasibility of mass production but also supports the commercialization of quantum processors, making quantum dots a promising candidate for future quantum technologies [94].

Photonic Qubits

Photonic qubits represent a significant advancement in quantum information science, where quantum information is encoded in the properties of photons, particularly their polarization. The polarization states of photons can be categorized as horizontal, vertical, or in a superposition of both, which allows them to function effectively as qubits. This encoding method provides photonic qubits with remarkable versatility, enabling them to travel long distances with minimal interaction with their environment, thus preserving quantum coherence more effectively than many other types of qubits [110, 111].

One of the primary advantages of photonic qubits is their suitability for quantum communication. Unlike matter-based qubits, which often require isolation from environmental noise to maintain coherence, photonic qubits can be transmitted through optical fibres or free space with minimal loss of quantum information. This characteristic makes them ideal candidates for developing quantum networks, where secure sharing of quantum information over long distances is paramount [112, 113]. The ability to leverage existing telecommunications infrastructure for quantum communications further enhances their practical applicability, paving the way for global quantum networks that could transform secure data exchange and national security communications [114, 115].

Photonic qubits play a crucial role in quantum cryptography, particularly in Quantum Key Distribution (QKD) protocols such as BB84 and E91. These protocols utilize the quantum properties of photons to establish secure cryptographic keys between parties. The fundamental principle of QKD is that any attempt to eavesdrop on a quantum-encrypted communication disturbs the quantum state, making such intrusions detectable. This inherent security feature ensures that photonic qubits can be utilized to create ultra-secure communication channels, establishing them as a cornerstone of quantum security applications [116, 117].

In addition to their applications in cryptography, photonic qubits are being explored for quantum computing. Optical quantum computing architectures utilize the interactions of photons through beam splitters, phase shifters, and nonlinear optical elements to perform quantum logic operations. Although challenges remain in achieving large-scale photonic quantum computation, the potential for all-optical quantum processors that do not require extreme cooling or vacuum conditions makes this an exciting area of research [118, 119]. The integration of photonic qubits into existing fibre-optic infrastructure not only enhances their practicality but also facilitates the development of robust quantum networks capable of supporting advanced quantum applications [116, 120].

Neutral Atom Qubits

Neutral atom qubits are increasingly recognized as a leading approach for large-scale quantum computing due to their unique properties and capabilities. These qubits utilize individual neutral atoms that are manipulated through advanced techniques such as laser cooling and optical lattices. The process of laser cooling reduces the thermal motion of the atoms to near absolute zero, allowing them to be trapped and arranged in a defined structure, which is essential for creating stable quantum states [121]. Optical lattices, formed by the interference of laser beams, provide a framework for positioning and holding these atoms in a grid-like array, enabling precise control over qubit interactions [122, 123].

One of the most significant advantages of neutral atom qubits is their long coherence times. Neutral atoms, being electrically neutral, exhibit reduced susceptibility to environmental noise and electromagnetic interference, which are critical factors contributing to decoherence in other qubit systems such as superconducting circuits and trapped ions [124, 125]. This inherent stability allows for the maintenance of quantum information over extended periods, making neutral atom qubits particularly suitable for complex quantum computations that require sustained coherence [126].

Scalability is another key benefit of neutral atom qubits. The ability to manipulate identical atoms in large numbers facilitates the construction of extensive qubit arrays with high connectivity. This contrasts sharply with superconducting qubits, which often necessitate intricate fabrication and calibration processes [127, 128]. The straightforward scalability of neutral atoms positions them as a viable candidate for developing quantum processors capable of addressing real-world problems across various domains, including optimization and materials science [121, 129].

Moreover, neutral atom qubits exhibit versatility in quantum operations, particularly through the use of Rydberg states. When atoms are excited to these high-energy states, they can interact strongly, allowing for the implementation of multi-qubit quantum gates essential for executing complex algorithms [130, 131]. The Rydberg blockade effect, which prevents multiple atoms from being excited simultaneously, is a pivotal mechanism that enables the realization of entangling gates, such as the controlled-NOT (CNOT) gate, which is fundamental for quantum computation [132, 133].

Leading companies in the quantum computing sector, such as QuEra and ColdQuanta, are actively exploring the potential of neutral atom quantum processors. Their efforts are focused on developing systems that can incorporate hundreds or even thousands of neutral atom qubits, thereby advancing the field towards achieving quantum advantage [134]. As research progresses, the promise of neutral atom qubits continues to grow, positioning them as a cornerstone of future quantum computing technologies.

Technology and Hardware Required to Implement Physical Qubits

The implementation of physical qubits, which serve as the fundamental building blocks of quantum computing, requires highly specialized technology and hardware. Unlike classical bits, which operate using voltage levels in traditional transistors, qubits leverage the principles of quantum mechanics, such as superposition and entanglement, to perform computations. This introduces unique challenges in terms of hardware design, control mechanisms, and environmental isolation. Different physical qubit modalities, including superconducting qubits, trapped ion qubits, quantum dots, photonic qubits, and neutral atom qubits, require specific technological infrastructure and hardware to function effectively.

1. Superconducting Qubits: Technology and Hardware

Superconducting qubits are among the most widely used qubit implementations in modern quantum computers. These qubits are constructed using Josephson junctions, which consist of two superconducting materials separated by a thin insulating layer. The key component of superconducting qubits is their ability to exhibit macroscopic quantum states when cooled to extremely low temperatures.

The essential hardware required for superconducting qubits includes:

- **Dilution Refrigerators**: Superconducting qubits operate at temperatures close to absolute zero (10-15 millikelvin) to achieve superconductivity. Dilution refrigerators are used to create these ultra-low temperatures, minimizing thermal noise and extending coherence times.
- **Microwave Control Systems**: Quantum operations in superconducting qubits are performed using microwave pulses. These pulses manipulate the quantum state of the qubits, enabling the execution of logic gates and computational tasks.
- **Superconducting Circuit Fabrication**: The qubits themselves are fabricated on semiconductor chips using standard lithographic techniques, similar to those used in the microelectronics industry. Materials such as niobium and aluminium are commonly used.
- **Cryogenic Wiring and Filtering**: Because superconducting qubits are highly sensitive to noise, specialized wiring and filtering techniques are necessary to minimize electromagnetic interference and thermal fluctuations.

Leading companies such as IBM, Google, and Rigetti Computing are heavily invested in the development of superconducting qubit technology. Their research focuses on improving coherence times, reducing error rates, and scaling up qubit numbers to achieve fault-tolerant quantum computing.

2. Trapped Ion Qubits: Technology and Hardware

Trapped ion qubits utilize individual charged atoms (ions) confined within electromagnetic fields inside a vacuum chamber. These qubits are manipulated using laser pulses, which allow precise control over quantum state transitions.

The hardware required for trapped ion quantum computers includes:

- **Ion Traps**: These are microfabricated electrode structures that generate electromagnetic fields to confine and position individual ions.
- **Ultra-High Vacuum Chambers**: Trapped ion systems require an environment free from unwanted gas molecules to prevent decoherence. These vacuum chambers maintain pressures lower than those found in outer space.
- **Laser Systems**: Multiple laser beams are required for cooling, state initialization, quantum gate operations, and readout of the trapped ions.
- **Optical and Electromagnetic Control Systems**: Trapped ion qubits interact via shared vibrational modes (phonons), which must be precisely controlled using sophisticated electromagnetic and optical systems.

Companies such as IonQ and Quantinuum (formerly Honeywell Quantum Solutions) are pioneering trapped ion quantum computing, leveraging its long coherence times and high-fidelity gate operations.

3. Quantum Dots: Technology and Hardware

Quantum dots are semiconductor nanostructures that confine single electrons, which serve as qubits. These qubits can be manipulated using electric and magnetic fields, making them compatible with silicon-based semiconductor technology.

The hardware required for quantum dot qubits includes:

- **Semiconductor Fabrication Facilities**: Quantum dots are created using advanced semiconductor manufacturing techniques, allowing integration with conventional silicon electronics.
- **Cryogenic Cooling Systems**: Although not as extreme as superconducting qubits, quantum dots require low temperatures (millikelvin range) to reduce thermal noise and improve coherence times.
- **Spin Control Electronics**: Quantum dot qubits operate using the electron spin degree of freedom, requiring precisely controlled electric and magnetic fields to perform quantum operations.

- **Charge Sensors**: Quantum dots are highly sensitive, requiring specialized charge detection systems to monitor quantum state transitions.

Quantum dots hold promise for scalable quantum computing because they can be integrated into existing semiconductor infrastructure, making them a potential solution for future large-scale quantum processors.

4. Photonic Qubits: Technology and Hardware

Photonic qubits encode quantum information in the polarization states of photons. These qubits are ideal for quantum communication, cryptography, and distributed quantum computing.

The essential hardware for photonic qubit systems includes:

- **Single-Photon Sources**: These devices generate individual photons with well-defined quantum states for use as qubits.
- **Beam Splitters and Phase Shifters**: Optical elements such as beam splitters and phase shifters enable quantum gate operations on photonic qubits.
- **Optical Fibers and Free-Space Optical Systems**: Photonic qubits can travel long distances through optical fibres, making them ideal for quantum networking and secure communication applications.
- **Single-Photon Detectors**: These highly sensitive detectors measure quantum states with high precision, enabling quantum key distribution (QKD) and quantum teleportation.

Organizations such as Toshiba, the University of Vienna, and the Chinese Academy of Sciences are at the forefront of photonic quantum computing and quantum networking, leveraging the unique properties of photonic qubits.

5. Neutral Atom Qubits: Technology and Hardware

Neutral atom qubits use individual atoms, typically held in place by optical tweezers, to perform quantum operations. These qubits benefit from long coherence times and excellent scalability.

The hardware required for neutral atom quantum computing includes:

- **Optical Lattices and Tweezers**: Neutral atoms are trapped and manipulated using laser beams that form an optical lattice, allowing for precise control over qubit positioning.
- **Laser Cooling Systems**: Atoms are cooled to near absolute zero using Doppler and sub-Doppler cooling techniques to reduce their motion and improve stability.

- **Rydberg State Control Systems**: Quantum operations are performed using Rydberg interactions, where atoms are excited to high-energy states that enable strong, controllable interactions between qubits.
- **Atom-Photon Interfaces**: These interfaces enable photonic control of neutral atom qubits, allowing for efficient quantum communication and processing.

Companies such as QuEra Computing and ColdQuanta are pioneering neutral atom quantum technology, which offers significant potential for large-scale quantum computing due to its natural scalability and long coherence times.

Implementing physical qubits requires a combination of highly specialized technology and infrastructure. Each qubit modality—whether superconducting, trapped ion, quantum dot, photonic, or neutral atom—requires different approaches to fabrication, control, and environmental isolation. While superconducting qubits currently dominate commercial quantum computing due to their compatibility with existing semiconductor technology, other modalities such as trapped ions and neutral atoms provide advantages in terms of coherence times and scalability. As quantum hardware continues to evolve, advancements in quantum error correction, cryogenic technology, and photonic interfaces will be crucial in developing the next generation of practical and fault-tolerant quantum computers.

Bloch Sphere Representation

The Bloch Sphere Representation of Qubit States

A qubit state can be visualized using the Bloch sphere, a geometric representation where each possible qubit state corresponds to a point on the sphere's surface. The Bloch sphere provides an intuitive way to describe quantum superposition and quantum gates.

Any pure qubit state can be represented in terms of two angles, the polar angle θ and the azimuthal angle ϕ:

$$|\psi\rangle = \cos\frac{\theta}{2}|0\rangle + e^{i\phi}\sin\frac{\theta}{2}|1\rangle$$

- $\boldsymbol{\theta}$ (theta) represents the angle relative to the positive z-axis.
- $\boldsymbol{\phi}$ (phi) is the angle measured counterclockwise in the xy-plane from the positive x-axis.

For example, when $\theta=0$, the qubit is at the north pole of the Bloch sphere, representing $|0\rangle$. When $\theta=\pi$, the qubit is at the south pole, representing $|1\rangle$. A qubit in a superposition state lies somewhere along the surface between these two poles.

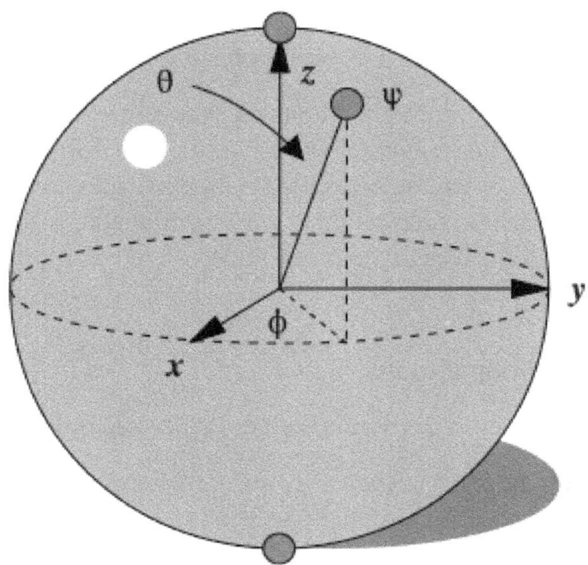

Figure 3: Bloch sphere diagram. CSTAR at English Wikipedia, CC BY-SA 3.0, via Wikimedia Commons.

The Bloch sphere representation allows us to understand quantum operations as rotations in a three-dimensional space, which is essential for quantum gates like the Hadamard gate, Pauli gates, and rotation gates.

The Bloch Sphere is a geometric representation of a single qubit state as a point on the surface of a unit sphere in a three-dimensional space. It provides an intuitive way to visualize quantum states and their evolution.

- The north pole of the Bloch Sphere represents the $|0\rangle$ state.
- The south pole represents the $|1\rangle$ state.
- Points on the equator represent equal superpositions of $|0\rangle$ and $|1\rangle$ with different phase factors.

The Bloch sphere provides an elegant and geometric way to interpret the properties of a qubit:

- **Pure States**: A pure qubit state (i.e., one that is in a definite quantum state without mixed states) lies on the surface of the Bloch sphere. Any point on the surface of the sphere corresponds to a unique pure state, which can be described by a specific combination of $|0\rangle$ and $|1\rangle$.

- **Superposition**: The superposition of |0⟩ and |1⟩ corresponds to all the points on the surface of the Bloch sphere, except for the poles. A qubit in a state like $\frac{1}{\sqrt{2}}(|0\rangle + |1\rangle)$ would be located on the equator of the Bloch sphere.
- **Coherent Rotation**: Quantum operations such as quantum gates (e.g., the Hadamard gate, Pauli gates) can be visualized as rotations of the qubit's state on the surface of the Bloch sphere. For example, applying a Hadamard gate to a qubit places it from |0⟩ to an equal superposition of |0⟩ and |1⟩, corresponding to a 90-degree rotation of the qubit state around the *y*-axis of the Bloch sphere.

Bloch Sphere and Quantum Operations

Quantum gates can be interpreted as rotations on the Bloch sphere. For example:

- **Pauli-X Gate (NOT Gate)**: This gate flips the state of the qubit, mapping |0⟩ to |1⟩ and vice versa. On the Bloch sphere, this operation corresponds to a 180-degree rotation around the *x*-axis.
- **Pauli-Y Gate**: This gate performs a 180-degree rotation around the *y*-axis, changing the qubit state in a manner similar to the Pauli-X gate, but with a phase shift.
- **Pauli-Z Gate**: This gate introduces a phase shift of π (or -1) to the |1⟩ state. It is represented by a 180-degree rotation around the *z*-axis.
- **Hadamard Gate**: The Hadamard gate transforms a state |0⟩ into a superposition of |0⟩ and |1⟩, and vice versa. Geometrically, it rotates the qubit state by 90 degrees around the axis halfway between the *x*- and *z*-axes.

In addition to pure states, the Bloch sphere can also represent mixed states, though these are represented as points inside the sphere, rather than on its surface. Mixed states are statistical mixtures of pure states and correspond to situations where the system is not fully coherent. As a result, the Bloch sphere representation for mixed states includes a radius *r* (ranging from 0 to 1) that indicates the level of coherence, with *r*=1 representing a pure state and *r*=0 representing a maximally mixed state (no coherence).

Special Quantum States on the Bloch Sphere

Some quantum states have particular significance in quantum computing due to their roles in quantum operations, superposition, and measurement outcomes. These states are categorized into computational basis states and superposition states, each occupying specific locations on the Bloch sphere.

The computational basis states correspond to the classical bit values, where the qubit is either in the |0⟩|0\rangle|0⟩ or |1⟩|1\rangle|1⟩ state. These states are the fundamental building blocks of quantum computation and serve as reference points for more complex superposition states.

- The state |0⟩ is located at the north pole of the Bloch sphere. This corresponds to the angles θ=0 and ϕ=0. When a qubit is in this state, measuring it will always return |0⟩ with 100% probability.
- The state |1⟩ is located at the south pole of the Bloch sphere, at θ=π and ϕ=0. Measuring a qubit in this state will always yield |1⟩.

These basis states align with the *z*-axis of the Bloch sphere. They represent classical binary values in quantum computing, but unlike classical bits, a qubit can exist in superposition between these states.

Superposition states are quantum states that are not confined to the poles but exist anywhere on the equatorial plane or other locations on the Bloch sphere. These states illustrate the fundamental difference between classical and quantum computing, as they enable qubits to exist in multiple states at once before measurement.

Equal Superposition States (X-basis States)

- The state **|+⟩** is an equal superposition of |0⟩ and |1⟩, defined as:

$$|+\rangle = \frac{|0\rangle + |1\rangle}{\sqrt{2}}$$

It is located at $\theta = \pi/2, \phi = 0$ on the equator of the Bloch sphere along the positive xxx-axis. This state is essential in many quantum algorithms, including the Hadamard transform, which places qubits into superposition.

The state **|−⟩** is another superposition state, but with a relative phase shift of π:

$$|-\rangle = \frac{|0\rangle - |1\rangle}{\sqrt{2}}$$

- It is located at $\theta = \pi/2, \phi = \pi$, on the opposite side of the equator, along the negative *x*-axis.

Both |+⟩ and |−⟩ states are used in quantum error correction and are crucial in quantum gates such as the Hadamard gate and the X-gate.

In addition to the real-number superposition states above, qubits can also exist in imaginary superposition states, which involve complex numbers (phases).

- The state |+i⟩ is a superposition where the |1⟩|1\rangle|1⟩ state has a phase factor of *i*:

$$|+i\rangle = \frac{|0\rangle + i|1\rangle}{\sqrt{2}}$$

It is located at $\theta = \pi/2, \phi = \pi/2$ on the equator, along the positive *y*-axis.

- The state |−i⟩ is a superposition where the |1⟩ state has a phase factor of −*i*:

$$|-i\rangle = \frac{|0\rangle - i|1\rangle}{\sqrt{2}}$$

- It is positioned at $\theta = \pi/2, \phi = -\pi/2$ on the equator, along the negative *y*-axis.

These states play a key role in quantum phase estimation and certain quantum Fourier transform operations. The presence of complex numbers (phases) in these states allows for interference effects, which are essential in many quantum computing algorithms.

Unlike classical bits, which can only be either 0 or 1, qubits can exist anywhere on the Bloch sphere, allowing for a vast range of possible states. This flexibility is what gives quantum computing its computational power. The ability to create and manipulate superposition states enables quantum computers to process multiple possibilities simultaneously, leading to speedups in algorithms such as Shor's algorithm (for factoring large numbers) and Grover's algorithm (for searching large databases).

By using quantum gates to rotate qubits on the Bloch sphere, we can transition between computational basis states and superposition states. Understanding the locations of these states helps in designing quantum circuits, controlling qubits efficiently, and implementing quantum algorithms.

The Bloch sphere representation of these special states illustrates why qubits are so powerful. Classical bits are confined to just two points (the poles), whereas qubits can explore the entire surface of the sphere, unlocking entirely new ways to process and compute information.

Quantum Gates and the Bloch Sphere

Quantum gates are the fundamental building blocks of quantum computing, manipulating qubits to perform operations in quantum algorithms. Unlike classical logic gates, which flip bits between 0 and 1 in a deterministic manner, quantum gates apply rotations to qubits around different axes of the Bloch Sphere. This allows quantum computers to exploit quantum mechanical properties like superposition, entanglement, and phase interference, enabling powerful computational abilities that surpass classical systems.

On the Bloch Sphere, a qubit is represented as a point on the unit sphere, with θ (theta) and ϕ (phi) determining its position. Quantum gates apply transformations that move the qubit's state to different locations on the sphere. These transformations are crucial in quantum computations, as they define how quantum information is manipulated to process data.

Pauli Gates (X, Y, Z)

Pauli gates are fundamental quantum operations that rotate the qubit 180° around different axes of the Bloch Sphere.

1. X-Gate (Bit Flip)

- The X-gate is equivalent to the classical NOT gate, flipping the qubit state.
- It performs a 180° rotation around the x-axis of the Bloch Sphere.
    - This means that applying an X-gate swaps |0⟩ and |1⟩:
      $$X|0\rangle = |1\rangle, \quad X|1\rangle = |0\rangle$$
- On the Bloch Sphere, this transformation moves the qubit from the north pole to the south pole and vice versa.

2. Y-Gate

- The Y-gate performs a 180° rotation around the y-axis.
- It swaps the computational basis states |0⟩|0⟩|0⟩ and |1⟩|1⟩|1⟩ but introduces a phase factor of i: $Y|0\rangle = i|1\rangle, \quad Y|1\rangle = -i|0\rangle$
- This phase difference is crucial for certain quantum algorithms where interference effects play a role.

3. Z-Gate (Phase Flip)

- The Z-gate does not change the probability amplitudes, but rather shifts the phase of the |1⟩ state.
- It performs a 180° rotation around the z-axis: $Z|0\rangle = |0\rangle, \quad Z|1\rangle = -|1\rangle$
- This gate is often used in quantum phase operations and quantum algorithms that rely on constructive and destructive interference.

Each of these gates can be visualized as a half-turn rotation along the corresponding axis of the Bloch Sphere, making them essential tools for basic quantum operations.

Hadamard Gate (H)

The Hadamard gate (H-gate) is one of the most crucial gates in quantum computing as it creates equal superpositions of |0⟩ and |1⟩, a fundamental requirement for many quantum algorithms.

- The Hadamard gate rotates the qubit 90° around the x+z diagonal axis of the Bloch Sphere.
- It transforms the computational basis states as follows:

$$H|0\rangle = \frac{|0\rangle + |1\rangle}{\sqrt{2}}, \quad H|1\rangle = \frac{|0\rangle - |1\rangle}{\sqrt{2}}$$

- This means that a qubit initialized in |0⟩ will now have an equal probability of being measured as |0⟩ or |1⟩.
- The Hadamard transformation is often applied to all qubits at the start of a quantum algorithm to create a superposition of all possible states, an essential step in quantum parallelism.

In terms of the Bloch Sphere, the Hadamard gate moves the qubit from the poles to the equator, placing it into an equal superposition state.

Phase Gates (S, T)

Phase gates modify the relative phase of a qubit state without changing its probabilities.

1. S-Gate (π/2 Phase Shift)

  - The S-gate applies a 90° (π/2) rotation around the z-axis.
  - It introduces a phase shift to the |1⟩ state:

$$S|0\rangle = |0\rangle, \quad S|1\rangle = i|1\rangle$$

- Used in phase-sensitive quantum algorithms, such as quantum Fourier transform.

2. T-Gate (π/4 Phase Shift)

- The T-gate is similar but applies a 45° (π/4) phase shift.
- It modifies the qubit state as follows:

$$T|0\rangle = |0\rangle, \quad T|1\rangle = e^{i\pi/4}|1\rangle$$

- This gate is important for fault-tolerant quantum computing, where precise phase control is needed.

Both S and T gates do not affect the qubit's position on the equator of the Bloch Sphere, but they rotate it around the z-axis, altering the phase relationship between states.

Rotation Gates (Rx, Ry, Rz)

Rotation gates provide **fine control** over qubit states by allowing arbitrary **θ (theta)** rotations around the Bloch Sphere.

1. Rx(θ) – Rotation Around the X-Axis

- Rotates the qubit by θ degrees around the x-axis:

$$R_x(\theta)|0\rangle = \cos(\theta/2)|0\rangle + i\sin(\theta/2)|1\rangle$$

- This gate is useful for precise state preparation and controlled bit-flip operations.

2. Ry(θ) – Rotation Around the Y-Axis

- Rotates the qubit by θ degrees around the y-axis:

$$R_y(\theta)|0\rangle = \cos(\theta/2)|0\rangle + \sin(\theta/2)|1\rangle$$

- Essential for changing the probability distribution of measurement outcomes.

3. Rz(θ) – Rotation Around the Z-Axis

- Rotates the qubit **by θ degrees around the z-axis**:

$$R_z(\theta)|0\rangle = e^{-i\theta/2}|0\rangle + e^{i\theta/2}|1\rangle$$

- This gate does not change measurement probabilities but alters the phase relationship between states.

Generalized single-qubit rotations combine these three rotations to move a qubit to any desired point on the Bloch Sphere, making them essential for universal quantum computation.

Implications for Quantum Computing

The ability of quantum gates to perform precise rotations on the Bloch Sphere is what enables quantum computers to perform complex computations that classical computers cannot. Since quantum states can exist in superpositions and quantum gates manipulate their positions on the Bloch Sphere, quantum circuits can execute operations on multiple values simultaneously.

Moreover, entanglement and controlled gate operations (such as CNOT) allow multiple qubits to interact, creating exponentially larger state spaces than classical computers. These properties are why quantum computers have the potential to solve problems intractable for classical computers, such as Shor's algorithm for factoring large numbers and Grover's algorithm for searching databases exponentially faster.

Understanding how quantum gates operate on the Bloch Sphere is crucial for designing efficient quantum algorithms, building quantum circuits, and implementing quantum computations that leverage the full power of quantum mechanics.

The Bloch Sphere is an essential tool in quantum computing for several reasons:

1. **Visualizing Qubit States**: It provides an intuitive way to understand quantum states, superposition, and phase relationships.
2. **Understanding Quantum Gates**: The effect of quantum gates on qubits can be seen as rotations on the Bloch Sphere, allowing easy interpretation of quantum logic operations.
3. **Quantum Error Correction**: The Bloch Sphere helps analyse errors and noise affecting qubits, guiding the development of quantum error correction techniques.
4. **Quantum Measurement Interpretation**: When a qubit is measured, it collapses onto either the $|0\rangle$ or $|1\rangle$ pole, reinforcing the probabilistic nature of quantum mechanics.

The Bloch Sphere demonstrates why quantum computing is fundamentally different from classical computing. Classical bits exist in two discrete states, represented as points on a line (0 or 1). In contrast, qubits exist anywhere on the Bloch Sphere, encoding much richer information through superposition and phase relationships. This enables quantum parallelism, where a quantum computer can explore multiple solutions at once.

While the Bloch Sphere is an excellent tool for visualizing single-qubit states, it does not scale well for multi-qubit systems. When multiple qubits are entangled, their state cannot be represented as independent Bloch Spheres, requiring more complex mathematical representations such as density matrices or tensor product spaces.

Decoherence and Error Correction Challenges

Quantum computing relies on delicate quantum states, such as superposition and entanglement, to perform computations that are infeasible for classical computers. However, one of the greatest challenges in building practical quantum computers is decoherence, which causes quantum information to degrade over time, leading to errors in calculations. Quantum error correction (QEC) aims to address these challenges, but it presents its own difficulties, such as scalability, resource requirements, and the need for redundancy.

Decoherence: The Loss of Quantum Information

Quantum decoherence is a fundamental challenge in quantum computing that occurs when a quantum system loses its quantum properties due to interactions with the environment. This loss of coherence causes quantum information to degrade, making computations unreliable and leading to errors. Decoherence is the process through which a quantum system transitions from a superposition state to a classical state, where quantum effects such as entanglement and interference are no longer observable. Understanding and mitigating decoherence is essential for the development of practical quantum computers.

Decoherence can be caused by several factors:

1. Environmental Interactions

- Any interaction between a qubit and its surrounding environment, such as thermal fluctuations, electromagnetic interference, or cosmic rays, can disturb its state.
- Unlike classical bits, which are stable in their 0 or 1 states, qubits must be isolated from external disturbances to maintain coherence.

2. Quantum Noise

- Quantum states are highly sensitive, and random fluctuations from external fields or imperfect control signals can introduce errors.
- Charge noise in superconducting qubits and laser fluctuations in trapped ion qubits are major sources of noise.

3. Decoherence Time (T1 and T2 Times)

- **T1 (Relaxation Time):** The time it takes for a qubit to decay from an excited state $|1\rangle$ to the ground state $|0\rangle$.
- **T2 (Dephasing Time):** The time over which quantum phase information is lost due to random interactions, even if the qubit remains in superposition.

- Long T1 and T2 times are crucial for stable quantum computation.

Because decoherence occurs rapidly (in microseconds for many qubit types), quantum computations must be completed before decoherence sets in, or error correction methods must be applied. The mechanisms of decoherence can be elucidated through several key aspects.

Environmental Interactions: Quantum systems are inherently coupled to their environments, which include electromagnetic fields, thermal vibrations, and other particles. These interactions introduce noise that disrupts the delicate superposition states of qubits, the fundamental units of quantum information. Decoherence is often described as a non-unitary evolution of the quantum state, where the environmental interactions lead to a loss of information from the system to the environment [135-137]. The coupling to the environment is an inevitable aspect of quantum systems, making decoherence a universal phenomenon that affects all quantum states [138].

Quantum Entanglement with the Environment: When a qubit interacts with its environment, it can become entangled with external particles. This entanglement results in the spreading of quantum information into the environment, rendering it inaccessible for computation. The entangled state can lead to a situation where the quantum information is effectively lost, as the system's state becomes correlated with the environment rather than remaining isolated [139]. This process is crucial for understanding how quantum information can degrade over time due to environmental influences.

Loss of Quantum Information: As decoherence progresses, the quantum state collapses into a classical state, effectively destroying the superposition and entanglement that are essential for quantum computation. The transition from a coherent quantum state to a classical mixture is a hallmark of decoherence, where the system loses its quantum characteristics and behaves classically [140, 141]. This loss of coherence is particularly detrimental in quantum computing, where maintaining the integrity of qubit states is paramount for reliable computation [142, 143].

Collapse of Superposition: The final stage of decoherence involves the irreversible loss of quantum superposition, where the qubit state transitions into a well-defined classical state. This transition signifies the breakdown of the probabilistic nature of quantum mechanics, as the system no longer exists in a superposition of states but rather in a definitive state that can be measured [144, 145]. The collapse of superposition is often accompanied by a significant reduction in the system's ability to exhibit quantum interference effects, which are essential for many quantum algorithms [146].

Time-Dependent Process: Decoherence is not an instantaneous event; rather, it is a gradual process that accumulates over time. The rate of decoherence can vary significantly depending on the strength of the interactions with the environment and the specific characteristics of the quantum system involved. Faster decoherence rates make it increasingly difficult to maintain stable quantum computations, as the qubits lose their coherence more rapidly [147, 148]. Understanding the time-dependent nature of decoherence is crucial for developing strategies to mitigate its effects in practical quantum computing applications [149].

The key causes of decoherence can be categorized into environmental interactions, quantum noise, and the inherent properties of qubits, specifically their decoherence times (T1 and T2).

Environmental Interactions: Quantum systems are highly sensitive to their surroundings, which can introduce noise and disrupt quantum states. Environmental disturbances such as thermal fluctuations, electromagnetic interference, and cosmic radiation play significant roles in this process. Thermal fluctuations, for instance, result from heat energy causing random motion in qubits, which alters their quantum states and leads to decoherence [150]. Electromagnetic interference from stray fields can also introduce noise that perturbs qubit coherence, as shown in studies focusing on superconducting qubits [151]. Additionally, cosmic rays and radiation can flip qubit states, further complicating the maintenance of quantum coherence [152]. Unlike classical computers, which store information in stable voltage states, quantum computers rely on maintaining fragile quantum states, making them particularly vulnerable to these environmental interactions [153].

Quantum Noise: Quantum noise is an intrinsic aspect of quantum mechanics that arises from fundamental uncertainties. Even in isolated systems, qubits experience fluctuations that can introduce errors. For example, charge noise in superconducting qubits is caused by fluctuations in nearby materials, leading to instability [150]. In trapped ion qubits, variations in laser intensity can result in errors during state manipulation, while photon loss in photonic qubits can lead to the loss of quantum information [153]. The unavoidable nature of quantum noise necessitates the implementation of quantum error correction methods to mitigate its effects [153].

Decoherence Time (T1 and T2): Decoherence is quantitatively described using two critical timescales: T1 (relaxation time) and T2 (dephasing time). T1 represents the time it takes for a qubit to decay from an excited state to the ground state, indicating how long a qubit can reliably store information [150]. T2, on the other hand, measures the time over which quantum phase information is lost due to environmental interactions, even if the qubit remains in superposition [154]. Long T1 and T2 times are essential for reliable quantum computations; for instance, superconducting qubits typically exhibit T1 times in the microsecond range, while trapped ion qubits can have coherence times measured in seconds [150]. The interplay between these timescales and environmental factors is crucial for understanding and improving quantum information processing [153].

Decoherence is a major limiting factor in the development of practical quantum computers. One of the most significant impacts of decoherence is on qubit stability. In order for quantum computations to be successful, qubits must maintain their quantum states long enough to perform complex calculations. However, decoherence disrupts these fragile quantum states, reducing the time in which qubits can remain coherent. This instability makes it difficult to execute long computations, as the loss of coherence leads to errors and prevents quantum parallelism from being fully utilized.

Another critical issue is quantum error rates. Decoherence introduces noise into quantum computations, making the results unreliable. Errors accumulate as qubits lose their quantum properties, leading to computational inaccuracies. High error rates significantly impact the performance of quantum algorithms, preventing them from reaching their theoretical advantages over classical computing. Without effective error correction techniques, decoherence makes it challenging to harness the full power of quantum systems.

Scalability challenges also arise as quantum computers grow in size. The more qubits that are added to a system, the greater the likelihood of decoherence affecting the computation. As the number of qubits increases, interactions between them and their environment become harder to control, resulting in greater instability. To scale quantum computers effectively, researchers must develop robust quantum error correction techniques that can counteract the effects of decoherence while maintaining the computational power of large quantum systems.

Decoherence also plays a fundamental role in the quantum-classical transition. In macroscopic systems, quantum effects are not observed because decoherence occurs so quickly that the system behaves classically. Large quantum systems, such as quantum computers, must be carefully isolated to prevent this transition. If decoherence occurs too rapidly, the system will lose its quantum behaviour, making it indistinguishable from a classical machine. Understanding and mitigating decoherence is crucial to preserving quantum properties at larger scales.

The performance of quantum algorithms is directly tied to decoherence. Many quantum algorithms rely on superposition and entanglement to outperform classical methods. However, if decoherence occurs too soon, these quantum effects are lost, reducing the efficiency and accuracy of the algorithm. As a result, decoherence sets a limit on the complexity of quantum computations that can be realistically implemented. Researchers are actively working on ways to extend coherence times and reduce errors, ensuring that quantum algorithms can operate effectively on real hardware.

Addressing decoherence is one of the most pressing challenges in quantum computing. While advancements in quantum error correction, cryogenic cooling, and novel qubit designs have helped improve coherence times, much work remains before quantum computers can operate reliably at scale.

To mitigate decoherence, researchers have developed several strategies that can be categorized into quantum error correction codes, environmental isolation, fast quantum operations, robust qubit designs, and dynamical decoupling.

Quantum Error Correction Codes (QECC) are fundamental in protecting quantum states from decoherence. Techniques such as the Shor code, Steane code, and Surface code encode logical qubits across multiple physical qubits, allowing for the detection and correction of errors that arise due to decoherence [155]. These codes are essential for maintaining the integrity of quantum computations, as they enable quantum systems to function effectively even when individual qubits experience decoherence [156]. The development of error correction strategies is critical, particularly for large-scale quantum computers, where the cumulative effects of decoherence can significantly hinder performance [157].

Environmental Isolation is another crucial strategy for mitigating decoherence. This involves techniques such as cryogenic cooling and electromagnetic shielding. Superconducting qubits, for instance, require dilution refrigerators that operate at millikelvin temperatures to minimize thermal noise, which is a significant source of decoherence [158]. Additionally, enclosing quantum hardware in shielded environments helps block stray electromagnetic interference, further protecting qubits from environmental disturbances [159]. These isolation techniques are vital for enhancing the coherence times of quantum systems, allowing for more reliable quantum computations.

Fast Quantum Operations are employed to execute quantum computations rapidly, thereby reducing the time during which decoherence can occur. Some quantum architectures prioritize speed over long coherence times, effectively minimizing the impact of noise on computations [160]. By completing operations quickly, researchers can circumvent the detrimental effects of decoherence, ensuring that quantum states remain intact for the duration of the computation.

Robust Qubit Designs are also being explored to enhance resistance to decoherence. Topological qubits, for example, are designed to be inherently robust against environmental noise, making them attractive for fault-tolerant quantum computing [156]. Similarly, neutral atom qubits and trapped ion qubits exhibit long coherence times, which are advantageous for maintaining quantum information over extended periods [158]. These robust designs are critical for the development of scalable quantum systems that can operate effectively in real-world environments.

Dynamical Decoupling is a technique that involves applying sequences of precisely timed control pulses to counteract environmental noise and extend coherence times. Methods such as the Carr-Purcell-Meiboom-Gill (CPMG) sequence are employed to protect qubits from random fluctuations caused by their surroundings [161]. This approach has shown promise in enhancing the fidelity of quantum computations by effectively mitigating the effects of decoherence [162]. By strategically controlling the quantum system, researchers can significantly improve the performance and reliability of quantum devices.

The Carr-Purcell-Meiboom-Gill (CPMG) sequence is a widely used dynamical decoupling technique in quantum computing and nuclear magnetic resonance (NMR). It is an extension of the Carr-Purcell (CP) sequence, designed to mitigate dephasing noise and extend the coherence time of qubits. The primary purpose of the CPMG sequence is to suppress low-frequency noise and improve the stability of quantum states, making it an essential tool for preserving quantum information in quantum computers.

Quantum systems, such as superconducting qubits and trapped ions, are highly sensitive to environmental interactions. These interactions cause decoherence, which results in the loss of quantum information over time. The CPMG sequence combats this issue by applying a carefully timed series of π (pi) pulses, also known as refocusing pulses, which effectively reverse the phase errors induced by external noise.

The CPMG sequence consists of:

1. An initial π/2 pulse that prepares the system into a superposition state.
2. A series of π pulses applied at regular intervals to refocus phase errors.
3. A final π/2 pulse for measurement.

These pulses work by flipping the quantum state at specific intervals, counteracting the effect of phase drift and extending the dephasing time (T_2) of the qubit.

The process can be understood step-by-step as follows:

1. Initialization and First π/2 Pulse:

- The system starts in its ground state (e.g., |0⟩).
- A π/2 pulse is applied, creating a superposition state (e.g., |+⟩ or |−⟩).
- This puts the qubit on the equator of the Bloch Sphere, where it is most susceptible to phase noise.

2. Application of π Pulses:

- After a certain time **τ**, a π pulse is applied, flipping the quantum state (e.g., from |+⟩ to |−⟩ or vice versa).
- If noise has caused a small phase shift before the π pulse, the phase error is reversed when the qubit is flipped.
- Multiple π pulses are applied at evenly spaced time intervals (τ, 2τ, 3τ...), ensuring that accumulated phase errors cancel out over time.

3. Final π/2 Pulse and Measurement:

  - After multiple π pulses, a final π/2 pulse is applied to rotate the state back into the measurement basis.

- This allows the system to recover a coherent quantum state with reduced decoherence effects.

The original Carr-Purcell (CP) sequence also uses a series of π pulses, but it suffers from phase errors accumulating over time. In 1958, Meiboom and Gill modified the CP sequence by adjusting the phase of the π pulses to match the spin echo alignment, preventing phase errors from compounding. This improvement led to the CPMG sequence, which is much more effective in prolonging coherence.

The CPMG sequence plays a critical role in enhancing qubit coherence and ensuring the stability of quantum operations. One of its primary applications is quantum error suppression, where it helps mitigate dephasing errors caused by environmental fluctuations. This is particularly beneficial for superconducting qubits and trapped ions, which are highly sensitive to phase noise. By applying the CPMG sequence, quantum hardware developers can extend the operational lifetime of qubits and improve the overall accuracy of quantum computations.

Another important application of the CPMG sequence is in quantum memory systems, where maintaining coherence for long durations is essential. Quantum memory is crucial for quantum repeaters, which enable long-distance quantum communication by storing and transmitting quantum states with minimal loss. The CPMG sequence helps preserve stored quantum information by counteracting the effects of noise, thereby extending the lifetime of quantum memory and improving its reliability.

In the field of quantum sensing, the CPMG sequence enhances the precision of quantum measurements. Quantum sensors, such as nitrogen-vacancy (NV) centres in diamonds, utilize the sequence to suppress noise and improve the accuracy of detecting magnetic and electric fields. By filtering out unwanted fluctuations, the CPMG sequence allows quantum sensors to achieve higher sensitivity, making them useful for applications in medical imaging, geophysics, and fundamental physics research.

The CPMG sequence also contributes to the advancement of quantum algorithms by allowing quantum processors to maintain coherence for extended periods. Many quantum algorithms, such as those used in optimization, cryptography, and materials simulation, require qubits to remain in a coherent state throughout the computation. By reducing decoherence, the CPMG sequence enables longer and more complex quantum computations, increasing the effectiveness of quantum algorithms.

Despite its advantages, the CPMG sequence has certain limitations and challenges that impact its effectiveness. One significant challenge is pulse imperfections. If the π pulses are not applied with perfect accuracy, they can introduce additional errors instead of correcting them. Hardware limitations, such as imprecise pulse shaping and control, can degrade the performance of the sequence, reducing its effectiveness in maintaining qubit coherence.

Another limitation is that the CPMG sequence is not effective for all types of noise. While it is highly efficient at counteracting low-frequency (1/f) noise, it is less effective against high-frequency noise sources. In such cases, additional techniques, such as nested dynamical decoupling sequences or alternative error mitigation strategies, may be required to achieve better noise suppression.

Implementing the CPMG sequence also increases the complexity of quantum control systems. The technique requires precise timing and accurate pulse sequences, which add to the computational and hardware demands of quantum processors. This complexity makes it challenging to integrate the CPMG sequence into large-scale quantum systems, particularly in architectures with many interacting qubits.

The Carr-Purcell-Meiboom-Gill (CPMG) sequence remains a fundamental tool for extending qubit coherence and improving the reliability of quantum computing. Its applications in quantum error suppression, quantum memory, quantum sensing, and algorithm optimization make it a valuable technique for advancing quantum technologies. Despite the challenges associated with pulse imperfections, noise limitations, and control complexity, ongoing research continues to refine the CPMG sequence and develop improved decoupling methods. These efforts contribute to the broader goal of building more robust and scalable quantum computers capable of solving complex problems beyond the reach of classical systems.

Decoherence is a fundamental obstacle in the path to large-scale quantum computing. It occurs due to environmental interactions, quantum noise, and fundamental instability in qubit designs. This loss of coherence causes quantum systems to behave classically, introducing errors in quantum computations. Managing decoherence is essential for building practical quantum computers, and various mitigation techniques, such as quantum error correction, cryogenic cooling, and robust qubit architectures, are actively being developed.

Despite these efforts, decoherence remains one of the most significant challenges in quantum computing. Future breakthroughs in materials science, quantum control techniques, and error correction methods will be necessary to achieve scalable, fault-tolerant quantum computation.

Quantum Error Correction (QEC)

Since qubits cannot be copied due to the no-cloning theorem, traditional error correction methods used in classical computing are ineffective. Quantum error correction (QEC) introduces redundancy and logical encoding to protect quantum information while adhering to the laws of quantum mechanics. Instead of directly copying qubits, QEC encodes quantum information across multiple physical qubits, ensuring that even if errors occur, the original quantum state can still be reconstructed.

Logical qubits and physical qubits are fundamental concepts in quantum error correction. Physical qubits are the actual hardware components that store and process quantum information, but they are highly susceptible to errors caused by decoherence and noise. To mitigate this, multiple physical qubits can be combined to create a logical qubit, which is an error-protected unit capable of maintaining quantum coherence longer than individual physical qubits. This encoding process requires significant redundancy, often needing dozens or even hundreds of physical qubits to construct a single logical qubit that can withstand errors.

To protect quantum information from errors, quantum error correction codes (QECC) have been developed. The Shor Code was the first QEC code and encodes one logical qubit into nine physical qubits, correcting both bit-flip and phase-flip errors. The Steane Code is a more efficient seven-qubit code that can also correct both types of errors. The Surface Code, one of the most scalable QEC approaches, encodes a logical qubit into a lattice of physical qubits and is widely used in modern quantum computing research due to its practical scalability.

A key aspect of QEC is syndrome measurement, which allows for error detection without collapsing the quantum state. Directly measuring a qubit would destroy its quantum superposition, so instead, ancilla qubits—auxiliary qubits entangled with the computational qubits—are used to detect and identify errors. These ancilla qubits extract information about the errors present in the system while preserving the quantum computation itself, making it possible to apply corrections without disturbing the original qubit states.

Despite its promise, quantum error correction faces significant challenges. One major issue is the high overhead of physical qubits. Current quantum processors from companies like IBM, Google, and IonQ operate with noisy physical qubits and have yet to demonstrate large-scale logical qubits. A fault-tolerant quantum computer may require millions of physical qubits, far beyond what current hardware can support. The challenge of error detection without disturbing computation further complicates QEC implementation. In classical computers, error detection is straightforward since bits can be measured without affecting their state. However, in quantum computers, measuring a qubit collapses its superposition, meaning errors must be identified indirectly through entangled qubits.

Another challenge is that decoherence limits computation time. Qubits have finite coherence times, meaning that quantum algorithms must execute before decoherence destroys the quantum state. Even with QEC, reducing errors in long computations remains an ongoing challenge, requiring improvements in qubit stability and correction efficiency. Additionally, scaling quantum error correction introduces complexities in syndrome measurements, entanglement management, and real-time error detection. While surface codes offer a scalable approach, they require highly connected quantum hardware, which is difficult to implement with current technologies.

Looking ahead, researchers are exploring various strategies to improve quantum error correction and achieve fault-tolerant quantum computing. One approach is improving qubit fidelity, which involves advancements in superconducting circuits, trapped ions, and neutral atom qubits. Better materials and enhanced qubit isolation techniques are being developed to extend coherence times and reduce intrinsic noise. Another focus is on developing more efficient QEC codes, such as low-overhead error correction schemes that require fewer physical qubits while maintaining fault tolerance. AI-driven QEC methods are also being investigated, leveraging machine learning to adapt error correction dynamically based on real-time noise conditions.

A promising area of research involves topological qubits and hardware-level error protection. Topological qubits, pursued by companies like Microsoft, aim to naturally resist decoherence by encoding quantum information in non-local degrees of freedom. If successfully implemented, topological qubits could dramatically reduce the need for complex QEC schemes, making large-scale quantum computing more feasible. Additionally, quantum feedback and active stabilization techniques, including real-time error correction loops and machine learning-based optimizations, are being explored to improve error mitigation.

Quantum error correction is essential for the future of quantum computing, enabling reliable and scalable quantum systems. While significant challenges remain, ongoing advancements in qubit fidelity, error correction codes, and topological qubits bring us closer to realizing fault-tolerant quantum computing, which will unlock new possibilities for solving problems beyond the capabilities of classical computers.

Examples of Quantum Error Correction in Practice

Quantum error correction (QEC) is a fundamental aspect of modern quantum computing, essential for protecting quantum information from errors induced by decoherence and noise. The significance of QEC is underscored by various real-world implementations and experimental demonstrations, which illustrate its critical role in advancing quantum computing technologies.

1. Logical Qubits vs. Physical Qubits

The distinction between logical qubits and physical qubits is pivotal in quantum computing. Logical qubits are constructed from multiple physical qubits, allowing for error correction and enhanced coherence times. IBM has made significant strides in this area, demonstrating a prototype logical qubit in 2023 that utilized 17 physical qubits arranged in a surface code architecture. This implementation notably reduced error rates compared to individual physical qubits, with the logical qubit exhibiting longer coherence times than its physical counterparts [163]. Similarly, Google's Sycamore quantum processor has explored logical qubits through surface codes, showing that increasing the number of physical qubits used to construct a logical qubit can effectively lower overall error rates, thus paving the way for scalable and fault-tolerant quantum computing [164].

2. Quantum Error Correction Codes (QECC)

Various quantum error correction codes have been successfully implemented in experimental quantum systems, demonstrating their effectiveness in preserving quantum information. The Shor code, for instance, has been implemented in trapped ion systems, where researchers utilized nine trapped ion qubits to protect a single logical qubit, successfully correcting both bit-flip and phase-flip errors [165]. Furthermore, Google's Quantum AI team conducted a surface code experiment in 2021 using 49 qubits, confirming that error rates decreased exponentially with the addition of more physical qubits, thus validating the scalability of the surface code for large-scale quantum computing [166]. The Steane code has also been employed by research groups at MIT and Harvard to enhance error rates in both superconducting and trapped-ion systems, requiring seven physical qubits to encode a logical qubit and correcting both types of errors [167].

3. Syndrome Measurement and Ancilla Qubits

Syndrome measurement is a crucial component of QEC, often utilizing ancilla qubits for error detection. In IBM's 127-qubit Eagle processor, specific qubits are designated as ancilla qubits, which assist in detecting and correcting errors without directly participating in computations [168]. IonQ has also implemented syndrome measurements using ancilla qubits in trapped ion systems, enabling error detection in multi-qubit entanglement experiments while preserving the quantum state [169]. This approach is vital for maintaining stable quantum operations in complex quantum systems.

4. High Overhead of Physical Qubits

The implementation of logical qubits necessitates a significant number of physical qubits, leading to high overhead in quantum computing architectures. IBM's roadmap includes the development of a 1,121-qubit quantum processor (Condor) by 2025, emphasizing the need for a large number of physical qubits to achieve practical, fault-tolerant quantum computing [168]. Similarly, Google's demonstration of a logical qubit using 72 superconducting qubits on their Bristlecone processor highlighted the challenges of controlling large qubit numbers, yet showed promising improvements in error rates [163].

5. Decoherence Limits Computation Time

Decoherence remains a formidable challenge in quantum computing, significantly limiting computation times. IBM has reported advancements in coherence times for superconducting qubits, achieving T1 times exceeding 300 microseconds, a substantial improvement over previous generations [168]. In contrast, Honeywell Quantum Solutions has demonstrated coherence times in trapped ion systems exceeding 10 minutes, making them strong candidates for long-running quantum algorithms, particularly in quantum cryptography and optimization tasks [167].

6. Scaling Quantum Error Correction

Scaling QEC is essential for the development of larger quantum systems. IonQ has proposed a modular approach that employs photonic links to connect multiple quantum processors, thereby distributing logical qubits across various physical qubit systems and improving error tolerance [168]. Concurrently, Google and Microsoft are exploring high-connectivity architectures that optimize the implementation of surface codes at scale, aiming to reduce the hardware overhead required for fault tolerance [167].

7. Future Approaches to Quantum Error Correction

Looking ahead, innovative approaches to QEC are emerging. Microsoft is investigating topological qubits, which inherently resist decoherence by encoding quantum information in non-local degrees of freedom, potentially minimizing the need for extensive QEC [168]. Additionally, AI-driven quantum error correction is gaining traction, with IBM and Google employing machine learning algorithms to adapt quantum circuits in real-time, enhancing error detection and correction capabilities [167]. These advancements are crucial for realizing the full potential of quantum computing, enabling the resolution of complex problems beyond the capabilities of classical computers.

Decoherence and quantum errors are significant barriers to the realization of practical quantum computing. In the current Noisy Intermediate-Scale Quantum (NISQ) era, quantum computers are limited by the high error rates of qubits, which are susceptible to decoherence from their environment. This phenomenon leads to the loss of quantum information, making it challenging to perform large-scale computations reliably. The need for robust quantum error correction (QEC) methods is paramount, as they provide a pathway toward achieving fault tolerance in quantum systems. However, implementing these error correction codes requires substantial resources and advancements in quantum hardware, which are still in development stages [170, 171].

Research indicates that decoherence is exacerbated by the entanglement present in quantum systems. For instance, while entangled states can enhance quantum advantages, they are also more vulnerable to decoherence effects [172]. This duality highlights the necessity for improved qubit designs that can withstand decoherence. Recent studies have shown that specific qubit architectures, such as superconducting qubits and quantum dots, can exhibit varying degrees of susceptibility to decoherence due to their unique physical properties [173, 174]. For example, superconducting qubits are particularly affected by magnetic flux noise, which can significantly degrade their coherence times [173].

Moreover, advancements in quantum error correction techniques are crucial for mitigating the effects of decoherence. The development of more efficient QEC codes and scalable quantum architectures is essential for achieving quantum advantage. Techniques such as encoding quantum information in decoherence-free subspaces have been proposed to protect against collective decoherence, thereby enhancing the reliability of quantum computations [175].

Additionally, the exploration of new materials and qubit designs, such as charge qubits and flux qubits, is ongoing to improve coherence times and reduce error rates [176, 177].

Overcoming decoherence through innovative qubit designs, efficient quantum error correction codes, and scalable architectures is critical for the future of quantum computing. The interplay between decoherence and quantum errors remains a central challenge, but ongoing research and technological advancements hold promise for realizing the full potential of quantum computing [171, 178].

Practical Applications and Implementation of Quantum Bits (Qubits) and Quantum Systems

The concepts outlined in this chapter play a critical role in the practical implementation of quantum computing.

Implementation of Qubit Principles in Quantum Computing

Developers of quantum computers use qubits to represent and process quantum information. Unlike classical bits, qubits leverage superposition, entanglement, and quantum interference to perform complex computations more efficiently than traditional computers. These properties allow quantum computers to solve problems such as optimization, cryptographic decryption, and quantum simulations exponentially faster.

For example, in superconducting quantum computers developed by IBM and Google, qubits are implemented using superconducting circuits cooled to near absolute zero to minimize decoherence. These superconducting qubits are controlled via microwave pulses, which manipulate their quantum states for computation.

In trapped ion quantum computing, companies like IonQ and Quantinuum use electromagnetic fields to trap and manipulate ions, allowing for highly stable and precise quantum operations. Unlike superconducting qubits, trapped ion qubits exhibit long coherence times, which makes them particularly suitable for quantum networking and fault-tolerant quantum computing.

Use of Superposition and Quantum Probability in Quantum Programming

Superposition is one of the fundamental quantum properties that developers exploit in programming quantum algorithms. It allows qubits to exist in multiple states simultaneously, which dramatically increases computational efficiency.

For example, the Hadamard gate (H-gate) is frequently used in quantum programming to put qubits into an equal superposition state. In IBM's Qiskit, developers use code like:

```
from qiskit import QuantumCircuit

qc = QuantumCircuit(1)
qc.h(0)   # Apply Hadamard gate to qubit 0
qc.measure_all()
print(qc.draw())
```

This prepares a qubit in superposition, allowing it to process multiple possibilities at once. This concept is foundational in quantum algorithms such as Shor's algorithm for factoring large numbers and Grover's algorithm for searching databases efficiently.

Entanglement and Quantum Communication

Quantum entanglement is essential for advanced quantum technologies such as quantum cryptography, quantum teleportation, and distributed quantum computing.

Developers create entangled qubit pairs using Controlled-NOT (CNOT) gates, which ensure that the state of one qubit is instantaneously correlated with another, no matter the physical distance between them. This is crucial in Quantum Key Distribution (QKD) protocols like BB84, which use entangled states to secure communications.

For example, in a quantum circuit using Qiskit, entanglement is created with the following code:

```
from qiskit import QuantumCircuit

qc = QuantumCircuit(2)
qc.h(0)   # Put the first qubit in superposition
qc.cx(0, 1)   # Apply a CNOT gate to entangle qubits
qc.measure_all()
print(qc.draw())
```

Such implementations are critical for secure quantum communication networks, enabling ultra-secure data transfer between distant locations.

Practical Use of the Bloch Sphere in Quantum Computing

The Bloch Sphere provides an intuitive way to visualize and manipulate qubit states. In practical quantum computing, Bloch Sphere transformations are implemented using quantum gates that rotate qubits along different axes.

Developers use Pauli gates (X, Y, Z), Hadamard gates (H), and rotation gates (Rx, Ry, Rz) to manipulate qubit states:

- **X-gate (bit flip)**: Swaps |0⟩ and |1⟩ by rotating the qubit 180° around the x-axis.
- **Z-gate (phase flip)**: Introduces a phase shift by rotating the qubit 180° around the z-axis.
- **Hadamard gate (H)**: Rotates the qubit 90° around the diagonal (x + z) axis, placing it into a superposition.
- **Rotation gates (Rx, Ry, Rz)**: Perform arbitrary rotations around different axes, allowing for precise quantum state control.

In quantum circuits, these gates are programmed as:

```
from qiskit import QuantumCircuit

qc = QuantumCircuit(1)
qc.h(0)  # Hadamard gate
qc.rz(3.14, 0)  # Rotate around Z-axis by π
qc.measure_all()
print(qc.draw())
```

This control over qubits enables the execution of complex quantum algorithms for cryptographic breaking, simulation, and AI applications.

Error Correction and the Challenges of Decoherence

One of the biggest practical challenges in quantum computing is decoherence, where qubits lose their quantum properties due to environmental interactions such as thermal noise, electromagnetic interference, and cosmic rays.

Developers use Quantum Error Correction (QEC) to mitigate these effects. Since qubits cannot be directly copied (due to the no-cloning theorem), QEC techniques encode logical qubits into multiple physical qubits. This redundancy allows for error detection and correction without collapsing the quantum state.

The Surface Code is one of the most scalable QEC techniques, encoding a logical qubit into a lattice of physical qubits to detect and correct bit-flip and phase-flip errors. Companies like Google and IBM are actively working on implementing fault-tolerant quantum computers using Surface Codes.

Example of error correction in quantum circuits:

```
from qiskit import QuantumCircuit

qc = QuantumCircuit(3)
qc.cx(0, 1)  # Apply a CNOT gate
qc.cx(0, 2)  # Redundant encoding for error detection
qc.measure_all()
print(qc.draw())
```

Such error correction is necessary for scaling quantum computers beyond the noisy intermediate-scale quantum (NISQ) era.

To achieve practical, large-scale quantum computing, developers are working on:

- **Improving Qubit Fidelity**: Researching topological qubits (Microsoft's approach) to naturally resist decoherence.
- **More Efficient QEC Codes**: AI-driven error correction that adapts dynamically to noise levels.
- **Hardware Innovations**: Using trapped ion qubits, neutral atom qubits, and photonic qubits to achieve long coherence times.
- **Quantum Networking**: Implementing quantum repeaters to link multiple quantum processors for distributed quantum computation.

The concepts from your document—qubits, superposition, entanglement, the Bloch Sphere, and quantum error correction—are practically implemented by developers to build reliable quantum computers. These principles guide the development of quantum hardware, algorithms, and error mitigation techniques. Companies like IBM, Google, Microsoft, and IonQ continue to refine these methods to scale quantum computing, making it a transformative technology for cryptography, material science, artificial intelligence, and optimization problems.

Chapter 3

Quantum Gates and Circuits

Having established the foundational principles of quantum mechanics in Chapter 1 and introduced the concept of qubits and quantum systems in Chapter 2, we are now ready to explore how quantum computers perform computations. Just as classical computers rely on logic gates to process information, quantum computers utilize quantum gates to manipulate qubits. However, unlike classical gates, which follow deterministic binary operations, quantum gates operate under the principles of quantum mechanics, enabling superposition, entanglement, and interference.

This chapter delves into the various types of quantum logic gates, explaining their role in quantum computation and how they differ from classical gates. We will explore single-qubit gates such as the Hadamard, Pauli, and phase gates, which manipulate individual qubits, as well as multi-qubit gates like the CNOT, Toffoli, and Swap gates, which enable interaction between qubits. These gates serve as the building blocks for quantum circuits, allowing us to construct algorithms that leverage the power of quantum parallelism.

Understanding quantum circuit design is essential for programming quantum computers and executing quantum algorithms efficiently. We will introduce key principles such as reversible computing, which ensures that quantum operations remain coherent and maintain information integrity. By the end of this chapter, you will have a firm grasp of how quantum circuits are structured and how they enable breakthroughs in fields such as cryptography, optimization, and artificial intelligence.

With a solid foundation in quantum gates and circuits, we will then proceed in Chapter 4 to examine quantum algorithms—powerful computational methods that harness quantum mechanics to solve problems that classical computers find infeasible. But before we can explore quantum speedups and groundbreaking algorithms, we must first understand the fundamental operations that make quantum computation possible.

Introduction to Quantum Logic Gates

Quantum logic gates are the fundamental building blocks of quantum circuits, just as classical logic gates are for traditional computing. However, unlike classical gates, which manipulate bits that are strictly 0 or 1, quantum gates operate on qubits—quantum bits that can exist in a superposition of both 0 and 1. This allows quantum computers to perform operations that are impossible for classical systems.

Quantum gates are unitary transformations, meaning they are reversible and preserve the total probability of a quantum state. These gates use the principles of superposition, entanglement, and quantum interference to execute computations that can exponentially outperform classical algorithms in certain tasks.

In classical computing, logic gates such as AND, OR, XOR, and NOT perform deterministic operations on binary data. These gates are irreversible, meaning information is lost after processing. For example, an AND gate takes two bits as input and produces one bit as output, which discards some information.

Quantum gates, on the other hand, must be reversible, meaning they can be undone by applying their inverse. They transform qubit states in a way that maintains quantum coherence, allowing for intricate operations such as parallel computation and quantum interference.

For example:

- A NOT gate flips a classical bit (0 → 1, 1 → 0).
- A Quantum X-gate (bit-flip gate) flips a qubit, but also allows superposition, meaning it can act on states that are mixtures of $|0\rangle$ and $|1\rangle$.

To understand quantum logic gates, it is essential to first examine their classical counterparts. Traditional computers process information using logic gates, which perform basic operations on bits—the fundamental units of classical data. Bits are represented as either 0 or 1, with logic gates manipulating them according to predefined rules.

For example, a NOT gate inverts a bit, turning 0 into 1 and vice versa. Other common logic gates include the AND gate, which outputs 1 only if both input bits are 1, and the OR gate, which outputs 1 if at least one input bit is 1. These classical logic gates serve as the building blocks of digital circuits, enabling everything from basic arithmetic to complex computations.

A crucial concept in classical computing is that any computation can be broken down into a sequence of universal gates—a set of fundamental logic operations that can be combined to execute any possible algorithm. This principle, first described by Alan Turing in 1936, underlies all modern computing systems. Classical computers, by chaining together sequences of AND, OR, and NOT gates, can perform diverse tasks ranging from web browsing to large-scale simulations.

Quantum computers operate fundamentally differently from classical computers. Instead of using classical bits, they manipulate qubits, which can be in a superposition of 0 and 1 at the same time. This property enables quantum computers to explore multiple solutions simultaneously, providing a potential speedup over classical systems for certain types of problems.

However, for a quantum computer to perform useful computations, qubits must be manipulated in precise ways to carry out quantum algorithms. These manipulations are achieved using quantum logic gates, which operate based on the principles of quantum mechanics.

Unlike classical gates, quantum gates are reversible, meaning they do not lose information and can be undone by applying their inverse operation. Quantum gates apply specific unitary transformations to qubits, rotating their states on the Bloch Sphere to execute complex quantum computations.

There are two main types of quantum logic gates:

- Single-qubit gates, which operate on individual qubits and are used to create superposition and perform basic transformations.
- Two-qubit gates, which allow qubits to interact, enabling quantum entanglement—a phenomenon where the state of one qubit is correlated with another, even across vast distances.

Quantum gates are fundamental components of quantum computing, enabling operations that leverage the unique properties of quantum mechanics, such as superposition and entanglement. These gates facilitate the execution of quantum algorithms, which can outperform classical algorithms in specific tasks, particularly in factoring large numbers and searching unsorted databases.

One of the primary functions of quantum gates is to create quantum superposition. The Hadamard gate (H), for instance, transforms a qubit from a definite state into a superposition of states, allowing quantum computers to explore multiple possibilities simultaneously. This property is essential for achieving quantum parallelism, which is a significant advantage over classical computing methods [179, 180]. The ability to manipulate qubits into superposition states is crucial for the efficiency of quantum algorithms, such as Shor's algorithm, which relies on the quantum Fourier transform (QFT) to factor integers exponentially faster than the best-known classical algorithms [181, 182].

Quantum entanglement is another critical aspect facilitated by quantum gates, particularly the controlled NOT (CNOT) gate. The CNOT gate creates entangled states between qubits, which is vital for various quantum protocols, including quantum teleportation and quantum cryptography [183, 184]. Entanglement allows for correlations between qubits that classical bits cannot achieve, enabling more complex quantum operations and enhancing the security of quantum communication systems [185]. The experimental realization of CNOT gates has been demonstrated, showcasing their importance in constructing universal quantum circuits [186, 187].

Furthermore, quantum gates contribute to the speedup of quantum algorithms. Shor's algorithm, which factors large integers, exemplifies how quantum gates can drastically reduce computation time compared to classical methods. The algorithm's efficiency stems from its reliance on quantum gates that perform operations like modular exponentiation and QFT, which are inherently faster in a quantum context [188, 189]. Similarly, Grover's search algorithm benefits from the use of quantum gates to achieve a quadratic speedup in searching unsorted databases [190, 191]. The design and optimization of these gates are ongoing areas of research, as advancements in quantum circuit design and fault-tolerant quantum computing are essential for realizing practical quantum computers [192, 193].

quantum gates are indispensable in quantum computing, enabling operations that exploit quantum mechanics' principles. Their unitary and reversible nature distinguishes them from classical gates, allowing for significant computational speedups. As research progresses, the refinement of gate designs and the development of fault-tolerant quantum processors will be crucial in unlocking the full potential of quantum technologies [194-196].

Quantum gates are represented using unitary matrices, which describe how they transform quantum states. A quantum state $|\psi\rangle$ is a vector in a Hilbert space, and quantum gates perform matrix multiplication to modify these states.

For example, the Pauli-X gate, which acts like a quantum NOT gate, is represented as:

$$X = \begin{bmatrix} 0 & 1 \\ 1 & 0 \end{bmatrix}$$

Applying this matrix to a qubit initially in state $|0\rangle$ gives:

$$X|0\rangle = \begin{bmatrix} 0 & 1 \\ 1 & 0 \end{bmatrix} \begin{bmatrix} 1 \\ 0 \end{bmatrix} = \begin{bmatrix} 0 \\ 1 \end{bmatrix} = |1\rangle$$

Similarly, applying it to $|1\rangle$ flips it back to $|0\rangle$.

Quantum gates can be broadly classified into **single-qubit gates** and **multi-qubit gates**:

1. Single-Qubit Gates

These gates operate on a single qubit at a time, modifying its state on the **Bloch Sphere**.

- **Pauli Gates (X, Y, Z):** Rotate a qubit 180° around different axes.
- **Hadamard Gate (H):** Creates superposition by transforming $|0\rangle$ into an equal mixture of $|0\rangle$ and $|1\rangle$.
- **Phase Gates (S, T):** Introduce a phase shift to a qubit's state.
- **Rotation Gates (Rx, Ry, Rz):** Perform arbitrary rotations around the x, y, or z-axis.

2. Multi-Qubit Gates

These gates operate on two or more qubits and enable quantum entanglement.

- **CNOT (Controlled-NOT) Gate:** Flips the second qubit's state if the first qubit (control qubit) is $|1\rangle$.
- **Toffoli Gate (CCNOT):** A three-qubit gate that flips the third qubit if both control qubits are $|1\rangle$.
- **SWAP Gate:** Swaps the states of two qubits.
- **CZ (Controlled-Z) Gate:** Applies a phase shift to the second qubit if the first is $|1\rangle$.

Operator	Gate(s)	Matrix
Pauli-X (X)	X	$\begin{bmatrix} 0 & 1 \\ 1 & 0 \end{bmatrix}$
Pauli-Y (Y)	Y	$\begin{bmatrix} 0 & -i \\ i & 0 \end{bmatrix}$
Pauli-Z (Z)	Z	$\begin{bmatrix} 1 & 0 \\ 0 & -1 \end{bmatrix}$
Hadamard (H)	H	$\frac{1}{\sqrt{2}}\begin{bmatrix} 1 & 1 \\ 1 & -1 \end{bmatrix}$
Phase (S, P)	S	$\begin{bmatrix} 1 & 0 \\ 0 & i \end{bmatrix}$
$\pi/8$ (T)	T	$\begin{bmatrix} 1 & 0 \\ 0 & e^{i\pi/4} \end{bmatrix}$
Controlled Not (CNOT, CX)		$\begin{bmatrix} 1 & 0 & 0 & 0 \\ 0 & 1 & 0 & 0 \\ 0 & 0 & 0 & 1 \\ 0 & 0 & 1 & 0 \end{bmatrix}$
Controlled Z (CZ)	Z	$\begin{bmatrix} 1 & 0 & 0 & 0 \\ 0 & 1 & 0 & 0 \\ 0 & 0 & 1 & 0 \\ 0 & 0 & 0 & -1 \end{bmatrix}$
SWAP		$\begin{bmatrix} 1 & 0 & 0 & 0 \\ 0 & 0 & 1 & 0 \\ 0 & 1 & 0 & 0 \\ 0 & 0 & 0 & 1 \end{bmatrix}$
Toffoli (CCNOT, CCX, TOFF)		$\begin{bmatrix} 1 & 0 & 0 & 0 & 0 & 0 & 0 & 0 \\ 0 & 1 & 0 & 0 & 0 & 0 & 0 & 0 \\ 0 & 0 & 1 & 0 & 0 & 0 & 0 & 0 \\ 0 & 0 & 0 & 1 & 0 & 0 & 0 & 0 \\ 0 & 0 & 0 & 0 & 1 & 0 & 0 & 0 \\ 0 & 0 & 0 & 0 & 0 & 1 & 0 & 0 \\ 0 & 0 & 0 & 0 & 0 & 0 & 0 & 1 \\ 0 & 0 & 0 & 0 & 0 & 0 & 1 & 0 \end{bmatrix}$

Figure 4: Common quantum logic gates by name, circuit form(s) and matrices. Rxtreme, CC BY-SA 4.0, via Wikimedia Commons.

Just as classical computers use universal logic gates to perform any computation, quantum computers can implement any quantum algorithm using a set of universal quantum gates. These include:

- **Pauli Gates (X, Y, Z):** Basic single-qubit gates that rotate qubits around different axes.
- **Hadamard Gate (H):** Places a qubit into an equal superposition of |0⟩ and |1⟩, enabling quantum parallelism.

- **Controlled-NOT (CNOT) Gate:** A two-qubit gate that flips the second qubit if the first qubit is $|1\rangle$, creating entanglement.
- **Phase and Rotation Gates (S, T, Rx, Ry, Rz):** These gates introduce phase shifts and enable precise quantum operations.

Through sequences of these gates, quantum computers can execute complex algorithms such as Shor's algorithm for factoring large numbers or Grover's algorithm for searching unsorted databases exponentially faster than classical methods.

One of the most promising quantum computing platforms is trapped ion quantum computing. In this approach, individual ions (charged atoms) are used to store qubits. The quantum information is encoded in the spin states of these ions, which are well-isolated from the environment, making them highly stable and resistant to errors.

To manipulate these qubits and implement quantum logic gates, trapped ion quantum computers use electromagnetic fields and laser pulses. Single-qubit gates are relatively straightforward—microwave pulses are applied to flip qubit states or create superpositions. By carefully adjusting the pulse duration and frequency, the qubit can be rotated to any desired state.

However, two-qubit gates are more complex. Unlike superconducting qubits, which use direct electrical connections, trapped ion qubits do not naturally interact. Instead, trapped ion computers exploit the Coulomb repulsion between ions—similar to the repulsion between two positively charged magnets—to create an artificial interaction. This interaction enables entanglement between qubits, which is essential for quantum computation.

There are two main approaches for implementing two-qubit gates in trapped ion systems:

1. High-power lasers are used to mediate interactions between qubits, inducing entanglement.
2. Specially shaped magnetic fields combined with microwave pulses can achieve the same effect without requiring complex laser control.

Some companies, like Universal Quantum, have developed quantum architectures that eliminate the need for lasers, making their approach more scalable. By avoiding lasers, these systems reduce errors and make it easier to build quantum processors with thousands or millions of qubits.

To realize the full potential of quantum computing, significant advancements in quantum gate fidelity, coherence times, and error correction techniques are essential. Currently, quantum processors operate within the Noisy Intermediate-Scale Quantum (NISQ) era, characterized by a limited number of qubits and high susceptibility to errors, which restricts the complexity of computations that can be performed [88, 197, 198]. The NISQ era, as defined by Preskill, encompasses quantum devices that typically feature around 50 to 100 qubits, with a universal but imperfect set of gates [88, 197]. This limitation underscores the need for ongoing research to enhance the performance of quantum systems.

Efforts to improve qubit connectivity are critical for enabling more efficient entanglement and computation. Enhanced connectivity allows for more complex quantum circuits and better utilization of qubits in computations. For instance, modular designs that facilitate entanglement across separate silicon dies have shown promise in improving qubit connectivity [199]. Additionally, the development of architectures that support any-to-any connectivity could significantly enhance the scalability and performance of quantum processors [200]. Such advancements are crucial as they directly impact the ability to implement fault-tolerant quantum error correction methods, which are vital for protecting quantum information from noise and errors [201, 202].

The development of fault-tolerant quantum error correction techniques is another area of active research. Current quantum systems are prone to errors due to decoherence and gate inaccuracies, which can severely limit computational capabilities [202, 203]. Error mitigation strategies are being explored to extend the computational reach of NISQ devices, allowing them to perform tasks that would otherwise be infeasible [202]. Techniques that reduce the impact of noise and improve gate fidelity are essential for achieving reliable quantum computations [88, 202]. For example, advancements in trapped ion systems have demonstrated high fidelity and long coherence times, making them suitable candidates for implementing effective error correction methods [204, 205].

Scaling up trapped ion and superconducting qubit systems is also a key focus area. Trapped ions, in particular, have shown exceptional performance in terms of coherence and gate fidelity, with some systems achieving coherence times exceeding 10 minutes and gate errors below the fault-tolerant threshold [204, 205]. These characteristics make trapped ion systems a leading platform for practical quantum applications. Similarly, superconducting qubits are being developed to enhance their connectivity and reduce error rates, which is critical for the realization of scalable quantum computing architectures [90, 206].

If these challenges are successfully addressed, quantum logic gates will enable the execution of powerful quantum algorithms, unlocking new capabilities across various fields, including materials science and secure communications. The transition from the NISQ era to a more robust quantum computing framework will hinge on overcoming these technical hurdles, allowing quantum processors to surpass classical computing capabilities in specialized tasks [90, 207].

Quantum Logic Gates in Gate-Based Quantum Computing

Quantum Computing

In quantum computing, quantum logic gates are the fundamental building blocks of quantum circuits. These gates perform unitary transformations on qubits, enabling quantum computations to be carried out efficiently. Unlike classical logic gates, which manipulate binary bits (0s and 1s), quantum logic gates operate on qubits, which can exist in superposition and be entangled with other qubits. The ability of quantum gates to transform qubit states through controlled rotations, phase shifts, and entanglement is what gives quantum computers their unique computational advantage.

In gate-based quantum computing, various sets of quantum gates are commonly used to express quantum operations. Each quantum gate has a specific mathematical representation in the form of a unitary matrix, ensuring that quantum transformations are reversible. This section details several key quantum logic gates, their properties, and their role in quantum computation.

Identity and Global Phase Gates

The Identity Gate (I) is the simplest quantum gate, performing a "no operation" (NOP) on a qubit. Mathematically, the identity gate is represented as:

$$I = \begin{bmatrix} 1 & 0 \\ 0 & 1 \end{bmatrix}$$

This gate does not change the qubit state, meaning that if a qubit is in state $|\psi\rangle$, applying the identity gate leaves it unchanged:

$$I|\psi\rangle = |\psi\rangle$$

Although the identity gate has no effect, it is useful in quantum circuits for defining waiting cycles or preserving qubits during computation.

The Global Phase Gate introduces a phase factor $e^{i\delta}$ to a quantum state:

$$\text{Global Phase} = e^{i\delta} I = e^{i\delta} \begin{bmatrix} 1 & 0 \\ 0 & 1 \end{bmatrix}$$

Since quantum states are defined up to a global phase, the global phase gate does not affect measurement outcomes. However, phase shifts become critical when qubits interact in multi-qubit systems.

Clifford Qubit Gates

Clifford gates form an essential subset of quantum logic gates. They are particularly important for error correction and can be efficiently simulated on classical computers.

Pauli Gates (X, Y, Z)

Pauli gates correspond to rotations by π radians around the X, Y, and Z axes of the Bloch Sphere.

1. Pauli-X (X Gate) – Bit Flip Gate

The X gate acts like a classical **NOT gate**, flipping the qubit state:

$$X = \begin{bmatrix} 0 & 1 \\ 1 & 0 \end{bmatrix}$$

If applied to |0⟩|0\rangle|0⟩, it flips it to |1⟩, and vice versa.

2. Pauli-Y (Y Gate) – Phase-Adjusted Bit Flip

The Y gate combines the X and Z operations:

$$Y = \begin{bmatrix} 0 & -i \\ i & 0 \end{bmatrix}$$

It introduces a phase factor of **i** when flipping states.

3. Pauli-Z (Z Gate) – Phase Flip Gate

The Z gate applies a π phase shift to |1⟩, leaving |0⟩ unchanged:

$$Z = \begin{bmatrix} 1 & 0 \\ 0 & -1 \end{bmatrix}$$

This is crucial in quantum phase algorithms like the Quantum Fourier Transform (QFT).

Hadamard and Phase Gates

1. Hadamard (H Gate) – Superposition Gate

The Hadamard gate transforms basis states into equal superpositions:

$$H = \frac{1}{\sqrt{2}} \begin{bmatrix} 1 & 1 \\ 1 & -1 \end{bmatrix}$$

Applying H to |0⟩ creates a superposition:

$$H|0\rangle = \frac{|0\rangle + |1\rangle}{\sqrt{2}}$$

2. S-Gate (π/2 Phase Shift)

The S gate applies a π/2 phase shift:

$$S = \begin{bmatrix} 1 & 0 \\ 0 & i \end{bmatrix}$$

3. T-Gate (π/4 Phase Shift)

The T gate introduces a π/4 phase shift, playing a key role in achieving quantum universality:

$$T = \begin{bmatrix} 1 & 0 \\ 0 & e^{i\pi/4} \end{bmatrix}$$

Controlled and Multi-Qubit Gates

1. Controlled-NOT (CNOT) – Entangling Gate

The CNOT gate flips the second qubit (target) only if the first qubit (control) is |1⟩:

$$CNOT = \begin{bmatrix} 1 & 0 & 0 & 0 \\ 0 & 1 & 0 & 0 \\ 0 & 0 & 0 & 1 \\ 0 & 0 & 1 & 0 \end{bmatrix}$$

2. Controlled-Z (CZ Gate)

This gate flips the phase of |11⟩:

$$CZ = \begin{bmatrix} 1 & 0 & 0 & 0 \\ 0 & 1 & 0 & 0 \\ 0 & 0 & 1 & 0 \\ 0 & 0 & 0 & -1 \end{bmatrix}$$

3. Toffoli (CCNOT) – Universal for Classical Computation

A three-qubit gate that flips the target qubit only if both control qubits are |1⟩:

$$CCNOT = \begin{bmatrix} 1 & 0 & 0 & 0 & 0 & 0 & 0 & 0 \\ 0 & 1 & 0 & 0 & 0 & 0 & 0 & 0 \\ 0 & 0 & 1 & 0 & 0 & 0 & 0 & 0 \\ 0 & 0 & 0 & 1 & 0 & 0 & 0 & 0 \\ 0 & 0 & 0 & 0 & 1 & 0 & 0 & 0 \\ 0 & 0 & 0 & 0 & 0 & 1 & 0 & 0 \\ 0 & 0 & 0 & 0 & 0 & 0 & 0 & 1 \\ 0 & 0 & 0 & 0 & 0 & 0 & 1 & 0 \end{bmatrix}$$

Rotation Gates (Rx, Ry, Rz)

Rotation gates provide fine control over quantum states:

$$R_x(\theta) = \begin{bmatrix} \cos(\theta/2) & -i\sin(\theta/2) \\ -i\sin(\theta/2) & \cos(\theta/2) \end{bmatrix}$$

$$R_y(\theta) = \begin{bmatrix} \cos(\theta/2) & -\sin(\theta/2) \\ \sin(\theta/2) & \cos(\theta/2) \end{bmatrix}$$

$$R_z(\theta) = \begin{bmatrix} e^{-i\theta/2} & 0 \\ 0 & e^{i\theta/2} \end{bmatrix}$$

These gates are essential for quantum state preparation and fine-tuned quantum operations.

Single-Qubit Gates (Hadamard, Pauli-X, Y, Z, Phase, T)

Pauli Gates in Quantum Computing

The Pauli Gates are fundamental quantum operations derived from the Pauli Matrices, named after the physicist Wolfgang Pauli. These gates are widely used in quantum computing for manipulating qubit states, performing quantum operations, and serving as building blocks for more complex quantum circuits.

In classical computing, logic gates such as AND, OR, and NOT operate on bits (0s and 1s). In quantum computing, operations are performed on qubits, which can exist in superposition states. The Pauli Gates (X, Y, and Z) are single-qubit quantum gates that transform qubit states in specific ways.

Each Pauli gate is represented as a 2×2 unitary matrix that operates on a qubit state |ψ⟩ in the Hilbert space. The Pauli matrices are:

1. Pauli-X Gate (Bit Flip)

$$X = \begin{bmatrix} 0 & 1 \\ 1 & 0 \end{bmatrix}$$

The X-gate is analogous to the classical NOT gate. It flips a qubit's state:

- $X|0\rangle = |1\rangle$
- $X|1\rangle = |0\rangle$

2. Pauli-Y Gate (Bit and Phase Flip)

$$Y = \begin{bmatrix} 0 & -i \\ i & 0 \end{bmatrix}$$

The Y-gate performs both a bit flip and a phase flip. It rotates a qubit 180° around the Y-axis on the Bloch Sphere:

- $Y|0\rangle = i|1\rangle$
- $Y|1\rangle = -i|0\rangle$

3. Pauli-Z Gate (Phase Flip)

$$Z = \begin{bmatrix} 1 & 0 \\ 0 & -1 \end{bmatrix}$$

The Z-gate flips the phase of a qubit, meaning it changes the sign of $|1\rangle$ but leaves $|0\rangle$ unchanged:

- $Z|0\rangle = |0\rangle$
- $Z|1\rangle = -|1\rangle$

These matrices satisfy the fundamental properties of the Pauli Group, which is important in quantum information theory.

Visualizing the Effect of Pauli Gates on the Bloch Sphere

The Bloch Sphere provides a powerful way to visualize the effect of Pauli gates on a qubit. It represents a qubit's state as a point on a unit sphere, where different quantum gates correspond to rotations around specific axes. The Pauli-X gate performs a 180° rotation around the X-axis, effectively swapping the basis states $|0\rangle$ and $|1\rangle$. This operation is equivalent to a classical bit-flip, meaning if a qubit starts in the $|0\rangle$ state, applying the X-gate moves it to $|1\rangle$, and vice versa.

The Pauli-Y gate also rotates the qubit by 180°, but around the Y-axis. Unlike the X-gate, this transformation introduces an imaginary phase shift, which changes the relative phase between the basis states. This means that applying a Y-gate to |0⟩ moves it to i|1⟩, and applying it to |1⟩ results in -i|0⟩. This unique combination of a bit-flip and phase shift makes the Y-gate essential in quantum algorithms that require precise control over phase.

The Pauli-Z gate performs a 180° rotation around the Z-axis, which does not affect the probability of measuring |0⟩ or |1⟩ but instead flips the phase of |1⟩. If a qubit is in the state |0⟩, applying the Z-gate leaves it unchanged, but if it is in |1⟩, the operation transforms it into -|1⟩. This phase-flip property is particularly useful in quantum phase estimation and error correction.

For instance, if a qubit starts in |0⟩ and the X-gate is applied, it moves to |1⟩. However, applying a Z-gate in this scenario will not change the measurement outcome but will alter the qubit's phase, making it -|1⟩. This distinction between state flipping and phase shifting is critical for understanding quantum superposition and interference, which are key to the power of quantum computing.

Practical Applications of Pauli Gates

Pauli gates are fundamental in quantum state preparation. The X-gate is frequently used to initialize a qubit in the |1⟩ state instead of the default |0⟩ state. This is crucial in quantum algorithms that require specific initial conditions. For example, in quantum algorithms like Grover's search, state initialization often involves Pauli gates to ensure that qubits start in the desired configuration before further transformations are applied.

In quantum error correction (QEC), Pauli gates play a critical role in detecting and correcting errors. Errors in quantum computing typically manifest as bit-flips (X errors), phase-flips (Z errors), or both (Y errors) due to decoherence and noise in the quantum system. Quantum error-correcting codes, such as the Shor code and Surface code, rely on syndrome measurements that identify the type of error and then apply an appropriate Pauli gate to restore the qubit to its correct state. These gates allow quantum computations to remain fault-tolerant even when errors occur.

In quantum teleportation, the transfer of quantum information between distant qubits relies on Pauli gates for state correction. After entangling two qubits and performing measurements on the sender's qubit, the receiver applies a Pauli X or Z gate to adjust their qubit based on the measurement outcomes. This ensures that the quantum state is successfully reconstructed at the destination, enabling secure quantum communication.

Superdense coding is another quantum communication protocol that utilizes Pauli gates to encode information. By applying Pauli X, Y, and Z operations to an entangled qubit, two classical bits of information can be transmitted using only a single qubit. This efficiency in quantum communication provides advantages in quantum networks and secure data transfer applications.

Pauli gates are also essential in constructing higher-order quantum gates. Complex quantum operations such as the Hadamard gate (H) and Controlled-NOT gate (CNOT) often rely on Pauli matrices as their fundamental building blocks. For example, the CNOT gate, which entangles qubits and enables quantum parallelism, can be decomposed into a combination of Pauli and rotation gates.

In simulating quantum systems, Pauli matrices are extensively used to describe quantum Hamiltonians in quantum mechanics. Many quantum simulations of molecules, condensed matter physics, and quantum materials rely on Pauli gates to model interactions between particles. Quantum computers, particularly those focused on quantum chemistry and material science, use Pauli operators to represent spin systems, electronic structures, and energy interactions, allowing researchers to study quantum effects that are otherwise difficult to simulate with classical computers.

The versatility of Pauli gates in state preparation, error correction, quantum communication, algorithm design, and quantum simulations makes them indispensable tools in quantum computing. Their applications extend from fundamental quantum mechanics to real-world implementations in quantum networks, cryptography, and advanced computational models.

Qiskit Implementation of Pauli Gates

To demonstrate Pauli gates in Qiskit, we can create a quantum circuit that applies X, Y, and Z gates to a qubit and observe the results.

Step 1: Import Qiskit and Define a Quantum Circuit

```
from qiskit import QuantumCircuit, Aer, execute
from qiskit.visualization import plot_histogram

# Create a quantum circuit with one qubit
qc = QuantumCircuit(1)

# Apply Pauli gates
qc.x(0)  # Apply X-gate (bit flip)
qc.y(0)  # Apply Y-gate (bit & phase flip)
qc.z(0)  # Apply Z-gate (phase flip)

# Visualize the circuit
print(qc)
```

Step 2: Simulate the Circuit

```
# Select a quantum simulator backend
backend = Aer.get_backend('qasm_simulator')

# Execute the circuit
job = execute(qc, backend, shots=1024)
result = job.result()

# Get and print the measurement results
counts = result.get_counts(qc)
print(counts)

# Plot the results
plot_histogram(counts)
```

The output of the Qiskit simulation will depend on the combination of Pauli gates applied. For example:

- Applying X-gate alone flips $|0\rangle$ to $|1\rangle$, so the histogram should show most results in the $|1\rangle$ state.
- Applying Y-gate introduces an i-phase, affecting quantum interference.
- Applying Z-gate does not change measurement probabilities but alters relative phases, which may not be visible in direct measurements.

The Pauli Gates (X, Y, and Z) are fundamental quantum gates that manipulate qubit states in specific ways. They serve as the basis for many quantum algorithms, error correction techniques, and quantum communication protocols. Their ability to flip states, introduce phase shifts, and generate superpositions makes them essential for quantum computation. Understanding and implementing these gates in frameworks like Qiskit is key to building practical quantum applications.

Hadamard Gate

The Hadamard gate (H-gate) is a fundamental quantum logic gate used in quantum computing. It operates on a single qubit and is primarily responsible for creating superposition, a crucial quantum property that enables quantum computers to perform parallel computations. Unlike classical bits, which can only exist in definite states 0 or 1, qubits can exist in a superposition of both states simultaneously, allowing quantum algorithms to achieve exponential speedups in computation.

The Hadamard transformation is mathematically represented by the Hadamard matrix, given by:

$$H = \frac{1}{\sqrt{2}} \begin{bmatrix} 1 & 1 \\ 1 & -1 \end{bmatrix}$$

When applied to a qubit in state |0⟩, the Hadamard gate transforms it into an equal superposition of |0⟩ and |1⟩:

$$H|0\rangle = \frac{1}{\sqrt{2}}(|0\rangle + |1\rangle)$$

Similarly, when applied to a qubit in state **|1⟩**, it transforms into:

$$H|1\rangle = \frac{1}{\sqrt{2}}(|0\rangle - |1\rangle)$$

This operation allows quantum computations to explore multiple possibilities at once, forming the basis for quantum parallelism.

Superposition is one of the key principles that give quantum computers their power. By applying a Hadamard gate to a qubit initially in state |0⟩, it becomes an equal probability mixture of |0⟩ and |1⟩. If this superposed qubit is measured, the outcome will be either 0 or 1 with equal probability (50%). When multiple qubits are put into superposition using Hadamard gates, quantum computers can simultaneously evaluate many possible outcomes, a feature leveraged in quantum algorithms like Shor's algorithm for factoring large numbers and Grover's search algorithm.

Applications of the Hadamard Gate

1. Quantum Superposition and Parallelism

The Hadamard gate is the most commonly used gate for creating superposition states. In quantum computing, superposition allows multiple computations to occur simultaneously. Many quantum algorithms start by applying Hadamard gates to initialize qubits into an equal superposition of all possible states.

Example:

If a quantum circuit has n qubits, applying the Hadamard gate to each qubit creates a superposition of 2^n possible states, which is a fundamental step in quantum computing.

2. Quantum Fourier Transform (QFT)

The Quantum Fourier Transform (QFT) is a quantum analogue of the classical Fourier Transform and is widely used in quantum algorithms for signal processing, quantum phase estimation, and Shor's factoring algorithm. The Hadamard gate is an essential component of QFT because it generates superpositions, enabling efficient quantum interference.

Example:

Shor's algorithm, which factors large numbers exponentially faster than classical methods, relies on the Hadamard gate for initial superposition and the Quantum Fourier Transform to extract periodicity information.

3. Quantum Teleportation

The Hadamard gate is a crucial part of quantum teleportation, a protocol that allows transferring quantum information from one qubit to another without physical movement. It ensures that entangled qubits are in the correct basis for measurement and transmission.

Example:

In a three-qubit teleportation circuit, a Hadamard gate is applied to the first qubit before Bell-state measurement is performed. This ensures that quantum information is transferred accurately to the target qubit.

4. Quantum Error Correction

Quantum error correction (QEC) protects quantum information from decoherence and noise. Many QEC protocols, such as the Steane code and Shor code, use Hadamard gates to create and manipulate logical qubits encoded in superpositions.

Example:

In the Steane code, Hadamard gates are used to convert error syndromes into measurable classical bits, enabling the correction of errors without collapsing quantum superposition.

5. Grover's Search Algorithm

Grover's algorithm provides a quadratic speedup for searching an unsorted database. The Hadamard gate plays a key role in its initialization and amplification steps, enabling constructive interference to amplify the correct solution.

Example:

In a 4-qubit Grover search circuit, Hadamard gates initialize all qubits into superposition, allowing the algorithm to search among 16 possibilities simultaneously instead of checking them one by one.

6. Bell State and Quantum Entanglement

Hadamard gates are fundamental in generating Bell states, which are maximally entangled two-qubit states. Quantum entanglement enables secure communication in quantum cryptography and is essential for quantum computing.

Example:

To create an entangled Bell state ($|\Phi+\rangle = (|00\rangle + |11\rangle) / \sqrt{2}$), a Hadamard gate is applied to one qubit, followed by a CNOT gate, linking the states of the two qubits.

Code Implementation in Qiskit

The Hadamard gate can be easily implemented in Qiskit, a quantum computing framework for Python. The following example demonstrates how to apply a Hadamard gate to a qubit and measure the results:

```
from qiskit import QuantumCircuit, execute, Aer

# Define a quantum circuit with one qubit
qc = QuantumCircuit(1, 1)

# Apply the Hadamard gate to the qubit
qc.h(0)

# Measure the qubit
qc.measure(0, 0)

# Execute the circuit on a quantum simulator
backend = Aer.get_backend('qasm_simulator')
job = execute(qc, backend, shots=1024)
result = job.result()

# Print the results
print(result.get_counts(qc))
```

Expected Output:

The results should show approximately 50% probability for "0" and 50% for "1", demonstrating the qubit's equal superposition.

Example Output:

```
'0': 512, '1': 512
```

This indicates that the measurement yields |0⟩ and |1⟩ with equal probability, confirming the expected behaviour of the Hadamard transformation.

The Hadamard gate is one of the most important quantum gates in quantum computing. It enables superposition, allowing qubits to exist in multiple states simultaneously. This property is the foundation of quantum parallelism, which powers many quantum algorithms. Applications of the Hadamard gate include quantum Fourier transform, quantum teleportation, Grover's search algorithm, quantum error correction, and entanglement generation. Its versatility and significance make it indispensable in quantum algorithm design and execution.

Phase Gate (S Gate) and T Gate

Quantum computing relies on various quantum logic gates to manipulate qubits and perform computations. Among these gates, the Phase Gate (S Gate) and the T Gate are fundamental single-qubit gates that introduce controlled phase shifts to quantum states. These gates play a crucial role in quantum algorithms, quantum state preparation, and quantum error correction.

Phase Gate (S Gate)

The S Gate, also known as the Phase Shift Gate, applies a π/2 (90°) phase shift to the |1⟩ state of a qubit while leaving the |0⟩ state unchanged. This gate is represented by the following unitary matrix:

$$S = \begin{bmatrix} 1 & 0 \\ 0 & i \end{bmatrix}$$

When applied to a qubit, it modifies the phase of the **|1⟩** state but does not affect its probability of measurement. The operation is mathematically defined as:

$$S|0\rangle = |0\rangle, \quad S|1\rangle = i|1\rangle$$

This means that if a qubit is in a superposition state, applying an S gate will shift the relative phase of |1⟩ by π/2 radians (90 degrees).

T Gate

The T Gate, also known as the π/8 Gate, applies a π/4 (45°) phase shift to the |1⟩ state while leaving the |0⟩ state unchanged. It is represented by the following unitary matrix:

$$T = \begin{bmatrix} 1 & 0 \\ 0 & e^{i\pi/4} \end{bmatrix}$$

Mathematically, the T gate transforms the qubit states as:

$$T|0\rangle = |0\rangle, \quad T|1\rangle = e^{i\pi/4}|1\rangle$$

Since $e^{i\pi/4}$ introduces a controlled phase shift of π/4 radians (45 degrees), the T Gate is sometimes called the "π/8 gate" because of its role in rotating quantum states on the Bloch sphere.

Both S and T gates rotate a qubit around the Z-axis of the Bloch sphere. Unlike the Pauli-X gate, which flips a qubit between |0⟩ and |1⟩, or the Hadamard gate, which creates superposition, phase shift gates alter the phase relationship between basis states without changing the measurement probabilities.

- **S Gate (π/2 shift):** Rotates the state 90° counterclockwise around the Z-axis.
- **T Gate (π/4 shift):** Rotates the state 45° counterclockwise around the Z-axis.

If a qubit is in a superposition state, applying S or T gates changes how the qubit interferes with itself, which is a key principle behind quantum algorithms.

The applications of S and T gates in quantum computing are pivotal across various domains, including quantum state preparation, quantum Fourier transform (QFT), quantum error correction, quantum teleportation, and quantum communication protocols.

Quantum State Preparation: The S and T gates are essential for the precise manipulation of qubits, enabling the preparation of specific quantum states necessary for quantum algorithms. These gates allow for fine-tuning of quantum phases, which is crucial for achieving desired quantum states. For instance, the S gate introduces a phase shift of π/2, while the T gate provides a phase shift of π/4, both of which are instrumental in creating superposition states needed for quantum computations [208, 209]. The ability to adjust qubit phases accurately is vital for the implementation of quantum algorithms that rely on specific quantum states.

Quantum Fourier Transform (QFT): The QFT is a fundamental component of many quantum algorithms, including Shor's algorithm for integer factorization. It requires precise phase adjustments, which are implemented using S and T gates. The QFT transforms quantum states into a frequency domain, where the S and T gates facilitate the necessary phase shifts to achieve the desired output states [210]. This transformation is crucial for efficiently solving problems that are intractable for classical computers, highlighting the importance of these gates in quantum algorithms.

Quantum Error Correction: Quantum error correction is vital for maintaining the integrity of quantum information against decoherence and operational errors. The S and T gates play a significant role in diagnosing and correcting phase errors within quantum error-correcting codes, such as the Steane Code and Surface Code. These gates help implement logical operations that are resilient to errors, thereby enhancing the reliability of quantum computations [211]. The ability to correct phase errors is particularly important given the susceptibility of quantum states to various types of noise.

Quantum Teleportation: Quantum teleportation involves transferring a quantum state between two distant qubits using entanglement. The S and T gates are crucial in adjusting the phase of the teleported qubit to match the original state. This process relies on the manipulation of quantum states through entangled pairs, where the S and T gates ensure that the phase information is preserved during the teleportation process (Mendes & Ramos, 2010; , Du & Qiao, 2012). The successful implementation of quantum teleportation is a testament to the effectiveness of these gates in practical quantum communication scenarios.

Clifford + T Universal Gate Set: The combination of S and T gates with other gates like Hadamard (H) and CNOT forms a universal gate set capable of performing any quantum computation. This universal set is foundational in quantum computing, as it allows for the approximation of any quantum operation. The S and T gates are particularly important in this context because they enable the implementation of non-Clifford gates, which are necessary for universal quantum computation [212]. Their role in this gate set underscores their significance in the broader landscape of quantum computing.

Superdense Coding and Quantum Cryptography: In quantum communication protocols such as superdense coding and quantum key distribution (QKD), the S and T gates are utilized to encode and manipulate quantum states securely. These gates facilitate the encoding of information into quantum states, allowing for enhanced communication efficiency and security [213]. The application of S and T gates in these protocols demonstrates their versatility and importance in quantum information theory.

Code Implementation in Qiskit

The following Qiskit implementation demonstrates the S and T gates in action.

```
from qiskit import QuantumCircuit, Aer, execute

# Define a quantum circuit with one qubit
qc = QuantumCircuit(1, 1)

# Apply the S gate
qc.s(0)
```

```
# Apply the T gate
qc.t(0)

# Measure the qubit
qc.measure(0, 0)

# Execute the circuit on a quantum simulator
backend = Aer.get_backend('qasm_simulator')
job = execute(qc, backend, shots=1024)
result = job.result()

# Print the results
print(result.get_counts(qc))
```

Multi-Qubit Gates (CNOT, Toffoli, Swap, Fredkin)

Multi-qubit gates are fundamental components in quantum computing that operate on multiple qubits simultaneously. Unlike single-qubit gates, which manipulate individual qubits, multi-qubit gates establish interactions between qubits, enabling quantum entanglement, conditional operations, and complex quantum computations. Some of the most important multi-qubit gates include the Controlled-NOT (CNOT) Gate, Toffoli Gate, SWAP Gate, and Fredkin Gate.

Controlled-NOT (CNOT) Gate: A Fundamental Two-Qubit Gate in Quantum Computing

The Controlled-NOT (CNOT) gate is one of the most fundamental two-qubit quantum gates. It performs a conditional NOT operation, meaning it flips the state of the target qubit if and only if the control qubit is in the state $|1\rangle$. If the control qubit is $|0\rangle$, the target qubit remains unchanged. This conditional behaviour makes the CNOT gate essential for creating quantum entanglement, performing quantum error correction, and enabling various quantum algorithms.

In quantum computing, logic gates are represented as unitary matrices that define their effect on qubit states. The CNOT gate is expressed as the following 4×4 unitary matrix:

$$CNOT = \begin{bmatrix} 1 & 0 & 0 & 0 \\ 0 & 1 & 0 & 0 \\ 0 & 0 & 0 & 1 \\ 0 & 0 & 1 & 0 \end{bmatrix}$$

This matrix represents the operation of the CNOT gate in the computational basis ($|00\rangle$, $|01\rangle$, $|10\rangle$, $|11\rangle$). The first two rows indicate that if the control qubit is $|0\rangle$, the target qubit remains unchanged. The last two rows show that if the control qubit is $|1\rangle$, the target qubit flips.

The behaviour of the CNOT gate can be understood by looking at how it transforms the input qubits:

Table 3: Behaviour of the CNOT gate.

Control Qubit (Input)	Target Qubit (Input)	Target Qubit (Output)
0	0	0
0	1	1
1	0	1
1	1	0

This transformation follows a simple rule:

- If the control qubit is $|0\rangle$, the target qubit remains unchanged.
- If the control qubit is $|1\rangle$, the target qubit flips ($0 \rightarrow 1$, $1 \rightarrow 0$).

For example, applying a CNOT gate to the state $|10\rangle$ will result in $|11\rangle$, while applying it to $|11\rangle$ will result in $|10\rangle$.

Applications of the CNOT Gate

1. Quantum Entanglement

One of the most important uses of the CNOT gate is in creating entanglement. Entanglement is a uniquely quantum mechanical property that links two or more qubits in such a way that the state of one qubit is dependent on the state of another, regardless of the physical distance between them.

The CNOT gate is used to generate Bell states, which are maximally entangled states and are fundamental to quantum teleportation, superdense coding, and quantum cryptography.

For example, applying a Hadamard (H) gate followed by a CNOT gate to the initial state $|00\rangle$ creates the Bell state:

$$|\Phi^+\rangle = \frac{|00\rangle + |11\rangle}{\sqrt{2}}$$

This is a superposition of |00⟩ and |11⟩, meaning if one qubit is measured to be |0⟩, the other will also be |0⟩, and if one qubit is |1⟩, the other will also be |1⟩—even if they are separated by a vast distance.

2. Quantum Error Correction

Quantum computers are highly susceptible to errors due to decoherence and noise from environmental interactions. Quantum error correction (QEC) is a technique that helps detect and correct these errors, and the CNOT gate is a crucial component of QEC codes.

One of the simplest quantum error correction schemes, the Shor code, uses CNOT gates to encode a single logical qubit into multiple physical qubits. By spreading quantum information across multiple qubits, errors can be detected and corrected without collapsing the quantum state.

For example, if a qubit experiences a bit-flip error, the CNOT gate enables redundancy so that the error can be identified by measuring auxiliary qubits and corrected using quantum error correction algorithms.

3. Quantum Algorithms

The CNOT gate is a key building block in many quantum algorithms. Some of the most notable applications include:

- Shor's Algorithm (Factoring Large Numbers)
    - The CNOT gate is essential for implementing modular exponentiation, a critical step in breaking classical RSA encryption using quantum computers.
- Grover's Algorithm (Quantum Search Algorithm)
    - CNOT gates are used to implement quantum oracle functions and amplitude amplification, which speed up searching an unsorted database.
- Quantum Fourier Transform (QFT)
    - QFT, a quantum analogue of the discrete Fourier transform, relies on CNOT gates to perform conditional phase rotations. It plays a crucial role in quantum chemistry simulations and optimization problems.
- Deutsch-Jozsa Algorithm (Quantum Speedup Over Classical Computation)
    - This algorithm determines whether a function is constant or balanced in a single quantum query, using CNOT gates to construct the function oracle.

Implementation of the CNOT Gate in Qiskit

Quantum gates can be implemented in Qiskit, a quantum computing framework provided by IBM. Below is an example of how to apply a CNOT gate to a pair of qubits.

```
from qiskit import QuantumCircuit, Aer, execute

# Create a quantum circuit with 2 qubits
qc = QuantumCircuit(2)

# Apply a Hadamard gate to the first qubit (creating
superposition)
qc.h(0)

# Apply the CNOT gate (q0 = control, q1 = target)
qc.cx(0, 1)

# Draw the quantum circuit
print(qc.draw())

# Execute the circuit on a quantum simulator
backend = Aer.get_backend('statevector_simulator')
job = execute(qc, backend)
result = job.result()

# Get the final quantum state
statevector = result.get_statevector()
print("Final Quantum State:", statevector)
```

Explanation of Code:

1. A Hadamard gate (H-gate) is applied to the first qubit q0 to create superposition.
2. The CNOT gate is applied, using q0 as the control qubit and q1 as the target qubit.
3. The circuit is simulated, and the final quantum state is printed.

The output state will be $|\Phi+\rangle = (|00\rangle + |11\rangle)/\sqrt{2}$, demonstrating quantum entanglement.

The Controlled-NOT (CNOT) gate is a critical quantum gate used for entanglement, error correction, and quantum algorithms. Its ability to conditionally flip a target qubit based on the control qubit's state enables a wide range of quantum computations that classical computers cannot efficiently perform. By leveraging the CNOT gate, quantum computers unlock new computational possibilities, including secure quantum communication, fast problem-solving, and powerful quantum simulations.

Understanding the CNOT gate and its applications is essential for quantum computing research and implementation. As quantum technologies continue to advance, the CNOT gate will remain a fundamental component in developing scalable, fault-tolerant quantum computers.

Toffoli Gate (Controlled-Controlled NOT Gate) in Quantum Computing

The Toffoli gate, also known as the Controlled-Controlled NOT (CCNOT) gate, is a fundamental three-qubit gate in quantum computing. It applies a NOT operation (bit-flip) to the target qubit if and only if both control qubits are in the $|1\rangle$ state. Otherwise, it leaves the target qubit unchanged. This gate is essential for reversible computing, error correction, and complex quantum circuits.

The Toffoli gate operates on three qubits and is represented by the following 8×8 unitary matrix:

$$Toffoli = \begin{bmatrix} 1 & 0 & 0 & 0 & 0 & 0 & 0 & 0 \\ 0 & 1 & 0 & 0 & 0 & 0 & 0 & 0 \\ 0 & 0 & 1 & 0 & 0 & 0 & 0 & 0 \\ 0 & 0 & 0 & 1 & 0 & 0 & 0 & 0 \\ 0 & 0 & 0 & 0 & 1 & 0 & 0 & 0 \\ 0 & 0 & 0 & 0 & 0 & 1 & 0 & 0 \\ 0 & 0 & 0 & 0 & 0 & 0 & 0 & 1 \\ 0 & 0 & 0 & 0 & 0 & 0 & 1 & 0 \end{bmatrix}$$

In this representation:

- The first six rows show that when the control qubits are not both $|1\rangle$, the target qubit remains unchanged.
- The last two rows indicate that when both control qubits are $|1\rangle$, the target qubit undergoes a bit-flip (NOT operation).

The Toffoli gate follows a specific rule:

- If both control qubits are $|1\rangle$, the target qubit flips (0 → 1, 1 → 0).
- In all other cases, the target qubit remains unchanged.

Table 4: Behaviour of the Toffoli Gate.

Control Qubit 1	**Control Qubit 2**	**Target Qubit (Input)**	**Target Qubit (Output)**
0	**0**	**0**	**0**
0	**0**	**1**	**1**
0	**1**	**0**	**0**

Control Qubit 1	Control Qubit 2	Target Qubit (Input)	Target Qubit (Output)
0	1	1	1
1	0	0	0
1	0	1	1
1	1	0	1
1	1	1	0

For example:

- If **q1 = 1** and **q2 = 1**, and the **target qubit is |0⟩**, it flips to |1⟩.
- If **q1 = 1** and **q2 = 1**, and the **target qubit is |1⟩**, it flips to |0⟩.
- If **either q1 or q2 is 0**, the target qubit remains **unchanged**.

The Toffoli gate is a fundamental component in quantum computing, with a wide range of applications spanning classical reversible computation, quantum error correction, cryptographic protocols, and quantum simulations. Its ability to conditionally flip a target qubit based on two control qubits makes it an essential tool for designing complex quantum circuits and implementing fault-tolerant quantum computing systems.

One of the most significant applications of the Toffoli gate is in classical reversible computation. Quantum computing must adhere to the principle of reversibility, meaning that operations must be invertible to prevent information loss. The Toffoli gate is universal for classical reversible logic, meaning it can simulate any Boolean function without losing information. In classical computing, logic operations such as NAND, NOR, and XOR gates serve as building blocks for computational circuits. The Toffoli gate can be used to construct these classical logic operations within a quantum framework. A NAND gate, which is a universal classical logic gate, can be implemented using Toffoli gates. This allows for the simulation of combinational logic circuits, including adders, multipliers, and Boolean function generators, directly within a quantum system. By leveraging the Toffoli gate's ability to perform classical logical functions in a reversible manner, quantum circuits can efficiently simulate classical computation while maintaining quantum coherence.

Another crucial application of the Toffoli gate is in quantum error correction (QEC). Quantum computers are highly susceptible to decoherence and noise, which introduce errors that can disrupt quantum computations. Toffoli gates are instrumental in quantum error-correcting codes such as the Steane Code and Shor Code, which are designed to protect quantum information from both bit-flip (X errors) and phase-flip (Z errors). In fault-tolerant quantum computing, logical qubits are encoded using multiple physical qubits to safeguard against errors. Toffoli gates are employed in error detection and correction protocols, ensuring that errors can be identified and rectified without collapsing the quantum state. Additionally, Toffoli gates are used in stabilizer circuits, which measure error syndromes by extracting information about errors while preserving the underlying quantum computation. By utilizing entangled qubits and ancilla qubits, the Toffoli gate facilitates real-time error correction, making it an essential component in the pursuit of large-scale, fault-tolerant quantum computing.

Beyond error correction, the Toffoli gate plays a pivotal role in quantum cryptography and simulation. Quantum cryptographic protocols, including Quantum Key Distribution (QKD), rely on controlled operations to ensure the secure exchange of cryptographic keys. The Toffoli gate enables complex multi-qubit operations that underpin the security of quantum cryptographic algorithms, making it a valuable tool in the development of secure quantum communication networks. In quantum simulations, the Toffoli gate is used to model controlled logical operations on multiple interacting quantum particles. Simulating complex molecular interactions in quantum chemistry requires precise control over multi-qubit states, which can be achieved using the Toffoli gate. For example, simulating atomic orbitals and their interactions in chemical reactions involves controlled operations that the Toffoli gate can efficiently implement. Additionally, quantum finance models use controlled unitary transformations, aided by the Toffoli gate, to optimize risk calculations and perform advanced quantum-based financial simulations.

The Toffoli gate is a foundational element in quantum computing with applications across classical reversible computation, quantum error correction, cryptography, and quantum simulations. Its ability to perform controlled multi-qubit operations makes it a critical tool for developing scalable and fault-tolerant quantum computing architectures. As quantum technology advances, the Toffoli gate will continue to be a central component in unlocking the potential of quantum computing for a broad range of applications.

Implementation of the Toffoli Gate in Qiskit

To implement a Toffoli gate (CCNOT) using IBM's Qiskit, we define a quantum circuit with three qubits and apply the ccx (Toffoli) gate.

```
from qiskit import QuantumCircuit, Aer, execute

# Create a quantum circuit with 3 qubits
qc = QuantumCircuit(3)

# Apply the Toffoli gate (q0 and q1 are control qubits, q2 is
the target)
qc.ccx(0, 1, 2)
```

```
# Draw the quantum circuit
print(qc.draw())

# Execute the circuit on a quantum simulator
backend = Aer.get_backend('statevector_simulator')
job = execute(qc, backend)
result = job.result()

# Get the final quantum state
statevector = result.get_statevector()
print("Final Quantum State:", statevector)
```

Explanation of Code

1. A three-qubit circuit is created.
2. The Toffoli gate (CCX gate) is applied, using q0 and q1 as control qubits and q2 as the target qubit.
3. The circuit is executed on a quantum simulator, and the final quantum state is displayed.

The output will show that the target qubit flips only when both control qubits are $|1\rangle$, demonstrating the expected behaviour of the Toffoli gate.

The Toffoli gate (CCNOT gate) is a key multi-qubit gate in quantum computing. Its ability to conditionally flip a target qubit based on two control qubits enables its use in reversible classical computation, quantum error correction, cryptography, and quantum simulations. As quantum hardware continues to advance, the Toffoli gate will remain a fundamental building block for scalable quantum computing.

SWAP Gate in Quantum Computing

The SWAP gate is a fundamental two-qubit quantum logic gate that exchanges the states of two qubits. Unlike other quantum gates that introduce superposition or entanglement, the SWAP gate purely permutes the quantum information stored in the qubits without altering their phase or introducing new quantum correlations.

In matrix form, the SWAP gate is represented as:

$$SWAP = \begin{bmatrix} 1 & 0 & 0 & 0 \\ 0 & 0 & 1 & 0 \\ 0 & 1 & 0 & 0 \\ 0 & 0 & 0 & 1 \end{bmatrix}$$

This matrix acts on a two-qubit system by exchanging the second and third elements in the state vector, which effectively swaps the quantum states between the two qubits.

The SWAP gate performs a direct exchange of quantum states between two qubits. If the input state of the two qubits is $|0\rangle$ and $|1\rangle$, applying the SWAP gate results in the first qubit becoming $|1\rangle$ and the second qubit becoming $|0\rangle$. Similarly, for an input state of $|1\rangle$ and $|0\rangle$, the qubits exchange their roles after the SWAP operation.

Table 5: Truth Table for the SWAP Gate.

Input Qubit 1	Input Qubit 2	Output Qubit 1	Output Qubit 2
0	0	0	0
0	1	1	0
1	0	0	1
1	1	1	1

From the table, it is evident that the SWAP gate only exchanges the quantum states of the two qubits, leaving the system otherwise unchanged. This property makes the SWAP gate particularly useful for quantum information routing in quantum circuits.

Applications of the SWAP Gate

In large-scale quantum processors, quantum communication between physically distant qubits is a significant challenge. Qubits may be positioned in a fixed layout, making direct interactions between distant qubits difficult. The SWAP gate enables the movement of quantum states between different parts of a quantum circuit, allowing qubits to be relocated for necessary interactions.

For example, in quantum networking, SWAP gates help move quantum information between nodes in a distributed quantum system. This is essential for quantum teleportation protocols, where entanglement must be distributed across different quantum systems for secure quantum communication.

Quantum processors have a physical layout that imposes constraints on qubit connectivity. Certain quantum algorithms require specific qubit interactions, but due to hardware limitations, qubits may not always be physically adjacent. The SWAP gate is used to reposition qubits within the processor to ensure that they can interact correctly.

For instance, in superconducting quantum computers (such as those developed by IBM and Google), qubits are arranged in a grid-like topology, and only adjacent qubits can interact directly. If an operation requires two distant qubits to interact, a sequence of SWAP gates can be applied to move one of the qubits into the correct position.

In variational quantum algorithms used for quantum machine learning, the SWAP gate plays an important role in reordering qubits to optimize computations. Many quantum neural networks and variational circuits rely on efficient qubit interactions to adjust quantum weights and correlations dynamically. By using SWAP gates, machine learning models can be structured to leverage optimal quantum state transfer, ensuring efficient execution of quantum feature encodings and parameterized transformations.

Example Implementation of the SWAP Gate in Qiskit

To better understand how the SWAP gate is implemented in quantum circuits, consider the following Qiskit implementation:

```
from qiskit import QuantumCircuit, Aer, execute

# Create a quantum circuit with 2 qubits
qc = QuantumCircuit(2)

# Apply the SWAP gate
qc.swap(0, 1)

# Execute the circuit on a simulator
backend = Aer.get_backend('qasm_simulator')
job = execute(qc, backend, shots=1024)
result = job.result()

# Print the results
print(result.get_counts(qc))

# Draw the circuit
print(qc.draw())
```

This quantum circuit swaps the quantum states of two qubits. If qubit 0 initially held $|1\rangle$ and qubit 1 held $|0\rangle$, after applying the SWAP gate, their states would be exchanged.

The SWAP gate is a crucial two-qubit operation that allows efficient movement of quantum information in a quantum system. Unlike other quantum gates, it does not introduce superposition or entanglement but serves as a tool for reordering quantum states within a quantum processor. Its primary applications include quantum communication, circuit optimization, and quantum machine learning, making it an indispensable component in the development of scalable quantum technologies.

Fredkin Gate (Controlled-SWAP) in Quantum Computing

The Fredkin gate, also known as the Controlled-SWAP (CSWAP) gate, is a fundamental three-qubit gate that conditionally swaps the states of two qubits based on the state of a control qubit. If the control qubit is in state $|1\rangle$, the gate swaps the states of the two target qubits. If the control qubit is $|0\rangle$, the target qubits remain unchanged.

This property makes the Fredkin gate particularly valuable in reversible computing, quantum data processing, and cryptographic applications where conditional logic and controlled operations are necessary.

The Fredkin gate can be represented by the following 8×8 unitary matrix, acting on a three-qubit system:

$$F = \begin{bmatrix} 1 & 0 & 0 & 0 & 0 & 0 & 0 & 0 \\ 0 & 1 & 0 & 0 & 0 & 0 & 0 & 0 \\ 0 & 0 & 1 & 0 & 0 & 0 & 0 & 0 \\ 0 & 0 & 0 & 1 & 0 & 0 & 0 & 0 \\ 0 & 0 & 0 & 0 & 1 & 0 & 0 & 0 \\ 0 & 0 & 0 & 0 & 0 & 0 & 1 & 0 \\ 0 & 0 & 0 & 0 & 0 & 1 & 0 & 0 \\ 0 & 0 & 0 & 0 & 0 & 0 & 0 & 1 \end{bmatrix}$$

In this matrix:

- The first four rows correspond to cases where the control qubit is $|0\rangle$, meaning the target qubits remain unchanged.
- The last four rows correspond to cases where the control qubit is $|1\rangle$, meaning the two target qubits swap places.

The Fredkin gate operates on three qubits: one control qubit and two target qubits. The behaviour of the gate depends on the control qubit:

- If the control qubit is $|0\rangle$, the target qubits remain unchanged.

- If the control qubit is $|1\rangle$, the states of the two target qubits are swapped.

Table 6: Truth Table for the Fredkin Gate.

Control Qubit	Target Qubit 1	Target Qubit 2	Output Qubit 1	Output Qubit 2
0	0	0	0	0
0	0	1	0	1
0	1	0	1	0
0	1	1	1	1
1	0	0	0	0
1	0	1	1	0
1	1	0	0	1
1	1	1	1	1

This conditional swapping mechanism makes the Fredkin gate valuable in quantum computing applications where controlled data manipulation and reversible logic are required.

The Fredkin gate is a universal gate for reversible computing, meaning that any classical Boolean function can be implemented using only Fredkin gates. Since quantum computing requires reversibility, this property makes it a key component in low-energy computational models that minimize energy dissipation. This is particularly useful for implementing logic functions such as AND, OR, and XOR in a reversible manner. In energy-efficient computing, where minimizing heat generation is crucial, the Fredkin gate ensures that computations can be performed without losing information, making it essential for classical logic circuit design in quantum computation.

In quantum computing, data sorting and reordering play a vital role in efficient data manipulation and parallel computation. The Fredkin gate allows controlled swapping of qubits, which makes it highly beneficial in quantum sorting networks, where large datasets must be arranged in a specific order. It is also valuable in data exchange operations, such as rearranging entangled qubits in a quantum register. Additionally, it contributes to quantum memory management, where certain operations require qubits to be stored or retrieved in an optimized sequence. For example, in quantum search algorithms, Fredkin gates can be used to dynamically rearrange quantum superposition states, optimizing performance in Grover's search algorithm.

The Fredkin gate plays a critical role in quantum cryptography and secure quantum communication by enabling controlled operations for encrypting and transmitting quantum information. In Quantum Key Distribution (QKD) protocols, Fredkin gates ensure secure swapping of quantum states, which protects cryptographic keys from eavesdropping. In quantum teleportation, the Fredkin gate assists in preserving quantum correlations while transferring quantum states between distant parties. It is also useful in error-resistant quantum communication, where conditional data swapping ensures entanglement fidelity during quantum state transfer. For instance, in the BB84 quantum key distribution protocol, conditional operations like the Fredkin gate help maintain secure entanglement channels, making QKD a practical alternative to classical encryption.

Example Implementation of the Fredkin Gate in Qiskit

To demonstrate the Fredkin gate in action, the following Qiskit code shows how to implement it using a quantum circuit:

```
from qiskit import QuantumCircuit, Aer, execute

# Create a quantum circuit with 3 qubits
qc = QuantumCircuit(3)

# Apply the Fredkin (Controlled-SWAP) gate
qc.cswap(0, 1, 2)

# Execute the circuit on a simulator
backend = Aer.get_backend('qasm_simulator')
job = execute(qc, backend, shots=1024)
result = job.result()

# Print the results
print(result.get_counts(qc))

# Draw the circuit
print(qc.draw())
```

Explanation of the Code:

- The circuit initializes three qubits where:
    - Qubit 0 acts as the control qubit.
    - Qubits 1 and 2 are the target qubits.

- The Fredkin gate (CSWAP) is applied using `qc.cswap(0, 1, 2)`, meaning that if qubit 0 is in state |1⟩, the states of qubits 1 and 2 will be swapped.
- The results are simulated using Qiskit's `qasm_simulator`.

This implementation allows for testing and visualizing the conditional swapping behaviour of the Fredkin gate in a quantum system.

The Fredkin gate (Controlled-SWAP) is a powerful three-qubit quantum gate that enables conditional swapping of quantum states. Unlike other gates, it preserves quantum information without introducing new entanglement, making it essential for quantum networking, data reordering, and cryptographic applications.

Its reversible nature makes it a cornerstone of low-energy computation models, and it is widely used in quantum algorithms, quantum sorting, and secure quantum communication. As quantum computing advances, the Fredkin gate continues to play a pivotal role in building fault-tolerant, scalable quantum circuits for a wide range of applications.

Quantum Circuit Design

A quantum circuit serves as the foundational framework for quantum computing, analogous to classical digital circuits in conventional computing. It comprises a series of quantum gates acting on qubits, the fundamental units of quantum information. Unlike classical bits, which can exist in one of two states (0 or 1), qubits can exist in a superposition of both states simultaneously, allowing for more complex computations [214, 215]. Quantum circuits can be designed to perform various operations, including quantum teleportation and the quantum Fourier transform, which are pivotal in quantum algorithms [216].

The structure of a quantum circuit is built around qubits and quantum gates. Qubits are manipulated through quantum gates, which are the quantum equivalent of classical logic gates. These gates, such as the Hadamard gate and the CNOT gate, facilitate operations like creating superposition and entanglement among qubits [214, 216]. Entanglement is a unique feature of quantum circuits, where the state of one qubit becomes dependent on the state of another, regardless of the distance separating them, thereby enabling complex correlations that classical systems cannot replicate [217].

Measurement in quantum circuits is crucial as it collapses the qubits' superposition into definite classical states, yielding the results of the computation. This process is inherently probabilistic, reflecting the fundamental principles of quantum mechanics [214, 215]. Quantum circuits excel in solving problems that are intractable for classical computers, such as factorization and optimization tasks, due to their ability to process vast amounts of information simultaneously [218].

Quantum Computing

DiVincenzo's Criteria, established by physicist David P. DiVincenzo in 2000, outlines essential requirements for practical quantum computing systems. These criteria emphasize the need for a scalable physical system capable of reliably creating and manipulating qubits, the availability of universal quantum gates, and the necessity for long qubit coherence times to maintain quantum states during computations [214]. Furthermore, high-fidelity qubit initialization and measurement are critical for ensuring accurate results, while effective quantum error correction schemes are vital for mitigating errors arising from decoherence and other quantum noise [219].

DiVincenzo's Criteria are a set of fundamental requirements to outline the essential conditions for building a practical quantum computer. These criteria ensure that a quantum system can store, manipulate, and process quantum information reliably while being scalable for real-world applications. Additionally, two extra criteria were proposed specifically for quantum communication systems to facilitate secure information transfer over long distances [220-222].

A quantum computer must have a scalable architecture where qubits can be reliably initialized, controlled, and measured. Qubits must be well-defined two-level quantum systems, such as superconducting circuits, trapped ions, or quantum dots, and should be physically realizable with minimal interference from external factors. Superconducting qubits in IBM and Google quantum computers, for example, are built using Josephson junctions, which allow precise control over quantum states.

Before computation begins, the qubits must be reliably initialized into a well-known state, typically $|0\rangle$. Proper initialization is crucial to prevent errors from propagating throughout the computation. In superconducting quantum computers, this initialization is achieved by cooling qubits to their ground state using dilution refrigerators, ensuring a clean starting point for quantum operations.

Quantum coherence refers to the time a qubit can maintain its quantum state before decohering due to noise or environmental interference. The system must maintain coherence for a duration long enough to perform quantum gate operations before the information is lost. Trapped-ion qubits exhibit long coherence times, often lasting several seconds, making them suitable for high-fidelity quantum computations.

To perform arbitrary quantum computations, the quantum system must support a universal set of quantum gates that allow operations on qubits. This includes single-qubit gates such as the Hadamard (H) and phase shift (S, T) gates, as well as multi-qubit gates such as the Controlled-NOT (CNOT) and Toffoli gates. With a universal gate set, any quantum algorithm can be decomposed into a sequence of quantum gates. Quantum algorithms like Shor's factoring algorithm, for example, rely on Hadamard, CNOT, and phase gates to manipulate qubits efficiently.

Measurement collapses a quantum state into a classical outcome, so quantum computers must allow precise qubit measurement without affecting the coherence of other qubits. This is essential for retrieving results from quantum computations. In superconducting quantum circuits, microwave resonators are used to non-destructively measure qubit states, allowing for reliable output without disturbing the overall quantum state.

Beyond computation, quantum communication requires additional capabilities. Quantum networks must enable the transfer of qubits between distant nodes. This means that quantum computers must be able to convert stationary qubits, such as superconducting or trapped-ion qubits, into flying qubits, such as photonic qubits, which can be transmitted over optical fibers or free-space links. Quantum teleportation, for instance, uses entangled photon pairs to transmit quantum information between distant locations.

For quantum networking and secure quantum communication, quantum states must be reliably transmitted over long distances without significant loss of coherence. This requires advanced error correction and quantum repeaters to extend transmission distances. Quantum key distribution (QKD) protocols, such as BB84, rely on the ability to send entangled photons over fibre-optic networks to ensure secure communication.

DiVincenzo's Criteria provide a benchmark for evaluating quantum hardware and determining whether a system is capable of scalable quantum computing. They have influenced the design of quantum computers and networks by highlighting the engineering challenges that must be overcome, including minimizing decoherence to extend qubit lifetimes, developing high-fidelity quantum gates to ensure accurate computations, and building scalable architectures that allow for error correction and fault tolerance. Most modern quantum computing platforms, including superconducting qubits from IBM and Google, trapped ions from IonQ, and photonic quantum computing from Xanadu and PsiQuantum, strive to meet these criteria.

As a roadmap for building functional quantum computers and quantum communication networks, DiVincenzo's Criteria play a critical role in guiding research and development. By addressing these seven fundamental requirements, researchers and engineers can develop scalable, fault-tolerant quantum systems capable of solving complex problems beyond the reach of classical computing. As quantum technology advances, ongoing improvements in qubit coherence, quantum error correction, and entanglement-based communication will bring us closer to practical, large-scale quantum computing.

Quantum circuit design is the process of constructing sequences of quantum gates to manipulate qubits and execute quantum computations. Just as classical circuits use logic gates to process binary data, quantum circuits employ quantum gates to transform qubit states according to quantum principles such as superposition, entanglement, and interference. Designing quantum circuits requires an understanding of quantum mechanics, linear algebra, and computational complexity.

A quantum circuit consists of several fundamental components, including qubits, quantum gates, and measurement operations. Qubits are the basic units of quantum information, analogous to classical bits but capable of existing in superposition states. Quantum gates perform operations on qubits, modifying their states in a reversible manner. Finally, measurement operations collapse qubit states into classical binary outcomes, providing the final result of the computation.

A quantum circuit is typically represented using a quantum circuit diagram, where qubits are depicted as horizontal lines, and quantum gates are represented as symbols applied to these lines. The progression of operations follows a left-to-right convention, indicating the sequence of gate applications.

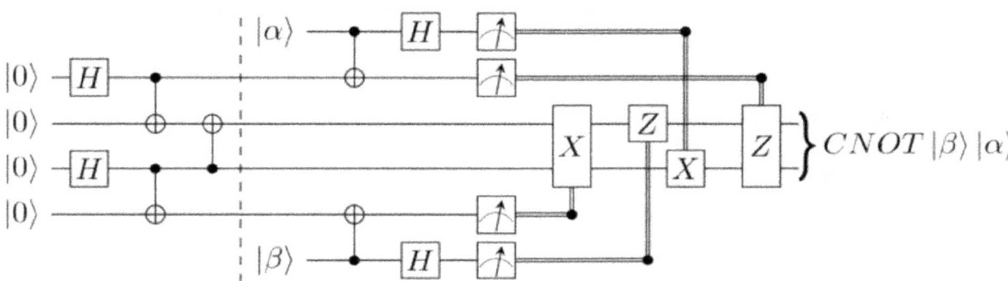

Figure 5: A circuit diagram for quantum teleportation. Vtomole, CC BY-SA 4.0, via Wikimedia Commons.

Quantum gates are the fundamental components of quantum circuits, allowing the manipulation of qubits to perform computations. These gates can be categorized based on their functionality and the number of qubits they operate on.

Single-qubit gates act on individual qubits and are crucial for basic quantum state manipulation. The Hadamard (H) gate is one of the most significant single-qubit gates, as it creates superposition by transforming the basis states into an equal combination of |0⟩ and |1⟩. The Pauli gates (X, Y, Z) serve various functions, including bit-flipping and phase shifts. The X gate, also known as the quantum NOT gate, inverts the qubit's state from |0⟩ to |1⟩ and vice versa. The Y gate combines a bit-flip and a phase-flip, while the Z gate applies a 180-degree phase shift to the |1⟩ state without affecting |0⟩. Additionally, rotation gates (Rx, Ry, Rz) provide fine-tuned control over qubits by rotating them around the Bloch sphere's X, Y, or Z axes, respectively.

Multi-qubit gates play a crucial role in enabling quantum entanglement and conditional operations, which are fundamental to quantum computing's advantage over classical systems. The Controlled-NOT (CNOT) gate is a two-qubit gate that flips the state of a target qubit only if the control qubit is in the |1⟩ state. This gate is essential for creating entanglement, a key resource in quantum algorithms and quantum cryptography. The Toffoli gate (CCNOT) is a three-qubit gate that applies a NOT operation to the target qubit only when both control qubits are in the |1⟩ state, making it useful for quantum error correction and classical reversible computing. The SWAP gate exchanges the states of two qubits, which is particularly useful in quantum networking and optimizing qubit arrangements in physical quantum hardware.

Phase shift gates modify the phase of a quantum state without affecting its probability amplitudes. The S gate, also called the Phase gate, introduces a π/2 phase shift, while the T gate, known as the π/8 gate, applies a π/4 phase shift. These gates are widely used in quantum algorithms, including error correction protocols and quantum cryptographic schemes.

A set of quantum gates is considered universal if it can be used to construct any possible quantum operation. The combination of Hadamard, CNOT, and T gates forms a universal gate set, enabling the execution of any quantum algorithm. Universal gate sets are essential for designing scalable quantum computers, as they ensure that complex quantum computations can be broken down into sequences of fundamental operations.

By leveraging these different types of quantum gates, researchers and engineers design quantum circuits that power quantum algorithms, quantum error correction, and quantum simulations, bringing practical quantum computing closer to reality.

Quantum circuit design plays a crucial role in the development of quantum algorithms and applications across various fields. One of its most impactful applications is in quantum cryptography, where quantum circuits are used to implement secure communication protocols such as the BB84 quantum key distribution protocol. This allows two parties to share encryption keys securely, leveraging the principles of quantum mechanics to detect eavesdroppers and prevent unauthorized access to sensitive information.

Another significant application is in quantum search algorithms, particularly Grover's algorithm, which enables a quantum computer to search an unsorted database much faster than classical algorithms. By utilizing quantum superposition and amplitude amplification, Grover's algorithm reduces the number of required search steps from *O(N)* in classical computing to $O(\sqrt{N})$, making it an essential tool for optimization problems.

Shor's algorithm for factoring is another example where quantum circuits outperform classical methods. This algorithm efficiently finds the prime factors of large numbers using quantum Fourier transforms, posing a significant threat to traditional encryption schemes such as RSA. The ability to break encryption systems that rely on integer factorization could revolutionize cybersecurity, making post-quantum cryptography a necessity.

In the field of quantum machine learning, quantum circuits are used to implement variational quantum algorithms (VQAs), which optimize functions and solve classification problems faster than classical methods. These circuits leverage quantum entanglement and superposition to explore complex datasets more efficiently, potentially revolutionizing artificial intelligence and data science.

Another critical application of quantum circuit design is in quantum simulation, where quantum circuits model the behaviour of complex quantum systems in physics and chemistry. Classical computers struggle to simulate quantum interactions accurately due to exponential growth in computational requirements, but quantum circuits can naturally model these interactions, enabling advancements in material science, drug discovery, and condensed matter physics.

Despite its transformative potential, quantum circuit design faces several significant challenges. One of the primary obstacles is noise and decoherence, where qubits are highly susceptible to external environmental interactions that cause them to lose their quantum states. Even minor disturbances can introduce errors in computations, limiting the reliability of quantum circuits.

Limited qubit connectivity is another challenge, as many quantum hardware platforms impose physical constraints on how qubits interact. In superconducting qubit systems, for example, not all qubits are directly connected, requiring the use of SWAP gates to reposition qubits for multi-qubit operations. This increases circuit complexity and execution time, leading to additional errors.

Scalability issues also pose a major challenge. As quantum circuits grow larger, error rates and computational complexity increase. Achieving fault-tolerant quantum computing requires improving qubit coherence times, reducing gate errors, and optimizing quantum hardware to support large-scale circuits.

To overcome these challenges, quantum error correction (QEC) is essential. QEC methods protect quantum information by encoding logical qubits across multiple physical qubits to detect and correct errors. However, implementing QEC requires significant additional resources, as a single logical qubit may need dozens or even hundreds of physical qubits. This increases circuit depth and hardware requirements, making fault-tolerant quantum computation a long-term goal rather than an immediate reality.

Despite these challenges, ongoing advancements in quantum circuit design, error correction techniques, and qubit technologies continue to push the boundaries of what is possible in quantum computing. With improvements in hardware and algorithms, quantum circuit design will play a vital role in unlocking the full potential of quantum technology.

Reading Quantum Circuit Diagrams

Quantum circuit diagrams are visual representations of quantum computations, with time flowing from left to right. Each component in the circuit corresponds to a quantum operation applied sequentially to qubits. These diagrams allow us to understand and track the transformations of quantum states as they progress through different quantum gates and measurements.

Take the following quantum circuit diagram as an example:

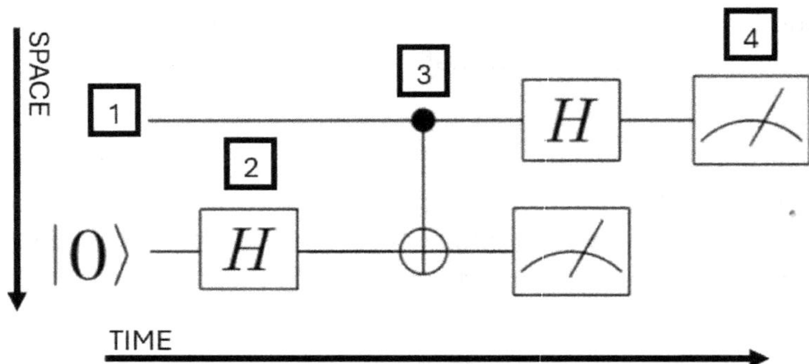

Figure 6: Sample Quantum Circuit.

1. A **qubit register** is represented by horizontal lines in a quantum circuit diagram, where each line corresponds to a single qubit. The qubits are labelled sequentially, with the topmost line representing qubit 0, the second line representing qubit 1, and so on.
2. A **quantum gate** represents an operation applied to one or more qubits, similar to how classical logic gates function in classical computing. These gates are depicted as labelled boxes placed along the qubit lines. For example, a Hadamard gate (H) is represented by a box labelled H, which applies a superposition operation to the qubit.
3. A **controlled gate** is a multi-qubit gate that performs an operation based on the state of a control qubit. One common example is the Controlled-NOT (CNOT) gate, where a black circle (●) represents the control qubit, and a plus sign inside a circle (⊕) represents the target qubit. The CNOT gate flips the target qubit if the control qubit is in the $|1\rangle$ state.
4. A **measurement operation** is denoted by a meter symbol and converts a qubit's quantum state into a classical bit. This operation collapses the qubit's superposition or entanglement, producing a definite classical output of either **0** or **1** based on the qubit's probability distribution.

In a quantum circuit, time progresses from left to right, meaning the leftmost gate is applied first. For example, in a given quantum circuit diagram, operations are performed sequentially in this order. However, matrix multiplication in quantum mechanics follows the opposite convention, where the rightmost matrix is applied first. This distinction between linear algebra notation and quantum circuit representation is crucial and can sometimes cause confusion. It is important to be aware of this difference when interpreting quantum circuits.

In a quantum circuit diagram, the horizontal lines, or wires, represent qubits that serve as inputs and outputs of quantum gates. When a qubit enters a gate, it undergoes a transformation, and the resulting qubit state exits the gate.

Since quantum operations are unitary and inherently reversible, the number of input qubits for a gate must always be equal to the number of output qubits. If a quantum gate had more outputs than inputs, it would contradict the unitary nature of quantum mechanics. As a result, every gate in a quantum circuit diagram must have the same number of incoming and outgoing qubit wires.

Multi-qubit operations in quantum circuits follow the same general structure as single-qubit gates. For example, a two-qubit unitary operation can be represented as a transformation applied to a two-qubit system. This can be viewed either as an operation on two independent qubits or as an operation on a single two-qubit register, depending on the circuit's context.

One of the key advantages of circuit diagrams is their abstract representation, allowing complex quantum algorithms to be visualized at a high level without requiring explicit compilation into fundamental gates. This enables a better understanding of the data flow within quantum algorithms without needing to focus on the specifics of individual subroutines.

Controlled quantum gates are multi-qubit gates where a designated control qubit determines whether a quantum operation is applied to a target qubit. The operation is performed only when the control qubit is in a specific state.

For example, a controlled quantum gate applies an operation to the target qubit only when the control qubit holds a particular value. This can be seen in the case of a Controlled-NOT (CNOT) gate, where a black circle (●) represents the control qubit, and a plus sign inside a circle ($\oplus$) represents the target qubit. The operation occurs only when the control qubit is in the $|1\rangle$ state.

More generally, circuit diagrams represent controlled gates using a vertical wire connecting the control qubit to the unitary operation applied to the target qubit. In special cases, specific notations are used for controlled versions of common gates, such as controlled-X (CNOT) and controlled-Z gates.

In addition to quantum-controlled operations, gates can also be controlled classically using measurement results. In such cases, the measurement outcome acts as a classical control bit, determining whether a gate is applied.

A classically controlled gate is represented in a circuit diagram using a double line that carries the classical bit from the measurement result. This double line serves as an input to the controlled gate, ensuring that the operation is applied only when the classical condition is met.

A measurement operation extracts classical information from a qubit by collapsing its quantum state into a definite classical state, either 0 or 1. The measurement is represented in circuit diagrams by a meter symbol, which takes a qubit register as input (solid line) and outputs a classical bit (double line).

In Q# and other quantum programming frameworks, the Measure operator is used to implement this measurement process. The measured classical values can then be used in subsequent quantum or classical computations.

By understanding these elements—gate applications, controlled operations, and measurements—one can effectively interpret quantum circuits and their role in executing quantum algorithms.

Example Quantum Circuits

Quantum Teleportation

The quantum circuit depicted in the image as Figure 7 represents quantum teleportation, a protocol that enables the transfer of an arbitrary quantum state from one qubit to another, using entanglement and classical communication.

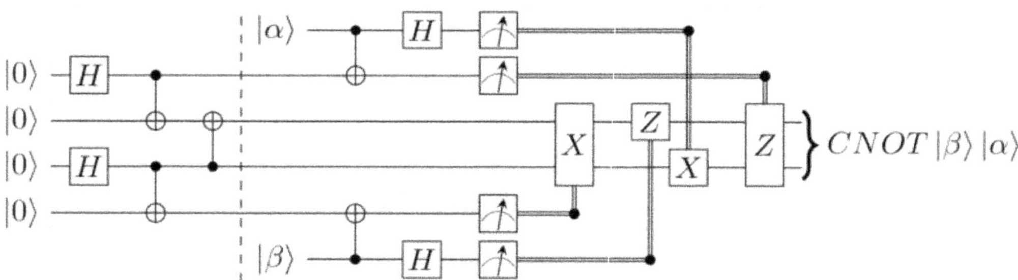

Figure 7: A circuit diagram for quantum teleportation. Vtomole, CC BY-SA 4.0, via Wikimedia Commons.

The quantum teleportation circuit consists of three qubits. The first qubit, represented by the top wire, carries the unknown quantum state |α⟩ that needs to be teleported. The second qubit, located on the middle wire, and the third qubit, represented by the bottom wire, form an entangled Bell pair. This entanglement is crucial for facilitating the teleportation process.

The circuit operates in three main stages. In the first stage, entanglement generation, Hadamard (H) gates are applied to the second and third qubits to place them in a superposition state. Controlled-NOT (CNOT) gates are then used to create an entangled Bell state between these two qubits, forming the quantum resource needed for teleportation.

In the second stage, Bell measurement, the first qubit, |α⟩, interacts with the second qubit through a CNOT gate. A Hadamard (H) gate is then applied to the first qubit, transforming the system into the Bell basis. Following this, both the first and second qubits are measured in the computational basis (|0⟩ or |1⟩), and their results are transmitted as classical bits.

The final stage involves conditional corrections based on the measurement results. Depending on the classical outcomes, the third qubit undergoes specific quantum operations: a Pauli X (bit flip) and/or a Pauli Z (phase flip) gate. These operations ensure that the third qubit accurately reconstructs the original quantum state |α⟩.

As a result, after applying the necessary corrections, the third qubit holds the original state |α⟩, achieving successful quantum teleportation. This process highlights how quantum information can be transmitted without the physical movement of qubits, relying instead on the principles of entanglement and classical communication.

Swap Test

Figure 8 is a swap test circuit, which is used to measure the similarity (or inner product) between two quantum states |ϕ⟩ and |ψ⟩. The circuit consists of three qubits: the first qubit (top wire) is an ancilla qubit initialized in the |0⟩|0\rangle|0⟩ state, while the second and third qubits contain the states |ϕ⟩ and |ψ⟩ that we want to compare.

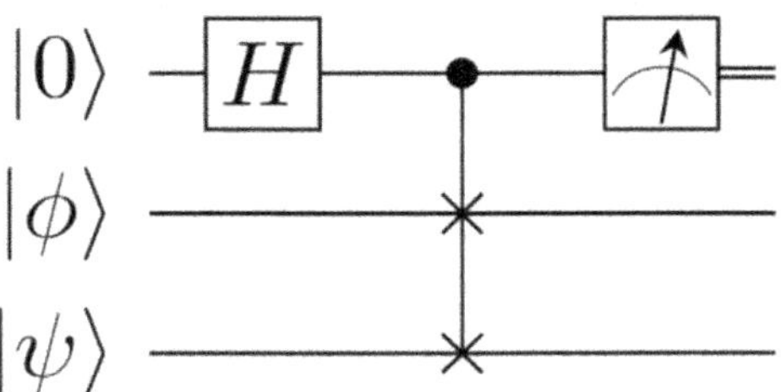

Figure 8: A quantum circuit for the Swap test. Vtomole, CC BY-SA 4.0, via Wikimedia Commons.

The circuit operates in the following steps:

1. Hadamard Gate on Ancilla Qubit: A Hadamard (H) gate is applied to the first qubit, creating a superposition state:

$$\frac{1}{\sqrt{2}}(|0\rangle + |1\rangle)$$

2. Controlled-SWAP (Fredkin) Gate: The ancilla qubit acts as a control for the SWAP operation on the second and third qubits. If the ancilla is in the |1⟩ state, the swap operation exchanges the states of |ϕ⟩ and |ψ⟩. If the ancilla is in |0⟩|0\rangle|0⟩, nothing changes.

3. Hadamard Gate on Ancilla Qubit Again: Another Hadamard gate is applied to the first qubit. This operation effectively interferes the two possible paths (swap vs. no swap), encoding the inner product information into the measurement outcome.

4. Measurement of Ancilla Qubit: The final measurement of the ancilla qubit determines the similarity between |ϕ⟩ and |ψ⟩. The probability of measuring |0⟩ is given by:

$$P(0) = \frac{1 + |\langle\phi|\psi\rangle|^2}{2}$$

This means that if |ϕ⟩ and |ψ⟩ are identical, the ancilla will always be measured as |0⟩|0\rangle|0⟩. If the two states are completely orthogonal, the ancilla will be measured as |1⟩ half of the time.

The swap test circuit is commonly used in quantum computing to compare quantum states without directly measuring them. This is useful in machine learning, quantum fingerprinting, and quantum cryptography, where the inner product between quantum states plays a key role in classification and verification.

Full Adder

The quantum circuit shown as Figure 9 represents a Quantum Full Adder, a fundamental quantum circuit used to perform binary addition on quantum bits (qubits). This circuit takes three input qubits $(|A\rangle, |B\rangle$, and $|C_{in}\rangle$, the carry-in bit) and produces two outputs: the sum $(|S\rangle)$ and the carry-out $(|C_{out}\rangle)$.

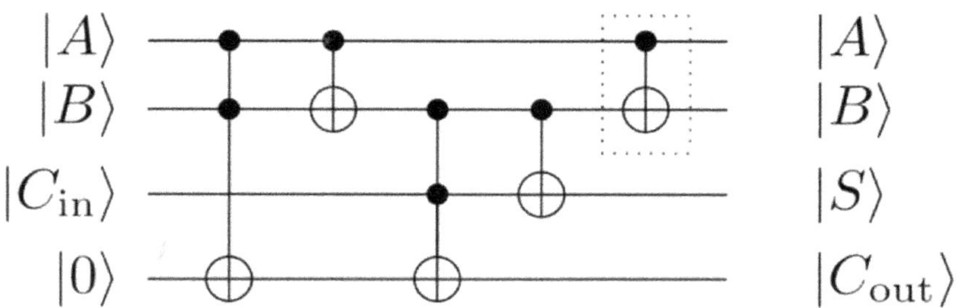

Figure 9: Quantum Full Adder, using Toffoli and CNOT gates. Omnissiahs hierophant, CC BY-SA 4.0, via Wikimedia Commons.

Breakdown of the Circuit:

- **Inputs:**
  - $|A\rangle$ and $|B\rangle$ represent the two qubit values being added.
  - $|C_{in}\rangle$ represents the carry bit from a previous computation.
  - An auxiliary qubit (|0⟩) is used to store the carry-out bit.
- **Controlled Gates:**
  - The circuit consists of multiple Toffoli gates (CCNOT gates) and CNOT gates, which are used to compute the sum and carry values.
  - **Toffoli Gates:** These are controlled-controlled-NOT (CCNOT) gates that flip the target qubit only when both control qubits are in the |1⟩.
  - **CNOT Gates:** These perform an XOR operation between two qubits.
- **Computation of Carry-Out (|Cout⟩):**
  - The carry-out bit is computed using Toffoli gates.
  - The first Toffoli gate sets the auxiliary qubit (initially |0⟩) if both |A⟩ and |B⟩ are |1⟩.
  - Another Toffoli gate processes the input carry (|Cin⟩) to account for additional carry propagation.
- **Computation of Sum (|S⟩):**

- The sum bit is calculated using CNOT gates, implementing an XOR operation: $S=A\oplus B\oplus Cin$
- This is achieved by applying a series of CNOT gates between |A⟩, |B⟩, and |Cin⟩.

- **Final Outputs:**
    - The circuit leaves |A⟩ and |B⟩ unchanged.
    - The third qubit (|S⟩) stores the sum of the three inputs.
    - The fourth qubit (|Cout⟩) holds the carry-out result.

This quantum full adder efficiently performs binary addition using quantum gates while preserving reversibility, a key requirement in quantum computing. The sum output follows classical binary addition logic, while the carry-out ensures proper cascading when used in larger computations. This type of quantum adder is an essential building block in quantum arithmetic operations, including quantum multiplication and quantum algorithms requiring arithmetic computations.

Quantum Fourier Transform

The quantum circuit shown in Figure 10 represents the Quantum Fourier Transform (QFT) on three qubits, a fundamental quantum operation used in quantum computing for applications like quantum phase estimation and Shor's algorithm.

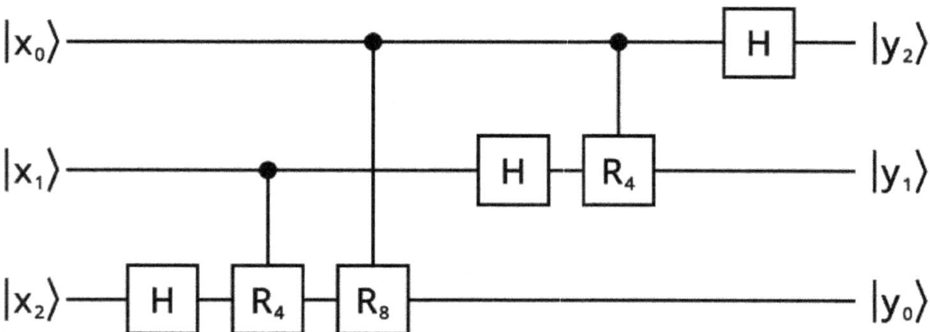

Figure 10: A quantum circuit of the quantum fourier transform on a 3-qubit system. Stefan Birkner (squizzz), CC BY-SA 3.0, via Wikimedia Commons.

The Quantum Fourier Transform (QFT) is the quantum counterpart of the classical discrete Fourier transform (DFT). It transforms quantum states into a frequency-based representation, which is crucial for many quantum algorithms.

This circuit consists of three qubits and performs the QFT using a combination of Hadamard gates (H) and controlled phase shift gates (R_k).

Step-by-Step Execution:

- **First Qubit Operations (Bottom Qubit in the Diagram)**
  - A Hadamard (H) gate is applied to the first qubit, placing it into a superposition state.
  - A controlled R_4 (π/4 phase rotation) gate is applied using the second qubit as control.
  - A controlled R_8 (π/8 phase rotation) gate is applied using the third qubit as control.
- **Second Qubit Operations (Middle Qubit in the Diagram)**
  - A Hadamard (H) gate is applied to the second qubit.
  - A controlled R_4 (π/4 phase rotation) gate is applied using the third qubit as control.
- **Third Qubit Operations (Top Qubit in the Diagram)**
  - A Hadamard (H) gate is applied to the third qubit.
- **Swap Gates (Not Explicitly Shown)**
  - The standard QFT includes swap operations to reverse the order of the qubits at the end.
  - In a full QFT implementation, qubits would be swapped to match the correct output order.

The Hadamard (H) gate is fundamental to the Quantum Fourier Transform as it places qubits into a superposition state. This allows quantum interference effects to take place, which are crucial for distributing probability amplitudes across different computational basis states. By applying Hadamard gates in combination with controlled phase shift gates, the circuit achieves the desired transformation into the frequency domain.

The controlled phase rotation (R_k) gates introduce conditional phase shifts that depend on the quantum state. These gates encode different frequency components, which are essential for the Fourier transform process. Mathematically, an R_k gate is defined as:

$$R_k = \begin{bmatrix} 1 & 0 \\ 0 & e^{2\pi i/2^k} \end{bmatrix}$$

For example, the R_4 gate (π/4 rotation) applies a phase shift of $e^{i\pi/4}$, while the R_8 gate (π/8 rotation) applies a smaller phase shift. These rotations help create interference patterns that map an input quantum state into its Fourier-transformed counterpart.

This quantum circuit effectively converts an input quantum state into its Fourier-transformed equivalent, encoding information in the frequency domain rather than the computational basis. The output of the Quantum Fourier Transform follows the general formula:

$$\frac{1}{\sqrt{2^n}} \sum_{k=0}^{2^n-1} e^{2\pi ikx/2^n} |k\rangle$$

where **n** represents the number of qubits involved in the circuit. This transformation is fundamental for various quantum algorithms, including quantum phase estimation and Shor's algorithm, as it enables efficient processing of periodic structures within quantum states.

Designing a Quantum Circuit

Step-by-Step Process

1. Initializing Qubits: At the beginning of a quantum computation, qubits are initialized in the **|0⟩** state. Specific gates may be applied to prepare the required input states.

2. Creating Superposition: To leverage quantum parallelism, quantum circuits often apply Hadamard gates to create superposition states. This allows a qubit to represent both **|0⟩** and **|1⟩** simultaneously.

3. Implementing Quantum Operations: Quantum gates are applied in sequence to manipulate the qubits and execute the desired computation. This may involve single-qubit transformations, multi-qubit interactions, and controlled operations.

4. Introducing Entanglement: Entanglement is a crucial resource in quantum computation. Multi-qubit gates such as the CNOT gate create entangled states, enabling correlations between qubits that classical systems cannot replicate.

5. Applying Quantum Interference: Quantum algorithms rely on interference effects to enhance correct solutions and cancel out incorrect ones. This is achieved using phase shift gates, Hadamard gates, and controlled operations.

6. Measurement and Output: At the end of the computation, qubits are measured, collapsing their states into classical outcomes. The measurement results are interpreted based on the specific problem being solved.

Example: Quantum Circuit for Creating a Bell State

A Bell state is a fundamental entangled state used in quantum computing and quantum communication.

Quantum Circuit:

1. Apply a Hadamard gate (H) to qubit **q0** to create superposition.
2. Apply a Controlled-NOT (CNOT) gate between **q0** (control) and **q1** (target) to entangle them.

```
from qiskit import QuantumCircuit, Aer, execute

# Create a quantum circuit with 2 qubits
qc = QuantumCircuit(2)

# Apply a Hadamard gate to q0
qc.h(0)

# Apply a CNOT gate between q0 and q1
qc.cx(0, 1)

# Visualize the circuit
print(qc.draw())

# Simulate and execute the circuit
backend = Aer.get_backend('qasm_simulator')
job = execute(qc, backend, shots=1024)
result = job.result()

# Print measurement results
print(result.get_counts(qc))
```

This circuit creates the Bell state:

$$|\Phi^+\rangle = \frac{1}{\sqrt{2}}(|00\rangle + |11\rangle)$$

which demonstrates quantum entanglement.

Universal Quantum Gates and Reversible Computing

Universal Quantum Gates

In classical computing, a set of universal logic gates, such as NAND or NOR, can be used to construct any Boolean function. Similarly, in quantum computing, a set of universal quantum gates can be used to implement any unitary transformation on qubits.

A universal set of quantum gates refers to a collection of gates that can be used to approximate any quantum operation with arbitrary precision. Just as classical computation relies on universal logic gates such as NAND or NOR, quantum computation depends on a set of fundamental gates that can be combined to perform any quantum algorithm.

One essential gate in this set is the Hadamard (H) gate, which creates superposition by transforming the computational basis states $|0\rangle$ and $|1\rangle$ into an equal combination of both states. This ability to create superpositions is fundamental to quantum computing, as it enables the parallelism that gives quantum computers their advantage over classical ones.

Phase gates, specifically the S and T gates, introduce controlled phase shifts to qubits. While they do not change the probability amplitude of the qubit states, they modify the relative phase, which is essential for certain quantum algorithms and interference-based computations.

The Pauli gates (X, Y, and Z) are quantum analogues of classical logic operations. The X gate functions as a quantum NOT gate, flipping $|0\rangle$ to $|1\rangle$ and vice versa. The Y and Z gates introduce phase shifts and rotations around different axes of the Bloch sphere, playing a key role in quantum state manipulation.

The CNOT (Controlled-NOT) gate is a two-qubit gate that flips the target qubit if the control qubit is in state $|1\rangle$. This gate is essential for entanglement, which is a core principle of quantum computing. The ability to entangle qubits allows quantum computers to perform complex computations that are infeasible for classical systems.

Controlled Phase (CZ) and Toffoli (CCNOT) gates extend multi-qubit control operations. The CZ gate applies a phase shift only when both qubits are in state $|1\rangle$, while the Toffoli gate (a three-qubit gate) flips the target qubit only when both control qubits are in $|1\rangle$. These gates enable more advanced quantum computations and are crucial for implementing fault-tolerant quantum circuits.

The T gate is a non-Clifford gate, meaning it cannot be expressed solely in terms of Clifford gates (H, S, CNOT). However, when combined with Clifford gates, it enables universal quantum computation. The inclusion of the T gate is necessary to achieve the full power of quantum computing.

Finally, the Quantum Fourier Transform (QFT) gate is a key component in many quantum algorithms, including Shor's algorithm for factoring large numbers. The QFT transforms quantum states into their frequency components, allowing for efficient solutions to problems that are computationally difficult for classical computers.

By combining single-qubit gates such as Hadamard, phase, and T gates with at least one two-qubit gate like CNOT, any quantum circuit can be constructed. This combination forms a universal quantum gate set, enabling the realization of arbitrary quantum computations.

Universal gate sets allow for the construction of any quantum circuit and enable the implementation of various quantum algorithms. Below are practical examples of how different gates from a universal set are used in quantum computing applications.

1. Quantum Teleportation Protocol (H, CNOT, and Measurement)

Gates Used: Hadamard (H), Controlled-NOT (CNOT), Measurement

Application: Secure quantum communication

Quantum teleportation allows the transfer of an unknown quantum state $|\psi\rangle$ between two parties (Alice and Bob) without physically transmitting the qubit. The protocol involves:

- Creating an entangled Bell state using a Hadamard gate (H) and a CNOT gate.
- Alice performing Bell state measurement on her qubits and sending the classical results to Bob.
- Bob applying conditional Pauli X and Z gates to recover the original state.

This protocol relies on the H and CNOT gates, demonstrating their fundamental role in quantum information transfer.

2. Grover's Search Algorithm (H, Oracle, and Diffusion Operator)

Gates Used: Hadamard (H), Pauli (X), Controlled-Z (CZ), and Multi-qubit Phase Gates

Application: Speeding up search problems

Grover's algorithm provides a quadratic speedup for searching an unsorted database. It operates as follows:

- Apply Hadamard gates (H) to initialize the quantum state in superposition.
- Use an oracle function (which involves multi-qubit phase gates like CZ and Pauli X) to mark the correct solution.

- Perform a diffusion operator, which includes H and X gates, followed by a controlled phase gate (CZ), to amplify the probability of the correct answer.
- Repeat the process $O(\sqrt{N})$ times for optimal results.
- This algorithm showcases the universality of H, X, and controlled gates in quantum search applications.

3. Shor's Algorithm for Factoring Large Numbers (QFT and Modular Exponentiation)

Gates Used: Hadamard (H), Controlled Phase (R_k), CNOT, Toffoli

Application: Breaking RSA encryption by factoring large numbers

Shor's algorithm uses Quantum Fourier Transform (QFT) to find the period of a function, which helps in integer factorization.

- The Hadamard and Controlled Phase (R_k) gates construct the QFT circuit, transforming states into their frequency domain representation.
- Modular exponentiation is implemented using CNOT and Toffoli gates to ensure efficient arithmetic operations on quantum states.
- Classical post-processing extracts factors from the periodicity information.

This algorithm highlights how QFT (built from H and R_k gates) plays a critical role in quantum cryptanalysis.

4. Quantum Error Correction (Toffoli and Ancilla Qubits)

Gates Used: Hadamard (H), CNOT, Toffoli (CCNOT)

Application: Ensuring fault-tolerant quantum computation

Quantum error correction codes protect qubits from decoherence and noise.

- The CNOT gate is used to create redundant encoding of logical qubits by spreading information across multiple physical qubits.
- The Toffoli gate helps in error detection and correction by implementing syndrome extraction and recovery operations.
- Hadamard gates are used to switch between computational and superposition bases, allowing detection of phase errors.

This example demonstrates the essential role of universal gate sets in fault-tolerant quantum computation.

5. Variational Quantum Eigensolver (VQE) for Chemistry Simulations

Gates Used: Hadamard (H), Pauli (X, Y, Z), CNOT

Application: Finding molecular ground-state energies

The VQE algorithm is used in quantum chemistry to approximate the lowest energy state of molecules.

- Parameterized quantum circuits (ansatz) use Pauli X, Y, and Z gates along with CNOT gates to entangle qubits.
- A Hadamard test is used to measure quantum expectation values, essential for energy calculations.
- The algorithm is hybrid, combining classical optimization with quantum circuits to minimize the energy function iteratively.

This showcases how universal gate sets enable near-term quantum applications in material science and chemistry.

Reversible Computing

Classical computation is inherently irreversible, meaning that information is lost when performing standard operations such as AND, OR, and XOR. These classical logic gates do not allow the reconstruction of their original inputs from their outputs. For example, in an AND gate, knowing the output alone is insufficient to determine the exact inputs. This loss of information results in increased energy dissipation due to Landauer's Principle, which states that erasing a bit of information leads to a fundamental energy cost.

In contrast, quantum computing operates using unitary transformations, which are inherently reversible. A unitary operation is a mathematical transformation that maintains all the information in a system, allowing computations to be reversed without any loss of data. Every quantum gate must be unitary, meaning that it has an inverse operation that can completely recover the original quantum state.

One key reversible gate used in quantum circuits is the Toffoli gate. This classical reversible gate takes three inputs and maps them to three outputs while preserving all input information. It can perform classical computations in a reversible manner, making it useful in quantum error correction and logical circuits that require reversibility.

Another important reversible gate is the Fredkin gate, also known as the controlled-swap (CSWAP) gate. This gate swaps two qubits only when a control qubit is in the |1⟩|1\rangle|1⟩ state, ensuring that the transformation can always be undone. The Fredkin gate is essential in both classical reversible computing and quantum computing, where reversibility is a fundamental requirement.

One major advantage of reversible computing is its energy efficiency. According to Landauer's Principle, erasing a bit of information leads to a small but inevitable loss of energy in the form of heat. Since quantum computation is fundamentally reversible, it does not erase information, thus avoiding this energy dissipation. This makes quantum computing potentially more energy-efficient compared to classical computation, especially for large-scale computations.

Reversible computing is crucial not only in quantum circuits but also in the development of low-power classical computing technologies, where minimizing heat dissipation is a key challenge. Since quantum algorithms naturally adhere to the principles of unitary evolution, they operate in a fully reversible manner, making quantum mechanics an essential foundation for the future of energy-efficient computation.

Below are some practical applications where reversible computing plays a key role.

1. Quantum Circuit Reversibility in Quantum Computing

Key Concept: Unitary transformations ensure reversibility

Application: Quantum algorithms, quantum error correction, and quantum simulations

Quantum computation is inherently reversible because quantum gates are unitary operations, meaning they can be inverted to retrieve original states. Practical applications include:

- **Shor's Algorithm for Factoring:** Uses Quantum Fourier Transform (QFT) and modular exponentiation, all reversible operations.
- **Grover's Algorithm for Search:** Iterative applications of oracle and diffusion operators must be reversible to ensure probability conservation.
- **Quantum Error Correction (QEC):** Detects and corrects errors in quantum computers using reversible logic gates like the Toffoli gate.

Since quantum information cannot be copied (no-cloning theorem), reversible transformations allow quantum computers to evolve states without losing data.

2. Reversible Logic in Energy-Efficient Classical Computing

Key Concept: Avoiding energy dissipation due to information loss

Application: Low-power computing, supercomputing, and embedded systems

Landauer's principle states that erasing a bit of information dissipates energy as heat. In conventional irreversible classical computing, gates like AND, OR, and XOR lose information because their inputs cannot always be recovered. This leads to:

- Heat dissipation in classical processors
- Power consumption issues in supercomputers

Reversible computing provides an alternative by using reversible logic gates like:

- **Toffoli Gate:** Implements classical AND logic but preserves all input bits.
- **Fredkin Gate (Controlled-SWAP):** Enables conditional swapping of bits without losing information.

These gates are used in low-power circuit designs, neuromorphic computing, and supercomputing architectures to reduce energy consumption.

3. Reversible Computing in Cryptographic Protocols

Key Concept: Secure, reversible transformations for cryptographic functions

Application: Quantum-safe encryption, reversible hash functions

Reversible logic gates are useful in cryptographic applications where information loss must be minimized for security. Examples include:

- **Quantum Cryptography:** Quantum key distribution (QKD) relies on reversible operations to ensure information security.
- **Hash Function Implementations:** Certain collision-resistant hash functions use reversible transformations to improve efficiency and security.
- **Homomorphic Encryption:** Some cryptographic techniques require reversible logic circuits to enable computations on encrypted data without decryption.

Reversible logic is particularly valuable in post-quantum cryptography, where quantum-resistant encryption is needed.

4. DNA Computing and Reversible Bioinformatics Processing

Key Concept: Biological systems use reversible reactions for information processing

Application: Simulating molecular interactions, DNA-based logic circuits

Biological systems naturally exhibit reversible computing principles, especially in DNA and protein synthesis:

- **DNA Strand Displacement:** Used for DNA-based computing, where reactions can be reversed to prevent loss of molecular information.
- **Reversible Chemical Reactions in Bioinformatics:** Many biochemical reactions obey reversibility constraints, making reversible computing ideal for modelling biological systems.

Reversible computing principles are applied in synthetic biology, where logic gates made from DNA molecules are used to control biological processes reversibly.

5. Fault-Tolerant Computing in Aerospace and Embedded Systems

Key Concept: Reversible computing reduces error accumulation

Application: Reliable computing for space missions, satellites, and autonomous systems

Reversible logic is advantageous in environments where:

- Heat dissipation is a concern (e.g., space missions)
- Energy-efficient processing is required
- Computational errors must be minimized

For example:

- **Fault-tolerant computing in spacecraft:** Reversible circuits minimize errors in embedded processors used in deep-space exploration.
- **Self-checking processors for critical applications:** Aerospace and military applications use reversible logic gates to ensure computations can be corrected in real-time.

Since error propagation is reduced in reversible computing, it is particularly useful in radiation-hardened computing for space applications.

6. Reversible Computing in Quantum Machine Learning (QML)

Key Concept: Reversible transformations for efficient quantum neural networks

Application: Optimization problems, data classification, quantum AI

Quantum machine learning algorithms often rely on reversible unitary operations:

- **Quantum Boltzmann Machines (QBMs):** Use reversible transformations to simulate probabilistic distributions.
- **Quantum Neural Networks (QNNs):** Employ reversible logic gates to preserve information and reduce computation overhead.
- **Quantum Data Encoding:** Uses reversible quantum circuits to compress and encode data efficiently.

Reversible logic ensures that quantum AI models are scalable and energy-efficient, reducing decoherence effects in quantum hardware.

Practical Applications of Quantum Gates and Circuits

The concepts discussed in this chapter are practically implemented in quantum computing in various ways. Developers of quantum computers use these principles to design, optimize, and execute quantum algorithms, ensuring that computations leverage the unique properties of quantum mechanics.

1. Implementation of Quantum Gates in Real Quantum Computers

Quantum gates form the building blocks of quantum circuits, similar to how logic gates function in classical computing. Developers use these gates to manipulate qubits, enabling complex computations that classical computers struggle with.

- **Hadamard Gate (H)**: Used in quantum algorithms like Grover's search and Shor's algorithm to create superposition states, allowing quantum computers to explore multiple possibilities simultaneously.
- **CNOT Gate**: Used for creating quantum entanglement, which is essential for quantum teleportation and quantum error correction.
- **Toffoli Gate**: Implemented in reversible computing, where classical logic operations (AND, OR, XOR) are executed without information loss.
- **Quantum Fourier Transform (QFT) Gate**: Used in Shor's algorithm for factoring large numbers exponentially faster than classical methods.

Developers implement these gates using superconducting circuits (IBM, Google), trapped ions (IonQ, Honeywell), or photonic systems (Xanadu), each with different approaches to achieving high-fidelity gate operations.

2. Quantum Circuit Design for Algorithm Execution

Quantum circuit design is essential for executing quantum algorithms. Developers build circuits that optimize qubit usage while minimizing errors and decoherence.

- **Quantum Parallelism**: Superposition and interference allow quantum computers to perform multiple computations simultaneously.
- **Quantum Entanglement**: Used in quantum cryptography (BB84 protocol) and distributed quantum computing.
- **Quantum Error Correction (QEC):** Implemented using codes like the Steane or Shor codes to protect against qubit decoherence.

Developers use quantum programming languages like Qiskit (IBM), Cirq (Google), and PennyLane (Xanadu) to implement these circuits and test their behaviour on quantum hardware or simulators.

Quantum programming languages provide tools and frameworks for writing, simulating, and executing quantum algorithms on real quantum hardware or simulators. The three major quantum programming languages—Qiskit (IBM), Cirq (Google), and PennyLane (Xanadu)—each have unique features and applications.

Qiskit (IBM)

Qiskit is an open-source quantum computing framework developed by IBM. It allows users to program quantum computers using Python and provides tools for designing, simulating, and running quantum circuits on IBM's quantum hardware.

Key Features:

- **Quantum Circuit Representation**: Qiskit uses a circuit-based model where users can create, modify, and visualize quantum circuits.
- **Quantum Algorithms**: Supports algorithms like Shor's algorithm, Grover's search, and variational quantum algorithms (VQAs).
- **Quantum Simulators**: Includes high-performance simulators for testing circuits before running them on real quantum hardware.
- **Error Mitigation and Noise Modelling**: Provides tools for mitigating errors and modelling noise in quantum processors.
- **Quantum Machine Learning and Optimization**: Qiskit supports quantum-enhanced ML models via the Qiskit Machine Learning module.

Use Cases:

- Developing and testing quantum algorithms on IBM Quantum Experience.

- Simulating quantum circuits before execution on a real quantum processor.
- Research in quantum cryptography, quantum chemistry, and quantum optimization.

Example Code:

```
from qiskit import QuantumCircuit, Aer, transpile, assemble,
execute

# Create a simple 2-qubit quantum circuit
qc = QuantumCircuit(2)
qc.h(0)  # Apply Hadamard gate
qc.cx(0, 1)  # Apply CNOT gate

# Simulate the quantum circuit
simulator = Aer.get_backend('statevector_simulator')
job = execute(qc, simulator)
result = job.result()
print(result.get_statevector())
```

Cirq (Google)

Cirq is a quantum programming framework developed by Google, optimized for near-term quantum devices. Unlike Qiskit, which focuses on gate-based universal quantum computation, Cirq is designed for variational quantum algorithms and near-term quantum applications.

Key Features:

- **Low-Level Quantum Circuit Control**: Cirq allows fine-grained control over quantum hardware, making it useful for quantum hardware research.
- **Tightly Integrated with Google's Sycamore Processor**: Used in Google's quantum supremacy experiment.
- **Supports Quantum Volume Benchmarking**: Measures how well a quantum processor performs complex computations.
- **Focus on Noise and Error Handling**: Provides tools for optimizing quantum circuits to reduce errors.

Use Cases:

- Research in quantum error correction, quantum machine learning (QML), and variational quantum algorithms (VQAs).
- Optimizing quantum circuits for Google's Sycamore quantum processor.

- Testing and validating quantum algorithms before deployment on real devices.

Example Code:

```
import cirq

# Create a simple 2-qubit quantum circuit
qubits = [cirq.GridQubit(0, 0), cirq.GridQubit(0, 1)]
circuit = cirq.Circuit()
circuit.append(cirq.H(qubits[0]))  # Hadamard gate
circuit.append(cirq.CNOT(qubits[0], qubits[1]))  # CNOT gate

# Simulate the circuit
simulator = cirq.Simulator()
result = simulator.simulate(circuit)
print(result)
```

PennyLane (Xanadu)

PennyLane is an open-source quantum machine learning (QML) framework developed by Xanadu, designed for hybrid quantum-classical computations. It integrates quantum computing with deep learning frameworks like TensorFlow and PyTorch.

Key Features:

- **Quantum Differentiable Programming**: Enables gradient-based optimization of quantum circuits, useful for quantum neural networks (QNNs).
- **Hybrid Quantum-Classical Workflows**: Combines quantum computing with classical machine learning techniques.
- **Supports Multiple Backends**: Works with Qiskit, Cirq, and hardware providers like IBM Quantum, Amazon Braket, and Rigetti.
- **Optimized for Variational Quantum Algorithms (VQAs)**: Useful in quantum chemistry, quantum finance, and quantum AI applications.

Use Cases:

- Quantum machine learning (QML) and quantum neural networks (QNNs).
- Implementing quantum finance models for portfolio optimization.
- Developing quantum-enhanced deep learning algorithms.

Example Code:

```
import pennylane as qml
```

```
import numpy as np

# Define a 2-qubit quantum circuit
dev = qml.device('default.qubit', wires=2)

@qml.qnode(dev)
def circuit(theta):
    qml.RX(theta, wires=0)
    qml.CNOT(wires=[0, 1])
    return qml.expval(qml.PauliZ(0))

theta = np.pi / 4
print(circuit(theta))
```

Table 7: Programming language Comparison Table.

Feature	Qiskit (IBM)	Cirq (Google)	PennyLane (Xanadu)
Best For	General quantum computing	Near-term quantum devices	Quantum machine learning
Programming Language	Python	Python	Python
Circuit Type	Universal, gate-based	Noisy Intermediate-Scale Quantum (NISQ)	Variational Quantum Circuits (VQCs)
Integration	IBM Quantum, Aer Simulator	Google Sycamore, XMON hardware	TensorFlow, PyTorch, JAX
Use Cases	Quantum algorithms, error correction, cryptography	Hardware optimization, variational circuits	QML, finance, quantum AI

3. Universal Quantum Gate Sets in Practical Computing

Universal gate sets (Hadamard, T, CNOT, and Phase gates) enable the implementation of any quantum algorithm. Developers use them to:

- Construct quantum compilers that translate high-level quantum programs into hardware-compatible gate sequences.
- Optimize quantum circuits to reduce gate count, improving execution fidelity.

- Implement fault-tolerant quantum computing, ensuring robust computations even with noisy qubits.

4. Reversible Computing in Quantum Hardware

Reversible computing is inherent in quantum computing because all quantum operations must be unitary and invertible.

- **Energy Efficiency**: Quantum gates do not lose information, avoiding the energy dissipation seen in classical irreversible gates.
- **Reversible Classical Logic Implementation**: Toffoli and Fredkin gates allow classical logic to be embedded in quantum circuits, which is useful for quantum simulation of classical algorithms.
- **Quantum Cryptographic Systems**: Reversible transformations ensure secure quantum key distribution without data loss.

5. Practical Use Cases for Developers

Quantum computing developers apply these concepts in real-world applications:

- **Optimization Problems**: Quantum annealing (D-Wave) and gate-based quantum computing (IBM, Google) solve logistics, finance, and scheduling problems more efficiently than classical methods.
- **Drug Discovery and Material Science**: Quantum simulations of molecules (using QFT and phase estimation circuits) accelerate the discovery of new drugs and materials.
- **Machine Learning and AI**: Quantum circuits are used in variational quantum algorithms (VQA) for AI applications.
- **Cryptography and Security**: Quantum circuits power post-quantum encryption algorithms and secure communication protocols.

The principles in this chapter guide the design and development of quantum computing hardware and software. Quantum developers use them to build efficient circuits, optimize quantum computations, and apply quantum mechanics to real-world problems. By leveraging universal gates, reversible computing, and entanglement-based operations, quantum computing moves toward scalable and fault-tolerant systems.

Chapter 4

Quantum Algorithms: The Power of Parallelism

Up to this point, we have explored the fundamental principles of quantum mechanics, the nature of qubits and quantum systems, and the design of quantum circuits. These foundations are crucial for understanding how quantum computers process information in ways that surpass classical computing. With qubits capable of existing in superposition and being entangled across vast distances, quantum systems have the potential to perform massively parallel computations. However, the true power of quantum computing emerges when these unique properties are harnessed within carefully designed algorithms.

This chapter delves into quantum algorithms—mathematical procedures designed specifically for quantum computers to solve complex problems more efficiently than classical counterparts. Quantum algorithms leverage superposition and entanglement to explore multiple solutions simultaneously, leading to exponential speedups for certain computational tasks. Some of the most well-known quantum algorithms, such as Shor's algorithm for integer factorization and Grover's algorithm for searching unsorted databases, illustrate the significant advantages quantum computing can offer in cryptography, optimization, and data analysis.

We will begin with an introduction to quantum algorithms, highlighting their role in unlocking the computational power of quantum systems. Then, we will explore key algorithms that demonstrate quantum advantage, including the Deutsch-Jozsa algorithm, which showcases the first example of quantum speedup, and the Quantum Fourier Transform, a critical component of many quantum algorithms. As we progress, we will examine the broader implications of quantum algorithms, particularly their applications in fields such as artificial intelligence, finance, and materials science.

By understanding how quantum algorithms operate and why they matter, we move one step closer to grasping the real-world potential of quantum computing. This chapter sets the stage for deeper discussions on error correction, quantum cryptography, and emerging quantum applications that will shape the future of computation.

Introduction to Quantum Algorithms

Quantum algorithms are computational procedures that leverage the principles of quantum mechanics—such as superposition, entanglement, and quantum parallelism—to solve problems more efficiently than classical algorithms. Unlike classical algorithms, which process information in a binary manner (0s and 1s), quantum algorithms use qubits, which exist in multiple states simultaneously, enabling exponentially faster computations in some cases.

Quantum algorithms fall into different categories based on their approach and application. Some of the most well-known include Shor's Algorithm, Grover's Algorithm, Quantum Fourier Transform (QFT)-based algorithms, and Variational Quantum Algorithms (VQAs).

Quantum algorithms harness the fundamental properties of quantum mechanics to achieve computational speedups over classical algorithms. Unlike classical computing, which processes information using bits (0s and 1s), quantum computing relies on qubits, which can exist in multiple states simultaneously. This enables quantum computers to perform operations in ways that classical computers cannot. The effectiveness of quantum algorithms is based on four core principles: superposition, entanglement, quantum parallelism, and interference.

Superposition: Superposition is one of the most fundamental principles of quantum computing. In classical computing, a bit can only exist in one of two states, either **0** or **1**, at any given time. In contrast, a qubit can exist in both states simultaneously in a quantum superposition. Mathematically, this means a qubit can be represented as:

$$|\psi\rangle = \alpha|0\rangle + \beta|1\rangle$$

where $\boldsymbol{\alpha}$ and $\boldsymbol{\beta}$ are probability amplitudes that determine the likelihood of measuring the qubit in the $|0\rangle$ or $|1\rangle$ state when observed. The key advantage of superposition is that it allows quantum computers to explore multiple possibilities at once, significantly increasing computational power for certain problems. For example, in Grover's search algorithm, a quantum computer can examine all possible solutions simultaneously before amplifying the correct answer, leading to a quadratic speedup over classical search methods.

Entanglement: Entanglement is a uniquely quantum phenomenon that allows two or more qubits to become strongly correlated, regardless of the distance between them. When qubits are entangled, the state of one qubit is immediately dependent on the state of the other, even if they are separated by vast distances. This property enables quantum computers to perform coordinated operations that classical computers cannot replicate efficiently.

For example, consider an entangled pair of qubits in the Bell state:

$$|\psi\rangle = \frac{1}{\sqrt{2}}(|00\rangle + |11\rangle)$$

If one qubit is measured and found to be $|0\rangle$, the other qubit is instantly known to be $|0\rangle$ as well. If the first qubit is $|1\rangle$, the second qubit must also be $|1\rangle$. This instantaneous correlation is what makes entanglement useful in quantum teleportation, quantum cryptography, and quantum networking.

Quantum algorithms, such as Shor's Algorithm for integer factorization, leverage entanglement to efficiently process large-scale computations that would take classical computers an impractical amount of time.

Quantum Parallelism: Quantum parallelism is the ability of quantum computers to evaluate multiple computations at the same time due to superposition. In classical computing, an operation on N bits requires N separate calculations, each executed one after another. However, in quantum computing, a quantum gate can operate on a superposition of states, performing multiple calculations in parallel.

For instance, in Grover's Algorithm, instead of searching for an item in a database one by one, a quantum computer can process all potential solutions in $\sqrt{N}$ time, providing an exponential advantage over classical methods. Similarly, in Shor's Algorithm, quantum parallelism enables the computation of periodicity in modular exponentiation, which is essential for factoring large numbers in polynomial time.

The ability to evaluate multiple possibilities at once is what makes quantum computing exceptionally powerful for applications such as optimization problems, cryptographic attacks, and machine learning.

Interference: Interference is a principle in quantum mechanics that allows quantum computers to manipulate probability amplitudes to steer a computation toward the correct answer while eliminating incorrect results. Unlike classical probability, where events are independent, quantum probability amplitudes can constructively or destructively interfere with each other.

A prime example of interference is Grover's Algorithm, where quantum gates modify the probability amplitudes of different computational paths. The algorithm amplifies the correct solution while simultaneously cancelling out incorrect possibilities, allowing the correct result to emerge with a much higher probability after just a few iterations.

Interference is achieved using quantum gates, such as the Hadamard gate (H), phase shift gates (S and T), and controlled quantum operations (CNOT, CZ, etc.). These gates adjust the quantum states in a way that enables quantum algorithms to find optimal solutions efficiently.

The principles of superposition, entanglement, quantum parallelism, and interference form the foundation of quantum algorithms. These principles enable quantum computers to solve complex problems in ways that classical computers cannot. By leveraging these properties, quantum algorithms such as Shor's Algorithm, Grover's Algorithm, and Quantum Fourier Transform (QFT)-based techniques provide significant speedups in areas such as cryptography, database searching, and optimization.

For developers working on quantum computing, understanding these principles is essential when designing and implementing quantum circuits. Quantum programming languages like Qiskit (IBM), Cirq (Google), and PennyLane (Xanadu) allow developers to experiment with these principles in real-world applications. As quantum hardware continues to evolve, these core principles will drive the advancement of quantum computing in various industries, from cryptography and materials science to machine learning and artificial intelligence.

Among these algorithms, Shor's Algorithm, Grover's Algorithm, Quantum Fourier Transform (QFT), Variational Quantum Algorithms (VQAs), and Quantum Phase Estimation (QPE) are particularly significant due to their diverse applications across fields like cryptography, optimization, and quantum chemistry.

Shor's Algorithm (Factoring Large Numbers)

Shor's Algorithm is renowned for its ability to efficiently factor large integers, posing a substantial risk to contemporary cryptographic frameworks such as RSA encryption. The classical complexity of integer factorization is exponential ($O(2^n)$), but Shor's Algorithm reduces this to polynomial time complexity ($O(n^3)$) primarily utilizing the Quantum Fourier Transform (QFT) to extract periodicity from modular exponentiation [223]. This algorithm effectively illustrates the quantum advantage, allowing for the rapid decomposition of integers into their prime factors, which is a task that grows increasingly difficult as the integer size increases [224, 225]. The QFT's central role further reinforces the interconnection between quantum algorithms and Fourier analysis, allowing Shor's method to leverage the periodicity in larger datasets efficiently [226, 227].

Shor's Algorithm holds significant implications for:

- **Cryptography:** Specifically, it threatens RSA encryption by enabling rapid factorization of public keys [228].
- **Number Theory:** It enhances the understanding of prime factorization and its applications in secure communication [229].

Grover's Algorithm (Quantum Search)

Grover's Algorithm, another seminal quantum algorithm, provides a quadratic speedup in searching unstructured databases. The classical search requires O(N) operations, while Grover's Algorithm performs the same task in $O(\sqrt{N})$ steps, utilizing quantum amplitude amplification to increase the probability of locating the desired result. This quadratic speedup makes it particularly useful in various optimization problems, including those that arise in logistics and finance [230, 231] as well as in cryptography for attacking symmetric key cryptosystems [232].

Grover's Algorithm is beneficial for:

- **Database Searching:** Efficiently retrieving information from large, unstructured datasets [230].
- **NP-Hard Problem Solving:** Offering strategies for approaching complex problems in polynomial time frames [231].

Quantum Fourier Transform (QFT) and its Applications

The QFT stands as a foundational operation in many quantum algorithms, including Shor's Algorithm and quantum phase estimation. It significantly enhances computational efficiency, transforming quantum states into their frequency domain with a complexity advantage over classical counterparts that require $O(N^2)$ operations; the QFT operates in $O(N \log N)$ time [227]. The application of Hadamard gates and controlled phase rotations is critical in realizing this transformation, which underlies several quantum protocols.

The QFT contributes to:

- **Shor's Algorithm:** Provides a crucial component in efficiently factoring numbers [223, 226].
- **Quantum Phase Estimation:** Helps in determining eigenvalues of unitary operators [233].
- **Quantum Signal Processing:** Serves roles in data analysis and manipulation within quantum computing frameworks [226].

Variational Quantum Algorithms (VQAs)

VQAs represent a hybrid approach, combining classical and quantum computing techniques to address optimization problems. Notable examples such as the Variational Quantum Eigensolver (VQE) and the Quantum Approximate Optimization Algorithm (QAOA) are designed to tackle tasks in quantum chemistry and provide solutions to combinatorial optimization problems [234]. These algorithms utilize parameterized quantum circuits and classical optimization techniques to iteratively refine solutions, making them versatile for practical applications.

VQAs find utility in:

- **Quantum Chemistry:** Simulating molecular systems to predict chemical properties [234].
- **Machine Learning:** Implementing quantum neural networks for advanced data inference [234].

Quantum Phase Estimation (QPE)

QPE is instrumental in estimating the eigenvalues of unitary operators with polynomial efficiency. It employs controlled quantum operations as well as the QFT to extract phase information, forming a backbone for various quantum algorithms, including Shor's Algorithm and many applications in quantum simulations [226, 235]. The ability to extract phase information efficiently translates into significant advances in precision measurement techniques across quantum computing disciplines.

QPE is crucial for:

- **Quantum Chemistry Simulations:** Allowing for efficient calculations of molecular properties [230, 231].
- **Precision Measurements:** Providing high accuracy in quantum states' characterization [224].

Ongoing Evolution of Quantum Algorithms

Quantum algorithms hold the potential to revolutionize computing by solving problems exponentially faster than classical algorithms. However, practical implementation faces challenges such as:

- **Quantum Hardware Limitations**: Current quantum computers suffer from noise and decoherence, making large-scale execution difficult.
- **Error Correction**: Quantum error correction techniques need improvement to maintain coherence in computations.
- **Scalability**: More qubits and stable quantum gates are required to perform meaningful computations on a large scale.

Despite these challenges, research in quantum algorithms continues to expand, with applications in cryptography, artificial intelligence, financial modelling, drug discovery, and materials science.

Quantum algorithms leverage the unique properties of quantum mechanics to outperform classical algorithms in certain problem domains. Shor's Algorithm, Grover's Algorithm, and Variational Quantum Algorithms (VQAs) are among the most widely studied, each with applications ranging from cryptography to optimization. As quantum hardware improves, these algorithms will play a crucial role in shaping the future of computation.

Deutsch-Jozsa Algorithm (First Quantum Speedup)

The Deutsch-Jozsa algorithm is one of the earliest quantum algorithms that demonstrates the advantage of quantum computing over classical computing. It solves a specific problem—determining whether a given function is constant or balanced—with exponential speedup compared to classical algorithms.

The Deutsch–Jozsa algorithm is a landmark deterministic quantum algorithm introduced by David Deutsch and Richard Jozsa in 1992, later advanced by Richard Cleve, Artur Ekert, Chiara Macchiavello, and Michele Mosca in 1998. This algorithm exemplifies a case where quantum computing demonstrates an exponential speedup over deterministic classical algorithms, serving as a focal point in comparative computational analysis.

The Deutsch–Jozsa problem is specifically structured to showcase the advantages of quantum computation. In contrast to classical deterministic algorithms that require exponentially many queries to determine whether a given function is constant or balanced, the Deutsch–Jozsa algorithm can ascertain this with a single query to the oracle [236, 237]. This characteristic illustrates how quantum algorithms exploit quantum superposition and parallelism to streamline computational tasks that would otherwise be difficult for classical counterparts [238-240].

The clear distinction in performance afforded by the Deutsch–Jozsa algorithm leads to an oracle separation between the class of problems solvable in polynomial time on quantum computers (EQP) and those solvable on classical computers (P) [241, 242]. While the problem can be tackled by probabilistic classical algorithms with a polynomial number of queries, it is critical to note that the Deutsch–Jozsa problem does not provide a clear oracle separation with respect to BPP (Bounded-error Probabilistic Polynomial Time) [243]. It highlights the unique capabilities of quantum computation in discerning functions in ways classical systems cannot achieve under deterministic constraints.

Moreover, while the Deutsch–Jozsa algorithm showed immense theoretical promise, its practical applications remain limited. The algorithm primarily serves as an instructive example within foundational studies of quantum computing rather than as a tool for practical computations [244, 245]. The elucidation of the Deutsch–Jozsa framework continues to refine researchers' understanding of the boundaries and capabilities of quantum computation, paving the way for more advanced algorithms and hybrid quantum-classical systems in the future [246].

The Function Classification Problem is at the core of the Deutsch-Jozsa algorithm, a quantum algorithm that demonstrates an exponential speedup over classical computation. The problem involves analysing a given function *f* and determining whether it is constant or balanced with the fewest possible function evaluations.

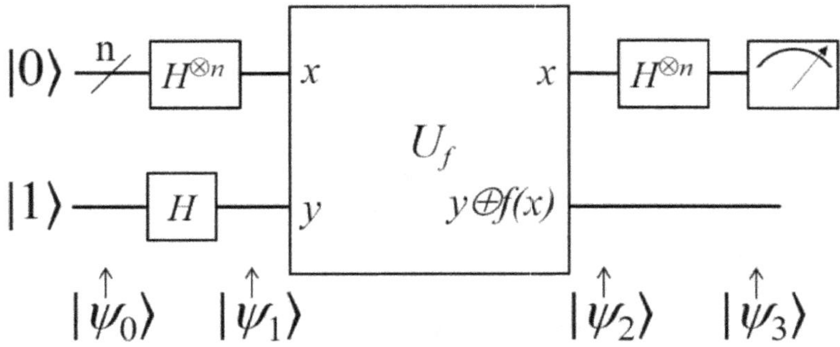

Figure 11: Deutsch-Jozsa algorithm's quantum circuit. Peplm, CC BY-SA 4.0, via Wikimedia Commons.

The quantum circuit shown as Figure 11 represents the Deutsch-Jozsa algorithm, which determines whether a given function $f : \{0,1\}^n \rightarrow \{0,1\}$ is constant (same output for all inputs) or balanced (equal number of 0s and 1s). This circuit showcases quantum parallelism and interference, enabling a solution with only one function evaluation, as opposed to the $O(2^n)$ queries needed classically.

Breakdown of the Quantum Circuit

Step 1: Input Initialization

- The circuit consists of **n+1 qubits**.
- The first n qubits (top wire, |0⟩ states) represent the input register *x*.
- The last qubit (bottom wire, initialized in the |1⟩ serves as an ancilla qubit used in function evaluation.

$$|\psi_0\rangle = |0\rangle^{\otimes n} \otimes |1\rangle$$

Step 2: Hadamard Transform (Creating Superposition)

- A Hadamard H gate is applied to each qubit:
  - On the first n qubits: Creates an equal superposition of all 2^n computational basis states.
  - On the ancilla qubit: Places it in an equal superposition of $|+\rangle = \frac{|0\rangle - |1\rangle}{\sqrt{2}}$.

Mathematically, this transforms the state to:

$$|\psi_1\rangle = \left(\frac{1}{\sqrt{2^n}} \sum_{x=0}^{2^n-1} |x\rangle \right) \otimes \frac{|0\rangle - |1\rangle}{\sqrt{2}}$$

This step ensures that all possible inputs are evaluated simultaneously when passed through the oracle.

Step 3: Oracle Function U_f(Evaluating the Function)

- The oracle U_f is a quantum black-box operation that applies the function *f(x)* by modifying the ancilla qubit:

$$U_f|x, y\rangle = |x, y \oplus f(x)\rangle$$

Since the ancilla qubit was in the $|-\rangle$ state, the effect of this operation is:

$$U_f \left(\frac{1}{\sqrt{2^n}} \sum_x |x\rangle \otimes \frac{|0\rangle - |1\rangle}{\sqrt{2}} \right)$$
$$= \frac{1}{\sqrt{2^n}} \sum_x (-1)^{f(x)} |x\rangle \otimes |-\rangle$$

The function value *f(x)* is encoded into the phase of the computational basis states in the first *n* qubits.

Step 4: Hadamard Transform (Interference)

- A second Hadamard transform is applied to the first *n* qubits.
- This creates interference between the quantum states.
- If the function is constant, all phase terms cancel out except for the $|0\rangle^{\otimes n}$ term.

- If the function is balanced, interference eliminates the |0⟩⊗n term, producing a nonzero measurement.

Step 5: Measurement

- The final measurement collapses the state into a computational basis.
- If the outcome is |0⟩⊗n, the function is constant.
- Otherwise, the function is balanced.

This circuit implements the Deutsch-Jozsa algorithm, efficiently classifying a function as constant or balanced using only one oracle call. It highlights key quantum computing principles such as superposition, interference, and phase encoding—core components of quantum advantage in algorithmic problems.

Problem Definition

We are given a function:

$$f : \{0,1\}^n \to \{0,1\}$$

where:

- The input to the function is an *n*-bit binary string (a sequence of *n* bits, each being either 0 or 1).
- The function produces a single-bit output, either 0 or 1.

Our goal is to determine whether *f(x)* is constant or balanced.

The function *f(x)* is guaranteed to be either constant or balanced, meaning it must follow one of the two conditions:

1. Constant Function:

  - The function returns the same output for all possible inputs.
  - That is, either:
    - *f(x)*=0 for all *x*
    - OR *f(x)*=1 for all *x*

2. Balanced Function:

  - The function returns **0** for exactly half of the possible inputs and **1** for the other half.

- That is, for a function with 2^n possible inputs, it outputs:
  - f(x)=0 for 2^{n-1} inputs
  - f(x)=1 for the remaining 2^{n-1} inputs

Examples:

For n=2 (i.e., two-bit input), the total number of possible inputs is $2^2=4$:

- Constant function example:
  - $f(00) = 0, f(01) = 0, f(10) = 0, f(11) = 0$
  - OR
  - $f(00) = 1, f(01) = 1, f(10) = 1, f(11) = 1$
- Balanced function example:
  - $f(00) = 0, f(01) = 0, f(10) = 1, f(11) = 1$
  - OR
  - $f(00) = 1, f(01) = 1, f(10) = 0, f(11) = 0$

The difficulty of the problem lies in determining whether *f* is constant or balanced with the fewest function evaluations.

Classical Approach

A classical deterministic algorithm must evaluate *f(x)* at least $2^{n-1}+1$ times to guarantee correctness:

- If any two outputs differ, we confirm *f* is balanced.
- If all outputs are the same for the first $2^{n-1}+1$ queries, we conclude *f* is constant.

This means that in the worst case, we need exponentially many function calls (e.g., for n=100, we might need $2^{99}+1$ evaluations).

Quantum Approach

The Deutsch-Jozsa algorithm solves this problem in just one function call by leveraging:

1. Superposition: Evaluating all possible inputs simultaneously.

2. Quantum Interference: Amplifying the probability of measuring the correct result.
3. Measurement: Extracting the global property (constant or balanced) from a single query.

This quantum advantage demonstrates the power of quantum parallelism, as the solution is obtained exponentially faster than classical methods.

The Function Classification Problem illustrates how quantum algorithms can solve problems that would otherwise take classical computers an impractical amount of time. By utilizing quantum superposition and interference, the Deutsch-Jozsa algorithm can distinguish between constant and balanced functions in a single function call, demonstrating one of the first examples of quantum supremacy in computation.

Grover's Algorithm (Quantum Search)

Grover's algorithm is a quantum algorithm designed to search an unsorted database or solve the unstructured search problem exponentially faster than any classical algorithm. It provides a quadratic speedup compared to classical brute-force search methods.

This algorithm was developed by Lov Grover in 1996 and is particularly useful for finding a specific item in an unsorted list, optimizing problems, and breaking cryptographic hash functions.

Specifically, it allows for the identification of a unique input within a black box function that produces a given output using only $O(\sqrt{N})$ evaluations, where N represents the size of the function's domain [247]. In contrast, classical algorithms tackle this problem with a linear complexity of $O(N)$, necessitating approximately N/2 evaluations on average [248]. This stark difference in efficiency highlights the advantages of quantum computational strategies such as Grover's, underscoring the power of quantum algorithms in efficiently solving specific classes of problems [249].

Bennett et al. established that any quantum algorithm designed for this problem must evaluate the function at least $\Omega(\sqrt{N})$ times, confirming Grover's algorithm as asymptotically optimal [250]. This optimality, however, comes with limitations; while Grover's algorithm showcases a quadratic speedup, it remains insufficient for resolving NP-complete problems in polynomial time, as the square root of an exponential remains exponential. Hence, it is recognized that Grover's approach cannot simplify the complexities associated with NP-complete tasks [251].

The applications of Grover's algorithm extend beyond mere theoretical interest. It plays a critical role in enhancing the performance of various algorithms that involve exhaustive search techniques, particularly in NP-complete problem domains [252]. For instance, variants of Grover's algorithm, such as those utilizing amplitude amplification, have been shown to optimize numerous computational tasks, enabling significant speedups in constraint satisfaction problems, including the well-known 3SAT problem [253]. However, it is crucial to note that the benefits of these speedups are strongly contingent upon the specific nature of the problems addressed, and there remains uncertainty about whether Grover's algorithm can improve upon the most effective classical algorithms currently implemented [249].

Moreover, Grover's algorithm has implications for cryptography, particularly in function inversion tasks where it facilitates brute-force attacks on symmetric cryptographic keys. For example, it allows for the effective brute-force cracking of a 128-bit key in approximately 2^{64} operations, and a 256-bit key in roughly 2^{128} operations. Despite these daunting numbers, experts argue that the risks posed by Grover's algorithm might not drastically overshadow those presented by classical algorithms, suggesting that evolving cryptographic standards may still maintain resiliency against quantum threats [254].

Problem Statement: Unstructured Search in Quantum Computing

Unstructured search refers to the problem of finding a specific item within an unsorted database of size $N=2^n$. The search process is guided by an oracle function *f(x)*, which provides binary feedback:

$$f(x) = \begin{cases} 1, & \text{if } x \text{ is the correct item} \\ 0, & \text{otherwise} \end{cases}$$

The objective is to identify the marked item (the item for which *f(x)*=1 as efficiently as possible, minimizing the number of function evaluations.

Classical Search Complexity

In a classical computing setting, there is no way to exploit structure if the database is unsorted. The only approach is a brute-force search, which requires sequentially checking each item one by one.

- In the worst case, a classical algorithm requires O(N) function evaluations, meaning every item may need to be examined before finding the correct one.
- If $N=2^n$, this means an exponential number of queries $O(2^n)$ are required for large datasets, making search operations computationally expensive.

For example, if there are one trillion (10^{12}) items, a classical search would require up to 10^{12} queries in the worst case.

Grover's Algorithm Complexity

Grover's algorithm provides a quantum advantage by significantly reducing the number of queries required to find the marked item. Instead of the classical *O*(N) evaluations, Grover's algorithm reduces the complexity to $O(\sqrt{N})$ function calls.

- Since $N=2^n$, Grover's algorithm requires only $O(2^{n/2})$ queries.
- This represents a quadratic speedup over classical search methods.

For instance:

- Searching through one trillion (10^{12}) items classically requires $O(10^{12})$ function evaluations.
- Grover's algorithm reduces this to just $O(10^6)$ function evaluations, which is a significant reduction in computational effort.

This improvement makes Grover's algorithm highly relevant for large-scale search problems, optimization, and database retrieval applications where brute-force search is infeasible.

Outline of Grover's Algorithm

Grover's algorithm is a quantum search algorithm designed to find marked items in an unstructured database with quadratic speedup over classical methods. The algorithm iterates through a quantum process that amplifies the probability of finding the correct solution while suppressing incorrect ones. Below is a step-by-step breakdown of the process:

1. Initialization of the Quantum Register

The algorithm begins with a register containing n qubits, which are initially set to the state |0⟩:

$$|0\rangle^{\otimes n}$$

This represents a system where all qubits are in the |0⟩|0\rangle|0⟩ state.

2. Creating a Uniform Superposition

To evenly distribute the probability across all possible states, a Hadamard gate (H) is applied to each qubit. This transforms the initial state into an equal superposition of all possible computational basis states:

$$H^{\otimes n}|0\rangle^{\otimes n} = \frac{1}{\sqrt{N}} \sum_{x=0}^{N-1} |x\rangle$$

where $N=2^n$ is the total number of possible states. This step ensures that each state has an equal likelihood of being measured.

3. Iterative Search Process

To amplify the probability of the correct solution, the following set of operations is applied k times, where k is the optimal number of iterations:

Step 1: Phase Oracle Operation

A quantum oracle function U_f is applied, which marks the correct solution(s) by introducing a conditional phase shift of −1 to those states. This is represented as:

$$U_f|x\rangle = \begin{cases} -|x\rangle, & \text{if } x \text{ is a valid solution} \\ |x\rangle, & \text{otherwise} \end{cases}$$

This step differentiates the correct solutions from incorrect ones by flipping their phase.

Step 2: Diffusion Operator (Amplification)

After marking the solution states, a diffusion transformation (sometimes called an inversion about the mean) is applied to increase the probability of measuring the correct item. This consists of:

1. Applying the Hadamard transform $H^{\otimes n}$ to return to the computational basis.
2. Applying a conditional phase shift of −1 to all states except |0⟩, represented by the unitary operation:

   $$U_s = 2|\psi\rangle\langle\psi| - I$$

   where |ψ⟩ is the uniform superposition state.
3. Applying another Hadamard transform $H^{\otimes n}$.

This process redistributes probability amplitudes, increasing the likelihood of measuring the correct solution while suppressing incorrect ones.

4. Measurement and Validation

Once the optimal number of iterations k is reached, the quantum register is measured. The result is an index corresponding to a high-probability solution.

- If the output corresponds to a correct solution, the search ends.
- If the solution is incorrect, the process is restarted.

The optimal number of iterations is given by:

$$k \approx \frac{\pi}{4}\sqrt{N/M}$$

where:

- $N=2^n$ is the total number of states.
- M is the number of marked solutions.

Grover's algorithm leverages quantum superposition and interference to enhance the probability of obtaining the correct solution. By applying an oracle function and a diffusion operator in an iterative process, the algorithm provides a quadratic speedup over classical brute-force search methods. Instead of requiring *O(N)* queries, it only requires $O(\sqrt{N})$, making it significantly more efficient for large search problems.

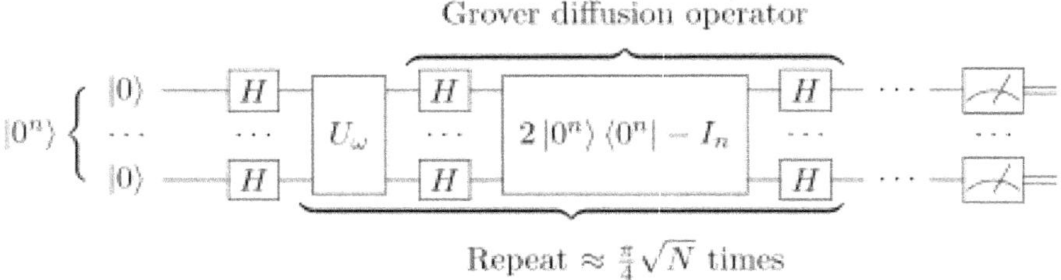

Figure 12: Circuit description of Grover's algorithm. Fawly, CC BY-SA 4.0, via Wikimedia Commons.

The Figure 12 quantum circuit represents Grover's Algorithm, a quantum search algorithm designed to efficiently find a marked item in an unsorted database. Let's break down the circuit step by step:

1. Initialization

- The circuit begins with n qubits initialized in the $|0\rangle$ state.
- A Hadamard (H) gate is applied to each qubit to create a uniform superposition over all possible states. Mathematically, this transforms the initial state:

$$|0\rangle^{\otimes n} \rightarrow \frac{1}{\sqrt{2^n}}\sum_{x=0}^{2^n-1}|x\rangle$$

- This ensures that each state has an equal probability amplitude.

2. Oracle Function U_f

- The oracle function is represented by the block labelled U_f.
- The oracle marks the correct solution(s) by flipping the phase of the state corresponding to the correct answer.
- This is achieved using a conditional phase shift:

$$U_f|x\rangle = \begin{cases} -|x\rangle, & \text{if } x \text{ is the correct solution} \\ |x\rangle, & \text{otherwise} \end{cases}$$

- In a quantum circuit, the oracle can be implemented using multi-qubit controlled-Z or Toffoli gates that recognize the correct solution.

3. Diffusion Operator U_s

- The next major block is the diffusion operator, also known as the amplitude amplification step.
- The purpose of the diffusion operator is to increase the probability amplitude of the marked solution while decreasing the amplitudes of incorrect solutions.
- The steps involved in the diffusion operator are:
    1. Apply Hadamard (H) gates to return to the computational basis.
    2. Apply a conditional phase shift (-1) to all states except $|0\rangle$.
    3. Apply Hadamard (H) gates again to bring the states back to superposition.

Mathematically, this operation is given by:

$$U_s = 2|\psi\rangle\langle\psi| - I$$

- where $|\psi\rangle$ is the uniform superposition state.

4. Repeated Iterations

- The oracle and diffusion operator are applied multiple times to amplify the probability of measuring the correct answer.
- The optimal number of iterations is:

$$k \approx \frac{\pi}{4}\sqrt{N/M}$$

- where:

- $N=2^n$ (total number of possible states)
- M (number of correct solutions)

- After the required number of iterations, the correct solution state has a significantly higher probability amplitude.

5. Measurement

- At the end of the algorithm, a measurement is performed in the computational basis.
- Due to the repeated amplification, the correct solution is measured with high probability.

Grover's Algorithm has several practical applications that make it valuable in quantum computing. One of its most well-known uses is solving unstructured search problems, where the goal is to locate a marked item within a large, unsorted database. Unlike classical search methods that require checking each item one by one, Grover's Algorithm achieves this in significantly fewer steps by leveraging quantum parallelism and amplitude amplification.

Another important application of Grover's Algorithm is in solving NP-complete problems, such as the Boolean satisfiability problem (SAT). In these types of problems, the challenge is to find a valid assignment of variables that satisfies a given Boolean formula. Since NP-complete problems are computationally difficult and grow exponentially with input size, Grover's Algorithm offers a potential quantum speedup in searching for valid solutions.

Grover's Algorithm also has significant implications in cryptographic attacks, particularly in breaking symmetric encryption schemes such as AES (Advanced Encryption Standard). Since symmetric cryptographic systems rely on brute-force key searches for security, classical computers would take O(N) steps to find the correct key, where N is the total number of possible keys. Grover's Algorithm reduces this search time to $O(\sqrt{N})$, effectively halving the security strength of encryption keys. For example, a 256-bit key, which would require 2^{256} classical operations to break, could be cracked in 2^{128} operations using quantum search, making it significantly more vulnerable.

Grover's Algorithm is a powerful quantum search method that dramatically reduces the number of queries required to find a solution in an unsorted dataset. It employs superposition, an oracle function, and amplitude amplification to enhance the probability of measuring the correct answer efficiently. While classical searching requires O(N) queries, Grover's Algorithm achieves the same result in just $O(\sqrt{N})$ steps, providing a substantial computational advantage in various domains, including database searches, optimization problems, and cryptography.

Shor's Algorithm (Factoring Large Numbers)

Shor's algorithm, developed by Peter Shor in 1994, represents a significant advancement in quantum computing due to its ability to efficiently factor large integers. This algorithm operates in polynomial time, which starkly contrasts the exponential time required by any known classical algorithm for integer factorization. Such efficiency poses a significant risk to the security of widely used cryptographic systems, including RSA encryption, which relies on the computational difficulty of factorizing large numbers to secure data [255, 256].

The practical implications of Shor's algorithm are profound. As noted by Ekerå and Håstad [255], Shor's algorithm not only efficiently factors integers but also plays a critical role in computing discrete logarithms. These operations are fundamentally linked to the security of commonly employed cryptographic algorithms. The ability to factor large integers quickly means that cryptographic protocols that assume such tasks are computationally intractable could be rendered insecure in the presence of quantum computers [256, 257].

Experimental implementations of Shor's algorithm have illustrated its potential. Notably, demonstrations utilizing various platforms, such as photonic qubits [256] and nuclear magnetic resonance [225], showcase how quantum systems can perform computations that classical computers cannot feasibly handle. For instance, a compiled version of Shor's algorithm executed on photonic chips was noted to factor the number 15, highlighting the foundational capabilities of quantum computing [225, 258]. Furthermore, these experiments emphasize that Shor's algorithm is instrumental not only for its theoretical implications but also for its applicability in real-world contexts [259].

The impact of Shor's algorithm extends beyond integer factorization; it represents a significant paradigm shift in computing. The advent of quantum computers capable of executing Shor's algorithm could usher in an era where current encryption standards are no longer viable, necessitating the development of quantum-resistant cryptographic algorithms [257, 260]. This urgency is reflected in various studies advocating for new cryptographic frameworks that can withstand quantum attacks, signalling both an academic and practical response to the challenges posed by quantum computing technologies [260, 261].

1. The Problem: Integer Factorization

Given a large integer N, the goal is to find its prime factors efficiently. Classical algorithms, like trial division and the general number field sieve (GNFS), require exponential time in the number of digits of N, making them impractical for very large numbers.

Shor's algorithm, however, solves this problem in polynomial time using a quantum computer.

2. The Core Idea: Period Finding

Shor's algorithm reduces the problem of integer factorization to a period-finding problem. The key mathematical observation is:

1. Select a random integer a such that $1<a<N$.
2. Compute $\gcd(a,N)$:
   - If $\gcd(a,N)\neq 1$, then we already found a factor of N.
   - If $\gcd(a,N)=1$, proceed to the next step.
3. Find the smallest integer r (called the order) such that: $a^r \equiv 1 \bmod N$
   - This means r is the period of the function $f(x)=a^x \bmod N$.
4. If r is even, compute:

$$\gcd(a^{r/2}-1,N)$$

$$\gcd(a^{r/2}+1,N)$$

   - At least one of these values is a nontrivial factor of N, which provides the desired factorization.

Finding the period r is difficult for classical computers but can be solved efficiently using quantum phase estimation on a quantum computer.

3. The Quantum Step: Finding the Period

Quantum computers use quantum Fourier transform (QFT) to efficiently find the period r.

1. **Superposition**: Prepare a quantum state that represents all possible values of x.
2. **Modular Exponentiation**: Compute $a^x \bmod N$ using quantum gates.
3. **Quantum Fourier Transform (QFT)**: Apply QFT to extract the period r.
4. **Classical Post-Processing**: Once r is found, use it to compute factors of N.

This step is what makes Shor's algorithm exponentially faster than classical methods.

4. Complexity of Shor's Algorithm

- **Classical Factorization Methods**: Exponential time $(O(e^{(\log N)^{1/3}(\log\log N)^{2/3}}))$.
- **Shor's Algorithm**: Polynomial time $(O((\log N)^3))$.

This speedup makes RSA encryption insecure against sufficiently powerful quantum computers.

5. Practical Considerations

- **Quantum Hardware**: Current quantum computers are too small and noisy to run Shor's algorithm for large numbers (e.g., 2048-bit RSA).
- **Quantum Error Correction**: Large-scale fault-tolerant quantum computers are needed to break real-world encryption.

Shor's algorithm is one of the most important breakthroughs in quantum computing, demonstrating that quantum computers can solve problems exponentially faster than classical ones. Its impact on cryptography is profound, prompting research into post-quantum cryptography to develop encryption schemes resistant to quantum attacks.

Quantum Fourier Transform

The Quantum Fourier Transform (QFT) is the quantum analogue of the Discrete Fourier Transform (DFT), a mathematical transformation widely used in classical computing for signal processing and solving periodicity-related problems. The QFT is a crucial component of many quantum algorithms, most notably Shor's algorithm, which relies on it to efficiently find periodicity in functions—a key step in factoring large numbers. Unlike classical Fourier transforms, which require computational resources that scale exponentially with input size, the QFT leverages quantum superposition and entanglement to perform the transformation exponentially faster.

In classical computing, the Discrete Fourier Transform (DFT) is a mathematical operation that transforms a sequence of values into a sum of sinusoidal components. For a vector $|x\rangle = (x_0, x_1, ..., x_{N-1})$ of length *N*, the DFT is defined as:

$$X_k = \frac{1}{\sqrt{N}} \sum_{j=0}^{N-1} x_j e^{2\pi ijk/N}, \quad k = 0, 1, ..., N-1$$

This transformation is fundamental in areas such as signal processing, data compression, and numerical solutions to differential equations. However, the classical computation of the DFT requires $O(N^2)$ operations, which becomes impractical for large datasets.

In quantum computing, the QFT performs a similar transformation but takes advantage of quantum mechanics to achieve an exponential speedup. The QFT transforms a quantum state |j⟩ into a superposition state:

$$QFT(|j\rangle) = \frac{1}{\sqrt{N}} \sum_{k=0}^{N-1} e^{2\pi ijk/N} |k\rangle$$

where $N=2^n$ for an n-qubit system. Unlike the classical Fourier transform, which processes data sequentially, the QFT leverages quantum parallelism to encode phase information across qubits simultaneously, significantly reducing computational complexity.

Quantum parallelism is a fundamental concept in quantum computing that allows quantum computers to perform multiple computations simultaneously. This unique capability arises from quantum superposition, a property of quantum bits (qubits) that enables them to exist in a combination of multiple states at once. Unlike classical bits, which can only be 0 or 1 at any given time, qubits can be in a superposition of both 0 and 1 simultaneously. This means that a quantum computer can evaluate a function for multiple inputs in a single computational step, vastly increasing efficiency for certain types of problems.

In classical computing, if we have a function *f(x)* and want to evaluate it for different inputs, we must do so sequentially or, at best, in parallel using multiple processors. However, in a quantum system, we can evaluate the function for many inputs simultaneously using a quantum register, which is a collection of qubits.

Consider a quantum computer with an n-qubit register, which can store all possible 2^n states at once due to superposition. If we prepare the register in the state:

$$\frac{1}{\sqrt{2^n}} \sum_{x=0}^{2^n-1} |x\rangle$$

and then apply a quantum function evaluation operator that maps each input $|x\rangle$ to its corresponding output $|f(x)\rangle$, we obtain:

$$\frac{1}{\sqrt{2^n}} \sum_{x=0}^{2^n-1} |x\rangle |f(x)\rangle$$

This result shows that the quantum system has evaluated the function for all possible inputs simultaneously in a single step. This is quantum parallelism—it allows a quantum computer to compute many values in parallel, something that is infeasible for classical computers.

The Quantum Fourier Transform (QFT) is a key element in quantum algorithms because of its ability to efficiently extract periodicity from quantum states. This capability is especially important in problems such as integer factorization, hidden subgroup problems, and quantum simulations. The main advantages of QFT include:

- **Efficient Period Finding**: QFT plays a crucial role in Shor's algorithm by allowing the discovery of the period of modular exponentiation, which is essential for factoring large integers efficiently.

- **Solving Hidden Subgroup Problems**: The QFT helps in solving problems related to group theory and cryptography, particularly in lattice-based problems and quantum cryptanalysis.
- **Exponential Speedup**: Compared to classical Fourier transforms, which require $O(N^2)$ operations, the QFT achieves the same results in just $O(N^2)$ operations, making it exponentially faster for large N.

The QFT circuit consists of a combination of Hadamard gates and Controlled Phase Shift gates, which together perform the Fourier transformation on quantum states.

Quantum Gates Used in QFT

1. **Hadamard Gate (H)** – Creates superpositions, enabling quantum parallelism.
2. **Controlled Phase Shift Gates (R_k)** – Introduces phase shifts based on the binary representation of the input state.

Steps to Perform QFT on an n-qubit System

1. **Apply a Hadamard gate** to the first qubit.
2. **Apply controlled phase shift gates** R_k to introduce phase shifts based on the binary representation of x.
3. **Repeat steps 1 and 2** for each remaining qubit in sequence.
4. **Apply swap operations** to correct the order of qubits since the QFT naturally reverses bit order.

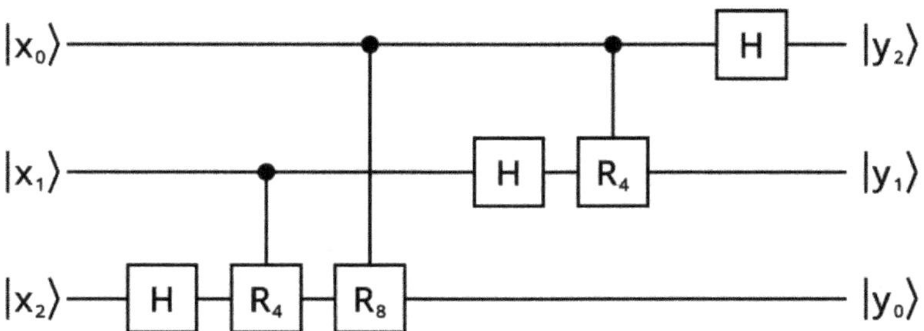

Figure 13: A quantum circuit of the quantum fourier transform on a 3-qubit system. Stefan Birkner (squizzz), CC BY-SA 3.0, via Wikimedia Commons.

where:

- H represents the Hadamard gate, responsible for creating superpositions.
- R_k represents the controlled phase shift gate, which applies a phase factor $e^{2\pi i/2^k}$.
- Swap operations are used to correct the qubit order at the end.

Unlike the classical Discrete Fourier Transform, which requires $O(N^2)$ operations, the QFT circuit needs only $O(n^2)$ quantum gates, making it exponentially more efficient.

Efficiency and Quantum Speedup

One of the primary advantages of the Quantum Fourier Transform (QFT) is its computational efficiency. The classical Discrete Fourier Transform (DFT) has a time complexity of $O(N^2)$, making it impractical for large values of N. In contrast, the QFT requires only $O(n^2)$ operations, where $N=2^n$.

Table 8: Comparison: Classical vs. Quantum Fourier Transform.

Method	Complexity
Classical DFT	$O(N^2)$
Quantum QFT	$O(n^2)$ (where $N=2^n$)

For instance, for a very large number such as $N=2^{1000}$, a classical Fourier transform would take years to compute, whereas a quantum Fourier transform could perform the same operation in milliseconds. This efficiency is why QFT is a cornerstone of quantum computing algorithms.

Applications of the Quantum Fourier Transform (QFT)

The Quantum Fourier Transform (QFT) has a wide range of applications in quantum computing, where it plays a crucial role in enabling efficient computations that are otherwise infeasible for classical computers. By leveraging the power of quantum superposition and interference, the QFT allows quantum computers to process information in ways that significantly outperform classical approaches. Some of its key applications include integer factorization, quantum phase estimation, quantum signal processing, and accelerating solutions to classical problems.

One of the most well-known applications of the QFT is in Shor's algorithm, which provides an efficient quantum method for factoring large numbers. This is of major importance in cryptography, particularly in breaking widely used encryption schemes such as RSA encryption, which relies on the difficulty of integer factorization. Shor's algorithm utilizes the QFT to find the period of a modular exponentiation function, a key step in determining the factors of a given integer. Since classical factoring algorithms require exponential time, while Shor's algorithm runs in polynomial time, the use of QFT in this process poses a significant threat to classical cryptographic security.

Another crucial application of the Quantum Fourier Transform is in Quantum Phase Estimation (QPE), a technique that allows quantum computers to determine the eigenvalues of unitary operators. QPE is fundamental in several quantum algorithms, including Shor's algorithm and applications in quantum simulations and chemistry. In fields like molecular modelling, quantum mechanics, and cryptography, QPE enables quantum computers to extract phase information efficiently, which is essential for understanding quantum states and solving complex mathematical problems. By utilizing the QFT, QPE achieves precise estimations of phase values, making it a powerful tool in quantum computation.

The QFT is also applied in quantum signal processing, which is used in error correction, quantum simulations, and quantum machine learning. In quantum error correction, QFT helps in identifying and correcting errors that arise due to quantum decoherence and noise, thereby improving the stability and reliability of quantum computations. Similarly, in quantum simulations, which model complex physical systems such as chemical reactions or condensed matter physics, QFT is used to transform and analyse quantum states efficiently. In quantum machine learning, QFT enables algorithms to process and manipulate quantum data structures, enhancing the ability of quantum computers to recognize patterns, optimize functions, and perform computations that are infeasible for classical systems.

Beyond quantum-specific applications, the Quantum Fourier Transform also provides substantial advantages in solving classical computational problems. Many problems in differential equations, number theory, and physics simulations require transformations similar to Fourier analysis, which are computationally expensive using classical methods. QFT can accelerate these transformations by performing them in polynomial time, significantly reducing computational complexity. This speedup makes quantum computers highly valuable in scientific research, engineering, and industrial applications, where solving such problems efficiently can lead to breakthroughs in areas such as material science, financial modelling, and climate simulations.

The Quantum Fourier Transform (QFT) is an important mathematical operation in quantum computing that enables efficient period finding, quantum phase estimation, and cryptanalysis. By leveraging quantum superposition and entanglement, QFT provides an exponential speedup over classical Fourier transforms, making it a fundamental component of Shor's algorithm and other quantum algorithms. As quantum computing technology advances, the QFT will play a central role in solving complex problems that are infeasible for classical computers, unlocking new possibilities in cryptography, simulations, and computational mathematics.

Variational Quantum Algorithms

Variational Quantum Algorithms (VQAs) are a class of quantum algorithms that leverage both quantum and classical computing to solve complex optimization and simulation problems. These algorithms are particularly useful for near-term quantum devices, known as Noisy Intermediate-Scale Quantum (NISQ) computers, which have limited qubit numbers and are prone to noise. VQAs offer a practical way to harness quantum computing power while mitigating hardware limitations by offloading certain computations to classical processors.

Variational Quantum Algorithms operate by iteratively optimizing a quantum circuit to minimize a given cost function. They follow a hybrid quantum-classical feedback loop, where a classical optimizer adjusts the parameters of a quantum circuit until an optimal or near-optimal solution is found. The general steps of a VQA are:

1. Prepare a Quantum State

- The quantum processor initializes a parameterized quantum circuit, often using variational quantum gates that depend on tunable parameters.

2. Apply a Cost Function

- The quantum circuit encodes a problem-dependent cost function, which measures the "quality" of the quantum state for solving the problem.

3. Measure the Quantum State

- The quantum computer measures the output of the circuit, providing data to update the parameters.

4. Classical Optimization

- A classical computer processes the measurement results and updates the quantum circuit parameters using optimization algorithms such as gradient descent or genetic algorithms.

5. Iterate Until Convergence

- Steps 1–4 are repeated until the cost function reaches an optimal value, meaning the algorithm has found the best possible solution within the given constraints.

The application of Variational Quantum Algorithms (VQAs), particularly the Variational Quantum Eigensolver (VQE) and the Quantum Approximate Optimization Algorithm (QAOA), has profound implications across domains such as quantum chemistry and combinatorial optimization. The integration of VQAs in these fields can potentially lead to groundbreaking advancements due to their unique problem-solving capabilities enabled by quantum mechanics.

The Variational Quantum Eigensolver (VQE) is notable for its ability to compute the ground state energy of quantum systems efficiently. VQE employs a hybrid quantum-classical optimization approach, wherein a parameterized quantum circuit generates a trial wavefunction. This circuit's parameters are adjusted by a classical optimizer based on the expectation value of the Hamiltonian, which represents the system's energy [262]. Given the complexity of solving the Schrödinger equation for large molecular systems, VQE offers a significant advantage, enabling the examination of systems that are computationally prohibitive for classical methods [262, 263].

The potential applications of VQE in quantum chemistry are vast, including drug discovery and materials science. By benchmarking its performance against classical methods, researchers can leverage VQE to achieve more accurate simulations of molecular interactions, leading to enhanced understanding in chemical engineering and the development of new materials [263, 264]. The algorithm's iterative nature allows it to converge efficiently to the ground state energy, making it a cornerstone in the field of quantum simulations [262, 263].

The Quantum Approximate Optimization Algorithm (QAOA) is designed for combinatorial problems, where the goal is to identify the optimal solution from a sizable configuration space, such as in the Traveling Salesman Problem or the Max-Cut Problem [265, 266]. QAOA operates by encoding the problem into a quantum circuit that applies alternating operations corresponding to a cost Hamiltonian and a mixing Hamiltonian, facilitating efficient exploration of the solution space [267].

QAOA has attracted interest due to its robustness in addressing various optimization challenges across fields such as finance, logistics, and artificial intelligence [265]. As studies continue to survey its performance on constrained problems and question the scalability of QAOA for more complex instances, its ability to provide approximate solutions more efficiently than classical algorithms is becoming increasingly evident [267, 268]. For example, the relationship between specific graph structures and QAOA performance has been thoroughly investigated, revealing insights that can guide future algorithm developments [269].

In the realm of quantum machine learning, VQAs are pivotal in the formulation of Quantum Neural Networks (QNNs). These networks utilize parameterized quantum circuits to model intricate data distributions, which can be advantageous in various learning tasks such as pattern recognition, anomaly detection, and financial modelling [266]. The utilization of quantum parallelism allows QNNs to potentially surpass classical networks in specific domains, presenting a significant opportunity for advancements in artificial intelligence [270].

Furthermore, the intersection of QNNs and VQAs indicates a promising future where quantum techniques could enhance machine learning processes, making them more efficient and effective compared to their classical counterparts [263, 264, 266]. Given that the architecture of quantum circuits plays a crucial role in determining the performance of these algorithms, ongoing research focuses on optimizing quantum architectures to improve the efficacy of QNNs and related VQAs [266, 271].

Variational Quantum Algorithms (VQAs) offer several advantages that make them well-suited for near-term quantum computing. One of their most important benefits is noise resilience. Quantum hardware today is prone to errors due to environmental noise and limited qubit stability, but VQAs mitigate this issue by relying on short-depth quantum circuits. These circuits minimize the number of quantum operations required, reducing the accumulation of errors and making the algorithm more robust on Noisy Intermediate-Scale Quantum (NISQ) devices. This characteristic allows VQAs to function effectively on existing quantum hardware without requiring full error correction, which is a major challenge in the development of large-scale quantum computers.

Another key advantage of VQAs is their hybrid efficiency, which results from the combination of quantum speedup with classical optimization. Instead of relying entirely on quantum resources, VQAs use a quantum processor to evaluate certain complex functions while leaving the optimization and refinement of parameters to a classical computer. This approach leverages the strengths of both quantum and classical computing, making it possible to run useful quantum algorithms with fewer qubits and lower error rates. By iterating between quantum measurements and classical updates, VQAs provide a practical way to extract meaningful results from quantum computations even with the hardware limitations of today's quantum devices.

The broad applicability of VQAs makes them valuable across multiple industries and scientific fields. They are widely used in quantum chemistry for solving molecular structures and predicting chemical properties, which has implications for drug discovery, material science, and pharmaceuticals. In finance, VQAs can optimize risk analysis, portfolio management, and financial modelling, offering potential advantages over classical optimization techniques. Similarly, in machine learning, logistics, and cryptography, these algorithms provide novel ways to solve complex optimization and classification problems that are computationally expensive for classical computers. This versatility makes VQAs one of the most promising approaches to unlocking quantum computing's practical benefits in real-world applications.

Unlike fully scalable quantum algorithms such as Shor's algorithm, which require fault-tolerant quantum computers, VQAs have the advantage of scalability on near-term quantum hardware. They are designed to work within the constraints of NISQ devices, which have a limited number of qubits and are prone to decoherence. Since VQAs rely on variational optimization and adaptive learning, they can still deliver meaningful results with relatively small quantum systems. This makes them one of the most realistic candidates for early quantum advantage, where quantum computers begin to outperform classical ones for specific tasks before full fault tolerance is achieved.

Despite these advantages, VQAs face several challenges that must be addressed to maximize their potential. One major issue is optimization difficulty, as classical optimizers may struggle with complex energy landscapes that contain local minima and barren plateaus. A barren plateau is a region where gradients vanish, making it difficult for optimization algorithms to make meaningful updates to circuit parameters. This problem can slow down or even prevent convergence, limiting the effectiveness of VQAs for large-scale problems.

Another limitation stems from hardware constraints. Current quantum computers have limited qubit connectivity, meaning that only certain pairs of qubits can interact directly. Additionally, short coherence times—the duration for which qubits maintain their quantum states—restrict how many operations can be performed before errors accumulate. Gate errors further introduce inaccuracies, reducing the reliability of quantum computations. These hardware limitations create practical challenges for implementing VQAs efficiently on existing quantum devices.

Scalability remains a concern, as VQAs perform well for small-scale problems but may face obstacles as system size increases. The increasing circuit depth and measurement overhead required for larger quantum computations could reduce the advantage of VQAs over classical approaches. As the number of qubits grows, noise and decoherence effects become more pronounced, potentially limiting the ability of VQAs to provide accurate and useful results in large-scale applications.

To overcome these challenges, researchers are actively exploring improved quantum-classical hybrid methods, better optimization techniques, and error-mitigation strategies. These efforts include gradient-free optimization methods to reduce the impact of barren plateaus, quantum error mitigation techniques to compensate for noise, and more efficient ansatz designs that optimize circuit depth and structure. Continued progress in these areas will be critical to making VQAs more effective and scalable for real-world applications.

Variational Quantum Algorithms (VQAs) represent a promising path toward practical quantum computing by combining the advantages of quantum mechanics and classical optimization. Their ability to solve problems in quantum chemistry (VQE), combinatorial optimization (QAOA), and quantum machine learning (QNNs) makes them highly valuable for industries that require significant computational power. While challenges such as optimization difficulties, hardware limitations, and scalability issues remain, ongoing advancements in quantum hardware, algorithm design, and optimization techniques are steadily improving their effectiveness. As quantum computing technology evolves, VQAs could become one of the first real-world applications of quantum computing, unlocking new possibilities in science, industry, and beyond.

Example of a Variational Quantum Algorithm: Variational Quantum Eigensolver (VQE)

One of the most well-known examples of a Variational Quantum Algorithm (VQA) is the Variational Quantum Eigensolver (VQE). This algorithm is designed to find the ground state energy of a given quantum system, making it particularly useful in quantum chemistry, material science, and condensed matter physics.

In many scientific fields, solving eigenvalue problems is essential for understanding the behaviour of quantum systems. Specifically, in quantum chemistry, determining the ground state energy of a molecule is crucial for predicting its chemical properties and reactivity. However, classical computers struggle with these calculations for larger molecules because the complexity of simulating quantum interactions grows exponentially with the system size.

The Variational Quantum Eigensolver (VQE) provides a hybrid quantum-classical approach to solving this problem efficiently. Instead of using a fully quantum algorithm, VQE leverages the power of both quantum computing for state preparation and measurement and classical optimization for parameter tuning.

The **Variational Quantum Eigensolver (VQE)** follows an iterative process that involves both **quantum and classical computations**:

1. **Hamiltonian Encoding:** The quantum problem is first formulated using a Hamiltonian HHH, which describes the energy of the system. For example, in quantum chemistry, the electronic Hamiltonian of a molecule is written in terms of fermionic operators, which are then mapped onto a qubit-based Hamiltonian using techniques like the Jordan-Wigner transformation.

2. **Parameterized Quantum Circuit (Ansatz):** A trial wavefunction (or quantum state) is prepared using a parameterized quantum circuit called an ansatz. The ansatz consists of quantum gates that depend on adjustable parameters θ. Different ansatz structures can be chosen based on the problem, such as hardware-efficient ansatz or chemically inspired ansatz.

3. **Expectation Value Measurement:** The quantum circuit is executed on a quantum computer, and the expectation value of the Hamiltonian is measured. This gives an estimate of the energy based on the current parameters θ. Since quantum measurements are probabilistic, multiple measurements are taken to obtain an accurate estimate.

4. **Classical Optimization:** The measured energy value is then fed into a classical optimizer (such as gradient descent, COBYLA, or SPSA), which updates the parameters θ to minimize the energy. The optimizer aims to find the lowest possible energy, corresponding to the ground state of the system.

5. **Iteration Until Convergence:** The process is repeated iteratively, adjusting the quantum circuit parameters until the energy converges to a minimum value. At this point, the quantum state stored in the quantum computer represents the approximate ground state of the system.

Mathematical Formulation

The goal of VQE is to minimize the expectation value of the Hamiltonian H with respect to a parameterized quantum state $|\psi(\theta)\rangle$:

$$E(\theta) = \langle\psi(\theta)|H|\psi(\theta)\rangle$$

By adjusting the parameters θ, the algorithm seeks to find the optimal wavefunction $|\psi(\theta*)\rangle$ that minimizes $E(\theta)$, representing the ground state energy of the system.

Example: Finding the Ground State Energy of the Hydrogen Molecule (H_2)

To illustrate how VQE works in quantum chemistry, consider the hydrogen molecule (H_2), one of the simplest molecules.

1. The Hamiltonian of H_2 is mapped onto a qubit representation. This can be done using quantum chemistry methods such as the Jordan-Wigner or Bravyi-Kitaev transformations, which convert the molecular electronic structure problem into a qubit-based Hamiltonian.
2. A variational ansatz (a trial wavefunction) is chosen, typically consisting of rotation gates (R_x, R_y, R_z) and entangling gates (CNOTs) to generate the correct quantum state.
3. The expectation value of the Hamiltonian is measured using a quantum processor, and a classical optimizer updates the parameters to minimize the energy.
4. The process is repeated until convergence, yielding an estimate of the ground state energy of H_2.

Practical Application and Implementation of Quantum Algorithms

The concepts outlined in this chapter, particularly superposition, entanglement, quantum parallelism, interference, Shor's Algorithm, Grover's Algorithm, the Quantum Fourier Transform (QFT), Variational Quantum Algorithms (VQAs), and Quantum Phase Estimation (QPE), have direct practical implementations in quantum computing. Below is a detailed explanation of how these concepts are applied in real-world quantum computing and how they are used by quantum computer developers.

Superposition is implemented in quantum computing using Hadamard gates (H-gates), which create an equal superposition of states. This ability allows quantum parallelism, enabling quantum computers to process multiple states at once.

- **Example in Practice**: In quantum algorithms like Grover's Algorithm, superposition allows a quantum system to examine all possible database entries simultaneously, rather than sequentially as in classical search algorithms.
- **Implementation by Developers**: Quantum developers use quantum programming languages like Qiskit, Cirq, and PennyLane to apply Hadamard gates to qubits, ensuring that quantum circuits leverage superposition effectively.

Entanglement is implemented using CNOT (Controlled-NOT) and CZ (Controlled-Z) gates, which create strong correlations between qubits.

- **Example in Practice**: Quantum teleportation and quantum key distribution (QKD) rely on entanglement to transmit information securely over long distances. In Shor's Algorithm, entanglement is used to efficiently process large-scale computations by correlating qubit states.
- **Implementation by Developers**: Developers create entangled qubit pairs using circuits designed with CNOT gates and Bell state preparation techniques. These entangled states are used in quantum cryptography, quantum networks, and distributed quantum computing.

Quantum interference is controlled using phase shift gates (S, T, and controlled phase gates), which adjust probability amplitudes to amplify correct solutions while cancelling out incorrect ones.

- **Example in Practice**: In Grover's Algorithm, interference is used to amplify the probability of measuring the correct answer while suppressing incorrect solutions.
- **Implementation by Developers**: Quantum developers fine-tune quantum gate sequences to control interference effects, ensuring that algorithms like Grover's search and quantum simulations achieve maximum efficiency.

Shor's Algorithm is implemented using quantum modular exponentiation and the Quantum Fourier Transform (QFT) to efficiently factor large numbers.

- **Example in Practice**: Shor's Algorithm threatens RSA encryption, which relies on the difficulty of factorization for security. A sufficiently powerful quantum computer running Shor's Algorithm could break RSA encryption by factoring large numbers exponentially faster than classical computers.
- **Implementation by Developers**: Quantum engineers use modular exponentiation circuits, phase estimation, and QFT-based subroutines to implement Shor's Algorithm

in quantum hardware. Although current quantum computers do not have enough qubits to break real-world cryptography, developers are working toward scalable fault-tolerant quantum systems.

Grover's Algorithm is implemented using Hadamard gates, quantum oracles, and amplitude amplification techniques.

- **Example in Practice**: Grover's Algorithm is useful for search problems, database retrieval, cryptanalysis (brute force attacks on symmetric cryptography), and combinatorial optimization.
- **Implementation by Developers**: Developers create oracle functions that mark the correct solution and use diffusion operators to amplify it. These are implemented in quantum programming frameworks like IBM Qiskit or Google Cirq to optimize search operations.

The QFT is a critical subroutine used in Shor's Algorithm and quantum phase estimation (QPE).

- **Example in Practice**: QFT is used in Shor's Algorithm to find periodicity in modular exponentiation, which is essential for factoring numbers. QPE is used in quantum chemistry simulations to determine molecular energy states.
- **Implementation by Developers**: Quantum engineers implement QFT using Hadamard gates and controlled phase shift gates, optimizing circuit depth to reduce quantum decoherence. QPE is used in quantum chemistry applications and material simulations.

VQAs like Variational Quantum Eigensolver (VQE) and Quantum Approximate Optimization Algorithm (QAOA) are used in quantum chemistry, finance, and machine learning.

- **Example in Practice:** VQE is used to simulate molecular structures for drug discovery, while QAOA is applied in logistics optimization and machine learning applications.
- **Implementation by Developers**: Quantum developers combine quantum circuits with classical optimizers (gradient descent, SPSA, COBYLA) to iteratively improve solutions. These are executed on cloud-based quantum platforms like IBM Quantum Experience and Google's Sycamore.

Quantum computing researchers and developers implement these principles using a variety of tools and techniques that bridge theoretical quantum mechanics with practical computing applications. One of the primary ways they achieve this is through quantum programming frameworks, which allow for the design and execution of quantum circuits on real or simulated quantum processors. IBM Qiskit, a Python-based framework, enables users to create and manipulate quantum circuits, simulate quantum algorithms, and run them on IBM's superconducting quantum hardware. Google Cirq is another widely used framework designed for near-term quantum devices, particularly those based on Google's Sycamore processor. Xanadu PennyLane, on the other hand, focuses on hybrid quantum-classical computing, making it especially useful for variational quantum algorithms (VQAs) and quantum machine learning applications.

Chapter 5

Quantum Error Correction and Noise Mitigation

As we progress through the exploration of quantum computing, we have established a foundation in quantum mechanics, examined the nature of qubits, and explored quantum gates, circuits, and algorithms. These discussions have highlighted the immense power of quantum computation, demonstrating its ability to process complex problems exponentially faster than classical computers. However, quantum computing is not without its challenges. The very quantum properties that grant quantum computers their power—superposition, entanglement, and quantum interference—also make them highly susceptible to errors and noise. This presents a fundamental obstacle to achieving practical, large-scale quantum computation.

Unlike classical computers, where information is stored in stable binary states (0s and 1s), quantum computers rely on qubits that exist in fragile quantum states. These states are highly sensitive to their environment, and even the slightest interference—such as temperature fluctuations, electromagnetic fields, or cosmic radiation—can cause qubits to lose their information. This phenomenon, known as decoherence, is one of the most significant barriers to building reliable quantum computers. Additionally, quantum operations introduce gate errors, where imperfect control over quantum gates results in computation inaccuracies. Without effective error correction, these errors accumulate, making it nearly impossible for quantum computers to execute long and complex computations reliably.

This chapter delves into the critical field of quantum error correction and noise mitigation, which seeks to overcome these challenges and make quantum computing viable on a large scale. We will explore the fundamental problem of decoherence and examine the key strategies developed to combat errors in quantum computations. The discussion begins with an overview of quantum error correction codes, including the Shor Code and Surface Codes, which enable quantum systems to detect and correct errors without disturbing the underlying quantum state. We will also examine the concept of fault-tolerant quantum computing, which aims to design quantum circuits that can function reliably despite inherent errors.

Additionally, we will discuss various noise mitigation strategies that researchers and engineers employ to reduce the impact of errors in near-term quantum devices. Techniques such as quantum error suppression, dynamical decoupling, and error-aware quantum algorithms have shown promise in improving quantum computation accuracy. Understanding these techniques is essential for appreciating the road ahead, as the success of quantum computing depends on our ability to build robust and scalable quantum systems.

As we move beyond this chapter, the concepts of quantum error correction and noise mitigation will remain central to discussions on quantum hardware, cryptography, and the future of practical quantum computing. By addressing the challenges of decoherence and errors, we edge closer to realizing the full potential of quantum computing in solving real-world problems.

The Problem of Decoherence

Quantum decoherence presents a significant challenge in the field of quantum computing, as it disrupts the delicate quantum states necessary for computations. Decoherence is defined as the process by which a quantum system transitions from a coherent superposition of states into a classical probabilistic mixture, primarily due to its interactions with the external environment [137, 159]. This phenomenon undermines the integrity of qubits—quantum bits that serve as the building blocks of quantum computers—making it increasingly difficult to maintain stable and reliable qubits over extended periods, which is vital for effective computational performance [272, 273].

The roots of decoherence can be traced back to environmental disturbances that introduce noise into the system. Such noise can manifest in various forms, including thermal fluctuations and electromagnetic interference, contributing to the loss of coherence in quantum states [159, 274]. Environmental interactions tend to displace the system from ideal quantum behaviour, often leading to an irreversible loss of quantum information [272]. Zurek extensively discusses the implications of decoherence in distinguishing between quantum and classical worlds, providing insight into how decoherent systems transition to classical behaviour, emphasizing the critical influence of these interactions [137, 273].

Decoherence not only affects the coherence of qubits but also the overall computational efficiency of quantum algorithms. For instance, Su et al. illustrate that in quantum walks, decoherence leads to a transition from an efficient quantum search algorithm to a suboptimal classical search mechanism, highlighting the importance of coherence in quantum computations [275]. As decoherence becomes increasingly prevalent, it serves as a barrier to achieving fault-tolerant quantum computation, which is essential for practical applications of quantum technology [276]. Strategies such as employing decoherence-free subspaces have been proposed to mitigate these challenges, although they rely on specific conditions regarding environmental noise [276, 277].

In an ideal quantum computer, qubits exist in a delicate state of superposition and entanglement, allowing them to perform multiple computations simultaneously. However, real-world quantum systems are not perfectly isolated; they are constantly interacting with their surrounding environment, including electromagnetic fields, vibrations, and thermal noise. These interactions cause qubits to lose phase information, meaning the quantum state collapses into a classical state before computations can be completed. This process effectively destroys the quantum advantage that quantum computers rely on.

Mathematically, decoherence can be understood in terms of density matrices. A pure quantum state $|\psi\rangle$ is represented by a density matrix $\rho=|\psi\rangle$. When decoherence occurs, the system evolves from a coherent superposition to a mixed state, losing its quantum interference properties. The time it takes for this loss to occur is characterized by decoherence time (T_2), which is a measure of how long a qubit can maintain its quantum properties before becoming classical.

Several factors contribute to decoherence in quantum systems. Environmental noise is a major issue, as qubits interact with unwanted electromagnetic radiation, cosmic rays, and other external influences. Thermal fluctuations also play a role, as temperature changes can disturb the quantum states of qubits, causing errors in computations. Additionally, material imperfections in quantum hardware, such as defects in superconducting circuits or ion-trap instability, introduce unwanted interactions that accelerate decoherence.

Another cause of decoherence is quantum measurement itself. The act of measuring a quantum state collapses it into a definite classical value, which is a fundamental limitation of quantum mechanics. This means that qubits must be carefully controlled to avoid unintended measurements during computation.

Decoherence is a major obstacle for building large-scale, fault-tolerant quantum computers. Because quantum operations must be completed before decoherence occurs, researchers need to design quantum circuits that operate within coherence time (T_2). If computations take too long, errors accumulate, rendering the results useless.

Decoherence also limits the number of qubits that can be effectively entangled in a quantum system. Since entanglement is crucial for quantum algorithms such as Shor's algorithm, Grover's search, and quantum error correction, losing coherence reduces the power of quantum computing.

To combat decoherence, researchers use quantum error correction (QEC) techniques. QEC involves encoding logical qubits using multiple physical qubits to detect and correct errors without disturbing the quantum computation. Techniques such as the Shor Code and the Surface Code are widely used to enhance qubit stability.

Another approach is to develop hardware with longer coherence times. Superconducting qubits (IBM, Google) and trapped-ion qubits (IonQ, Honeywell) have different decoherence characteristics, and researchers are optimizing qubit designs to reduce noise. Advanced materials and cooling techniques, such as operating at millikelvin temperatures in dilution refrigerators, also help extend coherence times.

Researchers are also exploring topological qubits, a theoretical form of qubits that rely on non-local quantum states, making them inherently more resistant to decoherence. Companies like Microsoft are investing in topological quantum computing as a potential long-term solution.

Topological Quantum Computing

Topological Quantum Computing (TQC) represents a significant advancement in the realm of quantum computing by exploiting topological states of matter for encoding and manipulating quantum information. This paradigm stands out from traditional quantum computation by providing inherent fault tolerance, which reduces the complexities involved in quantum error correction (QEC) that dominate conventional systems reliant on fragile qubits. Central to TQC is the concept of topological qubits, which employ non-local properties of quantum states to achieve this robustness, making them less sensitive to local disturbances such as noise and temperature variations [278, 279].

In traditional quantum computing frameworks, commonly used quantum bits (qubits) are vulnerable to decoherence, requiring extensive QEC strategies to maintain coherence. In stark contrast, topological qubits are defined by their dependence on the global properties of the system, allowing them to resist perturbations effectively. The utilization of exotic quasiparticles known as anyons—with non-abelian statistics—underpins this approach. These anyons, particularly Majorana fermions, enable a unique braiding process where quantum information is encoded into the configurations of the anyons rather than their individual states. This process not only facilitates fault-tolerant operations but also enhances the stability and coherence times of quantum computations [279-281].

Anyons are a unique class of quasiparticles that exist only in two-dimensional (2D) systems. Unlike ordinary particles, which are classified as fermions (such as electrons) or bosons (such as photons), anyons follow fractional statistics, meaning their behavior falls between those of fermions and bosons. This unusual property makes them particularly interesting for topological quantum computing, where they can be used to store and manipulate quantum information in a way that is naturally resistant to errors.

In quantum mechanics, particles are classified based on how they behave when swapped or exchanged:

- Bosons (e.g., photons, helium-4 atoms) obey Bose-Einstein statistics. When two bosons are exchanged, the quantum state remains the same:

$$|\psi_1, \psi_2\rangle = |\psi_2, \psi_1\rangle$$

- This symmetry allows bosons to "bunch together," leading to phenomena like Bose-Einstein condensation.
- Fermions (e.g., electrons, protons) obey Fermi-Dirac statistics and the Pauli exclusion principle, which prevents two identical fermions from occupying the same quantum state. When two fermions are exchanged, the quantum state acquires a negative sign:

$$|\psi_1, \psi_2\rangle = -|\psi_2, \psi_1\rangle$$

  This property is crucial for the structure of atoms and materials.

- Anyons, however, exist only in 2D systems and obey fractional statistics. When two anyons are exchanged, their wavefunction acquires a complex phase factor:

$$|\psi_1, \psi_2\rangle = e^{i\theta}|\psi_2, \psi_1\rangle$$

where $0<\theta<\pi$. This phase factor is neither 0 (like bosons) nor π (like fermions), meaning anyons exhibit intermediate behaviour.

Anyons come in two major types: Abelian anyons and non-Abelian anyons, with the latter being particularly useful for quantum computing.

Abelian Anyons: Abelian anyons follow fractional statistics, meaning that when they are exchanged, their quantum state is modified by a simple phase factor. These anyons have been experimentally observed in fractional quantum Hall systems, where the Hall conductance exhibits fractional values due to anyonic behavior.

Although Abelian anyons are interesting for fundamental physics, they do not provide the fault-tolerant quantum computational power that non-Abelian anyons do.

Non-Abelian Anyons: Non-Abelian anyons have more complex behavior: when two of them are exchanged (braided), they change the quantum state of the system in a way that depends on the entire history of braiding operations. This property allows them to store quantum information non-locally, making them highly resistant to decoherence and noise.

The most well-known theoretical example of non-Abelian anyons is the Majorana fermion, which emerges in certain superconducting systems. These particles obey non-Abelian braiding rules, meaning that moving them around each other performs a quantum computation.

The key property that makes non-Abelian anyons useful for topological quantum computing is their ability to perform computations through braiding. Unlike conventional qubits, which store information in the quantum states of individual particles, topological qubits rely on the global configuration of anyons within a system. The movement and exchange of these anyons create stable, error-resistant quantum states that can be manipulated to perform quantum computations.

In a topological quantum system, non-Abelian anyons exist as collective excitations that can be moved and exchanged in a controlled manner. The history of how these anyons are braided—meaning the way they are moved around each other—determines the final quantum state of the system. Because the quantum information is encoded in the braiding pattern rather than in the anyons themselves, it is naturally resistant to local disturbances, making topological quantum computing inherently more robust than other quantum computing approaches.

When two non-Abelian anyons are exchanged, their collective quantum state undergoes a unitary transformation, meaning that the system transitions from one quantum state to another in a predictable and deterministic way. This controlled transformation serves as the fundamental mechanism for quantum computations in topological quantum systems. Unlike classical qubits, which rely on individual particles to store quantum information, topological qubits encode information in the entire braiding configuration of multiple anyons. This structure makes them uniquely suited for error-resistant quantum computing.

Since quantum information is encoded non-locally in the braiding history, it becomes highly resistant to local errors. Small perturbations or environmental noise do not significantly affect the computation, making anyon-based quantum computing naturally fault-tolerant. In contrast to traditional quantum computers, which require extensive quantum error correction to maintain coherence, topological quantum computers benefit from intrinsic error protection due to the global nature of their quantum states.

Logical operations in a topological quantum computer are performed by physically braiding anyons in specific patterns. These braiding operations correspond to quantum logic gates, which are essential for executing quantum algorithms. Instead of relying on direct quantum gate manipulations, as in superconducting or trapped-ion quantum computers, topological quantum computers execute computations through the movement of anyons in carefully designed trajectories. These movements effectively act as quantum gates, allowing for complex operations with reduced error rates.

Once a computation is complete, the final quantum state of the system is determined by measuring the collective properties of the anyons rather than individual particle states. Because the information is stored in the global configuration of the system, the measurement process does not disturb the underlying quantum computation as much as in other quantum computing models. This non-local encoding and measurement strategy further enhances the robustness of topological quantum computing, making it a promising approach for building highly stable and scalable quantum processors.

The mechanism of using anyons for computation is predicated on their non-abelian statistics, which differ fundamentally from those of typical fermions or bosons. When anyons are manipulated through braiding—effectively exchanging their positions—their quantum states undergo changes that are governed by the histories of their movements. As a result, any disturbance originating from local noise will not significantly affect the overall computation, because the quantum information resides in the structure formed by the braiding rather than the specifics of individual particle states [279, 282]. This property yields a more fault-tolerant computational architecture than contemporary quantum systems based on standard qubits, fostering a promising pathway towards scalable quantum computing frameworks [280, 283].

One of the chief advantages of TQC is its potential for applications in practical quantum processors. The stability of topological qubits translates to longer coherence times, minimizing the detrimental effects common in traditional systems where qubits rapidly lose their state due to environmental interactions [279, 280]. Moreover, TQC architectures could potentially simplify hardware requirements, enabling a more efficient quantum processor design that does not necessitate intricate QEC protocols frequently used to maintain qubit functionality in classical quantum computers [284].

However, despite the promising theoretical framework, significant challenges remain in realizing and manipulating non-abelian anyons in practical applications. Notably, the experimental demonstration of Majorana zero modes (MZMs) and their manipulation through braiding remains a focal point of current research efforts. Institutions like Microsoft are at the forefront of these endeavours, aiming to develop practical topological quantum computers based on these principles [281, 285]. Yet, precise control over anyonic braiding and the integration of standard qubit systems with topological qubits to enhance computational reliability remains an ongoing investigation [279, 286].

Quantum Error Correction Codes (Shor Code, Surface Codes)

Quantum computing is highly sensitive to errors caused by decoherence, environmental noise, and imperfect quantum gate operations. Unlike classical bits, which can be protected using simple redundancy methods, quantum information is more fragile because qubits exist in superposition and can suffer from both bit-flip errors (changing 0 to 1 or vice versa) and phase-flip errors (changing the relative phase between quantum states). Quantum Error Correction (QEC) codes are designed to protect quantum information by encoding logical qubits into multiple physical qubits, allowing errors to be detected and corrected without disturbing the quantum computation. Two of the most important QEC codes are the Shor Code and Surface Codes.

Shor Code

The Shor Code is a 9-qubit code that protects against both bit-flip and phase-flip errors by encoding a single logical qubit into a distributed entangled state across nine physical qubits.

The Shor Code follows a two-step encoding process. First, a single logical qubit $|\psi\rangle=\alpha|0\rangle+\beta$ is encoded into three physical qubits using a bit-flip redundancy:

$$|0\rangle \rightarrow |000\rangle, \quad |1\rangle \rightarrow |111\rangle$$

This protects against bit-flip errors because if one qubit flips due to noise, a majority vote from the other two qubits can identify and correct the error.

Next, to protect against phase-flip errors, each of the three-qubit states is further encoded into three more qubits using an entanglement scheme that creates superpositions of quantum states:

$$|000\rangle \rightarrow \frac{1}{\sqrt{2}}(|000\rangle + |111\rangle)$$

$$|111\rangle \rightarrow \frac{1}{\sqrt{2}}(|000\rangle - |111\rangle)$$

This final step ensures that if a phase-flip occurs on one qubit, it can be detected and corrected using quantum measurements without collapsing the entire quantum state.

To correct errors, the Shor Code uses syndrome measurements, which involve measuring certain stabilizers that indicate whether an error has occurred without collapsing the quantum superposition. If an error is detected, a correction procedure is applied to restore the correct logical state.

While the Shor Code successfully protects against single-qubit errors, it requires 9 physical qubits per logical qubit, making it inefficient for large-scale quantum computations. Additionally, its error-correction procedures require complex quantum operations, which are difficult to implement on current quantum hardware. For practical quantum computing, more advanced error-correction techniques like Surface Codes have been developed.

Surface Codes

Surface Codes are one of the most promising quantum error correction schemes for real-world quantum computers. Unlike the Shor Code, which uses a direct redundancy approach, Surface Codes arrange physical qubits in a 2D grid and use stabilizer measurements to detect and correct errors. This structure allows for scalable, high-performance error correction that can protect against multiple types of quantum errors.

In a Surface Code, qubits are placed on a square lattice with two types of qubits:

1. Data qubits (which store quantum information)
2. Ancilla (measurement) qubits (which detect errors without disturbing data qubits)

Each data qubit is surrounded by multiple measurement qubits that monitor its state through stabilizer checks. These checks measure whether an error has occurred, but they do not collapse the quantum superposition, ensuring that quantum coherence is preserved.

Errors are detected using parity checks, where groups of ancilla qubits measure correlations between data qubits. If a measurement indicates an unexpected change, an error correction algorithm identifies the error's location and applies a correction operation to restore the original state.

One of the major advantages of Surface Codes is that they only require nearest-neighbour interactions, making them highly practical for superconducting qubit architectures like those developed by IBM, Google, and Rigetti. Unlike the Shor Code, which requires complex long-range interactions, Surface Codes use local interactions, which significantly reduce hardware complexity.

Surface Codes are also highly scalable. By increasing the size of the qubit lattice, the error correction becomes more robust. For example, a 5x5 grid of physical qubits can significantly reduce logical errors, and larger grids provide even stronger protection. This scalability makes Surface Codes the leading candidate for fault-tolerant quantum computing.

Surface Codes have a high error correction threshold, meaning that they can function effectively as long as the physical qubit error rate remains below a certain threshold (typically around 1% error per gate). This makes them more practical than earlier QEC methods, as they can tolerate realistic hardware imperfections.

Additionally, Surface Codes support logical qubit operations through lattice surgery, where logical qubits can be split, merged, or entangled by dynamically modifying the 2D grid structure. This feature is essential for implementing quantum algorithms on large-scale quantum processors.

Despite their advantages, Surface Codes still face practical challenges. They require a large number of physical qubits to encode a single logical qubit, with estimates suggesting that thousands of physical qubits may be needed to achieve fault-tolerant quantum computing.

Additionally, the implementation of Surface Codes relies on fast, high-fidelity quantum measurements, which remain a technological challenge in today's quantum hardware.

Applications

Quantum Error Correction is essential for building reliable, large-scale quantum computers, as quantum systems are prone to errors from decoherence and environmental noise. The Shor Code, as the first quantum error correction code, demonstrated how quantum information can be protected using entangled states, but its high qubit overhead limits its practical use. Surface Codes, in contrast, have become the leading approach for real-world quantum computing due to their scalability, high error threshold, and efficient use of local qubit interactions.

Table 9: Comparison: Shor Code vs. Surface Codes.

Feature	Shor Code	Surface Codes
Qubit Overhead	9 physical qubits per logical qubit	Large 2D lattice (hundreds-thousands of qubits)
Error Types Corrected	Bit-flip & phase-flip errors	Arbitrary local errors (bit-flip, phase-flip, depolarization)
Interaction Type	Long-range entanglement	Nearest-neighbour interactions
Scalability	Limited	Highly scalable
Implementation Feasibility	Difficult with current hardware	Actively being implemented (IBM, Google)

As quantum technology advances, Surface Codes are expected to play a central role in achieving fault-tolerant quantum computing, allowing for the development of powerful quantum processors capable of outperforming classical computers in critical applications such as cryptography, optimization, and quantum simulations.

Practical Examples of Quantum Error Correction (Shor Code and Surface Codes)

To illustrate these concepts, consider the following real-world analogies and practical implementations of Shor Code and Surface Codes in quantum computing.

Example 1: Shor Code – Protecting Against Bit-Flip and Phase-Flip Errors

Analogy: Voting to Correct Errors in a Message

Imagine you send a message over a noisy communication channel, but there is a risk that some letters will be misinterpreted. To prevent errors, instead of sending "A," you send "AAA" (triplicate encoding). If one of the letters is accidentally changed to "ABA," the receiver can use a majority vote and still correctly interpret the message as "A." This redundancy helps correct bit-flip errors.

Now, imagine that the font style of the message changes unpredictably (analogous to a phase-flip error in quantum computing). If each letter is stored as three copies, but their bold/italic formatting gets altered, this could introduce misinterpretation. Encoding the message using a second layer of redundancy, such as repeating entire words in a structured way, ensures that these phase-related alterations can also be detected and corrected.

Practical Implementation in Quantum Computing

The Shor Code applies a similar idea but in a quantum system. A single qubit is first encoded into three qubits to protect against bit-flip errors. For example, the quantum state $|0\rangle$ becomes $|000\rangle$, and $|1\rangle$ becomes $|111\rangle$. If one qubit flips due to noise, the system can still recover the original logical state by applying a majority vote.

To protect against phase-flip errors, each of these three qubits is further encoded into another three qubits, creating a 9-qubit entangled state. This means that instead of a single qubit carrying the quantum information, it is distributed across nine qubits, ensuring that both bit-flip and phase-flip errors can be corrected using syndrome measurements that detect errors without collapsing the quantum superposition.

This technique is experimentally implemented in trapped-ion quantum computers such as those developed by IonQ and Honeywell, where individual ions act as qubits and errors are corrected using multi-qubit entanglement and error-syndrome measurements. However, the Shor Code requires nine physical qubits per logical qubit, making it inefficient for large-scale quantum computers.

Example 2: Surface Codes – Scalable Error Correction in Modern Quantum Processors

Analogy: A City Grid with Traffic Monitors

Imagine a city grid where cars represent data qubits, and intersections have traffic monitors (analogous to ancilla measurement qubits). If a car deviates from its expected lane (an error), the traffic monitors at nearby intersections can detect the anomaly and signal a correction before a collision (data corruption) occurs. The key idea is that the monitors do not affect the traffic directly but instead observe the system holistically, ensuring that problems are fixed without disturbing the flow of traffic.

This concept is directly applied in Surface Codes, where error detection is done using ancilla qubits that monitor data qubits without directly measuring their quantum state.

Practical Implementation in Quantum Computing

Surface Codes encode logical qubits in a 2D lattice of physical qubits, where data qubits store quantum information, and ancilla (measurement) qubits detect errors. The lattice structure allows for nearest-neighbour interactions, making error correction more practical for superconducting qubit architectures used by companies like Google, IBM, and Rigetti Computing.

Google's Sycamore quantum processor, which achieved quantum supremacy in 2019, relies on Surface Code-based error correction to maintain stable quantum computations. In this system, a 5×5 grid of qubits can significantly reduce error rates, and larger grids (such as 10×10) provide even stronger protection.

Surface Codes are also implemented using superconducting qubits, where each qubit is coupled to multiple neighbours, allowing for real-time parity checks that detect errors. For instance, IBM's quantum computers use surface-code error correction to extend the coherence time of logical qubits, ensuring that quantum algorithms can run longer without errors accumulating.

The key advantage of Surface Codes over the Shor Code is their scalability. While the Shor Code requires 9 physical qubits per logical qubit, Surface Codes can protect a single logical qubit using a 2D lattice, making it the leading candidate for fault-tolerant quantum computing.

Fault-Tolerant Quantum Computing

Fault-Tolerant Quantum Computing (FTQC) refers to the ability of a quantum computer to perform long, complex computations despite the presence of errors caused by decoherence, environmental noise, and imperfect quantum gate operations. Because quantum information is extremely fragile, a fault-tolerant system must be capable of detecting, correcting, and managing errors in real-time without disturbing the quantum states needed for computation. The development of fault-tolerant quantum computers is a critical milestone in achieving scalable, reliable, and practical quantum computing that can outperform classical computers in solving real-world problems.

The emergence of fault-tolerant quantum computing (FTQC) represents a significant milestone in the pursuit of reliable quantum computations with low logical error rates. Achieving FTQC requires overcoming critical challenges related to error rates and the effective implementation of quantum error correction (QEC) strategies. A key requirement for FTQC is that the error correction threshold must be exceeded, ensuring that quantum information is adequately protected from environmental influences and that errors are managed locally within the qubit array [287, 288]. This capability is crucial since the current generation of quantum computers, operating with noisy intermediate-scale quantum (NISQ) devices, often produces results that are significantly impacted by errors, limiting their utility for real-world problem solving [289, 290].

Unlike classical computers, which can use simple redundancy (such as error-correcting codes in RAM or hard drives) to fix bit errors, quantum computing faces unique challenges. Qubits exist in superposition, meaning they store information in both 0 and 1 simultaneously. They also experience entanglement, where qubits share quantum correlations, making errors more complex.

Two major types of errors occur in quantum systems:

1. **Bit-Flip Errors** – A qubit in state $|0\rangle$ flips to $|1\rangle$ (or vice versa), similar to a bit-flip in classical computing.
2. **Phase-Flip Errors** – The relative phase between quantum states is altered, affecting quantum interference and making computation unreliable.

These errors arise due to quantum decoherence, which occurs when qubits interact with their surrounding environment, causing information loss. Since measurement collapses quantum states, we cannot directly copy qubits for redundancy (as in classical error correction). This makes Quantum Error Correction (QEC) essential for achieving fault-tolerance.

The importance of neutral atom quantum computing in advancing FTQC has been underscored recently. Neutral atoms are advantageous because they exhibit long coherence times compared to other qubit modalities, which enhances their noise resistance [291, 292]. Additionally, techniques like qubit shuttling facilitate mid-circuit measurements that support error correction processes, allowing for real-time checks and adjustments of potential errors during computations [293]. According to Bermúdez et al., the implementation of flag qubits can enhance error detection, which is pivotal for fault-tolerant operations [289].

In the pursuit of FTQC, various error suppression techniques are utilized, including derangement circuits, dynamic decoupling, and stabilizer codes like Steane's Code. This specific QEC methodology encodes one logical qubit using seven physical qubits, effectively addressing bit flip and phase flip errors; however, it requires further refinement to ensure its scalability and practical applicability [288, 294]. The development of error correction frameworks is critical, as not all classical error correction strategies are suitable for the quantum domain due to limitations such as the no-cloning theorem, which prohibits the duplication of unknown quantum states [288, 295]. The primary requirement for realizing FTQC remains the enhancement of the error-corrected fidelity of two-qubit entangling gates to over 99% accuracy, a benchmark achieved recently by a collaboration of researchers using 60 neutral atom qubits [292].

To be fault-tolerant, a quantum computer must meet the following criteria:

- **Error Detection & Correction:** It must identify and correct errors without measuring the quantum state directly (since measurement collapses quantum information).
- **Logical Qubits & Redundancy:** It must encode a single logical qubit into multiple physical qubits, ensuring that quantum information is spread across many qubits, making it more resilient to noise.
- **Threshold Theorem:** The system must maintain an error rate below a critical threshold, ensuring that error correction techniques can correct more errors than they introduce.

These principles are achieved through Quantum Error Correction (QEC) Codes and Fault-Tolerant Operations, ensuring that quantum computations can continue without data corruption.

Central to FTQC are techniques such as quantum error correction (QEC), fault-tolerant quantum gates, and the threshold theorem, which collectively mitigate the effects of decoherence and operational errors inherent in quantum systems.

1. Quantum Error Correction (QEC) Codes

Quantum error correction codes play an integral role in FTQC by encoding logical qubits into multiple physical qubits, thus enabling error detection and correction without collapsing quantum states. Among various QEC codes, the Shor Code stands out as it protects against both bit-flip and phase-flip errors by using a nine-qubit configuration to represent a single logical qubit [296]. The surface code is recognized as a leading architecture for scalable fault tolerance, utilizing a two-dimensional lattice of qubits where data qubits store information, while ancilla qubits detect errors via stabilizer measurements [297, 298]. Additionally, Bacon-Shor codes and colour codes are noteworthy QEC techniques designed to optimize performance for specific hardware setups, highlighting the adaptive nature of quantum error correction to differing technological contexts [299].

These QEC techniques are paramount as they facilitate the long-term stability of quantum computations necessary for practical applications. The robustness of these codes lies in their ability to address a wide range of errors in a systematic fashion, thereby enhancing the reliability of quantum operations over extended periods [300].

2. Fault-Tolerant Quantum Gates

The implementation of fault-tolerant gates is essential for preventing the propagation of errors during quantum computations. Transversal gates permit independent operations across multiple qubits, effectively isolating errors and preventing them from spreading—this method enables logical gates such as NOT operations to function without influencing other qubits adversely [301, 302]. Besides, magic state distillation is a vital technique for realizing non-Clifford gates (e.g., T-gate) fault-tolerantly; it employs specialized ancilla states that assist in executing more complex quantum gate operations while maintaining fidelity [303, 304].

The combination of QEC with fault-tolerant gates allows quantum computers to conduct extensive calculations while mitigating error accumulation, thus making progress towards practical quantum computation approaches [305, 306]. This synergy between error correction and fault-tolerant gate design is critical in addressing the challenges posed by noise and decoherence in real-world environments.

3. Threshold Theorem: The Key to Fault Tolerance

The Quantum Threshold Theorem posits that quantum error correction can outperform error accumulation as long as the error rate per quantum gate remains below a certain threshold, generally around 1% [301, 307]. For instance, Google's Sycamore processor exemplifies the theorem's implications, as it operates at an error rate of approximately 0.1%, underscoring the feasibility of large-scale, fault-tolerant quantum computing as advances in qubit coherence and gate fidelity continue to evolve [308].

Thus, the threshold theorem provides a theoretical foundation that supports the viability of QEC methods and fault-tolerant gates in practical quantum systems. By establishing a framework within which scalable quantum computation becomes attainable, the theorem paves the way for significant breakthroughs in realizing fault-tolerant quantum computing architectures [309].

Practical Examples of Fault-Tolerant Quantum Computing

Leading quantum companies like Google, IBM, and Microsoft are actively developing fault-tolerant quantum computing using surface codes, topological qubits, and advanced superconducting architectures. As technology advances, fault tolerance will be the key milestone that enables real-world applications in cryptography, AI, materials science, and financial modelling, bringing us closer to a future where quantum computing outperforms classical systems in solving complex problems. As examples:

- **Google's Surface Code-Based Error Correction:** Google has demonstrated surface-code error correction on its Sycamore quantum processor. In this system, logical qubits are encoded using a grid of superconducting qubits, and ancilla qubits continuously monitor for errors without disturbing computations. Their 2023 research showed that increasing the number of qubits reduced error rates, proving that fault-tolerant scaling is possible.

- **IBM's Quantum Error Correction on Eagle and Osprey Processors:** IBM has implemented QEC techniques on their superconducting qubit processors to extend coherence times. Their latest hardware, such as Eagle (127 qubits) and Osprey (433 qubits), is designed to integrate surface-code-based error correction to enable fault-tolerant quantum computations for enterprise applications.
- **Microsoft's Topological Qubits and Majorana Fermions:** Microsoft is developing topological qubits, which are inherently fault-tolerant due to their non-Abelian anyon properties. These qubits use Majorana fermions to encode quantum information non-locally, making them naturally resilient to errors without requiring complex QEC schemes.
- **Quantum Cryptography and Secure Communications:** Fault-tolerant quantum computers are crucial for quantum cryptography, where error-corrected qubits ensure secure transmission of quantum-encrypted information. For example, Quantum Key Distribution (QKD) systems use error correction to prevent eavesdropping while maintaining quantum-secure encryption.

Despite significant progress in fault-tolerant quantum computing, several challenges remain that must be overcome before large-scale quantum computers become practical. One of the most pressing issues is the hardware requirement. Fault tolerance demands encoding a single logical qubit into multiple physical qubits, often requiring thousands to millions of physical qubits for a reliable system. Given that current quantum processors only support a few hundred qubits at most, scaling up to the necessary levels presents significant engineering and fabrication challenges. The sheer complexity of maintaining coherence across such a vast number of qubits makes near-term implementation difficult.

Another major challenge is the need for high-fidelity quantum gates. For quantum error correction to be effective, quantum gates must operate with an error rate below a critical threshold (typically around 1%). While superconducting qubits and trapped-ion qubits have made significant strides in improving gate fidelity, they still suffer from short coherence times and noise from the surrounding environment. Even slight imperfections in gate operations can lead to error accumulation, making it difficult to maintain reliable quantum states over long computations. Developing qubits with longer coherence times and reducing gate operation errors are crucial for achieving fault-tolerant quantum computing.

Even with robust error correction mechanisms, quantum algorithms themselves must be optimized to work efficiently within fault-tolerant architectures. Certain quantum algorithms may require excessive computational resources, making them impractical even if fault-tolerant qubits are available. Researchers are actively refining existing quantum algorithms and developing error-resilient approaches that take full advantage of fault-tolerant systems while minimizing resource overhead.

To address these challenges, researchers are exploring alternative qubit designs that could offer improved stability and scalability. Silicon spin qubits, which leverage semiconductor technology, offer the potential for large-scale integration using existing fabrication techniques. Neutral atom qubits, which use laser-controlled atoms as qubits, provide promising long coherence times and scalability advantages. Photonic qubits, which encode quantum information in light particles, present another avenue for developing highly stable and easily interconnected quantum systems. Additionally, hybrid classical-quantum computing approaches are being explored to optimize quantum computation by integrating quantum processors with classical optimization techniques, improving efficiency while reducing error rates. These ongoing advancements are paving the way toward practical fault-tolerant quantum computing, bringing us closer to the realization of fully scalable and reliable quantum systems.

Noise Mitigation Strategies

Quantum computing is highly susceptible to noise, which arises from interactions with the environment, hardware imperfections, and control errors. This noise leads to decoherence, qubit instability, and gate errors, limiting the ability of quantum computers to perform reliable computations. Unlike classical computers, where error correction is relatively straightforward, quantum systems require specialized noise mitigation strategies to preserve quantum states and enhance computational accuracy. Researchers and engineers have developed various methods to reduce, correct, or compensate for noise, making quantum computing more practical.

Quantum Error Correction (QEC)

One of the most critical noise mitigation strategies is Quantum Error Correction (QEC). Because qubits cannot be directly copied due to the no-cloning theorem, classical redundancy techniques do not work. Instead, QEC encodes a logical qubit into multiple physical qubits, distributing quantum information across a larger system to detect and correct errors without measuring the quantum state directly.

Different QEC codes are used depending on the hardware and application. Surface codes, one of the most promising error correction methods, organize qubits in a 2D lattice, where additional ancilla qubits measure errors while preserving superposition states. Another well-known QEC technique, the Shor Code, encodes a logical qubit into nine physical qubits, protecting against both bit-flip and phase-flip errors. These methods significantly enhance quantum coherence and are essential for fault-tolerant quantum computing.

While QEC is powerful, it requires a large number of physical qubits per logical qubit, making it impractical for near-term quantum devices. As a result, alternative error mitigation techniques are necessary for today's Noisy Intermediate-Scale Quantum (NISQ) devices.

Implementing Quantum Error Correction (QEC) requires designing quantum circuits that encode logical qubits, detect errors, and apply corrections without directly measuring the quantum state. Since qubits cannot be copied due to the no-cloning theorem, QEC relies on encoding a single logical qubit into multiple physical qubits in a way that allows for error detection and correction while preserving quantum coherence. Two widely used QEC methods are the Shor Code and Surface Codes, which provide different approaches to protecting quantum information from decoherence and noise.

The Shor Code is a fundamental QEC scheme that protects a single logical qubit from both bit-flip errors and phase-flip errors by encoding it into nine physical qubits. It performs error detection using syndrome measurements and applies corrections as needed. The encoding process starts by introducing bit redundancy, where a logical qubit is triplicated to protect against bit-flip errors. For example, $|0\rangle_L$ is represented as $|000\rangle$. If one qubit flips due to noise, majority voting can be used to detect and correct the error. To protect against phase-flip errors, each of these triplets is further encoded into three additional qubits, creating an entangled state:

$$|0\rangle_L = \frac{1}{\sqrt{2}}(|000\rangle + |111\rangle)$$

$$|1\rangle_L = \frac{1}{\sqrt{2}}(|000\rangle - |111\rangle)$$

This ensures that phase errors can be corrected by checking the relative phase among the three sets.

To detect bit-flip errors, syndrome measurements check the parity of neighbouring qubits using stabilizer operators such as $Z_1Z_2, Z_2Z_3, Z_4Z_5, Z_5Z_6, Z_7Z_8, Z_8Z_9$. If any measurement result differs from the expected value, an error is detected. For phase-flip errors, a Hadamard transformation is applied to all qubits before measuring parity in the X-basis using stabilizers like $X_1X_2, X_2X_3, X_4X_5, X_5X_6, X_7X_8, X_8X_9$. If a deviation from expected parity is found, the error's location is identified. Once detected, errors are corrected by applying X-gates (for bit-flip errors), Z-gates (for phase-flip errors), or both if necessary.

The Shor Code has been successfully demonstrated using trapped-ion qubits (IonQ, Honeywell) and superconducting qubits (IBM, Google, Rigetti). Quantum circuits implementing the Shor Code are built using quantum programming frameworks such as Qiskit

(IBM), Cirq (Google), and PennyLane (Xanadu). However, despite its effectiveness, the Shor Code requires nine physical qubits per logical qubit, making it challenging to scale for larger quantum systems.

To address scalability, Surface Codes provide a more practical and efficient approach to QEC by arranging qubits in a 2D lattice structure. A Surface Code consists of two types of qubits: Data Qubits, which store quantum information, and Ancilla (Measurement) Qubits, which detect errors without collapsing quantum states. Each data qubit is surrounded by stabilizer qubits that monitor for errors. The lattice follows a checkerboard pattern, where white squares represent Z-stabilizers (detecting bit-flip errors) and black squares represent X-stabilizers (detecting phase-flip errors).

To detect errors, ancilla qubits perform parity checks across neighbouring data qubits. If any parity check deviates from the expected value, an error is detected. The stabilizers used in Surface Codes are:

$$Z\text{-stabilizers: } Z_1 Z_2 Z_3 Z_4$$

$$X\text{-stabilizers: } X_1 X_2 X_3 X_4$$

These stabilizers ensure that errors can be detected in real-time without disturbing the quantum computation.

Once an error is detected, it is corrected by applying appropriate quantum gates. A bit-flip error is corrected using an X-gate, a phase-flip error is corrected using a Z-gate, and multiple errors are corrected using a decoder algorithm, such as the Minimum Weight Perfect Matching (MWPM) algorithm. This approach makes Surface Codes highly robust for large-scale quantum processors.

Surface Codes have been implemented in superconducting quantum processors, including Google's Sycamore processor, IBM's Eagle processor, and Rigetti's quantum systems. These codes are programmed using quantum software like Qiskit, Cirq, and cloud-based quantum platforms, making them a cornerstone of current fault-tolerant quantum computing research. Unlike the Shor Code, Surface Codes require a large number of physical qubits but scale more efficiently by organizing them into a structured lattice, making them the leading approach for scalable error correction in modern quantum computing.

Dynamical Decoupling (DD)

Dynamical Decoupling (DD) is a technique that mitigates decoherence by applying a sequence of fast, carefully timed quantum gate pulses to counteract noise effects. This method is inspired by spin echo techniques in nuclear magnetic resonance (NMR), where rapid spin flips help cancel out unwanted interactions with the environment.

In quantum computing, DD sequences are applied to qubits at specific intervals to reverse phase errors and maintain coherence for longer durations. Some of the most commonly used DD pulse sequences include Carr-Purcell-Meiboom-Gill (CPMG) and Uhrig Dynamical Decoupling (UDD), which effectively protect qubits from low-frequency noise.

The Carr-Purcell-Meiboom-Gill (CPMG) sequence is a dynamical decoupling technique used in quantum computing and nuclear magnetic resonance (NMR) to extend the coherence time of quantum states by mitigating the effects of decoherence and low-frequency noise. It is an improvement over the Carr-Purcell (CP) sequence, designed to counteract spin dephasing and phase errors by applying a series of π (pi) pulses to refocus quantum states periodically.

CPMG is widely used in trapped-ion quantum computers, superconducting qubits, and solid-state spin qubits (such as nitrogen-vacancy centres in diamonds). It plays a crucial role in noise mitigation by reducing the impact of unwanted interactions between qubits and their environment, thus maintaining quantum coherence for longer periods.

The CPMG pulse sequence is based on repeatedly applying π (pi) pulses around the Y-axis, which corrects phase errors and reverses the effects of dephasing. The general structure of the CPMG sequence follows these steps:

1. **Initialization** – Prepare the qubit in a superposition state (e.g., using a Hadamard or microwave pulse).
2. **π/2 Pulse (First Rotation)** – Apply an initial π/2 pulse (typically around the Y-axis) to create a superposition state.
3. **Repeated π Pulses** – Apply a series of π pulses at regular time intervals τ to continuously refocus the qubit's state and counteract errors.
4. **Final π/2 Pulse (Readout Preparation)** – After applying N refocusing π pulses, perform a final π/2 pulse to bring the state back to the measurement basis for readout.

Mathematically, the pulse sequence can be represented as:

$$(\pi/2)_y - (\tau - \pi_x - \tau)_N - (\pi/2)_y$$

where:

- π/2 represents a 90-degree rotation (first and last pulses).
- πx represents 180-degree refocusing pulses around the X-axis.
- τ is the time between pulses.
- *N* is the number of refocusing pulses.

The key idea is that each π pulse inverts the phase errors, so by the end of the sequence, the qubit is effectively "refocused," minimizing decoherence.

To implement the Carr-Purcell-Meiboom-Gill (CPMG) pulse sequence in quantum hardware, a structured approach is required to ensure qubit coherence is maintained for an extended period. The process begins by preparing the qubit in a superposition state. Initially, the qubit is set in either the $|0\rangle$ or $|1\rangle$ state, and a Hadamard gate (H-gate) or a π/2 pulse around the Y-axis is applied to transform it into a superposition state:

$$|0\rangle \rightarrow \frac{1}{\sqrt{2}}(|0\rangle + |1\rangle)$$

This step ensures that the qubit is in a state where it is vulnerable to dephasing effects, making it suitable for applying the CPMG refocusing technique to counteract such errors.

Once the qubit is in superposition, a π/2 pulse is applied around the Y-axis using a microwave control pulse. Mathematically, this is represented as:

$$R_y(\pi/2)$$

This rotation brings the quantum state into the XY-plane, making it more susceptible to decoherence and phase noise. The purpose of the CPMG sequence is to mitigate these effects by applying periodic refocusing pulses.

After initialization, the main refocusing sequence is executed by applying a series of π pulses at evenly spaced time intervals (τ). These pulses are performed around the X-axis and take the form:

$$R_x(\pi) - \tau - R_x(\pi) - \tau - R_x(\pi) - \tau \ldots$$

Each π pulse flips the qubit's phase, effectively reversing any accumulated dephasing errors caused by environmental noise. The effectiveness of this correction depends on choosing an optimal time interval (τ), which is determined based on the specific decoherence dynamics of the qubit.

After the sequence of π pulses, a final π/2 pulse around the Y-axis is applied to bring the quantum state back into a form that can be measured:

$$R_y(\pi/2)$$

This last step ensures that the final quantum state aligns with the computational measurement basis, allowing for an accurate readout of the qubit's state.

The final step in implementing the CPMG sequence is measuring the qubit in the computational basis. A standard quantum measurement is performed, and if the CPMG sequence was successful, the qubit's coherence time should be extended compared to an experiment where no dynamical decoupling was applied. By mitigating dephasing and reducing environmental noise effects, CPMG allows quantum systems to maintain quantum coherence for longer periods, significantly improving the stability of quantum computations.

CPMG sequences are implemented using quantum programming frameworks like IBM Qiskit, Google Cirq, and Rigetti's PyQuil. Below is an example of implementing a CPMG sequence in Qiskit:

```
from qiskit import QuantumCircuit, Aer, execute
from qiskit.providers.aer import AerSimulator

# Create a quantum circuit with 1 qubit
qc = QuantumCircuit(1, 1)

# Apply π/2 pulse (Hadamard gate simulates Ry(π/2))
qc.h(0)

# Apply N refocusing π pulses
N = 4  # Number of π pulses
for _ in range(N):
    qc.barrier()
    qc.rx(3.1415, 0)  # π (pi) pulse around X-axis

# Apply final π/2 pulse
qc.h(0)

# Measure the qubit
qc.measure(0, 0)

# Run the circuit on a quantum simulator
simulator = AerSimulator()
job = execute(qc, simulator, shots=1000)
result = job.result()
counts = result.get_counts()

# Print results
print(counts)
```

This Qiskit program applies the CPMG sequence to a single qubit, helping to extend its coherence time by refocusing phase errors through repeated π pulses.

Uhrig Dynamical Decoupling (UDD) is an advanced dynamical decoupling (DD) technique designed to extend the coherence time of qubits by mitigating the effects of decoherence and environmental noise. It was introduced by Götz Uhrig in 2007 as an optimization over Carr-

Purcell-Meiboom-Gill (CPMG) sequences, offering a more efficient way to suppress low-frequency and high-frequency noise in quantum systems.

Unlike standard equidistant pulse sequences such as CPMG, which apply π pulses at uniform time intervals, UDD places these pulses at non-uniform, strategically optimized time intervals. This non-uniform spacing enables UDD to better counteract high-frequency noise, making it particularly effective in superconducting qubits, trapped-ion qubits, and solid-state spin qubits (such as NV centres in diamonds).

UDD applies N refocusing π pulses, but instead of being evenly spaced, the pulses occur at positions determined by the formula:

$$t_j = T \sin^2\left(\frac{\pi j}{2N+2}\right)$$

where:

- T is the total duration of the decoupling sequence.
- N is the number of refocusing pulses.
- T_j is the time at which the j-th π pulse is applied.
- The spacing follows a sinusoidal distribution, concentrating more pulses near the beginning and end of the sequence, which helps counteract both slow and fast noise components.

By optimizing pulse placement, UDD enhances coherence protection against a broader noise spectrum than standard dynamical decoupling techniques.

To implement the Uhrig Dynamical Decoupling (UDD) sequence, the process begins by preparing the qubit in a superposition state. The qubit is first initialized in either the |0⟩ or |1⟩ state. A Hadamard gate (H-gate) or a π/2 pulse around the Y-axis is then applied to transform it into a superposition state:

$$|0\rangle \rightarrow \frac{1}{\sqrt{2}}(|0\rangle + |1\rangle)$$

This step ensures that the qubit is in a state where it is vulnerable to dephasing, making it suitable for Uhrig Dynamical Decoupling, which aims to mitigate such effects.

Once the qubit is in superposition, a π/2 pulse is applied around the Y-axis using a microwave control pulse:

$$R_y(\pi/2)$$

This rotation moves the quantum state into the XY-plane, where phase errors accumulate over time. By applying a carefully optimized sequence of π pulses, UDD counteracts these errors, preserving the quantum state.

After initialization, the Uhrig-optimized π pulses are applied at non-uniform time intervals, unlike the standard CPMG sequence, which spaces pulses evenly. The precise timing of the pulses is determined by the formula:

$$t_j = T \sin^2 \left(\frac{\pi j}{2N+2} \right)$$

where *T* is the total duration of the sequence, *N* is the number of π pulses, and t_j represents the time at which each pulse is applied. These carefully spaced pulses effectively reverse accumulated dephasing errors caused by environmental fluctuations.

For an implementation with *N* = 5 pulses, the five π pulses are applied at the following non-uniform time fractions of the total sequence duration *T*:

$$t_1 = T \sin^2 \left(\frac{\pi}{12} \right), \quad t_2 = T \sin^2 \left(\frac{2\pi}{12} \right), \quad t_3 = T \sin^2 \left(\frac{3\pi}{12} \right), \quad t_4 = T \sin^2 \left(\frac{4\pi}{12} \right), \quad t_5 = T \sin^2 \left(\frac{5\pi}{12} \right)$$

These optimized pulse placements provide better noise suppression, particularly against high-frequency dephasing noise, allowing for longer qubit coherence.

Once all π pulses have been applied, a final π/2 pulse around the Y-axis is executed to return the quantum state to the measurement basis:

$$R_y(\pi/2)$$

This step ensures that the qubit is properly aligned for measurement, allowing the system to extract meaningful results from the UDD sequence.

Finally, a quantum measurement is performed to determine the final state of the qubit. If UDD was effective, the coherence time of the qubit should be significantly extended compared to an experiment where no dynamical decoupling was applied. This demonstrates the power of UDD in mitigating decoherence and maintaining quantum information over longer timescales, making it a valuable technique for improving the reliability of quantum computations.

UDD sequences can be implemented in IBM Qiskit, Google Cirq, and Rigetti PyQuil. Below is an example of a UDD sequence in Qiskit:

```
from qiskit import QuantumCircuit, Aer, execute
from qiskit.providers.aer import AerSimulator
import numpy as np

# Define the number of refocusing pulses
```

```
N = 5
T = 10  # Total evolution time (arbitrary units)

# Compute the UDD pulse timings
pulse_times = [T * (np.sin((np.pi * j) / (2 * N + 2)))**2 for j
in range(1, N + 1)]

# Create a quantum circuit with 1 qubit
qc = QuantumCircuit(1, 1)

# Apply π/2 pulse (Hadamard gate simulates Ry(π/2))
qc.h(0)

# Apply the Uhrig-decoupled π pulses
for t in pulse_times:
    qc.rx(np.pi, 0)  # π pulse around X-axis
    qc.barrier()

# Apply final π/2 pulse
qc.h(0)

# Measure the qubit
qc.measure(0, 0)

# Run the circuit on a quantum simulator
simulator = AerSimulator()
job = execute(qc, simulator, shots=1000)
result = job.result()
counts = result.get_counts()

# Print results
print(counts)
```

This code implements the Uhrig Dynamical Decoupling sequence by applying π pulses at non-uniform intervals, as determined by the UDD formula. The barriers ensure proper spacing between pulses for better visualization and hardware execution.

Dynamical Decoupling is particularly useful in trapped-ion quantum computers and superconducting qubits, where external electromagnetic noise can cause significant phase decoherence. By periodically refocusing qubits, DD extends the lifetime of quantum states, improving computation reliability.

Quantum Error Mitigation (QEM)

Quantum Error Mitigation (QEM) refers to a set of techniques that reduce the impact of errors without full error correction, making them ideal for near-term quantum devices. Unlike QEC, which requires many additional qubits, QEM techniques focus on correcting errors at the algorithmic level after computation.

One of the most widely used QEM methods is Zero-Noise Extrapolation (ZNE). ZNE works by deliberately amplifying noise in a controlled way, running quantum circuits multiple times with different noise levels, and then extrapolating back to a zero-noise estimate of the correct result. This technique is particularly useful in variational quantum algorithms such as VQE (Variational Quantum Eigensolver) and QAOA (Quantum Approximate Optimization Algorithm), where small improvements in accuracy can significantly enhance performance.

Unlike Quantum Error Correction (QEC), which requires encoding logical qubits into multiple physical qubits, ZNE works by estimating the ideal noiseless result from a set of noisy computations. It is particularly useful for Noisy Intermediate-Scale Quantum (NISQ) devices, where full fault-tolerant quantum computing is not yet feasible due to hardware limitations.

The core idea behind ZNE is to intentionally increase the noise level in a controlled way, observe how the quantum system behaves, and then use mathematical extrapolation to estimate what the result would be if there were no noise at all. This process allows quantum computers to recover more accurate results without the need for additional qubits or complex error correction codes.

ZNE consists of three main steps:

1. Execute the quantum circuit at its normal noise level to obtain a baseline result.
2. Increase the noise level multiple times in a controlled manner (for example, by stretching quantum gates or adding artificial noise).
3. Use extrapolation techniques (such as linear or polynomial regression) to estimate the zero-noise result based on the noisy measurements.

By repeating the computation at different noise levels, a mathematical model can predict what the computation's outcome would be if no noise were present. The effectiveness of ZNE depends on how well the noise scaling is controlled and how accurately the extrapolation is performed.

The first step in implementing Zero-Noise Extrapolation (ZNE) is to execute the quantum circuit at its normal noise level without any modifications. Running the circuit on a noisy quantum processor provides a baseline measurement of the quantum computation under the influence of existing noise. This initial result serves as a reference for the extrapolation process that follows.

To generate additional data points for extrapolation, the noise level is artificially increased in a controlled manner. There are several ways to achieve this. Gate folding is a common approach where the effective noise is increased by repeating quantum gate operations in a way that does not change the outcome of the computation. For example, replacing a single X gate with three consecutive X gates (since XXX=X) artificially increases the noise level without altering the logical operation. Another method involves stretching pulse durations, which is applicable in quantum hardware that allows continuous control over gate operations. In this case, quantum gates are extended in duration, increasing their exposure to environmental noise. Some quantum platforms also support artificial noise injection, where users can add noise to specific gates or qubits to simulate higher error rates. By applying these techniques, multiple noisy results are obtained at different levels of noise amplification, creating a dataset for extrapolation.

Once the results from different noise levels are collected, an extrapolation function is applied to estimate the computation's outcome in a zero-noise scenario. Several mathematical techniques can be used for extrapolation, depending on the expected noise behaviour. Linear extrapolation is the simplest approach and is applied when the noise impact is assumed to be proportional to the error rate, fitting a straight-line model. Polynomial extrapolation is used when noise effects are more complex, requiring a higher-degree polynomial function to capture the variations. Exponential extrapolation is applied when the noise follows an exponential decay pattern, allowing for a more precise estimation of the zero-noise limit. By fitting these extrapolation models to the noisy data, the function is evaluated at zero noise, providing an improved estimate of the true quantum computation result that would have been obtained in an ideal noise-free environment.

Below is an example of how to implement Zero-Noise Extrapolation (ZNE) using Qiskit and the Mitiq library, which is widely used for quantum error mitigation.

```
from qiskit import QuantumCircuit, Aer, execute
from mitiq.zne import execute_with_zne, exponential_factory
from mitiq.interface.mitiq_qiskit import qiskit_executor

# Define a simple quantum circuit
qc = QuantumCircuit(1, 1)
qc.h(0)  # Hadamard gate
qc.measure(0, 0)

# Define the quantum simulator
simulator = Aer.get_backend("qasm_simulator")

# Define the executor function for Mitiq
def executor(qc):
    job = execute(qc, simulator, shots=1000)
    return job.result().get_counts().get('0', 0) / 1000  #
Probability of measuring |0>
```

```
# Apply Zero-Noise Extrapolation (ZNE) using an exponential fit
zne_result = execute_with_zne(qc, executor,
exponential_factory(scale_factors=[1, 2, 3]))

print("Zero-Noise Extrapolated Result:", zne_result)
```

This code runs a simple Hadamard gate circuit and applies ZNE using an exponential extrapolation model. The Mitiq library scales the noise using different scale factors (1, 2, 3) and then extrapolates the final zero-noise result.

Another effective QEM method is Probabilistic Error Cancellation (PEC), which estimates the noise present in a system and applies a probabilistic correction to cancel out errors. This approach is inspired by classical Monte Carlo techniques and is used in IBM's quantum cloud services to improve the fidelity of quantum computations.

Probabilistic Error Cancellation (PEC) is an advanced quantum error mitigation technique that corrects errors by probabilistically applying inverse operations to counteract noise. Unlike Quantum Error Correction (QEC), which requires encoding logical qubits using multiple physical qubits, PEC operates by reconstructing an error-free quantum process using a combination of noisy quantum operations and classical post-processing. This makes PEC particularly useful for Noisy Intermediate-Scale Quantum (NISQ) devices, where full fault-tolerant quantum computing is not yet feasible.

The core idea behind PEC is to model the noise affecting a quantum circuit and use statistical techniques to apply corrections probabilistically. By carefully choosing a set of noisy quantum operations that, when combined, approximate the inverse of the noise process, PEC allows quantum computers to recover results that are closer to the ideal error-free computation. However, because this technique relies on probabilistic sampling, it requires a large number of quantum circuit executions (shots) to reduce statistical uncertainty.

PEC follows a three-step process:

1. Characterize the Noise Model – The first step is to accurately characterize the noise affecting the quantum system. This is done by quantum process tomography, which determines the error channel affecting each gate in the circuit. The noise model is expressed in terms of a linear combination of noisy operations.

2. Decompose Noisy Operations into a Set of Corrective Operations – Once the noise is characterized, a linear inverse map is constructed using a set of quantum operations that, when probabilistically sampled, approximate the inverse of the noise process. These operations are chosen so that their weighted sum cancels out the errors introduced by noise.

3. Run the Circuit Multiple Times with Probabilistic Corrections – Instead of applying a deterministic correction, the inverse operations are sampled according to the probability distribution derived from the noise model. After running the circuit many times, a classical post-processing step statistically reconstructs the noise-free result.

Since the noise-inverted operations are applied stochastically, the expectation value of the measurement results converges to the ideal value as the number of circuit runs increases.

The first step in implementing Probabilistic Error Cancellation (PEC) is to characterize the noise model affecting the quantum hardware. This is achieved through quantum process tomography, which involves running calibration circuits specifically designed to measure how quantum gates deviate from their ideal behaviour. The noise is typically modelled as a completely positive trace-preserving (CPTP) map, which describes how an input quantum state transforms due to errors introduced by the quantum system.

Many quantum platforms, such as IBM Qiskit, Google Cirq, and Rigetti PyQuil, provide built-in tools for noise characterization. These include techniques like gate set tomography (GST) and randomized benchmarking (RB), which help estimate error rates and gate fidelity. Once the noise model is extracted, it is mathematically decomposed into a sum of quantum operations that can be used to cancel errors.

Once the noise model is well understood, the next step is to decompose the noisy operations into a linear combination of corrective operations. This process involves constructing an inverse map that represents the error-free quantum operation using a weighted sum of noisy operations. The ideal quantum gate $\mathcal{U}$ can be expressed as:

$$\mathcal{U} = \sum_i c_i \mathcal{N}_i$$

where $\mathcal{U}$ is the target error-free operation, $\mathcal{N}_i$ are the available noisy quantum operations, and c_i are the coefficients that determine the probability of selecting each operation.

Since the inverse map contains a weighted sum of quantum operations, the system randomly samples operations based on the probabilities $| c_i |$ and adjusts the final measurement outcomes accordingly. Over many circuit runs, the expected result converges to the ideal zero-noise computation, effectively cancelling the errors introduced by the quantum hardware.

The final step in PEC is to apply probabilistic corrections and perform measurement post-processing. The quantum circuit is executed multiple times, with each run applying different corrective operations based on the probabilistic distribution derived from the noise model. Since these probabilistic corrections introduce a bias, the final measurement results must be rescaled using classical post-processing to recover the true, error-free expectation values.

Because PEC relies on statistical sampling, it requires a significantly larger number of quantum executions to achieve high accuracy. However, it provides a powerful method for mitigating noise without requiring additional qubits for error correction. By leveraging a detailed understanding of the noise model and applying probabilistic techniques, PEC enables high-accuracy quantum computations even on noisy quantum hardware, making it one of the most promising error mitigation strategies for near-term quantum devices.

Implementing PEC in Qiskit Using Mitiq - The Mitiq library provides built-in support for Probabilistic Error Cancellation (PEC) in quantum circuits. Below is an example of how to apply PEC using Qiskit and Mitiq:

```
from qiskit import QuantumCircuit, Aer, execute
from mitiq.pec import execute_with_pec, complete_meas_cal,
apply_correction
from mitiq.interface.mitiq_qiskit import qiskit_executor

# Create a simple quantum circuit
qc = QuantumCircuit(1, 1)
qc.h(0)  # Apply Hadamard gate
qc.measure(0, 0)  # Measure the qubit

# Define the quantum simulator
simulator = Aer.get_backend("qasm_simulator")

# Define an executor function for Mitiq
def executor(qc):
    job = execute(qc, simulator, shots=1000)
    return job.result().get_counts().get('0', 0) / 1000  #
Probability of measuring |0>

# Apply Probabilistic Error Cancellation (PEC)
pec_result = execute_with_pec(qc, executor)

print("Probabilistic Error Cancellation Result:", pec_result)
```

This code constructs a quantum circuit, runs it with probabilistic error cancellation, and retrieves an improved result with reduced noise effects. The Mitiq library automatically determines the linear decomposition of the noisy operations, performs the stochastic sampling, and applies post-processing to correct the measured results.

Quantum Control Techniques

Precise quantum control techniques help minimize noise by optimizing how qubits are manipulated. Advanced calibration methods ensure that quantum gates operate with maximum accuracy while reducing crosstalk and unintended interactions between qubits.

One commonly used approach is pulse shaping, where the control pulses applied to qubits are fine-tuned to reduce gate errors. Instead of using simple square pulses, researchers use Gaussian, DRAG (Derivative Removal by Adiabatic Gate), and composite pulse sequences, which minimize unwanted energy leakage and phase errors.

Additionally, active qubit feedback control allows real-time monitoring and correction of quantum states. In this approach, machine learning algorithms analyse qubit behaviour and dynamically adjust gate operations to compensate for drift and noise, improving quantum computation accuracy over time.

Since quantum gates are implemented using electromagnetic pulses (such as microwave or laser pulses), imperfections in these pulses can lead to unwanted energy leakage, phase errors, and decoherence, reducing the accuracy of quantum operations. By optimizing the shape of these pulses, quantum systems can achieve higher gate fidelity, making quantum computations more reliable.

In superconducting qubits, trapped-ion qubits, and other hardware platforms, quantum gates are not applied instantaneously but rather as continuous control pulses. If these pulses are not properly shaped, they can cause over-rotation or under-rotation of qubits, increased susceptibility to noise, and crosstalk between qubits. Pulse shaping techniques help mitigate these errors by adjusting the amplitude, duration, and phase of the pulses in a way that reduces distortions.

A quantum gate is typically implemented by applying a control pulse that drives the evolution of a qubit's quantum state. The simplest approach is to use a square pulse, where the control field is turned on at a constant amplitude for a fixed duration. However, square pulses introduce high-frequency noise, which can lead to unwanted excitations and reduce gate fidelity.

Instead of using a simple square pulse, pulse shaping techniques apply smooth, optimized waveforms that reduce high-frequency components and minimize errors. Commonly used pulse shaping functions include:

- Gaussian Pulses: Smoothly varying pulses that minimize spectral leakage and reduce noise effects.
- DRAG (Derivative Removal by Adiabatic Gate) Pulses: Used in superconducting qubits to correct leakage into higher energy states by applying a derivative component to compensate for unwanted phase errors.

- Sinusoidal and Cosine Pulses: Reduce high-frequency components while maintaining efficient qubit control.
- Composite Pulses: Sequences of pulses that collectively correct systematic errors, such as BB1 and CORPSE pulses in quantum control.

By optimizing the pulse shape, quantum gates can be applied with higher accuracy, reducing the impact of hardware imperfections and increasing quantum coherence times.

The first step in implementing pulse shaping is to design an optimized waveform for quantum gate operations. In superconducting qubits, an X-gate (π rotation on the Bloch sphere) is typically implemented using a Gaussian-modulated microwave pulse instead of a simple square pulse. The use of Gaussian pulses helps minimize spectral leakage and unwanted excitations that degrade quantum gate fidelity.

A Gaussian pulse is mathematically described as:

$$A(t) = A_0 e^{-\frac{(t-t_0)^2}{2\sigma^2}}$$

where A_0 is the peak amplitude, t_0 is the centre of the pulse, and σ controls the pulse width. This smooth waveform reduces abrupt transitions that could introduce high-frequency noise and errors.

To further improve quantum gate performance, DRAG (Derivative Removal by Adiabatic Gate) correction is applied to counteract phase errors and leakage into higher energy states. The DRAG pulse includes an additional derivative term:

$$A_{DRAG}(t) = A(t) + \alpha \frac{dA}{dt}$$

where α is a tunable parameter that helps suppress phase errors. This correction is particularly useful in superconducting qubits, where excess energy can drive the qubit into unwanted higher states, leading to computation errors.

In IBM Qiskit, pulse shaping is implemented using the Qiskit Pulse module, which allows users to define custom pulse waveforms for qubit control. Below is an example of a Gaussian pulse applied to a qubit using Qiskit:

```
from qiskit import pulse
from qiskit.pulse.library import Gaussian
from qiskit.test.mock import FakeHanoi  # Example quantum
processor

backend = FakeHanoi()  # Load a mock superconducting qubit
backend
dt = backend.configuration().dt  # Get system time resolution
```

```
# Define Gaussian pulse parameters
duration = 128  # Pulse duration in units of dt
sigma = 16  # Standard deviation of Gaussian pulse
amp = 0.1  # Pulse amplitude

# Create a Gaussian pulse
gaussian_pulse = Gaussian(duration=duration, amp=amp,
sigma=sigma)

# Create a pulse schedule
schedule = pulse.Schedule()
with pulse.build(schedule) as sched:
    pulse.play(gaussian_pulse, pulse.DriveChannel(0))  # Apply
pulse to qubit 0

# Display the schedule
sched.draw()
```

This Qiskit program defines and applies a Gaussian pulse to a superconducting qubit, ensuring smooth transitions and reducing the likelihood of unwanted qubit transitions. The pulse schedule is built using Qiskit Pulse, specifying the duration, standard deviation, and amplitude to fine-tune the waveform.

After defining a pulse, fine-tuning its parameters is crucial to achieving high-fidelity quantum gate operations. Calibration experiments are performed by running the shaped pulse on quantum hardware and measuring its effect on qubit states. Adjustments to the pulse width, amplitude, and phase offset help minimize errors and optimize gate performance.

Advanced optimization techniques, including machine learning-based approaches, can further refine pulse parameters. Reinforcement learning or gradient-based algorithms analyse experimental results and dynamically adjust pulse parameters to maximize quantum gate fidelity. This ensures that the control pulses applied to qubits are precisely tailored to counteract noise and imperfections, leading to more reliable quantum computations.

Hardware-Based Noise Suppression

Many noise sources originate from the physical structure of quantum hardware, making material improvements and environmental control essential for reducing noise.

In superconducting qubits, noise is often caused by fluctuations in material properties, such as two-level system (TLS) defects in insulating layers. Researchers are experimenting with higher-purity superconducting materials and better fabrication techniques to reduce these defects and improve qubit coherence.

For trapped-ion and neutral atom qubits, laser and electromagnetic field noise can introduce errors. Improved laser stabilization and advanced ion-trapping techniques, such as cryogenic ion traps, help minimize these issues.

Shielding quantum processors from external electromagnetic interference, such as cosmic rays and temperature fluctuations, is also critical. Google, IBM, and other quantum research labs use ultra-low-temperature dilution refrigerators, operating at millikelvin temperatures, to extend qubit coherence times.

Fault-Tolerant Quantum Computing (FTQC)

The ultimate goal of noise mitigation is to achieve fault-tolerant quantum computing, where qubits can operate indefinitely without accumulating errors. Fault-tolerant architectures rely on high-threshold error correction codes, such as Surface Codes, to ensure quantum computations remain stable even with noisy qubits.

Leading quantum computing companies, such as Google, IBM, Microsoft, and Rigetti, are actively working on scalable fault-tolerant designs. Google's Sycamore processor demonstrated error correction scaling, showing that increasing qubit redundancy improved overall computational accuracy. IBM's quantum roadmap aims to implement fault-tolerant logic by integrating thousands of physical qubits into a stable computational unit.

Topological quantum computing, pursued by Microsoft, provides another potential pathway toward fault tolerance by using Majorana fermions to create naturally error-resistant qubits.

Threshold Theorem and Scalability

The Quantum Threshold Theorem is a significant finding in the field of quantum error correction, establishing that reliable fault-tolerant quantum computing is achievable if the physical error rate of qubits remains below a critical threshold. This enhances the understanding of quantum computing by positing that, despite the inherent noise in quantum systems, it is possible to maintain fidelity in computations through appropriate error correction mechanisms.

In classical computing, error correction often involves redundancies, such as encoding a single bit across multiple copies. However, the no-cloning theorem in quantum mechanics prohibits this straightforward approach. Instead, Quantum Error Correction (QEC) methods, such as Shor's Code and Surface Codes, are employed to protect quantum information by encoding it across multiple qubits. These codes allow for error detection and correction without directly measuring the quantum states involved, which would otherwise collapse them [310-312].

Mathematically, the fault-tolerant threshold is expressed as a critical error probability p_{th}. If the physical gate error rate per operation is below p_{th}, then logical qubits (protected by QEC) can be made arbitrarily reliable by increasing the number of physical qubits used for encoding. For many QEC codes, the threshold is around 10^{-2} to 10^{-3} per gate operation, meaning that quantum computers must achieve error rates below 1% per gate to be scalable [313, 314].

The implications of the Quantum Threshold Theorem extend to the scalability of quantum computers. Once hardware achieves an error rate below the critical threshold, it becomes feasible to add more qubits and utilize deeper layers of error correction, which enhances the reliability and complexity of quantum computations permissible within a reasonable timeframe [311, 312, 314]. This foundational theorem supports the belief in the potential for large-scale quantum computing applications, as it assures that errors can be managed effectively as long as certain technical conditions are met [310, 313].

Scalability refers to the ability of a quantum computer to increase in size while maintaining performance and reliability. A truly scalable quantum computer must be able to increase the number of qubits and operations while keeping error rates manageable. The Quantum Threshold Theorem provides the theoretical guarantee that scalability is possible, but practical challenges remain in implementing large-scale quantum error correction and maintaining coherence.

One of the biggest challenges to scalability is hardware fidelity. For quantum computers to scale, they must achieve gate fidelities above 99.9%, meaning that only one error occurs per thousand operations or less. This requires high-quality qubits, precise control over quantum gates, and minimal decoherence. Different quantum computing platforms—such as superconducting qubits, trapped-ion qubits, neutral atoms, and photonic qubits—are actively being developed to achieve these high fidelities.

Another key factor in scalability is error correction overhead. Fault-tolerant quantum computing requires thousands of physical qubits per logical qubit to effectively suppress errors. For example, in Surface Codes, a single logical qubit may require hundreds or even thousands of physical qubits. This means that a practical quantum computer capable of outperforming classical supercomputers (e.g., for breaking RSA encryption or simulating quantum chemistry) would require millions of physical qubits, far beyond today's capabilities.

To address scalability, researchers are developing hardware-efficient error correction methods that reduce qubit overhead, such as low-overhead QEC codes, hardware-level error suppression, and improved qubit connectivity. Emerging techniques like topological quantum computing using non-Abelian anyons may also provide a more scalable approach by naturally protecting quantum information.

Scalability also depends on quantum control architectures that can handle thousands or millions of qubits efficiently. Classical control systems must be able to manage qubit interactions, apply error correction, and synchronize quantum operations across large-scale systems. Approaches such as distributed quantum computing, cryogenic control electronics, and photonic interconnects are being explored to overcome these challenges.

Implementing the Quantum Threshold Theorem in Quantum Computing

1. Achieving Error Rates Below the Threshold

The Quantum Threshold Theorem states that fault-tolerant quantum computing is achievable if the physical qubit error rate remains below a critical threshold. To implement this theorem in practice, quantum computers must first reach error rates below the threshold (typically 10^{-2} to 10^{-3} per gate operation). This requires high-fidelity quantum hardware with precise control over quantum gates.

Different quantum hardware platforms—including superconducting qubits, trapped-ion qubits, neutral atoms, and photonic qubits—are actively being optimized to minimize gate errors, qubit decoherence, and environmental noise. Superconducting qubits, for example, are enhanced through high-coherence transmon qubits and improvements in microwave pulse shaping, while trapped-ion systems benefit from long-lived qubit states and high-fidelity laser-based quantum gates.

To improve qubit fidelity, quantum devices also use error suppression techniques such as dynamical decoupling, which applies periodic control pulses to reduce decoherence, and pulse shaping, which optimizes gate implementations to reduce gate errors.

2. Implementing Quantum Error Correction (QEC) for Fault Tolerance

Once error rates are below the threshold, Quantum Error Correction (QEC) must be implemented to protect quantum information. Unlike classical error correction, quantum systems cannot use simple redundancy due to the no-cloning theorem. Instead, quantum information is encoded across multiple physical qubits using error correction codes like Shor's Code and Surface Codes.

Shor's Code encodes one logical qubit into nine physical qubits, using a combination of bit-flip and phase-flip protections. While effective, this method requires a large overhead. More scalable is the Surface Code, which arranges qubits in a 2D lattice and uses stabilizer measurements to detect and correct errors.

To implement Surface Codes, physical qubits are divided into data qubits (which store quantum information) and ancilla qubits (which detect errors). Error detection is performed through syndrome measurements, which check the parity of qubit states without disturbing the quantum computation. If an error is detected, corrective operations (such as Pauli gates X and Z) are applied to restore the logical qubit state.

Practical implementations of Surface Codes require hardware with high connectivity, fast syndrome extraction, and low measurement noise. Leading quantum hardware providers, such as IBM, Google, and Rigetti, are developing hardware architectures optimized for Surface Code error correction, enabling fault-tolerant quantum computations.

3. Scaling Up Quantum Computers for Practical Applications

The Quantum Threshold Theorem ensures that once error rates are below the threshold, it becomes feasible to scale up quantum computers by adding more qubits and applying deeper layers of error correction. However, practical scalability requires overcoming several technical challenges.

One of the biggest challenges is hardware fidelity. To build a scalable quantum computer, gate fidelities must exceed 99.9%, meaning that only one error occurs per thousand operations or less. This demands high-quality fabrication techniques, precise quantum gate control, and improved qubit connectivity.

Another critical challenge is error correction overhead. A single logical qubit in a Surface Code typically requires hundreds or thousands of physical qubits for protection. This means that to build a practical quantum computer capable of tasks such as breaking RSA encryption or simulating quantum chemistry, millions of physical qubits are needed. To address this, researchers are developing low-overhead QEC codes, optimizing qubit layouts, and improving the efficiency of syndrome extraction circuits.

Scalability also depends on efficient quantum control architectures. Classical systems must manage thousands or millions of qubits, synchronize operations, and apply real-time error correction. Approaches such as distributed quantum computing, where multiple quantum processors are linked together, and cryogenic control electronics, which allow efficient control of superconducting qubits, are being explored. Additionally, photonically linked quantum processors offer a potential way to scale quantum computers by enabling communication between separate quantum chips.

Chapter 6

Hardware Approaches in Quantum Computing

Quantum computing is more than just an abstract concept; it requires physical systems that can implement quantum principles in a controlled and scalable manner. In previous chapters, we explored the theoretical foundations of quantum computing, including the fundamental principles of quantum mechanics, the role of qubits, quantum gates, and the power of quantum algorithms. While these concepts illustrate the extraordinary computational potential of quantum computers, realizing this potential in practice depends on overcoming significant hardware challenges.

Chapter 6 delves into the different hardware approaches used to build quantum computers, examining the technologies that enable the manipulation and control of qubits. Unlike classical computers, which rely on well-established silicon-based transistor technology, quantum computers require specialized architectures that can maintain and manipulate fragile quantum states. This chapter explores the leading qubit implementations, including superconducting qubits, which dominate the field with contributions from companies like IBM, Google, and Rigetti; trapped ion qubits, pioneered by IonQ and Honeywell; topological qubits, a promising but still experimental approach championed by Microsoft; and photonic quantum computing, which leverages light-based qubits for scalable and energy-efficient quantum processing.

A major challenge in quantum hardware is error correction and decoherence, as quantum states are highly sensitive to environmental noise. This chapter also addresses how different hardware platforms attempt to mitigate these issues and the trade-offs between various qubit implementations. Additionally, quantum annealers, such as those developed by D-Wave, offer a specialized approach to solving optimization problems, providing a distinct hardware model that differs from universal quantum computing architectures.

As quantum hardware continues to advance, researchers are working to scale up quantum processors, integrate quantum computing with classical computing infrastructure, and improve qubit coherence times. This chapter provides a comprehensive overview of where we stand today in quantum hardware development and what lies ahead in the race to build scalable, fault-tolerant quantum computers. Understanding these hardware approaches is crucial for appreciating the technical challenges that must be overcome to bring quantum computing into mainstream applications.

Superconducting Qubits (IBM, Google, Rigetti)

Superconducting qubits have emerged as a predominant technology in the development of quantum computers, integrated by major companies such as IBM, Google, and Rigetti. The backbone of these qubits is the Josephson junction, which allows for the manipulation and control of quantum states using microwave pulses. This control mechanism is particularly potent as superconducting qubits operate within a cryogenic environment, typically below 10 millikelvin, ensuring that quantum coherence is maintained long enough to facilitate meaningful computational tasks [83, 315].

The architecture of superconducting quantum processors utilizes superconducting circuits to encode quantum states. This design enables rapid gate operations and enhances scalability, making integration with classical control electronics feasible [316, 317]. Among the various types of superconducting qubits, the transmon qubit stands out, having been developed to reduce sensitivity to charge noise, thus improving performance. The transmon architecture aids in achieving higher coherence times, an attribute crucial for the realization of quantum computations [97, 318]. Recent advancements have led to the development of improved fabrication techniques for these qubits, which further bolster coherence and operational fidelity [319].

Despite the technological advances with superconducting qubits, challenges remain, particularly concerning quantum error correction (QEC). While qubit lifetimes and gate fidelity have significantly improved, achieving error correction for large-scale quantum processors is essential. Current strategies in error correction like the surface code show promise but require further enhancements to fully integrate quantum data within a fault-tolerant framework [315, 320, 321]. Notably, recent experimental setups have demonstrated techniques for detecting and correcting errors in qubits, showcasing the viability of superconducting systems in practical applications of quantum computing [320, 321].

IBM's Superconducting Qubit Technology

IBM is a pioneer in superconducting qubit technology and has been developing quantum computers for over two decades. IBM's Quantum Experience platform allows researchers and developers to run quantum circuits on real quantum hardware via the cloud.

IBM's superconducting qubits are based on transmon architecture, which reduces noise sensitivity by increasing the capacitor size of Josephson junctions. The company has developed high-fidelity two-qubit gates, achieving error rates below 1% per gate operation.

IBM uses a heavy-hexagonal qubit layout, which optimizes connectivity while minimizing crosstalk between qubits. The company has been steadily increasing the number of qubits in its processors, with Eagle (127 qubits), Osprey (433 qubits), and Condor (1,121 qubits) planned for future deployment. IBM is also developing Quantum Error Correction (QEC) techniques using Surface Codes to enable fault-tolerant quantum computing.

IBM's quantum software stack is built on Qiskit, an open-source framework that provides tools for quantum circuit design, simulation, and quantum hardware execution.

Key Achievements of IBM:

- Developed a 127-qubit quantum processor (Eagle) and a roadmap for 1,000+ qubit devices.
- Achieved two-qubit gate fidelities above 99.9%.
- Provides cloud-based quantum computing access through IBM Quantum Experience.
- Actively working on fault-tolerant quantum computing with Surface Code error correction.

Google's Superconducting Qubit Technology

Google has been at the forefront of quantum computing research, particularly in demonstrating quantum supremacy. Google's quantum processors use superconducting transmon qubits, similar to IBM, but focus on high-connectivity qubit architectures and improved error correction techniques.

Google's biggest achievement came in 2019, when its Sycamore processor (53 qubits) demonstrated quantum supremacy by performing a computational task in 200 seconds, which would take a classical supercomputer 10,000 years. The Sycamore processor uses high-fidelity single-qubit and two-qubit gates, achieving error rates below 0.1% for single-qubit gates and ~0.3% for two-qubit gates.

Google is now working on Quantum Error Correction by implementing logical qubits using repetition codes and Surface Codes. In 2023, Google demonstrated error suppression in logical qubits, showing that adding more physical qubits reduces logical error rates. This is a critical milestone toward fault-tolerant quantum computing.

Google's quantum computing research is supported by Cirq, an open-source quantum computing framework optimized for running quantum circuits on Google's hardware.

Key Achievements of Google:

- Demonstrated quantum supremacy with the 53-qubit Sycamore processor.
- Developed high-fidelity quantum gates with error rates below 0.1% for single-qubit gates.
- Achieved error suppression in logical qubits, advancing toward fault-tolerant quantum computing.
- Leading research in quantum error correction and scalable quantum architectures.

Rigetti Computing's Superconducting Qubit Technology

Rigetti Computing is a quantum-first company focused on developing superconducting qubit processors and hybrid quantum-classical computing platforms. Unlike IBM and Google, which are large tech giants, Rigetti is a startup specializing in quantum computing.

Rigetti's superconducting qubits are based on transmon technology, but the company focuses on modular architectures to make quantum computing more scalable. Rigetti's Aspen series processors range from 8-qubit to 80-qubit devices and are accessible via Rigetti's Quantum Cloud Services (QCS).

Rigetti differentiates itself by working on hybrid quantum-classical computing, where quantum processors are tightly integrated with classical systems to perform complex calculations. Rigetti is also exploring low-overhead quantum error correction methods to reduce the number of physical qubits required for fault tolerance.

The company's quantum programming environment, PyQuil, is an open-source framework that integrates with the Forest quantum computing platform, enabling researchers to design and execute quantum algorithms efficiently.

Key Achievements of Rigetti:

- Developed the Aspen series quantum processors with up to 80 qubits.
- Focuses on hybrid quantum-classical computing for practical applications.
- Provides cloud-based quantum access through Rigetti Quantum Cloud Services (QCS).

- Exploring modular superconducting qubit architectures for improved scalability.

Superconducting Quantum Processors

Superconducting quantum processors are among the most advanced quantum computing platforms, utilizing superconducting circuits to create and manipulate quantum bits (qubits). These processors are built using Josephson junctions, which enable quantum behaviour at macroscopic scales and allow precise control of qubit states.

Superconducting qubits function at millikelvin temperatures (near absolute zero), where superconducting materials exhibit zero electrical resistance. This extreme cooling environment is necessary to maintain coherence, reduce thermal noise, and enable long-lived quantum states suitable for computation.

The operation of superconducting quantum processors involves several key components:

1. Superconducting Qubit Design (Transmon Qubits)
2. Quantum Gates and Control Mechanisms
3. Qubit Readout and Measurement
4. Quantum Error Correction and Scalability
5. Cryogenic Environment and Quantum Coherence

1. Superconducting Qubit Design (Transmon Qubits)

Superconducting qubits are fabricated using Josephson junctions, which consist of a thin insulating layer sandwiched between two superconducting electrodes. Unlike classical superconducting circuits, which follow standard electronic principles, Josephson junctions introduce nonlinearity into the circuit, allowing qubits to form well-defined quantum states.

The most commonly used type of superconducting qubit is the transmon qubit, an improved version of earlier charge qubits. Transmons reduce sensitivity to charge noise by adding a large capacitor in parallel with the Josephson junction, effectively lowering the energy-level spacing fluctuations caused by environmental disturbances.

A transmon qubit behaves as a nonlinear quantum oscillator, where the two lowest energy levels ($|0\rangle$ and $|1\rangle$ states) form a quantum bit. Unlike a classical oscillator, the anharmonicity introduced by the Josephson junction prevents the system from accidentally transitioning to higher energy states ($|2\rangle$, $|3\rangle$, etc.), which would degrade computation accuracy.

Mathematically, the Hamiltonian of a superconducting transmon qubit is:

$$H = 4E_C(n - n_g)^2 - E_J \cos(\varphi)$$

where:

- E_C is the charging energy (related to the capacitor size),
- E_J is the Josephson energy (determined by the junction properties),
- n is the Cooper pair number (discrete charge on the island),
- n_g is the gate charge,
- φ is the phase difference across the junction.

By tuning E_C and E_J, a stable qubit can be created with long coherence times (~100 μs) and high gate fidelities (>99.9%).

2. Quantum Gates and Control Mechanisms

Superconducting qubits are manipulated using microwave pulses to induce quantum gate operations. These pulses interact with the qubit's resonant frequency, driving transitions between the qubit's states |0⟩ and |1⟩.

Single-Qubit Gates

Single-qubit gates are implemented using resonant microwave pulses applied to a dedicated drive line. The most common single-qubit gates are:

- X Gate (π rotation on the Bloch sphere around the X-axis)
    - Implemented using a microwave pulse resonant with the qubit's frequency.
- Y Gate (π rotation around Y-axis)
    - Similar to the X gate but phase-shifted.
- Z Gate (Phase shift by θ without energy exchange)
    - Achieved virtually via frame rotation in the control software, without applying a physical pulse.

A general Rabi oscillation drive is described by:

$$H = \frac{\hbar}{2}\Omega(e^{-i\omega t}|0\rangle\langle 1| + e^{i\omega t}|1\rangle\langle 0|)$$

where Ω is the Rabi frequency, which controls the qubit rotation speed.

Two-Qubit Gates (Entanglement)

Two-qubit gates are necessary for universal quantum computing. The most commonly used entangling gate is the Controlled-Z (CZ) gate, which creates entanglement between two qubits by exploiting coupling interactions via a shared resonator or a tunable coupler.

The Hamiltonian for two coupled qubits with a coupling strength g is:

$$H = \hbar\omega_1 a_1^\dagger a_1 + \hbar\omega_2 a_2^\dagger a_2 + \hbar g(a_1^\dagger a_2 + a_1 a_2^\dagger)$$

where:

- ω_1, ω_2 are the qubit resonance frequencies,
- g is the coupling strength,
- $a^\dagger$ and a are creation and annihilation operators.

By tuning the interaction strength g, a CZ gate or an iSWAP gate can be performed, allowing controlled quantum entanglement.

3. Qubit Readout and Measurement

Qubit measurement is performed using dispersive readout via a superconducting resonator coupled to each qubit. A microwave probe signal is sent into the resonator, and the qubit state alters the resonator's frequency, shifting the reflected signal phase.

The Hamiltonian of a qubit-resonator system is given by the Jaynes-Cummings model in the dispersive limit:

$$H = \hbar\omega_r a^\dagger a + \frac{\hbar\omega_q}{2}\sigma_z + \hbar g(\sigma_+ a + \sigma_- a^\dagger)$$

where:

- ω_r is the resonator frequency,
- ω_q is the qubit frequency,
- g is the qubit-resonator coupling strength,
- σ_+, σ_- are Pauli operators.

By measuring the phase shift in the reflected microwave signal, the qubit state is inferred without directly collapsing the wavefunction, preserving coherence in multi-qubit systems.

4. Quantum Error Correction and Scalability

Superconducting quantum processors are actively developing Quantum Error Correction (QEC) using Surface Codes. In these implementations:

- Logical qubits are encoded using multiple physical qubits.
- Ancilla qubits perform syndrome measurements to detect and correct errors.
- Repeated measurements and feedback stabilize logical states over long computations.

IBM, Google, and Rigetti are integrating QEC into their quantum chips, with Google demonstrating error suppression using logical qubits in 2023.

5. Cryogenic Environment and Quantum Coherence

Superconducting qubits must operate at millikelvin temperatures using dilution refrigerators to prevent thermal noise from destroying quantum coherence. The cooling process involves:

- Liquid helium-based precooling to ~4K.
- Helium-3/Helium-4 dilution to 10-20 millikelvin.
- Cryogenic shielding to reduce electromagnetic noise and blackbody radiation.

This extreme cooling allows superconducting quantum states to persist long enough (~100 μs) for computations, enabling high-fidelity quantum gates.

Hardware and Environment Requirements

Superconducting quantum processors require a highly specialized hardware infrastructure and extreme environmental conditions to function reliably. Unlike classical processors, which operate at room temperature with conventional electrical signals, superconducting quantum processors must be cooled to millikelvin temperatures, shielded from electromagnetic interference, and precisely controlled using microwave signals. The key requirements for these quantum systems include:

1. Superconducting Qubit Fabrication and Design
2. Cryogenic Cooling and Dilution Refrigerators
3. Microwave Control and Signal Processing
4. Qubit Readout and Measurement Systems

5. Electromagnetic and Mechanical Shielding
6. Classical Control Electronics and Error Correction Infrastructure

Superconducting Qubit Fabrication and Design: Superconducting qubits are built using Josephson junctions, which consist of a thin insulating barrier sandwiched between two superconducting electrodes. These junctions are embedded in superconducting circuits, allowing quantum states to be created and manipulated. The most commonly used qubit design is the transmon qubit, which is optimized for longer coherence times and reduced charge noise.

The materials and fabrication techniques used in superconducting quantum processors must meet strict requirements:

- Superconducting Materials: Typically niobium (Nb) or aluminium (Al) thin films are used because they exhibit superconductivity at cryogenic temperatures.
- Substrate and Dielectric Selection: Sapphire, silicon, or silicon-on-insulator (SOI) substrates are chosen to minimize dielectric loss.
- Josephson Junction Fabrication: Requires electron beam lithography (EBL) and double-angle evaporation to precisely define the junction thickness (~1-2 nm).
- Coherence Time Optimization: Surface treatments and low-loss dielectric materials help improve coherence times (~100 μs or more).

Each qubit is coupled to a microwave resonator, which allows external control signals to manipulate and read out the qubit states. On-chip capacitors and tunable couplers are also integrated to enable precise qubit frequency tuning.

Cryogenic Cooling and Dilution Refrigerators: Superconducting qubits require ultra-low temperatures (~10–20 millikelvin) to function correctly. At these temperatures, superconducting materials exhibit zero electrical resistance, and thermal noise is minimized, preserving quantum coherence.

To achieve such extreme cooling, superconducting quantum processors are housed inside a dilution refrigerator, a cryogenic system that uses a mixture of Helium-3 and Helium-4 isotopes to reach millikelvin temperatures. The cooling process involves multiple temperature stages:

Table 10: Cooling temperatures and temperature stages.

Cooling Stage	Temperature Range	Purpose
Room Temperature	300 K (ambient)	Standard electronics & signal processing
Liquid Nitrogen Stage	77 K	Initial cooling to reduce thermal load
Liquid Helium Stage	4 K	Pre-cooling before dilution
Cold Plate	~800 mK	Further temperature reduction
Mixing Chamber	~10-20 mK	Final stage for quantum coherence

Key requirements for cryogenic cooling:

- Vibration isolation to prevent mechanical noise from disturbing qubit states.
- Low-noise coaxial cables and thermal anchors to minimize heat dissipation.
- Thermal shielding to block blackbody radiation from interfering with qubits.

The dilution refrigerator provides a thermally stable environment, allowing qubits to maintain coherence for long enough to execute quantum operations (~100 μs or more).

Microwave Control and Signal Processing: Superconducting qubits are manipulated using microwave pulses that drive quantum gates. Each qubit has a unique resonance frequency (~5–7 GHz), and control pulses must be precisely shaped to execute operations such as X, Y, Z rotations and entangling gates (CZ, iSWAP).

To generate and control the signals used in superconducting quantum processors, specialized hardware is required. Arbitrary waveform generators (AWGs) are used to shape microwave pulses, allowing precise control over qubit state transitions. IQ mixers enable phase and amplitude modulation, ensuring that the microwave signals can be fine-tuned to execute quantum gate operations accurately. Low-noise amplifiers (LNAs) are essential for boosting weak readout signals, improving the reliability of qubit measurements. Additionally, cryogenic circulators and attenuators are incorporated to prevent backscattering noise, which could interfere with qubit coherence and overall system stability.

The microwave control electronics must maintain phase stability to prevent unwanted quantum state drift, which could lead to computational errors. High time resolution, typically on the order of nanoseconds, is required for executing fast and precise gate operations. The system must also ensure low power dissipation to minimize thermal disturbances that could disrupt the superconducting state and reduce coherence times.

Each qubit is connected to a dedicated drive line for microwave signals, which is responsible for executing quantum gate operations. In addition, a flux bias line is used for dynamically tuning the qubit frequency, allowing control over qubit-qubit interactions and enabling entangling operations. These signals must be precisely synchronized to ensure reliable quantum operations, requiring sophisticated control algorithms and low-latency hardware to maintain system performance at scale.

Qubit Readout and Measurement Systems: Qubit states are measured using dispersive readout, where each qubit is coupled to a superconducting resonator. The readout process involves:

1. Sending a probe microwave signal into the resonator.
2. Detecting the phase shift in the reflected signal, which depends on the qubit state ($|0\rangle$ or $|1\rangle$).
3. Amplifying the signal using Josephson parametric amplifiers (JPAs) to avoid measurement errors.

The resonator's frequency is designed so that the qubit state slightly shifts the resonator's response, allowing non-destructive measurement.

The key hardware components for qubit readout in superconducting quantum processors include high-Q superconducting resonators, low-noise quantum amplifiers, and fast analogue-to-digital converters (ADCs). High-Q superconducting resonators, typically operating in the 5–7 GHz range, are used to couple with qubits and enable dispersive readout. These resonators shift their frequency based on the qubit state, allowing for non-destructive measurement while preserving quantum coherence.

To amplify the weak qubit signals without introducing excessive noise, low-noise quantum amplifiers such as Josephson Parametric Amplifiers (JPAs) and Traveling Wave Parametric Amplifiers (TWPAs) are employed. These amplifiers operate at cryogenic temperatures and provide high-gain, low-noise amplification, ensuring that qubit readout is accurate and reliable.

Fast analogue-to-digital converters (ADCs) are essential for real-time signal processing. These components capture the amplified readout signals and convert them into digital data for further analysis and quantum state determination. The speed and precision of ADCs directly impact the ability to perform high-fidelity measurements and execute quantum algorithms efficiently.

Efficient qubit readout is crucial for implementing error detection and quantum error correction (QEC) in large-scale quantum processors. By continuously monitoring qubit states, the system can identify and correct errors before they propagate, significantly improving the stability and scalability of quantum computations.

Electromagnetic and Mechanical Shielding: Superconducting qubits are extremely sensitive to electromagnetic interference (EMI), stray magnetic fields, and mechanical vibrations. To maintain coherence, processors are housed in a multi-layer shielding system, including:

- Cryogenic magnetic shielding using mu-metal enclosures to block external magnetic fields.
- Faraday cages around the entire cryogenic system to shield against RF interference.
- Vibration isolation platforms to reduce mechanical noise that could affect qubit frequencies.

These shielding mechanisms are crucial for preventing fluctuations in qubit energy levels and ensuring high-fidelity quantum operations.

Classical Control Electronics and Error Correction Infrastructure: Scaling up superconducting quantum processors requires large-scale classical control infrastructure to manage qubit operations. The control system includes:

- FPGAs (Field-Programmable Gate Arrays) for real-time qubit control.
- Cryogenic control chips to reduce wiring complexity and improve scalability.
- Error correction processors to implement Quantum Error Correction (QEC) codes.

A major challenge in scaling quantum computers is managing thousands of qubits with minimal latency. Innovations such as distributed quantum computing architectures and photonic interconnects are being explored to overcome this limitation.

Trapped Ion Qubits (IonQ, Honeywell)

Trapped-ion qubits represent one of the most promising technologies in the realm of quantum computing. Utilizing individual ions (charged atoms) as qubits, these systems leverage electromagnetic fields to confine and suspend the ions in a vacuum, facilitating precise control through laser manipulation to execute quantum gate operations. The unique characteristics of trapped ions, such as their natural homogeneity and stability due to atomic identicality, provide significant advantages over other qubit technologies, such as superconducting qubits.

Trapped ions are distinguished by their long coherence times, which contribute to the stability and reliability of quantum operations. Specifically, trapped-ion quantum systems showcase exceptional fidelity in single and multi-qubit operations, capitalizing on the Coulomb interaction that allows for high-connectivity between qubits, a critical aspect of scalable quantum computing architectures [322, 323]. Furthermore, the manipulation of these qubits is predominantly achieved through laser-based operations, which are characterized by high precision and robustness against noise, thus ensuring the integrity of quantum processes [201, 324].

In comparison to superconducting qubits, which depend on Josephson junctions and microwave controls, trapped-ion qubits maintain a more straightforward and inherently stable structure. This is particularly evidenced by their ability to perform entangling gates with minimal error, as demonstrated in studies where high-fidelity gates were achieved [325, 326]. The architectural developments in trapped-ion systems, such as modular configurations allowing for scalable ion-trap quantum computations, are valuable for advancing the realization of practical quantum computing [327].

Among the frontrunners in the field of trapped-ion quantum computing are IonQ and Honeywell Quantum Solutions, both recognized for their contributions to developing viable quantum computing frameworks based on trapped-ion technology. Their efforts are further evidenced by articles outlining the effective application of trapped-ion systems in quantum information processing and their potential to implement complex algorithms with high fidelity [323, 324, 326].

Trapped-ion quantum processors operate by isolating and trapping atomic ions using electromagnetic fields in a vacuum chamber. The most commonly used ions for quantum computing include Ytterbium (Yb^+) and Barium (Ba^+), which have internal energy levels suitable for encoding qubits.

Each qubit is represented by two stable electronic states of the ion. Quantum gates are performed using precisely tuned laser pulses, which control the qubit states and generate entanglement between ions. Qubit connectivity is achieved through phonon-mediated interactions, where the ions' collective vibrational modes serve as a quantum bus for performing two-qubit gates.

Measurement is done via state-dependent fluorescence, where a laser selectively excites one of the ion's states, causing it to emit light if it is in a specific quantum state. This allows high-fidelity readout of qubit states.

Key advantages of trapped-ion qubits include:

- Long coherence times (up to minutes, compared to microseconds in superconducting qubits).
- All-to-all connectivity, meaning any qubit can be entangled with any other qubit without requiring nearest-neighbour interactions.
- High-fidelity gate operations, with single-qubit gate fidelities exceeding 99.9% and two-qubit gate fidelities above 99%.

However, challenges remain, including slow gate speeds (~1 ms per operation) and the complexity of scaling up trapped-ion architectures due to the need for precise laser control and stable trapping conditions.

IonQ's Trapped-Ion Quantum Processors

IonQ is a leading company in trapped-ion quantum computing, originally a spin-off from the University of Maryland and Duke University. IonQ has built quantum processors based on Ytterbium ions (Yb^+), which are trapped and manipulated using radiofrequency (RF) Paul traps and laser-based quantum gates.

IonQ's trapped-ion quantum computers feature:

- Fully connected qubits, allowing flexible quantum circuit designs.
- High-fidelity operations, with single-qubit fidelities of 99.97% and two-qubit fidelities of ~99.3%.
- Scalable quantum architecture, with continuous improvements in the number of trapped ions and laser gate control.

One of IonQ's breakthroughs is the use of reconfigurable ion chains, where qubits can be dynamically rearranged to optimize computation. IonQ also develops hybrid quantum-classical algorithms to integrate its quantum processors with classical computing resources.

IonQ has commercialized its quantum computing services through cloud platforms like Amazon Braket, Microsoft Azure Quantum, and Google Cloud, allowing external researchers and companies to run quantum experiments.

Honeywell's Trapped-Ion Quantum Processors (Now Quantinuum)

Honeywell Quantum Solutions, now part of Quantinuum, has developed high-performance trapped-ion quantum computers with a strong focus on error correction and industrial applications. Honeywell's trapped-ion systems are based on Ytterbium (Yb^+) and Barium (Ba^+) ions, utilizing a unique QCCD (Quantum Charge-Coupled Device) architecture to shuttle ions between different processing zones for optimized computation.

Honeywell's key innovations in trapped-ion quantum computing include:

- QCCD Architecture, which allows ions to be physically moved between different zones, enabling parallel quantum operations.
- High-fidelity gates, with two-qubit fidelities exceeding 99.5%.

- Integrated mid-circuit measurement, which enables real-time quantum error correction and advanced quantum algorithms.

Quantinuum's H-Series quantum computers, such as the H1 and H2 models, leverage these technologies to achieve some of the highest-performing quantum processors available. The company is heavily focused on developing fault-tolerant quantum computing using logical qubits and quantum error correction (QEC).

Honeywell/Quantinuum has integrated its trapped-ion quantum systems into various cloud services and enterprise applications, targeting industries such as pharmaceuticals, finance, and materials science.

Table 11: Comparison: IonQ vs. Honeywell (Quantinuum).

Feature	IonQ	Honeywell (Quantinuum)
Qubit Type	Ytterbium (Yb^+) ions	Ytterbium (Yb^+) and Barium (Ba^+) ions
Qubit Connectivity	Fully connected (all-to-all)	Fully connected (via ion shuttling)
Quantum Gate Control	Laser-based single and two-qubit gates	QCCD-based gate operations with ion transport
Fidelity	99.97% (single-qubit), ~99.3% (two-qubit)	>99.5% (two-qubit gates)
Error Correction Focus	Developing logical qubits	Strong focus on QEC with mid-circuit measurement
Scalability Approach	Reconfigurable ion chains	QCCD architecture with ion transport
Cloud Access	Available on AWS Braket, Microsoft Azure Quantum, and Google Cloud	Available via cloud access for enterprise applications
Industry Focus	Hybrid quantum-classical applications	Error-corrected quantum computing for industry

Hardware and Environment Requirements

Trapped-ion quantum computing processors require highly specialized hardware and controlled environmental conditions to operate effectively. Unlike superconducting qubits, which use electrical circuits, trapped-ion processors rely on individual atomic ions that are electromagnetically trapped in a vacuum and manipulated using laser-based control systems. The hardware setup must maintain the stability of ion qubits, allow precise quantum gate operations, and minimize decoherence from environmental disturbances.

The essential hardware and environmental requirements for trapped-ion quantum computing include:

1. Ion Trapping System and Vacuum Chamber
2. Cryogenic or Ultra-High Vacuum (UHV) Environment
3. Laser Systems for Qubit Manipulation
4. Optical and Electrostatic Ion Control
5. Qubit Readout and Measurement System
6. Classical Control Electronics and Error Correction

Ion Trapping System and Vacuum Chamber: Trapped-ion qubits are individual charged atoms that are confined within a vacuum chamber using electromagnetic fields. These qubits are held in place using specialized ion traps that ensure their stability and isolation from environmental disturbances. There are two main types of ion traps used in trapped-ion quantum processors. Radiofrequency (RF) Paul Traps use oscillating electric fields to confine ions in a linear chain, allowing for precise manipulation of qubits using laser pulses. Penning Traps, on the other hand, use a combination of static magnetic and electric fields to create a stable trapping environment for the ions.

The trapping system must maintain high stability to ensure that ions remain suspended without experiencing unwanted motion, which could introduce errors in quantum gate operations. To achieve this, the system incorporates a carefully designed set of electrode structures that generate the required electric fields to hold ions in place. Additionally, electro-optical components provide fine control over individual ions, ensuring that each qubit can be addressed independently with high precision. Another critical element of the system is the ion-loading mechanism, where a neutral atom is ionized using a laser to create an individual qubit. This process ensures that the qubit can be initialized into a well-defined quantum state, ready for computation.

Trapped-ion processors require sub-micrometre precision in ion positioning, as even the smallest vibrations or fluctuations in electric fields can disrupt qubit coherence and interfere with entanglement. Maintaining this level of stability is essential for executing high-fidelity quantum operations and scaling up trapped-ion quantum computing systems. The ability to precisely manipulate and control individual ions in these traps is one of the key advantages of trapped-ion technology, contributing to its long coherence times and high gate fidelities.

Cryogenic or Ultra-High Vacuum (UHV) Environment: To preserve ion coherence and minimize collisions with residual gas molecules, trapped-ion processors must operate in an ultra-high vacuum (UHV) environment, with pressures as low as 10^{-11} torr. Maintaining such low pressure prevents unwanted interactions between the trapped ions and stray gas molecules, which could introduce quantum errors by causing decoherence or state transitions. The vacuum chamber plays a crucial role in ensuring the long coherence times and high-fidelity quantum operations that make trapped-ion systems effective for quantum computing.

Unlike superconducting qubits, which require millikelvin temperatures inside dilution refrigerators, trapped-ion systems do not always require extreme cryogenic cooling. These systems can function at room temperature in a well-maintained UHV environment. However, in some implementations, cryogenic cooling is used to enhance performance by reducing thermal motion of the ions, which improves qubit stability and reduces noise. Cooling also minimizes blackbody radiation, which can lead to spontaneous state transitions in ion qubits, affecting their coherence. Additionally, cryogenic conditions help improve laser stability and optical coherence, ensuring precise quantum gate operations.

Different trapped-ion quantum computing platforms employ varying approaches to temperature control. Companies like Quantinuum (formerly Honeywell Quantum Solutions) use cryogenic cooling to enhance qubit performance and increase gate fidelity. In contrast, IonQ operates its trapped-ion systems at room temperature while maintaining UHV conditions to achieve similar levels of stability and precision. The choice between room-temperature and cryogenic operation depends on factors such as scalability, qubit coherence requirements, and the complexity of maintaining an ultra-cold environment.

Laser Systems for Qubit Manipulation: Quantum gates in trapped-ion processors are executed using laser pulses that drive transitions between the ion's internal electronic states. These lasers must be highly stable, precisely tuned, and phase-coherent to ensure accurate quantum operations. Even the slightest fluctuation in laser intensity or phase can introduce errors, affecting gate fidelity and overall computational reliability. The laser system used in trapped-ion quantum computing consists of several specialized components, each serving a distinct function in qubit manipulation, cooling, and measurement.

Qubit initialization lasers are used to prepare the ion in a well-defined initial quantum state, either $|0\rangle$ or $|1\rangle$, before computation begins. These initialization steps ensure consistency in quantum operations and enable accurate processing of quantum algorithms. Quantum gate lasers are responsible for executing both single-qubit and two-qubit gate operations, typically operating in the ultraviolet (UV) or visible light spectrum. These laser pulses precisely control the quantum states of individual ions, allowing the implementation of logic gates necessary for quantum computations.

To maintain qubit stability and minimize errors due to ion movement, cooling lasers are employed. Techniques such as Doppler cooling and resolved-sideband cooling help reduce ion motion, preventing unwanted decoherence and improving overall gate fidelity. Finally, qubit readout lasers are used to measure the ion's quantum state through state-dependent fluorescence detection, where the ion either emits photons or remains dark depending on its final state.

Trapped-ion quantum processors require laser pulses with nanosecond-scale precision to ensure high-fidelity gate operations. Any instability in laser intensity, phase, or timing can lead to computational errors, reducing the accuracy of quantum algorithms. Advanced optical systems, including acousto-optic modulators (AOMs) and electro-optic modulators (EOMs), are used to fine-tune laser pulses in real time, ensuring optimal qubit control and minimizing gate errors. The precision and stability of these laser systems are crucial for maintaining the exceptional coherence times and connectivity that make trapped-ion quantum computing a leading platform for scalable quantum processors.

Optical and Electrostatic Ion Control: Since ions are manipulated using laser pulses, a high-precision optical system is required to focus and direct laser beams onto individual ions with extreme accuracy. The ability to control these beams with nanometre-scale precision ensures that quantum gates are executed with minimal errors and that each qubit can be independently addressed without unintended interference.

The optical system includes a network of high-resolution lenses and mirrors that precisely direct laser beams onto targeted ions. These components must be carefully aligned to ensure that the laser pulses interact with the correct ion in a chain or array. To control the timing and characteristics of these laser pulses, acousto-optic modulators (AOMs) and electro-optic modulators (EOMs) are used. These modulators allow for rapid changes in laser pulse frequency, amplitude, and phase, enabling fast and precise quantum gate operations. Additionally, beam steering systems ensure that individual ions in an array can be addressed selectively, allowing complex multi-qubit interactions to be performed with high fidelity.

For large-scale quantum computing, researchers are developing microfabricated ion traps that integrate optical control elements directly into the trap architecture. By embedding optical components such as waveguides and beam splitters within the ion-trap chip itself, these systems reduce the reliance on bulky free-space optics. This approach significantly improves scalability by minimizing alignment issues and reducing optical losses, paving the way for larger, more stable trapped-ion quantum processors capable of supporting hundreds or even thousands of qubits.

Qubit Readout and Measurement System: Trapped-ion qubit states are measured using state-dependent fluorescence detection, a highly accurate method that allows for precise quantum state readout. When a laser is applied to an ion, the ion will either fluoresce, emitting photons if it is in one quantum state, or remain dark if it is in another. This fluorescence pattern provides a direct way to determine the qubit's final state after a computation, enabling measurement with exceptional reliability.

To capture and analyse the emitted photons, high-efficiency photodetectors such as photomultiplier tubes (PMTs), charge-coupled device (CCD) cameras, or avalanche photodiodes are used. These detectors are capable of counting individual photons with extreme sensitivity, ensuring accurate qubit state determination. High-numerical-aperture (NA) lenses are employed to maximize the collection of emitted photons, reducing signal loss and improving measurement precision.

The detected photon signals are processed using real-time signal processing electronics, which perform high-speed analysis to distinguish between bright and dark states. This rapid processing allows for near-instantaneous quantum state readout, which is critical for implementing feedback-based quantum error correction protocols.

Measurement fidelity in trapped-ion systems can exceed 99.9%, making them one of the most accurate quantum measurement techniques available. This high precision ensures that quantum computations remain reliable and that errors in quantum algorithms are minimized, contributing to the overall stability and scalability of trapped-ion quantum processors.

Classical Control Electronics and Error Correction: Trapped-ion quantum computers require a sophisticated classical control system to manage the precise execution of laser pulses, ion trapping fields, and qubit measurement operations. Unlike classical computers, where operations are performed using electrical signals, trapped-ion systems rely on laser-driven interactions that must be carefully orchestrated. The control electronics must generate precisely timed laser pulses to execute quantum gates with nanosecond accuracy. These pulses must be synchronized across multiple laser beams to ensure that qubits in an array are manipulated in a coordinated manner. Additionally, the system must be capable of real-time analysis of fluorescence signals to determine qubit states with high accuracy during measurement.

For fault-tolerant quantum computing, trapped-ion processors must incorporate quantum error correction (QEC) to protect quantum information from noise and decoherence. This requires the ability to detect errors using ancilla qubits and stabilizer measurements, allowing the system to identify deviations in quantum states without disturbing computational qubits. Once errors are detected, quantum feedback control is applied dynamically to correct them before they accumulate and compromise the computation. Scaling up error correction requires implementing advanced QEC codes, such as Surface Codes and Bacon-Shor codes, which distribute quantum information across multiple physical qubits to enhance fault tolerance.

Leading trapped-ion quantum computing platforms, including those developed by IonQ and Quantinuum, are integrating real-time quantum error correction techniques to improve the stability of quantum operations and extend computation times. These advancements bring trapped-ion processors closer to achieving scalable fault-tolerant quantum computing, where logical qubits can perform computations indefinitely without succumbing to environmental noise or gate errors.

Table 12: Comparison: Trapped-Ion vs. Superconducting Qubit Hardware Requirements.

Feature	Trapped-Ion Qubits	Superconducting Qubits
Qubit Type	Individual atomic ions (e.g., Yb^+, Ba^+)	Josephson junctions (transmon qubits)
Environment	Ultra-high vacuum (~10^{-11} torr)	Cryogenic (~10 mK, dilution fridge)
Gate Control	Laser pulses (visible/UV light)	Microwave pulses (~5 GHz)
Qubit Readout	Fluorescence detection with photodetectors	Dispersive readout via superconducting resonators
Connectivity	All-to-all connectivity	Nearest-neighbour connectivity
Error Correction	Surface codes, QCCD-based logical qubits	Surface codes, superconducting logical qubits
Scalability Challenges	Precision laser control, ion trapping stability	Qubit coherence time, fabrication scalability

Challenges and Future Directions of Trapped-Ion Qubits

While trapped-ion qubits offer exceptional coherence times and high-fidelity operations, there are significant challenges to scaling up these systems for large-scale quantum computing. One of the biggest limitations is gate speed, as laser-controlled quantum gates are significantly slower (millisecond-scale) compared to superconducting qubits (nanosecond-scale).

To overcome these challenges, companies like IonQ and Quantinuum are focusing on:

- Parallel gate execution to increase processing speed.
- Improved ion-trapping techniques for larger qubit arrays.

- Advanced quantum error correction (QEC) methods to create fault-tolerant logical qubits.
- Hybrid quantum architectures that combine trapped ions with other quantum technologies for improved scalability.

Despite these challenges, trapped-ion quantum computing remains one of the most promising approaches due to its high qubit uniformity, low error rates, and all-to-all connectivity, making it a strong contender for future fault-tolerant quantum computing systems.

Topological Qubits (Microsoft's Majorana Fermions)

Topological qubits, as a theoretical class of qubits, harness topological states of matter to effectively encode and manipulate quantum information. Distinct from traditional qubits—such as superconducting or trapped-ion qubits, which represent quantum information through discrete particle states—topological qubits utilize non-Abelian anyons. These exotic quasiparticles arise in specific low-dimensional quantum systems and are characterized by their unique exchange statistics, which confer robustness to the qubit states against local noise and perturbations, thereby enabling them to exhibit intrinsic fault tolerance [278, 328, 329]. The fault tolerance of topological qubits comes from their information being stored in non-local degrees of freedom, which are less susceptible to localized errors than conventional systems [330].

Microsoft has emerged as a frontrunner in the pursuit of topological quantum computing, focusing primarily on the development of qubits informed by Majorana fermions—quasiparticles that serve as their own antiparticles. Majorana fermions are inherently linked to non-Abelian statistics and are expected to provide a highly stable framework for quantum computation due to their unique properties that guard against decoherence [331-333]. The connection between Majorana fermions and topological qubits signifies the potential for fault-tolerant quantum computing, since the braiding of these particles enables the implementation of topologically protected operations [334-336]. Such operations facilitate the manipulation of quantum information while maintaining a defence against the types of errors that typically challenge quantum computational systems.

A critical aspect of Microsoft's exploration lies in the ability to create stable Majorana fermion states within topological superconductors. Research has shown that these fermions can be employed not only in theoretical frameworks but also in practical, experimental setups, providing a pathway toward realizing topologically protected qubits that could notably advance the field of quantum information processing [337-339]. Moreover, the proposal of various operational schemes using Majorana fermions illustrates their versatility as qubits, integrated into diverse physical systems ranging from superconducting circuits to trapped-ion architectures [340, 341].

Majorana Fermions and Their Role in Topological Quantum Computing

Majorana fermions, first proposed by Ettore Majorana in 1937, have emerged as significant players in the realm of condensed matter physics, particularly in the context of topological quantum computing. They are not elementary particles but rather arise as quasiparticle excitations, particularly in topological superconductors, where they manifest as Majorana zero modes (MZMs). These modes are characterized by their non-Abelian statistics, which allows for the manipulation of quantum information in a fundamentally different way from conventional qubits, offering inherent resistance to decoherence and errors due to environmental noise [286, 342, 343].

MZMs exist as localized zero-energy states bound to defects or boundaries in a topological superconductor. This localization is crucial; it permits the encoding of quantum information with high fidelity [286, 342, 344]. The theoretical underpinnings suggest that by properly tuning the parameters of these systems, it is possible to not only observe but also manipulate these modes for practical quantum computation [286, 344].

The role of braiding MZMs is particularly noteworthy due to their non-Abelian exchange statistics, which means that the outcome of exchanging two MZMs depends solely on their braiding history. This feature is essential for implementing topological quantum gates, wherein quantum information is stored in the states of MZMs rather than their specific locations. This paradigm results in a computational model that is more resilient to certain types of quantum errors compared to traditional quantum computing models [345-347].

Recent research has expanded our understanding of MZMs in various topological superconductors, including those classified as second-order topological superconductors (SOTS), which host MZMs as localized quasiparticles [345]. These advancements not only reinforce the theoretical framework but also open avenues for experimental validation and practical applications in quantum computing. Critical results indicate that Majorana-based systems can achieve scalability—one of the essential requirements for practical quantum computing—due to the inherent non-locality and topological protection offered by the MZMs [342, 348].

Key advancements have been made in the design and implementation of devices intended to host MZMs. These include engineered nanowire-superconductor hybrid systems, promoted by companies like Microsoft, which focus on harnessing MZMs for robust and fault-tolerant computational operations [347, 349, 350]. Observations such as the fractional ac Josephson effect have provided experimental evidence supporting the existence of MZMs, further solidifying their role in quantum computation [343].

Microsoft's topological quantum computing efforts rely on nanowire-superconductor hybrid systems, where Majorana zero modes can theoretically form at the ends of semiconducting nanowires coupled with a superconducting layer. These systems are designed to create and manipulate topological qubits in a way that minimizes decoherence and gate errors.

The process involves:

1. Creating a Topological Superconductor

- A semiconducting nanowire (such as indium arsenide (InAs) or indium antimonide (InSb)) is placed on top of a superconducting material (such as aluminium).
- A strong magnetic field is applied along the nanowire to induce a topological superconducting phase.

2. Emergence of Majorana Zero Modes

- Under the right conditions, Majorana zero modes appear at both ends of the nanowire.
- These Majorana modes can be braided to perform quantum operations.

3. Encoding Quantum Information

- Unlike conventional qubits, where quantum states are stored in a single particle or system, topological qubits store quantum information non-locally across multiple Majorana zero modes.
- This delocalization makes the qubit immune to many types of noise, providing inherent error protection.

4. Performing Quantum Computation

- Quantum gates are performed by braiding Majorana fermions, meaning physically moving them around each other in a controlled way.
- This results in topologically protected quantum operations, where the system's evolution depends only on the braiding path, not on the details of the local environment.

Topological qubits are theoretically superior to other qubit technologies due to their built-in fault tolerance and robustness against decoherence. Unlike superconducting and trapped-ion qubits, which require active error correction, topological qubits are naturally resistant to many types of quantum noise. This could significantly reduce the overhead needed for quantum error correction, making large-scale quantum computing more feasible.

Because quantum gates in topological systems depend only on the global braiding of Majorana fermions, they are immune to local disturbances, such as electrical noise, charge fluctuations, and magnetic field variations. This could enable more stable and scalable quantum processors compared to current approaches.

Table 13: Comparison: Topological Qubits vs. Other Qubit Technologies.

Feature	Topological Qubits (Majorana Fermions)	Superconducting Qubits	Trapped-Ion Qubits
Qubit Encoding	Non-local (spread across multiple Majorana modes)	Local (single qubit states)	Local (individual trapped ions)
Error Protection	Intrinsic fault tolerance	Requires active QEC (Surface Codes)	Requires active QEC (Surface Codes)
Gate Operations	Braiding of Majorana fermions	Microwave control	Laser-based control
Scalability Potential	High (if realized)	Medium (limited by coherence)	High (all-to-all connectivity)
Development Stage	Theoretical and experimental phase	Fully operational (IBM, Google, Rigetti)	Fully operational (IonQ, Quantinuum)
Key Challenge	Experimental confirmation of Majorana fermions	Decoherence and noise	Slow gate speed

Despite their theoretical promise, topological qubits have not yet been fully realized experimentally, and Majorana zero modes are still under intense research. Microsoft and its research partners have been working to detect and confirm Majorana fermions in nanowire-superconductor hybrid systems. However, recent studies have shown conflicting results, leading to further investigations into whether the observed signals are truly from Majorana fermions.

Some key challenges include:

- Experimental Verification: Majorana zero modes have not been conclusively demonstrated, and their signatures can sometimes be confused with other quantum phenomena.

- Fabrication Complexity: Creating stable topological superconductors with reliable Majorana modes requires precise material engineering and extremely low temperatures.
- Scalability and Control: Even if Majorana-based qubits are realized, developing a practical system for braiding operations and large-scale quantum computation remains an open challenge.

Microsoft has taken a long-term approach to quantum computing, investing heavily in theoretical and experimental research to refine the materials and architectures needed for topological qubits. While superconducting qubits (IBM, Google, Rigetti) and trapped-ion qubits (IonQ, Quantinuum) are already being used in real quantum processors, Microsoft's approach is still in the research phase, aiming for a breakthrough in fault-tolerant quantum computing.

If Microsoft succeeds in building a scalable Majorana-based quantum processor, it could revolutionize the field of quantum computing by dramatically reducing the need for active quantum error correction. This would make quantum computers far more stable and efficient, allowing for deeper quantum circuits and more complex computations.

While other quantum computing platforms are progressing with superconducting and trapped-ion qubits, Microsoft remains committed to its long-term vision of topological quantum computing, banking on the potential breakthrough that Majorana fermions could provide. If successful, this approach could make fault-tolerant quantum computers a reality, unlocking new frontiers in cryptography, materials science, and complex simulations.

For now, theoretical and experimental research continues, with ongoing efforts to conclusively prove the existence of Majorana fermions and refine the fabrication techniques needed to create a stable topological qubit. If these challenges can be overcome, topological quantum computing could emerge as a game-changing technology in the quest for practical, large-scale quantum computers.

Processor Hardware and Environment Requirements

Topological quantum computing, specifically Microsoft's approach based on Majorana fermions, requires a highly specialized hardware infrastructure that is fundamentally different from superconducting or trapped-ion quantum processors. The goal of this approach is to create topological qubits, which store quantum information in a way that is intrinsically fault-tolerant, reducing the need for active quantum error correction.

Building a quantum processor based on Majorana zero modes (MZMs) involves several key hardware and environmental requirements, including:

1. Topological Superconductors and Nanowire-Based Qubit Structures

2. Ultra-Low Temperature Cryogenic Systems
3. High-Precision Magnetic Field Control
4. Semiconductor-Superconductor Hybrid Materials
5. Braiding and Measurement Systems for Majorana Qubits
6. Classical Control Electronics for Quantum Operations

Topological Superconductors and Nanowire-Based Qubit Structures: Unlike superconducting qubits that rely on Josephson junctions or trapped-ion qubits that use electromagnetic confinement, topological qubits are created within engineered nanowires coupled with superconductors. These nanowires host Majorana zero modes, which appear at the ends of the wire under specific conditions.

To fabricate such qubits, semiconducting nanowires such as indium arsenide (InAs) or indium antimonide (InSb) are deposited on top of a superconducting material like aluminum (Al) or niobium (Nb). This semiconductor-superconductor hybrid system is the foundation for Majorana-based quantum computing. The interaction between the semiconductor and the superconductor allows the system to enter a topological phase, in which Majorana fermions emerge at the wire's edges.

To achieve a reliable topological quantum processor, the nanowires must have:

- Atomically smooth interfaces between the semiconductor and the superconductor to minimize defects.
- Precise control over electron density within the nanowire using electrostatic gate voltages.
- Minimal disorder and impurities, as any disruption in the material can destroy the fragile topological phase.

The layout of the qubits is another critical aspect. In a Majorana-based quantum computer, qubits are non-locally encoded in pairs of Majorana zero modes, meaning that multiple nanowires must be interconnected in a scalable architecture that supports braiding operations, where Majorana fermions are moved around each other to implement quantum gates.

Ultra-Low Temperature Cryogenic Systems: Majorana fermions only emerge in specific superconducting materials at extremely low temperatures. Microsoft's Majorana-based qubits require dilution refrigerators to cool the system to below 20 millikelvin (mK), a temperature close to absolute zero. This is necessary to:

- Maintain superconductivity in the aluminium or niobium layers.

- Prevent thermal noise from disrupting the delicate quantum states of Majorana fermions.
- Enable long coherence times by minimizing unwanted excitations in the system.

The cryogenic system used is similar to those used for superconducting qubits but must also incorporate advanced shielding techniques to prevent thermal radiation from destroying the topological phase required for Majorana qubits.

High-Precision Magnetic Field Control: A significant requirement for realizing Majorana qubits is the application of a strong and precisely controlled magnetic field along the nanowires. The presence of this field induces a topological superconducting phase, which is essential for the formation of Majorana zero modes.

The magnetic field strength must be finely tuned, typically in the range of 0.5 to 2 Tesla, depending on the material properties of the nanowire-superconductor system. The field must be:

- Uniform across the entire chip to ensure consistent behaviour of all qubits.
- Precisely aligned with the nanowires to avoid breaking the topological protection.
- Stable over time to prevent fluctuations that could destroy Majorana fermions.

Specialized superconducting magnets or nano-engineered magnetic field sources are integrated into the quantum hardware to generate these controlled fields while minimizing stray magnetic interference.

Semiconductor-Superconductor Hybrid Materials: The successful realization of topological qubits depends on the quality of the semiconductor-superconductor hybrid system. Key material properties include:

- Strong spin-orbit coupling in the semiconductor (InAs or InSb) to enable the formation of Majorana fermions.
- High superconducting gap in the superconductor (Al, Nb) to maintain quantum coherence.
- Clean interfaces with minimal disorder, which requires advanced material fabrication techniques such as molecular beam epitaxy (MBE) or atomic layer deposition (ALD).

The challenge of material synthesis and fabrication is one of the biggest hurdles in developing Majorana-based qubits, as even small imperfections can destroy the topological phase needed for quantum computation.

Braiding and Measurement Systems for Majorana Qubits: Unlike other quantum systems, where quantum gates are performed using microwave or laser pulses, quantum computation with Majorana qubits relies on braiding operations. In this approach:

- Majorana zero modes are physically moved around each other, changing the quantum state of the system in a topologically protected way.
- The history of braiding paths determines the computational outcome, meaning that quantum gates depend on the order in which Majoranas are exchanged.

To implement this, the quantum processor must include nano-electrodes and gate voltage controls that allow precise movement of Majorana fermions within the nanowire network. These operations must be:

- Fast enough to enable practical computation.
- Precisely controlled to avoid unintended state transitions.
- Scalable so that multiple qubits can be braided efficiently.

Measuring the state of a Majorana qubit is also different from conventional quantum measurement techniques. Instead of direct state readout, parity measurements are performed by detecting the presence or absence of quantum states through specialized superconducting charge sensors.

Classical Control Electronics for Quantum Operations: Even though Majorana qubits are expected to require less active error correction, a sophisticated classical control system is still needed to:

- Manage electrostatic gate voltages that control qubit formation and movement.
- Synchronize Majorana braiding operations for quantum gate execution.
- Perform parity measurements to read out the quantum state of the system.

Microsoft's Azure Quantum platform is expected to integrate classical control electronics with cloud-based quantum computing, enabling hybrid quantum-classical processing for large-scale applications.

Photonic Quantum Computing

Photonic quantum computing is emerging as a compelling approach in the field of quantum computing, utilizing photons as qubits. This method presents several advantages over traditional qubit systems, such as trapped ions or superconducting circuits. Key benefits include high-speed operation, functionality at room temperature, and scalability through optical networks. The intrinsic properties of photons allow for low decoherence rates due to their minimal interaction with the environment, facilitating long-distance transmission and enhancing the viability of photonic systems for quantum communication and distributed quantum computing applications [351, 352].

Recent developments highlight the advantages of photonic systems further. Photons exhibit naturally low decoherence rates compared to solid-state systems and can maintain coherence even at room temperature, making them especially suitable for real-world applications in quantum technologies [351]. Moreover, photonic quantum states enable robust long-distance communication due to their resilience against environmental noise [351]. These characteristics position photon-based systems as promising candidates for both quantum computing architectures and advanced communication protocols, leveraging the capabilities of integrated photonic circuits to facilitate direct applications in quantum information processing [353].

Several leading organizations are pioneering photonic quantum computing technologies, specifically PsiQuantum, Xanadu, and QuiX Quantum. These companies are exploring various architectures for optical quantum computation, focusing on distinct photonic technologies such as linear optics and squeezed states. For example, Xanadu's work on silicon-based quantum circuits aims to create scalable photonic systems that integrate quantum dots and nonlinear optics [353]. QuiX Quantum is advancing research on high-capacity photonic systems that use entangled states to increase computational efficiency through optical interconnects, reinforcing the potential of photonic platforms in scaling quantum computing capabilities [354].

The challenges associated with photonic quantum computing primarily involve the implementation of reliable quantum gates and the integration of photonic circuits at scale. High-fidelity quantum gates are essential for achieving computational reliability; however, current photonic implementations often rely on probabilistic methods, which can limit the effectiveness of quantum operations [352]. Ongoing research aimed at overcoming these hurdles includes innovations in quantum logic gates, such as hyper-controlled NOT gates and deterministic gate designs, indicating a path forward for implementing robust photonic quantum computing systems [355, 356].

Instead of relying on matter-based qubits, photonic quantum computing encodes quantum information into the properties of light. Photons, as quantum particles of light, can be manipulated to represent qubits in multiple ways, offering a versatile approach to quantum information processing. One common method is polarization encoding, where the horizontal and vertical polarization states of a photon correspond to the quantum states $|0\rangle$ and $|1\rangle$. This approach is widely used due to its simplicity and compatibility with existing optical technologies.

Another technique is path encoding, where a single photon travels through different optical paths to represent quantum states. By carefully controlling the interference between these paths, quantum superpositions and entanglement can be achieved. Time-bin encoding provides another way to encode qubits, where quantum information is stored in the photon's arrival time at different time slots. This method is particularly useful for fibre-optic quantum communication since time-bin qubits remain stable during transmission over long distances.

A more advanced approach is squeezed light states, which are used in continuous-variable (CV) quantum computing. Instead of using discrete qubits, quantum information is stored in the phase and amplitude of light waves. This method leverages principles from quantum optics to perform computations in a way that differs from traditional qubit-based approaches.

To process quantum information, optical components such as beam splitters, phase shifters, waveguides, and nonlinear crystals are used to implement quantum gate operations. These elements manipulate photons, creating superposition, interference, and entanglement, which are essential for quantum computation. Measurement is performed using photon detectors, which determine the presence or absence of photons in different optical modes. The detection process collapses the quantum state, allowing the extraction of computational results. This unique approach enables high-speed, low-decoherence quantum computing, making photonic quantum computers a promising platform for scalable quantum processing.

There are two primary approaches to photonic quantum computing:

1. Gate-Based Photonic Quantum Computing (PsiQuantum, QuiX Quantum)

This model follows the standard quantum circuit model, where quantum gates (such as Hadamard, CNOT, and phase gates) are implemented using linear optical elements such as beam splitters and phase shifters. Qubits are encoded in single photons, and entangling operations are performed using quantum interference at beam splitters.

The challenge of this approach is that photons do not naturally interact with each other, making two-qubit gates difficult to implement. One solution is to use measurement-based quantum computing (MBQC), where entanglement is pre-generated in a large optical cluster state, and computations are carried out via measurements.

2. Continuous-Variable (CV) Photonic Quantum Computing (Xanadu, NTT)

In CV photonic quantum computing, quantum information is encoded in the quadratures of the electromagnetic field, rather than discrete photon states. This model uses squeezed light states and applies operations through Gaussian transformations and non-Gaussian gates.

Xanadu's Borealis processor, for example, implements Gaussian boson sampling, a photonic quantum algorithm used for complex combinatorial problems such as molecular simulations and graph optimization.

Gaussian Boson Sampling (GBS) is a quantum computing paradigm that leverages the unique properties of continuous-variable (CV) quantum optics to perform computations that are extremely difficult for classical computers. It is a specialized quantum algorithm that falls under the category of sampling problems, meaning that it is designed to generate probability distributions that are infeasible to simulate using classical methods. GBS is considered an extension of Boson Sampling, first proposed by Aaronson and Arkhipov in 2011, but instead of using single-photon input states, it employs Gaussian states, which are more practical to generate in optical quantum systems.

This approach has attracted significant attention due to its potential to demonstrate quantum advantage, particularly in problems related to combinatorial optimization, graph theory, and quantum chemistry.

GBS is based on the manipulation of squeezed states of light, a special type of quantum optical state where the uncertainty in one quadrature (amplitude or phase) is reduced at the expense of increased uncertainty in the other. These squeezed states serve as the input to a linear optical network, which consists of beam splitters, phase shifters, and interferometers that mix the photons before they are measured.

The process can be broken down into the following steps:

- **State Preparation**
  - A series of Gaussian squeezed vacuum states are generated using optical parametric down-conversion (PDC) sources or optical squeezers. These states contain pairs of correlated photons and follow a Gaussian probability distribution in phase space.
- **Linear Optical Interference**
  - The squeezed states are fed into a linear optical network (LON) consisting of beam splitters and phase shifters. This network causes interference between the photons, creating a complex quantum superposition of possible photon detection patterns.
- **Photon Detection and Sampling**
  - After passing through the optical network, the photons are detected using photon-number-resolving detectors (PNRDs), which count the number of photons at each output mode. The measurement results form a sample from a probability distribution that is computationally difficult to simulate classically.

The underlying mathematical structure of GBS is based on Hafnians, a function that describes the probability amplitudes of bosonic quantum states. Computing Hafnians for large matrices is known to be a computationally hard problem for classical computers, which is why GBS is believed to be capable of quantum advantage over classical algorithms.

A Hafnian is a mathematical function used to compute certain properties of symmetric matrices, particularly those related to bosonic quantum systems. It plays a crucial role in Gaussian Boson Sampling (GBS), where it determines the probability of detecting specific photon distributions in an optical quantum circuit. The Hafnian function is closely related to the permanent of a matrix, which appears in classical Boson Sampling, but is computationally even more complex in many cases.

Hafnians arise naturally when dealing with Gaussian states in quantum optics, where quantum information is represented in continuous variables such as phase space quadratures. Their complexity makes them computationally hard to calculate for large matrices, which is why Gaussian Boson Sampling is considered a strong candidate for quantum computational advantage—meaning it can solve problems intractable for classical computers.

Mathematically, the Hafnian of a 2N × 2N symmetric matrix A is defined as:

$$\mathrm{Haf}(A) = \sum_{\text{all perfect pairings } P} \prod_{(i,j)\in P} A_{i,j}$$

This means that the Hafnian sums over all possible ways to pair up the indices of the matrix and multiplies the corresponding matrix elements for each pairing.

For example, given a 4 × 4 symmetric matrix:

$$A = \begin{bmatrix} 0 & a & b & c \\ a & 0 & d & e \\ b & d & 0 & f \\ c & e & f & 0 \end{bmatrix}$$

The Hafnian is computed by summing over all possible perfect pairings:

$$\mathrm{Haf}(A) = ad + af + ae + bf + be + cf$$

For small matrices, this can be calculated directly, but for large matrices, computing Hafnians quickly becomes intractable because the number of terms in the sum grows exponentially with matrix size.

In Gaussian Boson Sampling (GBS), the output probabilities of photon detection events in an optical circuit are directly related to the Hafnian of the covariance matrix of the Gaussian state. The Hafnian describes the probability amplitude for bosons (photons) to occupy certain output modes after passing through a linear optical network.

If the covariance matrix encodes a quantum state of squeezed photons, the probability of detecting a particular photon-number pattern at the output ports is proportional to the absolute value of the Hafnian squared:

$$P(\text{photon pattern}) \propto |\text{Haf}(A)|^2$$

Since calculating Hafnians for large matrices is #P-hard (a complexity class even harder than NP-hard), classical computers struggle to efficiently simulate GBS experiments, reinforcing the belief that quantum photonic processors have a computational advantage in specific tasks.

Computing Hafnians is exponentially hard for classical computers. While small Hafnians can be computed using direct summation, large matrices require advanced numerical algorithms, such as Monte Carlo sampling or tensor network methods.

- The best known classical algorithms for Hafnians have exponential time complexity $O(2^N)$, making large-scale calculations infeasible.
- This is one of the reasons why Gaussian Boson Sampling is believed to provide a clear demonstration of quantum advantage—because simulating it on classical hardware would require computing Hafnians of matrices with thousands of dimensions.

Table 14: Comparison: Gaussian Boson Sampling vs. Other Quantum Computing Models.

Feature	Gaussian Boson Sampling (GBS)	Gate-Based Quantum Computing	Superconducting Qubits
Qubit Type	Photons (squeezed states)	Qubits (superposition states)	Josephson junctions
Computational Model	Sampling-based	Universal (circuit model)	Universal (circuit model)
Error Resilience	Naturally resilient to decoherence	Requires quantum error correction	Requires quantum error correction
Scalability	Limited by photon loss	Scalable with error correction	Scalable with error correction
Best Use Cases	Optimization, graph theory, chemistry	General-purpose quantum computing	General-purpose quantum computing

Hardware and Environment Requirements for Photonic Quantum Processors

Unlike superconducting and trapped-ion qubits, which require ultra-low temperatures and vacuum systems, photonic quantum processors can operate at room temperature, which makes them highly scalable and suitable for integration into existing optical fibre networks. The key hardware components include:

Single-Photon Sources: A reliable source of indistinguishable single photons is critical for photonic quantum computing. These can be generated using:

- **Spontaneous Parametric Down-Conversion (SPDC):** A nonlinear crystal splits a high-energy photon into two entangled photons.
- **Quantum Dot Emitters:** Semiconductor quantum dots generate single photons on demand.
- **Silicon Photonics Sources:** Integrated photonics chips generate single photons via nonlinear optical effects.

Linear Optical Quantum Gates: Quantum gates in photonic systems are implemented using passive and active optical components, such as:

- **Beam Splitters:** Used to create superpositions by splitting a photon into two paths.
- **Phase Shifters:** Modify the phase of a photonic qubit to perform quantum gate operations.
- **Waveguides:** Integrated optical pathways that direct photons through a circuit.
- **Nonlinear Optical Crystals:** Enable entanglement between photons by mediating interactions.

Quantum Entanglement and Measurement: Since photons do not interact easily, entanglement is generated using quantum interference effects at beam splitters, such as the Hong-Ou-Mandel (HOM) effect. To read out quantum information, highly efficient single-photon detectors are required, including:

- **Superconducting Nanowire Single-Photon Detectors (SNSPDs):** Ultra-sensitive detectors operating at cryogenic temperatures.
- **Silicon Avalanche Photodiodes (Si-APDs):** Room-temperature detectors with high efficiency.

Integrated Photonics and Optical Quantum Chips: To scale up photonic quantum processors, companies are developing integrated photonic chips, where thousands of optical components are fabricated onto a single silicon wafer. This enables compact, stable, and scalable quantum processors that can be mass-manufactured using CMOS-compatible fabrication techniques.

Table 15: Comparison of Photonic Quantum Computing vs. Other Approaches.

Feature	Photonic Qubits	Superconducting Qubits	Trapped-Ion Qubits
Qubit Type	Photons (light particles)	Josephson junctions	Trapped ions
Temperature Requirement	Room temperature	< 10 mK (cryogenic)	Room temperature / UHV
Gate Operations	Linear optics, phase shifts	Microwave control	Laser-based control
Error Protection	Some natural robustness	Requires active QEC	Requires active QEC
Connectivity	Naturally networked	Nearest-neighbour	All-to-all
Scalability Potential	High (integrated photonics)	Medium	High (scalable via ion transport)
Key Challenges	Photon loss, two-qubit interactions	Decoherence, fabrication complexity	Slow gate speed

Advantages and Challenges of Photonic Quantum Computing

One of the primary advantages of photonic quantum computing is its room-temperature operation. Unlike superconducting and trapped-ion qubits, which require cryogenic or ultra-high vacuum conditions, photonic qubits can function at ambient temperatures. This makes them far more practical for large-scale deployment and integration into existing infrastructure, reducing the complexity and cost of quantum computing systems.

Another key benefit is low decoherence. Photons, being massless particles, do not interact with their environment in the same way that matter-based qubits do. This means that photonic qubits experience longer coherence times and are inherently less susceptible to noise, reducing the need for complex error correction.

Photonic quantum processors also enable fast quantum operations since optical quantum gates operate at the speed of light. This makes photonic-based quantum systems inherently faster than other qubit architectures, providing significant advantages for applications requiring rapid computation.

Additionally, photonic quantum computing offers natural integration with quantum networks. Photonic qubits are ideally suited for quantum communication and distributed quantum computing, forming the backbone of the future quantum internet. Since photons can travel long distances through fibre-optic cables with minimal loss, they provide a seamless interface between quantum processors, enabling secure quantum communication and scalable quantum networks.

Despite its advantages, photonic quantum computing faces significant challenges, one of which is photon loss. Optical components such as beam splitters, fibre optics, and detectors introduce losses that degrade quantum computation fidelity. Unlike matter-based qubits, where qubit loss is rare, the probability of losing a photon in a photonic quantum circuit can be high, requiring redundancy and error mitigation techniques.

Another major limitation is the low probability of two-qubit gates. Unlike superconducting and trapped-ion qubits, which can be directly entangled through electromagnetic interactions, photons do not naturally interact with each other. Implementing two-qubit gates requires complex techniques such as heralded entanglement or measurement-based quantum computing (MBQC), which introduce additional computational overhead.

Scalability remains an open challenge, as large-scale photonic quantum computing requires integrated photonic chips with precise control over thousands or millions of optical components. While integrated quantum photonics is a promising avenue, developing large-scale quantum error correction (QEC) methods for photonic systems is still an ongoing area of research. Overcoming these challenges will be essential for realizing fault-tolerant photonic quantum computing and achieving the full potential of this technology.

Quantum Annealers (D-Wave and Optimization Problems)

Quantum annealing represents a distinct approach within the field of quantum computing, specifically tailored to address optimization problems. This paradigm leverages quantum superposition, tunnelling, and entanglement to identify the lowest energy state of complex systems, thereby providing solutions to combinatorial optimization challenges more efficiently than classical methods [357, 358]. Unlike universal gate-based quantum computers, which execute a series of quantum gates to perform computations, quantum annealers employ an adiabatic process to gradually transition from a simple Hamiltonian to a more complex problem-specific Hamiltonian. According to the adiabatic theorem, this method enables the system to remain in its ground state, which ultimately leads to an effective sampling of solutions from the problem landscape [358, 359].

D-Wave Systems stands as a leader in the commercialization of quantum annealing technology, having developed several generations of quantum annealers that exemplify this technique. The company's systems are optimized specifically for solving optimization problems, contrasting with other major players in the quantum computing space like IBM and Google, whose architectures aim for general-purpose quantum computing. D-Wave's quantum annealers are particularly adept at addressing a range of applications, including route optimization, financial portfolio management, drug discovery, and machine learning [360-362]. Notably, research has demonstrated the effective application of D-Wave's technology in real-world scenarios such as vehicle routing and traffic signal optimization, underscoring its practical utility in a variety of complex optimization problems [360, 362, 363].

Furthermore, the D-Wave quantum annealers are designed to handle Quadratic Unconstrained Binary Optimization (QUBO) problems natively, which are prevalent across diverse fields of computational research [364, 365]. These systems operate by encoding QUBO parameters using superconducting qubits, enabling the retrieval of the lowest energy configuration through an annealing process that exploits quantum mechanics principles such as superposition and entanglement [365, 366]. As the technology evolves, enhancements in qubit connectivity and count are anticipated, allowing D-Wave systems to tackle even more complex optimization issues with shorter computation times and greater effectiveness [360, 362, 363].

Quantum annealing works by representing an optimization problem as a quantum system where the solution corresponds to the lowest energy state of that system. The process involves:

1. Encoding the Problem in a Hamiltonian

- The optimization problem is translated into a mathematical function called the Ising model or Quadratic Unconstrained Binary Optimization (QUBO) problem.
- This function is mapped onto a network of superconducting qubits arranged in a lattice, where each qubit represents a variable in the problem.

2. Initializing the Quantum State

- The system is initialized in a superposition of all possible states, meaning that all potential solutions exist simultaneously.

3. Adiabatic Evolution and Quantum Tunnelling

- The system is gradually evolved from a simple initial Hamiltonian (where the solution is easy to determine) to the problem Hamiltonian (which encodes the optimization problem).
- During this evolution, the system explores many possible configurations, leveraging quantum tunnelling to move through energy barriers that would be difficult to escape in classical optimization methods.

4. Finding the Optimal Solution

- As the quantum system settles into its lowest-energy configuration, it collapses into the optimal solution for the given problem.
- The final state of the qubits provides the best solution (or a close approximation) to the optimization problem.

This process allows quantum annealers to solve complex combinatorial optimization problems faster and more efficiently than traditional classical algorithms, especially for problems with large solution spaces.

D-Wave has developed multiple generations of quantum annealers, including the D-Wave 2000Q, Advantage, and Advantage2, each incorporating an increasing number of qubits and improved connectivity. Unlike gate-based superconducting quantum computers, such as those developed by Google and IBM, D-Wave's quantum processors are specifically designed for optimization problems. They are built on superconducting flux qubits, which naturally behave like artificial spin systems, making them well-suited for solving combinatorial optimization challenges.

The latest D-Wave Advantage quantum annealer represents a significant step forward in quantum hardware capabilities. It features over 5000 qubits, offering a substantial increase in computational power. Each qubit has 15-way connectivity, allowing for more complex interactions and enhancing the system's ability to solve optimization problems more efficiently. The hardware also incorporates lower noise levels, which improve the accuracy of results by reducing the impact of environmental interference. Additionally, D-Wave integrates hybrid quantum-classical solvers, enabling users to combine quantum annealing with classical optimization algorithms, which is crucial for practical applications that require a balance of quantum and classical computing power.

These advancements make D-Wave's quantum annealers highly effective for tackling real-world industrial problems in logistics, finance, pharmaceuticals, and artificial intelligence. The ability to efficiently optimize large-scale systems—such as route planning, financial portfolio management, drug discovery, and machine learning models—positions D-Wave's technology as a valuable tool for industries that require high-performance computing solutions. As quantum annealing technology continues to evolve, its role in solving large-scale optimization problems is expected to expand, offering new possibilities for businesses and research institutions worldwide.

Quantum annealers are particularly well-suited for solving combinatorial optimization problems, where a system must evaluate an enormous number of possible configurations to determine the best solution. These problems arise in a variety of fields, including logistics, finance, pharmaceuticals, and artificial intelligence, making quantum annealing a valuable computational tool.

One of the most prominent applications is route optimization and logistics, where quantum annealers can improve efficiency in transportation and supply chain management. In the Traveling Salesperson Problem (TSP), quantum annealing helps find the shortest possible route that allows a traveller to visit all required locations and return to the starting point. This problem is fundamental in logistics, delivery systems, and manufacturing processes, where optimizing travel routes can significantly reduce costs and fuel consumption. Traffic flow optimization is another critical area, where quantum annealers determine the best way to minimize congestion in urban transportation networks by analysing and optimizing road usage patterns. Similarly, supply chain optimization benefits from quantum annealing by improving warehouse logistics, delivery schedules, and inventory management, ensuring that goods are transported efficiently with minimal delays.

In financial portfolio optimization, quantum annealers assist in asset allocation, helping investors determine the best mix of assets to maximize returns while minimizing risk. The risk analysis capabilities of quantum annealing allow financial institutions to evaluate large-scale financial risk scenarios using quantum-enhanced simulations, identifying potential weaknesses and optimizing investment strategies. These capabilities make quantum annealing particularly valuable for hedge funds, banks, and insurance companies dealing with complex financial markets.

Another major application is in drug discovery and molecular simulation, where quantum annealers help solve problems that require analysing vast molecular structures. Protein folding, a problem critical to drug discovery, involves predicting the three-dimensional structure of proteins, which determines their biological function. Understanding protein folding can accelerate the development of new drugs and treatments. Additionally, quantum annealers improve molecular similarity searches, where pharmaceutical researchers identify compounds with similar chemical structures, speeding up the discovery of potential drug candidates.

In the field of machine learning and artificial intelligence, quantum annealers assist with feature selection, a crucial step in training AI models. Feature selection involves identifying the most relevant variables in a dataset, which improves model accuracy and efficiency. Quantum annealers also enhance graph neural networks (GNNs), which are used in AI applications that rely on graph-based learning, such as social network analysis, fraud detection, and recommendation systems. By optimizing the structure of these networks, quantum annealers contribute to more powerful and efficient AI models.

As quantum annealing technology advances, its ability to solve large-scale optimization problems across multiple industries will continue to expand, offering businesses and researchers a competitive advantage in tackling complex computational challenges.

Table 16: Comparison: Quantum Annealing vs. Gate-Based Quantum Computing.

Feature	Quantum Annealing (D-Wave)	Gate-Based Quantum Computing (IBM, Google)
Purpose	Optimization and combinatorial problems	General-purpose quantum computing
Qubit Type	Superconducting flux qubits	Superconducting transmon qubits / Trapped ions
Quantum Operations	Adiabatic evolution, tunnelling	Discrete gate-based logic operations
Error Correction	Not required for optimization tasks	Requires quantum error correction
Best Use Cases	Logistics, finance, drug discovery, AI	Cryptography, quantum simulation, material science
Scalability	Large number of qubits (>5000)	Limited by qubit connectivity and error rates

Despite its advantages in optimization, quantum annealing faces several limitations that impact its broader applicability. One of the most significant constraints is that D-Wave's quantum annealers are not universal quantum computers. Unlike gate-based quantum processors, which can execute a wide range of quantum algorithms, quantum annealers are specialized for optimization problems and cannot perform arbitrary quantum computations, such as Shor's algorithm for factoring large numbers or quantum simulations of physical systems. This makes quantum annealing a valuable but domain-specific tool rather than a general-purpose quantum computing solution.

Another challenge is susceptibility to noise. While quantum annealers do not require as much active error correction as gate-based quantum computers, they are still affected by thermal noise and environmental fluctuations, which can impact the accuracy of solutions. Since quantum annealing relies on adiabatic evolution to guide the system to its lowest-energy state, any interference from external noise can cause the system to settle in suboptimal states, reducing the effectiveness of the computation. Ensuring a high degree of coherence and stability remains a significant challenge in scaling quantum annealers for real-world applications.

Quantum annealing also has limited speedup for certain problems. While it has demonstrated advantages over classical methods for some combinatorial optimization problems, there are cases where classical optimization algorithms, such as simulated annealing and quantum-inspired heuristics, can achieve comparable or even superior performance. This means that quantum annealing does not always provide an exponential speedup and must be carefully evaluated against state-of-the-art classical algorithms to determine when it offers a genuine advantage.

D-Wave is actively working on addressing these challenges by developing next-generation quantum annealers with improved capabilities. One of the key areas of focus is increasing qubit connectivity, which allows for solving more complex problems by improving interactions between qubits in the processor. Enhancing coherence times is another priority, as this directly impacts solution accuracy and reliability. Additionally, D-Wave is advancing hybrid quantum-classical algorithms, which combine quantum annealing with classical computing methods to maximize computational efficiency. By integrating classical optimization techniques with quantum-enhanced search processes, these hybrid approaches aim to deliver practical performance improvements in industrial applications.

With continuous advancements in quantum hardware and software, quantum annealers are expected to play an increasingly important role in high-performance optimization tasks. While they may not replace classical or universal quantum computing models, they serve as a highly practical tool for solving complex optimization problems that are infeasible for classical algorithms alone. As quantum annealing technology matures, it has the potential to provide immediate commercial and scientific benefits across multiple industries, including logistics, finance, artificial intelligence, and materials science.

Costs of Quantum Computers

The cost of quantum computers can significantly differ based on their capabilities, qubit count, and specific applications they are designed for. High-end quantum computers such as those developed by IBM and D-Wave are priced in the tens of millions of dollars and are often sold as part of comprehensive service contracts that span several years. For example, IBM's quantum computing systems, as observed in various reports, reflect this pricing strategy, where multi-million dollar contracts include extensive service and operational agreements [367, 368]. D-Wave Systems, a pioneer in quantum computing technology, introduced its first commercially available quantum annealer, the D-Wave One, priced around $10 million in

2011, and subsequently launched the D-Wave 2000Q priced at $15 million in 2017 [369]. More recent offerings, such as the Advantage quantum computer, have not disclosed specific pricing but are expected to align with these high-value contracts [367, 369].

In contrast, the emerging market for mid-range quantum systems presents a more accessible pricing structure. Companies like SpinQ have launched desktop quantum computers available for around $5,000, which are aimed at educational institutions and small-scale research projects. These systems provide a more feasible entry point into quantum computing, particularly for academic purposes [367, 370]. Furthermore, cloud services for quantum computing represent yet another pricing model, where users can access powerful quantum processors through a pay-per-use framework. IBM's Pay-As-You-Go plan is illustrative of this trend, starting at $96 per minute, allowing customers to utilize quantum processing capabilities without committing to substantial upfront investments [367, 368].

The economic landscape of quantum computers showcases a broad spectrum of costs, ranging from approximately $5,000 for educational models to tens of millions for advanced industrial-grade systems. Cloud-based services enhance accessibility further by offering flexible pricing structures based on actual usage, catering to diverse consumer needs within the quantum computing realm [367, 368, 370].

Chapter 7

Quantum Cryptography and Secure Communications

As quantum computing continues to advance, one of its most significant and widely discussed implications lies in the realm of cryptography and secure communications. While classical cryptographic systems have provided the backbone for modern security protocols, they rely on mathematical problems that quantum computers could potentially solve in a fraction of the time. This means that encryption methods currently protecting sensitive data, online transactions, and secure communications could become vulnerable in the age of quantum computing.

Up to this point, this book has explored the foundations of quantum mechanics, the nature of qubits and quantum gates, the power of quantum algorithms, and the challenges of error correction and hardware development. These chapters have laid the groundwork for understanding the capabilities of quantum computers, including their strengths and limitations. Now, as we transition into the domain of quantum cryptography, we turn our attention to how quantum mechanics can be leveraged not only as a tool for breaking encryption but also as a means to create fundamentally secure communication channels.

This chapter delves into quantum cryptographic protocols, such as Quantum Key Distribution (QKD) and the BB84 protocol, which use the principles of quantum mechanics to enable unbreakable encryption. It also explores post-quantum cryptography, which seeks to develop classical cryptographic algorithms that remain secure even against quantum adversaries. Additionally, we examine the concept of quantum random number generation, an essential tool for enhancing security in encryption schemes. Finally, this chapter discusses the looming threat that quantum computing poses to classical cryptographic systems, particularly those based on RSA and elliptic curve cryptography (ECC), and how researchers are actively working to develop new security solutions.

By understanding quantum cryptography, we gain insight into how quantum technologies are reshaping cybersecurity, from secure communication networks to the future of internet privacy. As quantum computing progresses toward real-world applications, these cryptographic advances will play a crucial role in protecting digital infrastructures and ensuring data security in the quantum era.

Quantum Key Distribution (QKD) and BB84 Protocol

Quantum Key Distribution (QKD) is a cryptographic technique that enables two parties to securely share encryption keys by leveraging the principles of quantum mechanics. Unlike classical key exchange protocols, which rely on mathematical complexity for security, QKD guarantees security through the laws of quantum physics, making it resistant to attacks from both classical and quantum computers.

The key advantage of QKD is its ability to detect eavesdropping. Any attempt by an attacker to intercept the quantum key introduces quantum disturbances, which can be detected by the communicating parties. If an intrusion is detected, the key exchange is aborted, ensuring that only secure keys are used for encryption.

BB84 Protocol

The BB84 protocol, proposed by Charles Bennett and Gilles Brassard in 1984, is the first and most widely used QKD protocol. It allows two parties, traditionally called Alice (sender) and Bob (receiver), to generate a shared secret key that can be used for secure encrypted communication.

The protocol operates by encoding quantum information into the polarization states of single photons, which are transmitted over a quantum channel. Since quantum states cannot be measured without disturbance, any attempt by an eavesdropper (Eve) to intercept the key can be detected.

The BB84 protocol consists of four main steps:

1. Quantum State Preparation (Encoding the Key)

- Alice randomly chooses a sequence of classical bits (0s and 1s), which will form the key.

- Each bit is then encoded into one of four possible polarization states of a photon:
    - Rectilinear Basis (|, —) → 0 is vertical (|), 1 is horizontal (—)
    - Diagonal Basis (/ , \ ) → 0 is 45° (), 1 is 135° (/)
- Alice randomly selects either the rectilinear or diagonal basis to prepare each photon.

2. Transmission Over a Quantum Channel

- The encoded photons are sent to Bob through a quantum communication channel, such as an optical fibre or free-space link.
- Since quantum states cannot be cloned, Eve cannot copy or intercept the photons without introducing errors.

3. Basis Selection and Measurement

- Bob randomly chooses a measurement basis (either rectilinear or diagonal) for each photon.
- If Bob chooses the correct basis (matching Alice's), he measures the correct bit.
- If Bob chooses the wrong basis, his measurement results are random (introducing 50% uncertainty).

4. Key Reconciliation and Error Detection

- After the transmission, Alice and Bob publicly compare the bases they used for each bit (but not the actual bit values).
- They discard the bits measured in the wrong basis, keeping only the correctly measured bits.
- This results in a shared raw key between Alice and Bob.
- Alice and Bob then perform error correction and privacy amplification to refine the key and remove any possible information that Eve might have obtained.

Once this process is complete, Alice and Bob share a secure cryptographic key, which can be used for encrypting messages using classical encryption schemes (e.g., AES encryption).

Working through these steps, Alice (the sender) aims to securely share a private cryptographic key with Bob (the receiver) over a quantum communication channel. The security of this process is based on the fundamental principles of quantum mechanics, specifically the no-cloning theorem and the disturbance of quantum states upon measurement.

Encoding the Quantum Key

Alice begins by generating two random bit strings, denoted as a and b, each containing n bits. The bit string a represents the raw key data that will eventually be shared with Bob, while the bit string b determines the encoding basis for each qubit. The combined values of a_i and b_i (where i is the index of a bit) dictate the quantum state of each qubit Alice prepares. The complete quantum state, representing n qubits, is given by:

$$|\psi\rangle = \bigotimes_{i=1}^{n} |\psi_{a_i b_i}\rangle$$

where each qubit $|\psi_i\rangle$ is determined by the values of $\mathbf{a_i}$ and $\mathbf{b_i}$ according to the following mapping:

- If $\mathbf{b_i = 0}$, Alice uses the computational basis:
  - $a_i = 0 \rightarrow |\psi_{00}\rangle = |0\rangle$
  - $a_i = 1 \rightarrow |\psi_{10}\rangle = |1\rangle$
- If $\mathbf{b_i = 1}$, Alice uses the Hadamard (diagonal) basis:
  - $a_i = 0 \rightarrow |\psi_{01}\rangle = |+\rangle = (|0\rangle + |1\rangle) / \sqrt{2}$
  - $a_i = 1 \rightarrow |\psi_{11}\rangle = |-\rangle = (|0\rangle - |1\rangle) / \sqrt{2}$

The key element here is that b_i determines whether the qubit is encoded in the computational basis ($|0\rangle$, $|1\rangle$) or the Hadamard basis ($|+\rangle$, $|-\rangle$). Since the bases are not mutually orthogonal, an observer who does not know b cannot reliably determine a without introducing errors.

Transmission Over the Quantum Channel

After encoding the qubits in the selected bases, Alice sends the quantum state $|\psi\rangle$ to Bob over a public but authenticated quantum communication channel denoted as E. This channel can be affected by environmental noise and potential eavesdropping from a third party, commonly referred to as Eve. The quantum state that Bob receives is $E(\rho) = E(|\psi\rangle\langle\psi|)$, meaning it may have been altered due to transmission imperfections or malicious interference.

Because Alice has kept her encoding basis b secret, neither Bob nor Eve can immediately determine the exact states of the received qubits. Eve cannot intercept and copy the qubits without disturbing them due to the no-cloning theorem, which states that an unknown quantum state cannot be perfectly duplicated. If Eve attempts to measure the qubits, she introduces errors with a 50% probability if she selects the incorrect measurement basis, making her presence detectable.

Bob's Measurement and Basis Selection

Upon receiving the qubits, Bob randomly selects a measurement basis for each qubit by generating his own random bit string b' of length n. This means Bob measures each qubit either in the computational basis (|0⟩, |1⟩) or in the Hadamard basis (|+⟩, |−⟩), independent of Alice's choices. As a result of this randomness:

- If Bob's basis b'_i matches Alice's basis b_i, his measurement yields the correct bit a_i with high probability.
- If Bob's basis b'_i differs from Alice's basis b_i, his measurement results are completely random, introducing uncertainty into the key.

Once Bob completes his measurements, he obtains his own raw bit string a', which contains some correctly received bits and some errors due to incorrect basis choices or noise.

Public Basis Reconciliation and Key Sifting

Bob now publicly announces that he has received Alice's transmission. Alice then publicly announces her choice of bases b over a classical authenticated communication channel. Bob compares b with b' and identifies which of his measurements were performed in the correct basis.

Both Alice and Bob discard the bits from their respective bit strings a and a' where $b_i \neq b'_i$, keeping only the bits where $b_i = b'_i$. This process is called key sifting and reduces the key length from n to approximately n/2, as Bob randomly chooses the correct basis about 50% of the time.

Error Checking and Eavesdropping Detection

After obtaining the sifted key of k bits, Alice and Bob need to check for errors, which could have been introduced by noise or eavesdropping. Alice randomly selects k/2 bits and discloses their values over the public channel. Bob checks whether his corresponding bits match Alice's disclosed bits.

- If the error rate is below a predefined threshold, Alice and Bob conclude that the quantum channel was secure, and they proceed to generate a final cryptographic key.
- If the error rate is too high, this suggests that Eve may have intercepted and measured the qubits, introducing disturbances. In this case, Alice and Bob abort the key exchange and restart the protocol.

This process ensures that any eavesdropping attempt is detected before the key is used for encryption.

Final Key Generation and Security Enhancements

To further improve the security of the shared key, Alice and Bob apply two additional steps:

1. Information Reconciliation: A classical error correction algorithm is applied to remove any remaining discrepancies between Alice's and Bob's keys, ensuring they hold identical bit strings.

2. Privacy Amplification: If Eve has obtained partial information about the key, Alice and Bob use a hash function to compress the key, eliminating any leaked information and producing a final shorter but more secure secret key.

After these steps, Alice and Bob now share a secure, identical cryptographic key, which can be used for symmetric encryption (e.g., AES encryption) to securely transmit messages.

Security of BB84 and Eavesdropping Detection

The security principles embodied in the BB84 quantum key distribution (QKD) protocol primarily rely on two key concepts: the no-cloning theorem and the principle of quantum measurement disturbance. The no-cloning theorem asserts that it is impossible to create an identical copy of an arbitrary quantum state. This property is crucial for security, as it implies that if an eavesdropper, traditionally named Eve, attempts to intercept and measure the quantum states being transmitted between the legitimate parties, Alice and Bob, she cannot create perfect copies of the quantum information, thus introducing errors into the transmission due to her limited knowledge about the encoding basis used by Alice [371-373].

Eve's attempts to measure the states lead to inevitable disturbances in the quantum channel, violating the delicate superposition and polarization states that are used to encode information. This disturbance is inherently linked to the Heisenberg uncertainty principle, which states that the act of measuring a quantum state alters it, leading to alterations in the key that Alice and Bob are trying to establish. The principle of quantum measurement disturbance ensures that Alice and Bob can detect discrepancies introduced by Eve's interference when they compare a portion of their shared key [374, 375]. If the error rate detected exceeds a predefined threshold, they recognize the potential presence of an eavesdropper and may choose to abort the key exchange, thus preserving the security of their communication [376, 377].

Furthermore, the robust framework provided by the BB84 protocol allows Alice and Bob to ascertain the fidelity of their key. The quantum bit error rate (QBER) serves as a vital parameter; consistent monitoring of this value allows them to determine the integrity of their communication. Research has indicated that the BB84 protocol can tolerate a maximum error rate of about 11% before secure key generation becomes impossible [377, 378]. In the presence of Eve, the detectable anomalies translate into errors that can be quantified and analysed, thus leveraging the quantum mechanical principles to ensure the integrity and secrecy of the transmission [374]. Hence, the combination of the no-cloning theorem and the measurement disturbance principle forms the backbone of the security landscape surrounding the BB84 protocol, making quantum key distribution a resilient approach against eavesdropping attempts.

Advantages, Challenges and Applications of QKD and BB84

Quantum Key Distribution (QKD) offers unbreakable security based on the fundamental laws of quantum physics rather than computational complexity. Unlike classical encryption methods, which rely on mathematical difficulty (such as factoring large numbers), QKD remains secure even against quantum computers capable of breaking traditional cryptographic algorithms. The security of QKD is ensured by the no-cloning theorem and quantum measurement disturbance, making it fundamentally resistant to unauthorized interception.

Another critical advantage of QKD is eavesdropping detection. Any attempt by an adversary to intercept or measure the quantum key inevitably introduces detectable disturbances in the quantum states being transmitted. This allows the communicating parties to identify potential security breaches and discard any compromised keys, ensuring the integrity of their communication.

QKD also provides long-term security, which is a crucial advantage over classical cryptographic systems. As advancements in quantum computing threaten the security of traditional encryption methods, QKD remains a future-proof solution. Since its security does not rely on computational assumptions, even the most powerful quantum computers cannot break a QKD-secured communication channel. This makes QKD an essential technology for organizations that require long-term data protection, such as governments, financial institutions, and defence agencies.

Despite its advantages, QKD faces several practical challenges and limitations. One of the most significant issues is transmission loss. When photons travel through fibre-optic cables, they experience attenuation, which limits the effective communication range to a few hundred kilometres. Environmental noise, scattering, and imperfections in optical components can further degrade the signal, making long-distance quantum communication difficult without additional technology.

To overcome these distance limitations, quantum repeaters are required. Quantum repeaters work by extending the range of quantum communication by using entanglement swapping and quantum teleportation to relay quantum states over large distances. However, practical and scalable quantum repeaters are still under development, making long-range QKD networks an ongoing research challenge. Without these repeaters, QKD is limited to point-to-point communication over moderate distances, restricting its widespread implementation.

Another major challenge is the high cost of QKD infrastructure. Unlike classical encryption systems that rely on software-based algorithms, QKD requires specialized quantum optics hardware, including single-photon sources, highly sensitive detectors, and dedicated quantum communication channels. The deployment and maintenance of QKD systems are expensive and complex, making them impractical for widespread consumer use at this stage. Only organizations with critical security needs, such as government agencies and financial institutions, can justify the investment in QKD technology.

Despite these challenges, QKD is already being commercially deployed in high-security applications where the need for long-term encryption protection is paramount. Financial institutions use QKD to ensure secure transactions and data transfers, particularly for high-value financial exchanges where the risk of cyberattacks is significant. Banks and stock exchanges are beginning to explore QKD as a means of securing their communication channels from future quantum threats.

Government communications also rely on QKD to protect classified information from cyber threats. Many countries are investing in quantum communication networks to secure their diplomatic, military, and intelligence communications. The ability to detect eavesdropping in real-time ensures that sensitive information remains protected from adversaries attempting to intercept or decrypt classified transmissions.

The development of a quantum-secured internet is another major application of QKD. Countries such as China, the United States, and members of the European Union are working on building global QKD networks. China's Micius satellite demonstrated the feasibility of space-based QKD, allowing secure key exchange over thousands of kilometers. Similarly, the European Quantum Communication Infrastructure (EuroQCI) aims to create a secure quantum network across Europe. These initiatives are laying the foundation for a future global quantum-secured internet, where communications can be protected from both classical and quantum cyber threats.

Quantum Key Distribution (QKD), particularly the BB84 protocol, provides a highly secure method for exchanging encryption keys using the principles of quantum mechanics. By leveraging the fundamental laws of quantum physics, QKD ensures that any eavesdropping attempt is detectable, making it a critical technology for post-quantum cryptography.

While technical challenges remain, including scalability, transmission losses, and infrastructure costs, continuous advancements in quantum networks and quantum repeaters are expected to expand the reach and practicality of QKD. As quantum technology evolves, QKD is likely to become an essential component of future secure communication systems, ensuring that sensitive data remains protected in an era where quantum computers could render classical encryption obsolete.

Practical Implementation of Quantum Key Distribution (QKD) and BB84 Protocol

Implementing Quantum Key Distribution (QKD) using the BB84 protocol requires both a quantum communication channel (such as fibre optics or free-space optical links) and a classical communication channel for basis reconciliation and error correction. Below is a step-by-step practical guide on how to implement BB84 QKD using real-world hardware.

Step 1: Setting Up the QKD System

To implement QKD in practice, we need the following components:

- **Quantum Transmitter (Alice)**: A device that generates and transmits quantum-encoded photons.
- **Quantum Receiver (Bob)**: A device that receives and measures the transmitted photons.
- **Quantum Channel**: A medium for transmitting single photons, such as fibre optics or free-space optical links.
- **Classical Channel**: A standard, authenticated network connection (e.g., TCP/IP) for basis reconciliation and error correction.
- **Single-Photon Sources**: A laser or attenuated light source capable of producing individual photons.
- **Single-Photon Detectors (SPDs)**: Highly sensitive detectors capable of measuring single photons.
- **Polarization Controllers and Beam Splitters**: Optical components to prepare and manipulate photon states.
- **Synchronization Electronics**: Hardware to coordinate timing between Alice and Bob.

Step 2: Generating Quantum States (Alice's Transmission)

1. Random Bit Generation

- Alice generates two random bit strings:
    - **Key bits (a)**: Represents the raw key values (0s and 1s).
    - **Basis bits (b)**: Determines whether each bit is encoded in the computational basis (**|0⟩, |1⟩**) or the Hadamard basis (**|+⟩, |−⟩**).

- Example:

```
a = 10110010
b = 01100101
```

2. Encoding Photons in Quantum States

- Alice prepares each photon according to the following mapping:
    - If b_i **= 0** → Computational Basis
        - **a_i = 0 → |0⟩ (Vertical Polarization)**
        - **a_i = 1 → |1⟩ (Horizontal Polarization)**
    - If b_i **= 1** → Hadamard Basis
        - **$a_i = 0 \rightarrow |+\rangle = (|0\rangle + |1\rangle) / \sqrt{2}$**
        - **$a_i = 1 \rightarrow |-\rangle = (|0\rangle - |1\rangle) / \sqrt{2}$**

3. Photon Transmission

- Alice uses a **polarized light source** (such as a laser and polarization controller) to generate and send each photon along the quantum channel (optical fibre or free space).

Step 3: Receiving and Measuring Photons (Bob's Detection)

1. Random Basis Selection

- Bob generates a random bit string **b'** to decide which basis to use for each measurement.
- Example:

```
b' = 11000110
```

2. Photon Measurement

- Using polarization beam splitters and single-photon detectors, Bob measures each photon in either the computational basis or the Hadamard basis, based on his random choice **b'**.
- If Bob's basis choice b'_i matches Alice's basis b_i, he measures the correct bit a_i.

- If Bob's basis choice **b'_i** does not match Alice's basis **b_i**, his measurement result is **random**.

3. Raw Key Generation

  - Bob records his measurement results as a raw key **a'**, which contains some correct and some incorrect bits due to basis mismatch.

Step 4: Basis Reconciliation Over a Public Channel

1. Alice and Bob Announce Basis Choices

- Over the public classical channel, Alice and Bob compare their basis choices **b** and **b'**.
- They discard all bits where **$b_i \neq b'_i$**, keeping only the correctly measured bits.
- Example:

```
Alice: a  = 10110010
Alice: b  = 01100101
Bob:   b' = 11000110
Bob:   a' = 10010110
```

- Matching basis indices: **b = b' at positions [3, 6, 7]**
- Final sifted key:

```
Alice: a  = 110
Bob:   a' = 110
```

2. Eavesdropping Check (Error Detection)

- Alice and Bob randomly select a subset of the key (e.g., half the bits) and compare them publicly.
- If the error rate is too high, it suggests an eavesdropper (Eve) interfered, and they discard the session.
- If the error rate is below a threshold, they proceed to key distillation.

Step 5: Error Correction and Privacy Amplification

1. Error Correction (Information Reconciliation)

- Alice and Bob correct any discrepancies in their sifted keys using classical error correction codes (e.g., Cascade or Winnow algorithms).

2. Privacy Amplification

- To remove any partial information Eve might have gained, Alice and Bob apply a hash function (e.g., SHA-256) to produce a final, shorter secure key.

Step 6: Using the Shared Key for Secure Communication

1. Key Storage

- The final shared key is stored securely in Alice and Bob's encryption systems.

2. Message Encryption

- The key is used in a symmetric encryption algorithm such as AES (Advanced Encryption Standard) or One-Time Pad to encrypt and decrypt messages.

3. Message Transmission

- Alice encrypts a message using the key and sends it over a public classical network.
- Bob decrypts the message using the same key, ensuring unbreakable security.

Table 17: Example Hardware Setup for Implementing BB84 QKD.

Component	Description
Single-Photon Source	Weak coherent laser or spontaneous parametric down-conversion (SPDC) source
Polarization Modulator	Liquid crystal or electro-optic modulator to encode quantum states
Beam Splitters	Splits photons based on polarization
Single-Photon Detectors	Avalanche photodiodes (APDs) or superconducting nanowire detectors
Optical Fibers	Quantum channel for transmitting photons
Classical Network	TCP/IP-based authenticated channel for basis reconciliation

Post-Quantum Cryptography

Post-Quantum Cryptography (PQC) refers to cryptographic algorithms that are designed to be secure against attacks from quantum computers. Unlike classical cryptographic methods, which rely on the computational difficulty of mathematical problems like integer factorization (RSA) and discrete logarithms (ECC), PQC is built on problems that remain hard even for powerful quantum computers.

Quantum computers pose a major threat to traditional cryptographic systems because algorithms such as Shor's algorithm can efficiently break widely used encryption schemes, including RSA, Diffie-Hellman, and Elliptic Curve Cryptography (ECC). To counter this, researchers are developing quantum-resistant cryptographic algorithms that will remain secure even in a world where large-scale quantum computers exist.

The emergence of quantum computing represents a significant shift in computing paradigms, with profound implications for information security. Current encryption standards, which rely on the mathematical difficulty of certain problems for their security, face an existential threat from quantum technologies. Quantum computers, especially through the implementation of Shor's algorithm, can efficiently solve problems like integer factorization and discrete logarithms, which underpin common encryption methods such as RSA and ECC [21, 379, 380]. This capability poses severe risks across various domains, including financial transactions, government communications, personal privacy, and broader internet security.

The reliance on asymmetric encryption for secure online banking and credit card transactions renders these systems vulnerable to quantum attacks. As conventional algorithms like RSA become susceptible, the safety of electronic payments and secure financial communications is under threat [381, 382]. If quantum computers achieve sufficient capability, they may decrypt data that has been stored securely under current standards, thus compromising sensitive financial information [383, 384].

Classified communications, essential for national security, also rely heavily on current encryption standards. The quantum threat could lead to significant vulnerabilities within government data integrity, revealing sensitive information to adversaries who could use quantum computing to break existing encryption schemes [381, 385]. The anticipated arrival of sufficiently powerful quantum computers underscores the urgent need for standardization of post-quantum cryptography methods to protect sensitive information [386, 387].

Secure messaging applications, email encryption, and electronic health records are increasingly threatened by the capabilities of quantum computing [381, 388]. Current encryption techniques that safeguard personal information, such as AES and ECC, are at risk due to the potential for quantum computers to exploit their underlying vulnerabilities, rendering personal data susceptible to unauthorized access and misuse [389].

Quantum Computing

Protocols that secure internet communications, including HTTPS and VPNs, are at risk of obsolescence. Quantum computing can undermine digital certificates and the integrity of secure communications established through classical encryption, thus exposing vulnerabilities that could be exploited through advanced quantum-driven attack techniques [380, 390]. The transition towards quantum-safe methods is critical to ensure ongoing security in digital spaces [383, 384].

The concept of "Harvest Now, Decrypt Later" emphasizes the pressing need for immediate action as quantum technologies continue to evolve. Attackers can seize encrypted data today with the intention of decrypting it in the future when quantum computing capabilities mature [382]. As a proactive measure, the adoption of post-quantum cryptography (PQC) is crucial for safeguarding against these future vulnerabilities. There is a consensus in the literature that transitioning to PQC will be fundamental for preserving the confidentiality and integrity of sensitive data against quantum threats [385, 387, 389].

Quantum Attacks on Classical Cryptography

Classical cryptographic systems rely on the computational difficulty of certain mathematical problems to ensure security. However, the development of quantum computing poses a significant threat to these systems due to the efficiency of quantum algorithms that can solve these problems exponentially faster than classical methods. Two of the most critical quantum algorithms that threaten classical encryption are Shor's algorithm and Grover's algorithm.

Shor's algorithm, developed by Peter Shor in 1994, is a quantum algorithm that can efficiently factor large integers. This poses a direct threat to the RSA encryption scheme, which relies on the difficulty of factoring large numbers as its security foundation. Additionally, Shor's algorithm can solve the discrete logarithm problem, which is the basis of Diffie-Hellman key exchange and Elliptic Curve Cryptography (ECC). With a sufficiently powerful quantum computer, these cryptographic protocols can be completely broken, exposing encrypted communications and digital signatures to quantum attacks. A quantum computer with thousands of fault-tolerant qubits could factor numbers large enough to break RSA encryption within minutes, rendering current public-key cryptographic systems obsolete.

Grover's algorithm, on the other hand, does not completely break symmetric encryption but reduces the time complexity of brute-force attacks. In classical computing, finding a secret key in an n-bit key space requires $O(2^n)$ operations. Grover's algorithm reduces this to $O(2^{n/2})$, effectively halving the security level of symmetric encryption schemes. This means that AES-128 encryption would offer the same level of security as AES-64, making it vulnerable to quantum attacks. To maintain security in a post-quantum world, cryptographers recommend using AES-256, which remains secure even under quantum attacks since Grover's speedup still requires exponential time.

These quantum attacks demonstrate the urgent need for Post-Quantum Cryptography (PQC). Classical encryption methods that are vulnerable to Shor's and Grover's algorithms must be replaced with quantum-resistant cryptographic algorithms. Researchers are actively developing new cryptographic techniques, such as lattice-based, code-based, and hash-based cryptography, to ensure secure communication and data protection in the age of quantum computing. The transition to PQC is critical for safeguarding sensitive information from future quantum threats.

Types of Post-Quantum Cryptographic Algorithms

In recent years, the surge in quantum computing capabilities has necessitated the exploration and development of post-quantum cryptographic (PQC) algorithms. These algorithms are pivotal as they aim to secure communications against potential quantum threats. The field consists of various classes of algorithms, notably including lattice-based cryptography, code-based cryptography, multivariate polynomial cryptography, hash-based cryptography, and isogeny-based cryptography. Each class presents distinct mathematical foundations, advantages, and challenges, leading to diverse applications in securing digital information.

Lattice-Based Cryptography: Lattice-based cryptography is predicated on the computational intractability of problems such as Learning With Errors (LWE) and the Shortest Vector Problem (SVP). These problems remain challenging even for quantum computers, making lattice-based constructions a prominent area in post-quantum research. Notable implementations include Kyber for encryption and Dilithium for digital signatures, which have gained attention during the NIST PQC standardization process [391, 392]. Additionally, the algorithms in this category are often noted for their efficiency and speed, contributing to their growing popularity [393].

Code-Based Cryptography: Code-based cryptography employs the complexity of decoding error-correcting codes, with Classic McEliece being one of the earliest known implementations. Although this category has a long history of demonstrating security against both classical and quantum attacks, its main drawback is the large key sizes required for robust security [394, 395]. Recent advancements aim to mitigate size-related challenges while still leveraging the established security properties of code-based systems [394].

Multivariate Polynomial Cryptography: Multivariate polynomial cryptography is based on the difficulty of solving systems of multivariate quadratic equations. Algorithms such as Rainbow and GeMSS exemplify this class, which is primarily tailored for digital signatures [396]. This approach is appreciated for its rapid computation and relatively small signature sizes, making it a favourable option for certain applications where efficiency is paramount [397].

Hash-Based Cryptography: Hash-based cryptographic approaches utilize hash functions to create secure signatures resistant to quantum attack, as illustrated by schemes like SPHINCS+ and XMSS. These methods benefit from a well-understood foundation, relying heavily on the properties of cryptographic hashes such as those in SHA-2 and SHA-3 [398, 399]. The advantages include low computational overhead and robustness for long-term data integrity applications [400]. Moreover, the hash-based schemes have been formally standardized and recognized for their security resilience against quantum threats [399, 400].

Isogeny-Based Cryptography: Relying on the mathematical properties of elliptic curves, isogeny-based cryptography represents a newer field within PQC. While notable examples like SIKE and CSIDH show promise, there are concerns regarding vulnerabilities under quantum analysis [401]. The challenges inherent in this approach include solvable problems by current quantum techniques, though ongoing research continues to refine the scope and application of isogeny-based constructs for secure communications [401, 402].

NIST Post-Quantum Cryptography Standardization

The National Institute of Standards and Technology (NIST) has initiated a critical process to standardize Post-Quantum Cryptography (PQC) algorithms to secure cryptographic systems against potential quantum computing threats. This effort is part of a broader strategy to transition from classical cryptographic methods, such as RSA and ECC, which are vulnerable to quantum attacks, to robust, quantum-resistant algorithms. The ongoing NIST PQC Standardization process began in December 2017, and it has progressed through multiple rounds of evaluation, ultimately culminating in the announcement of four finalists in July 2022: Kyber, Dilithium, FALCON, and SPHINCS+ [403]. These candidates were selected from an initial pool of 69 algorithms based on their security and performance metrics.

The four finalists represent distinct cryptographic approaches: Kyber, a lattice-based algorithm designed for public-key encryption and key encapsulation; Dilithium and FALCON, both lattice-based algorithms focused on digital signatures; and SPHINCS+, which employs a hash-based method for digital signatures. The lattice-based schemes (Kyber, Dilithium, and FALCON) are particularly noted for their resilience against quantum attacks due to their reliance on hard mathematical problems related to lattice structures, which are currently considered intractable for quantum computers [404, 405].

Kyber has been specifically tailored for public-key encryption scenarios, while Dilithium aims to provide digital signature functionalities with strong security assurances based on lattice problem hardness [406, 407]. FALCON also focuses on digital signatures through a secure hash-and-sign approach rooted in lattice problems. Researchers emphasize that these algorithms must not only be resilient to quantum attacks but also exhibit efficiency in both software and hardware implementations [406, 408]. This dual focus on security and performance is vital as the migration to these new standards, expected around 2024-2025, aims to ensure that cryptographic infrastructures remain secure against evolving technological threats [405, 408].

Moreover, the implications of the NIST standardization process extend beyond the algorithms themselves, influencing various applications, particularly in sectors leveraging critical digital infrastructure, such as automotive systems and mobile communications [409]. The anticipated adoption of these algorithms underscores the need to prepare existing cryptographic frameworks to incorporate PQC standards, ultimately providing a robust defence against emerging quantum threats [410].

Transitioning to Post-Quantum Cryptography

The transition from classical cryptography to Post-Quantum Cryptography (PQC) is a complex process that requires careful planning and implementation to ensure security while maintaining compatibility with existing systems. Since classical encryption methods such as RSA, ECC, and Diffie-Hellman key exchange are vulnerable to quantum attacks, organizations must begin integrating quantum-resistant cryptographic algorithms before large-scale quantum computers become a reality.

One of the first steps in this transition is the adoption of hybrid cryptographic systems. Many organizations are implementing hybrid encryption, which combines classical cryptographic methods with PQC algorithms to provide backward compatibility while ensuring security against future quantum threats. This approach allows existing infrastructure to continue functioning while introducing quantum-resistant protection. For example, TLS 1.3 Hybrid Mode integrates traditional RSA and ECC encryption with Kyber, a lattice-based key exchange protocol that provides quantum-resistant security for internet communication. Hybrid encryption serves as a temporary safeguard during the transition to full quantum security.

Another critical aspect of the transition is upgrading cryptographic infrastructure. Governments, financial institutions, and corporations must replace vulnerable encryption protocols with quantum-resistant alternatives to protect sensitive data. Financial institutions, for instance, are already preparing to update secure transactions using Kyber-based key exchange, which offers robust protection against quantum attacks. This transition will require updating hardware security modules (HSMs), cryptographic libraries, and communication protocols to support PQC algorithms while maintaining performance and reliability.

Quantum-resistant digital signatures are also an essential component of the transition. Digital certificates, identity verification systems, and blockchain-based authentication mechanisms rely heavily on cryptographic signatures that could become insecure once quantum computers break RSA and ECC. To address this, organizations are adopting post-quantum digital signatures such as SPHINCS+, a hash-based digital signature scheme that remains secure against quantum attacks. SPHINCS+ can replace traditional RSA-based certificates, ensuring that digital authentication, document signing, and software integrity verification remain protected in a post-quantum world.

Securing cloud computing and VPN services is another major challenge in the transition to PQC. Cloud service providers like Google and Microsoft are actively testing post-quantum secure VPNs using PQC algorithms to protect sensitive data transmissions. Since quantum computers could eventually decrypt encrypted cloud storage, emails, and virtual private network (VPN) connections, it is crucial to integrate PQC protocols into these services to maintain long-term data security. Ensuring that cloud security frameworks, encrypted backups, and secure remote access solutions remain quantum-resistant is a priority for technology companies and cybersecurity experts worldwide.

As the development of large-scale quantum computers progresses, organizations must act now to transition from classical cryptographic systems to PQC. The implementation of hybrid encryption, infrastructure upgrades, quantum-resistant digital signatures, and secure cloud services will ensure that data remains protected against quantum threats. The shift to quantum-safe cryptography is not just a theoretical concern—it is a necessary step to future-proof digital security in an era of rapidly advancing quantum technology.

Post-Quantum Cryptography (PQC) is becoming a critical security solution across various industries as organizations prepare for the potential threat of quantum attacks. One of the most significant areas where PQC is being integrated is banking and finance. Financial institutions rely on encryption to protect online transactions, prevent fraud, and ensure the security of customer data. With quantum computers posing a risk to RSA-based digital signatures, banks are beginning to adopt quantum-resistant cryptographic algorithms to secure financial transactions and digital signatures on critical documents. This transition ensures that sensitive financial information remains safe from potential quantum decryption threats.

Government and defence agencies are also prioritizing PQC to protect classified communications and national security data. Governments store and transmit vast amounts of sensitive information, including diplomatic communications, military intelligence, and defence strategies. If adversaries were to intercept and store encrypted government data today, they could potentially decrypt it in the future when large-scale quantum computers become available. To prevent this, national cybersecurity agencies are developing quantum-safe encryption protocols to safeguard government communications from future quantum threats.

In the healthcare and medical records sector, PQC is crucial for long-term data security and patient privacy compliance. Medical records, genomic data, and research databases must remain secure for decades, as unauthorized access to such information can lead to serious privacy violations and ethical concerns. With quantum computing posing a future risk to traditional encryption, hospitals, research institutions, and healthcare providers are beginning to implement quantum-resistant cryptographic solutions to protect patient records and maintain compliance with privacy regulations such as HIPAA and GDPR.

Telecommunications and internet security are also heavily impacted by the need for quantum-resistant encryption. Internet communications, cloud data storage, and virtual private networks (VPNs) rely on encryption protocols to protect user data and online transactions. Leading technology companies are actively working on PQC-based encryption standards to ensure that cloud computing platforms and global communication networks remain secure against quantum-enabled cyberattacks. The transition to quantum-safe TLS (Transport Layer Security) and quantum-secure VPNs will help protect businesses and consumers from data breaches in the post-quantum era.

Blockchain and cryptocurrencies are another area where PQC is playing a crucial role. Blockchain networks rely on public-key cryptography to secure transactions, verify digital signatures, and prevent unauthorized access to digital assets. If quantum computers become capable of breaking elliptic curve cryptography (ECC), existing blockchain systems could become vulnerable to attacks, allowing malicious actors to forge transactions and steal digital assets. To counter this threat, blockchain developers are exploring quantum-resistant cryptographic techniques to ensure the long-term security of cryptocurrencies, decentralized finance (DeFi) platforms, and smart contracts. Some blockchain projects are already experimenting with lattice-based cryptographic schemes to prepare for a quantum-secure future.

As quantum computing continues to advance, industries across banking, government, healthcare, telecommunications, and blockchain are actively integrating post-quantum cryptographic solutions to protect sensitive data and secure digital communications. The transition to quantum-resistant encryption is a crucial step in maintaining cybersecurity resilience against the evolving threats posed by quantum technology.

Post-Quantum Cryptography (PQC) is a critical response to the growing threat of quantum computing. Since Shor's algorithm can break existing encryption methods, organizations must transition to quantum-resistant cryptographic algorithms before large-scale quantum computers become operational.

The NIST PQC standardization process is leading the development of lattice-based, hash-based, and code-based encryption methods that will replace vulnerable public-key cryptography. Governments, businesses, and cybersecurity professionals must start planning now to adopt PQC algorithms to ensure long-term data security and resilience against future quantum threats.

Quantum Random Number Generation

Quantum Random Number Generation (QRNG) represents a significant advancement in the production of truly random numbers, leveraging the inherent unpredictability of quantum mechanics rather than the deterministic algorithms utilized by classical random number generators (pseudo-random number generators, PRNGs). Unlike PRNGs, which can be traced back to an initial state or seed, QRNGs derive randomness from quantum phenomena such as superposition, quantum measurement, and the probabilistic nature of photon behaviour [411, 412]. This fundamental difference enables QRNGs to produce sequences that are completely unpredictable, making them ideal for applications requiring high levels of security and true randomness, such as quantum key distribution (QKD) and secure communications [413, 414].

The significance of QRNG lies in its ability to eliminate the vulnerabilities associated with classical RNGs. Classical RNGs, despite employing complex algorithms, are ultimately deterministic, thus potentially susceptible to reverse engineering if the seed or algorithm is compromised [415]. The peculiarities of quantum mechanics ensure that outcomes cannot be predetermined, allowing QRNGs to generate numbers that are fundamentally unrepeatable, thereby providing enhanced security for cryptographic systems [416, 417]. These properties render QRNGs crucial for applications where traditional algorithms may expose systems to significant risks, including cryptographic key generation and secure session tokens [418, 419].

Recent research highlights advancements in QRNG technology, which improve both the quality and generation rates of random numbers, enhancing their applicability across multiple domains such as finance and statistics [420, 421]. The use of randomness extractors in conjunction with QRNG systems minimizes any potential noise or predictability stemming from the quantum source, thereby solidifying the reliability of the generated randomness [422, 423]. Overall, QRNG serves as a robust foundation for securing digital information systems, providing theoretical guarantees drawn from quantum mechanics and practical implementations that address the limitations of classical RNG paradigms [423].

Quantum Random Number Generators leverage various quantum phenomena to generate randomness. Some of the most commonly used physical processes include:

1. Single-Photon Detection

In this method, a light source (usually a laser or LED) emits single photons that hit a beam splitter. A beam splitter is an optical device that sends each photon down one of two possible paths with a 50% probability. The photon is then detected by one of two single-photon detectors, with each detector corresponding to a binary output (0 or 1). Since quantum mechanics dictates that the path a photon takes is completely random, this setup generates a truly random sequence of bits.

Example Process:

- A single photon is emitted.
- It encounters a beam splitter, which randomly directs it toward Detector A or Detector B.
- If the photon reaches Detector A, the output is 0.

- If the photon reaches Detector B, the output is 1.

Repeating this process many times produces a random bit stream.

2. Quantum Vacuum Fluctuations

Quantum mechanics predicts that even in a vacuum, fluctuations in electromagnetic fields occur due to the Heisenberg Uncertainty Principle. QRNGs can use these fluctuations as a source of randomness by measuring the vacuum noise in a quantum system.

Example Process:

- A laser or an electronic device detects quantum vacuum fluctuations.
- These fluctuations are converted into random electrical signals.
- The signals are digitized into a random number sequence.

3. Electron Spin or Nuclear Decay

Another approach involves measuring electron spin states or detecting radioactive nuclear decay events. Since quantum mechanics dictates that these events occur randomly, they can be used to generate unpredictable random numbers.

Example Process:

- A quantum system is prepared in a superposition state.
- When measured, the state collapses into either spin-up (0) or spin-down (1) randomly.
- The measurement results form a random binary sequence.

One of the key advantages of Quantum Random Number Generators (QRNGs) is their ability to produce true randomness. Unlike classical random number generators (RNGs), which rely on deterministic algorithms to simulate randomness, QRNGs leverage quantum mechanical principles such as superposition and quantum measurement to generate completely unpredictable sequences. This ensures that the randomness is fundamental and not reliant on computational methods that could be reverse-engineered.

QRNGs also provide exceptional security and cryptographic strength, making them ideal for applications where high-level encryption is required. Traditional Pseudo-Random Number Generators (PRNGs) can be biased, compromised, or even reproduced if an attacker gains access to the algorithm or initial seed value. In contrast, QRNGs generate randomness from quantum processes that cannot be cloned or predicted, making them a critical component for quantum cryptography, secure encryption, and authentication systems.

Another significant advantage of QRNGs is that they exhibit no periodicity. PRNGs eventually repeat their sequences over time due to their algorithmic nature, which can introduce vulnerabilities in systems requiring long-term security. QRNGs, however, do not rely on algorithmic cycles and instead generate continuous, non-repeating random outputs, making them superior for long-term security applications such as cryptographic key generation and secure communications.

Furthermore, QRNGs offer resistance to reverse engineering. Since they rely on quantum mechanics rather than computational algorithms, their output cannot be predicted, simulated, or reconstructed by any classical or quantum computer. This makes QRNGs particularly valuable for cybersecurity, military applications, and financial systems that demand absolute unpredictability in their encryption processes.

Despite their advantages, QRNGs face several challenges that limit their widespread adoption. One of the primary challenges is hardware complexity. Unlike classical RNGs, which can be implemented using simple software algorithms or standard electronic circuits, QRNGs require specialized quantum hardware such as beam splitters, photon detectors, and quantum noise sensors. This adds complexity to the design, manufacturing, and maintenance of QRNG systems.

Another limitation is the need for calibration and noise reduction. While quantum processes are inherently random, external factors such as detector inefficiencies, environmental noise, and system imperfections can introduce unwanted biases. To maintain true randomness, QRNGs must undergo precise calibration and error correction to eliminate these disturbances.

The cost and accessibility of QRNGs also remain a barrier to mainstream adoption. High-precision quantum hardware is expensive to produce, and QRNG technology is currently not widely available for consumer applications. However, companies such as ID Quantique, QuintessenceLabs, and Toshiba are actively developing commercial QRNG devices, and ongoing advancements in quantum hardware may eventually make QRNGs more affordable and scalable.

Cryptography and Cybersecurity are among the most critical applications of QRNG. In Quantum Key Distribution (QKD), QRNGs are used to generate truly random cryptographic keys that ensure secure encryption. Since QKD allows parties to detect eavesdropping attempts, QRNG-based encryption is being deployed in banks, government agencies, and military networks to protect sensitive communications from both classical and quantum cyber threats.

In online gambling and gaming, QRNG ensures fair and unbiased randomness in applications such as digital lotteries, slot machines, and card games. Many gambling systems rely on provable randomness to maintain trust and integrity, and QRNGs provide a secure, tamper-proof method of generating unpredictable gaming outcomes.

Scientific fields such as quantum physics, artificial intelligence, and climate modelling benefit from QRNG's ability to generate unbiased random data for complex simulations. Monte Carlo simulations, used in finance, physics, and engineering, require high-quality randomness for accurate modelling of probabilistic systems. QRNG enhances these simulations by ensuring that no algorithmic biases are present in the data.

Blockchain and cryptocurrencies also rely on secure randomness for wallet key generation, smart contract security, and blockchain consensus mechanisms. Since quantum computers could eventually break traditional cryptographic keys, QRNG is being explored as a method to create post-quantum secure blockchain systems that are resistant to quantum attacks.

For national security and military applications, QRNG plays a vital role in secure communications, intelligence encryption, and cybersecurity defence strategies. Governments and defence agencies use QRNG-generated encryption keys to protect classified military operations and diplomatic communications, ensuring that adversaries cannot compromise their security infrastructure.

Quantum Random Number Generation (QRNG) is a breakthrough technology that provides true, unbreakable randomness based on the fundamental principles of quantum mechanics. By leveraging photon behaviour, quantum vacuum fluctuations, and electron spin measurements, QRNG produces completely unpredictable sequences that are essential for secure encryption, cryptographic key generation, and scientific applications.

As the world prepares for quantum computing advancements, QRNG is emerging as a critical tool for cybersecurity, blockchain, AI, and national defence. While challenges such as hardware complexity, calibration, and cost remain, continued research and commercialization efforts are making high-quality quantum randomness more accessible for the future of secure communications, AI modelling, and next-generation encryption protocols.

The Threat of Quantum Computing to Classical Cryptography (Breaking RSA & ECC)

Classical cryptographic systems, such as RSA (Rivest-Shamir-Adleman) encryption and Elliptic Curve Cryptography (ECC), rely on the computational difficulty of mathematical problems that are practically infeasible to solve with classical computers. These encryption methods secure internet transactions, financial communications, government data, and digital identities. However, with the advent of large-scale quantum computers, these cryptographic algorithms face a significant existential threat.

Quantum computers use quantum bits (qubits) and take advantage of quantum parallelism, superposition, and entanglement to solve certain mathematical problems exponentially faster than classical computers. One of the most severe consequences of this capability is the ability to break public-key cryptography through Shor's Algorithm, which can efficiently factor large integers and compute discrete logarithms—rendering RSA, ECC, and Diffie-Hellman key exchange insecure.

RSA encryption is one of the most widely used cryptographic protocols, securing everything from secure email, HTTPS connections, VPNs, banking transactions, and government communications. Its security is based on the mathematical challenge of integer factorization—the difficulty of breaking a product of two large prime numbers. The strength of RSA depends on the fact that classical computers require an infeasible amount of time to factorize large numbers.

1. Classical RSA Security

In RSA, encryption keys are generated as follows:

- Choose two large prime numbers, p and q.
- Compute their product N = p × q, which serves as the public modulus.
- Select a public exponent e (usually 65537 for efficiency).
- Compute a private exponent d using modular arithmetic, which remains secret.

The security of RSA depends on the fact that given N, it is extremely difficult for classical computers to recover p and q through factorization. For example, factoring a 2048-bit RSA key would take billions of years using the best-known classical algorithms.

2. Shor's Algorithm and Quantum Factorization

In 1994, Peter Shor developed Shor's Algorithm, a quantum algorithm that can factor large integers exponentially faster than classical algorithms. While classical computers require sub-exponential or exponential time for factorization, Shor's Algorithm solves it in polynomial time.

Using a sufficiently large and error-corrected quantum computer, Shor's Algorithm can break RSA-2048 encryption in minutes or hours, rendering all communications encrypted with RSA completely insecure. This means that encrypted emails, banking transactions, digital signatures, and SSL/TLS security mechanisms could be decrypted by adversaries once large-scale quantum computers become available.

Elliptic Curve Cryptography (ECC) is another widely used public-key encryption method that is more efficient than RSA because it provides the same level of security with much smaller key sizes. ECC is used in Bitcoin, blockchain security, digital signatures, and secure authentication protocols such as TLS, SSH, and ECDSA.

1. Classical ECC Security

ECC security is based on the Elliptic Curve Discrete Logarithm Problem (ECDLP), which is significantly harder to solve than traditional discrete logarithms. Given an elliptic curve equation:

$$y^2 = x^3 + ax + b$$

A public key is generated as:

- Choose a private key d (a random integer).
- Compute the public key Q = dG, where G is a predefined base point on the curve.

The security of ECC relies on the fact that, given G and Q, it is infeasible to determine d (the private key) due to the difficulty of solving the discrete logarithm problem. For classical computers, breaking ECC is harder than RSA, which is why ECC-256-bit keys provide security equivalent to 3072-bit RSA keys.

2. Shor's Algorithm and ECC

Shor's Algorithm can also solve the discrete logarithm problem efficiently, breaking ECC in polynomial time. This means that once a quantum computer with thousands of logical qubits becomes available, it will be able to compute private keys from public keys almost instantly, completely breaking ECC-based encryption.

Since ECC is widely used in blockchain technology, cryptocurrency wallets, smart contracts, and authentication mechanisms, the impact of breaking ECC is enormous. Attackers with access to a quantum computer could steal cryptocurrency funds, forge digital signatures, and bypass authentication protocols, leading to massive security breaches across multiple industries.

Currently, the realm of quantum computing has made significant strides; however, the largest quantum computers developed by organizations such as IBM, Google, and IonQ still face limitations in capacity and reliability. Presently, these quantum systems possess dozens to hundreds of qubits, which are inherently noisy and susceptible to errors, thus hindering their capability to run critical quantum algorithms such as Shor's Algorithm effectively. Shor's Algorithm is particularly important as it has the potential to factor large integers efficiently, posing a significant threat to traditional cryptographic systems like RSA and ECC. However, existing quantum computers do not yet have the necessary fault tolerance and error correction needed to successfully execute this algorithm on a substantial scale [88, 424, 425].

The theoretical consensus suggests that to break RSA-2048 or ECC-256, a fault-tolerant quantum computer would require several thousand logical qubits, which corresponds to millions of physical qubits due to the required error correction mechanisms [11, 156, 425]. While researchers project that such large-scale, fault-tolerant quantum computers could be realized within 10 to 20 years [426], the current state of quantum hardware remains firmly entrenched in what is termed the Noisy Intermediate-Scale Quantum (NISQ) era. These NISQ devices are characterized by significant noise levels, limiting their practical utility for reliable quantum computations [88, 427].

Anticipating the implications of quantum computing on cybersecurity, experts assert that governments and intelligence agencies worldwide are increasingly investing in quantum research to mitigate potential threats posed by future quantum computers capable of decrypting sensitive information. This foresight leads to emerging strategies among adversaries, notably the "Harvest Now, Decrypt Later" approach, where encrypted data is stored with the intent to decrypt it later once quantum computing technology has matured sufficiently to mitigate the defences of classical encryption [156, 425, 426]. The proactive measures being put in place underline the rapid evolution and strategic importance of quantum computing research as it moves toward operational maturity.

Given the looming threat of quantum attacks, governments, cybersecurity experts, and major tech companies are actively working on transitioning to Post-Quantum Cryptography (PQC)—cryptographic algorithms that remain secure even against quantum computers.

The National Institute of Standards and Technology (NIST) has been leading the effort to standardize quantum-resistant encryption algorithms, with lattice-based cryptography, hash-based cryptography, and code-based cryptography emerging as promising solutions. Algorithms like Kyber, Dilithium, and SPHINCS+ are expected to replace RSA and ECC in the coming years.

Governments, financial institutions, and enterprises must begin transitioning to quantum-safe encryption now to protect data that needs to remain secure for decades. This involves:

- Deploying hybrid encryption schemes that combine classical and post-quantum cryptography.
- Updating digital certificates and authentication mechanisms to quantum-resistant standards.
- Implementing Quantum Key Distribution (QKD) for ultra-secure communication.
- Researching blockchain security upgrades to withstand quantum attacks.

Emerging Quantum Security Solutions

As quantum computing continues to progress, it presents a substantial risk to traditional cryptographic techniques, including widely-used methods such as RSA, ECC, and Diffie-Hellman key exchange. To mitigate these threats, researchers and industry experts are actively pursuing various emerging quantum security solutions. These solutions primarily include Post-

Quantum Cryptography (PQC), Quantum Key Distribution (QKD), Quantum Random Number Generation (QRNG), and quantum-resistant blockchain security.

PQC consists of cryptographic algorithms specifically designed to withstand quantum attacks. Unlike traditional methods, which rely on mathematical problems solvable by quantum computers—such as those tackled by Shor's algorithm—PQC hinges on problems that remain challenging even for quantum computers [428]. The National Institute of Standards and Technology (NIST) is spearheading standardization efforts for PQC, having selected four algorithms in July 2022: Kyber, Dilithium, FALCON, and SPHINCS+ [429]. These algorithms accommodate key exchange, digital signatures, and stateless signatures, thus ensuring long-term data security [429]. As organizations prepare for a future with quantum capabilities, adopting PQC is viewed as a critical measure for safeguarding sensitive information [430].

Given the lengthy phase-in period required for full adoption of PQC, many organizations are opting for hybrid cryptographic systems that merge classical and post-quantum encryption. This hybrid approach aims to maintain compatibility with existing systems while enhancing security against future quantum threats. For example, the TLS 1.3 Hybrid Mode incorporates PQC algorithms, such as Kyber, alongside traditional methods like RSA or ECC, thereby facilitating a transition to quantum resistance without sacrificing legacy compatibility [429]. Such hybrid frameworks serve as a vital bridge during the transition to a quantum-secure infrastructure while addressing operational security needs [429].

QKD represents a significant advancement in cryptographic practices, utilizing principles of quantum mechanics to secure key exchanges. This method ensures that any attempts at eavesdropping are detectable in real-time, which significantly enhances security [431]. The most prominent QKD protocol, BB84, effectively allows two parties—often referred to as Alice and Bob—to generate a shared key using the quantum states of photons [431]. Major QKD networks, such as those established by China's Micius Satellite, push the boundaries of global quantum-secure communication [431]. Nevertheless, practical challenges persist, including distance limitations due to photon loss in optical fibres. Researchers are exploring satellite-based QKD, quantum repeaters, and Measurement-Device-Independent QKD (MDI-QKD) to mitigate these challenges and enhance system security [431].

QRNG technologies leverage quantum phenomena to produce true randomness, which is essential for cryptographic applications. Unlike classical random number generators that can be deterministic, QRNGs utilize quantum states to generate sequences that are fundamentally unpredictable [432]. This unpredictability makes them highly suitable for cryptographic key generation and secure communications across various sectors, including banking and government services [432].

With the rise of cryptocurrencies and blockchain technology, the urgency of maintaining security against quantum threats is paramount. Quantum computers could potentially compromise the integrity of blockchain by exposing vulnerabilities in public key infrastructure [430]. To counter these risks, developers are creating quantum-secure blockchain solutions utilizing PQC and quantum-safe consensus mechanisms, thus adopting algorithms like those proposed in NIST's PQC standardization efforts [429]. Projects like Ethereum 3.0 and Hyperledger are exploring quantum-resistant protocols to ensure the viability of blockchain in post-quantum scenarios [429].

As the understanding of quantum security evolves, it is expected that new cryptographic protocols, enhanced quantum hardware, and fortified secure communication networks will emerge. Future advancements could include combining Fully Homomorphic Encryption (FHE) with PQC to secure data in cloud computing environments against quantum threats [433]. The focus will increasingly be on developing quantum-secure IoT devices, advanced quantum authentication protocols, and other technologies that balance security and performance under the pressures of quantum decryption capabilities [433].

Chapter 8

Quantum Machine Learning and AI

The intersection of quantum computing and artificial intelligence (AI) represents one of the most promising frontiers in modern computational science. Up to this point in the book, we have explored the foundational principles of quantum mechanics, the structure and function of qubits, quantum logic gates, algorithms, and the challenges posed by noise and error correction. We have also examined different hardware approaches and the implications of quantum cryptography for secure communications. Now, we turn our attention to how quantum computing can revolutionize AI and machine learning.

Traditional machine learning techniques rely on vast amounts of computational power to process large datasets, train models, and optimize decision-making. As datasets grow exponentially and AI models become more complex, classical computing systems are struggling to keep up with the demand for higher processing speeds and efficiency. Quantum computing offers a new paradigm, enabling the acceleration of machine learning tasks through quantum parallelism, faster optimization algorithms, and quantum-enhanced data structures.

This chapter introduces the fundamental concepts of quantum machine learning (QML), covering how quantum computers can encode and process data in novel ways. We will explore quantum neural networks, quantum support vector machines, and variational quantum algorithms, examining how they differ from their classical counterparts and where they offer advantages. Additionally, we will discuss the implications of quantum computing for big data analysis, optimization problems, and AI-driven applications across various industries.

By the end of this chapter, readers will gain a clear understanding of how quantum computing is poised to reshape machine learning, the current challenges facing QML research, and the potential breakthroughs that lie ahead. As quantum hardware continues to evolve, so too will its impact on AI, paving the way for a new era of intelligent systems that leverage quantum-enhanced computation to solve problems beyond the reach of classical methods.

Quantum Data Encoding and Representation

Quantum Machine Learning (QML) represents a significant interdisciplinary merger between quantum computing and machine learning, utilizing quantum principles to enhance classical machine learning algorithms. One of the primary challenges within this emerging field is the efficient encoding of classical data into quantum states suitable for processing by quantum hardware. Quantum computers utilize qubits, which leverage superposition and entanglement to represent and manipulate information in fundamentally different ways compared to classical bits. Effective data encoding is crucial because it determines the capacity of quantum algorithms to execute various tasks—including classification, clustering, regression, and pattern recognition—effectively and efficiently [434, 435].

Encoding techniques in QML fundamentally shift how data is represented and processed. Quantum data encoding refers to mapping classical data into the quantum state space, which must be optimized to extract meaningful information while maximizing efficiency [436, 437]. In particular, various quantum encoding methods, such as those based on Hamiltonian simulation, have been proposed to enhance the expressivity and computational efficiency of quantum algorithms [438]. For instance, methods utilizing quantum convolutional neural networks demonstrate strong advantages in handling tasks like image classification by capitalizing on hybrid quantum-classical learning frameworks [439, 440].

Another priority in quantum data encoding is addressing the "Quantum Information Gap" (QIG), an issue that can significantly impact the efficacy of QML algorithms. It occurs due to the disparity between classical and quantum feature representations, posing a challenge that researchers are actively working to minimize through innovative approaches like the Quantum Information Preserving (QIP) loss function [441]. By refining the encoding of data within quantum states, quantum machine learning algorithms can improve their performance, allowing for better utilization of the unique characteristics offered by quantum computation.

The selection of an appropriate encoding scheme directly affects the scalability and predictive power of QML approaches. Studies have highlighted that kernel methods—the methodologies that allow quantum models to match the capacity of classical models—are intimately tied to the way classical data gets encoded into quantum states [442, 443]. This makes the exploration of new and more efficient quantum transformations essential for furthering the potential advantages of quantum machine learning algorithms [444].

Types of Quantum Data Encoding

There are several methods to encode classical data into quantum states, each with unique advantages and computational implications. The most commonly used encoding techniques in Quantum Machine Learning (QML) are Basis Encoding, Amplitude Encoding, Angle Encoding, and Hamiltonian Encoding.

1. Basis Encoding

Basis Encoding represents classical data using the computational basis states of qubits. Each classical bitstring corresponds to a quantum state in the computational basis. For example, a classical dataset with binary values (0s and 1s) can be mapped to quantum states:

$$\text{Classical } \{00, 01, 10, 11\} \quad \longrightarrow \quad \text{Quantum } \{|00\rangle, |01\rangle, |10\rangle, |11\rangle\}$$

This encoding is simple and does not require superposition, but it has a major limitation: it requires one qubit per classical bit, which makes it impractical for encoding large datasets due to the limited number of qubits in today's quantum hardware.

2. Amplitude Encoding

Amplitude Encoding is one of the most efficient encoding techniques, where classical data values are embedded into the amplitude coefficients of a quantum state. A classical vector *x* is encoded as a quantum state |ψ⟩:

$$|\psi\rangle = \sum_i x_i |i\rangle$$

For example, a two-dimensional classical vector $x = (x_1, x_2)$ can be encoded as:

$$|\psi\rangle = x_1|0\rangle + x_2|1\rangle$$

The advantage of amplitude encoding is that it allows exponential data compression, meaning that an ***N*-dimensional classical dataset can be encoded using only $log_2(N)$ qubits. However, preparing amplitude-encoded quantum states requires complex quantum operations, making it challenging to implement on near-term quantum devices.

3. Angle Encoding

Angle Encoding embeds classical data into the rotation angles of quantum gates. Given a classical feature x_i, it is encoded into a quantum state using a rotation gate (Rx, Ry, Rz):

$$|\psi\rangle = R_y(x_i)|0\rangle = \cos(x_i)|0\rangle + \sin(x_i)|1\rangle$$

If we have multiple features, each feature is encoded into a separate qubit by applying rotation gates:

$$R_y(x_1)R_y(x_2)R_y(x_3)\ldots|00\ldots0\rangle$$

Angle encoding is simple to implement and works well with variational quantum machine learning models, such as Quantum Neural Networks (QNNs) and Quantum Support Vector Machines (QSVMs). However, it requires multiple qubits for high-dimensional data, making it less efficient for large datasets.

4. Hamiltonian Encoding (Quantum Feature Maps)

Hamiltonian Encoding is used when data needs to be encoded into quantum operators for quantum optimization tasks. Classical data values are mapped into Hamiltonians (energy operators), which evolve over time:

$$U(x) = e^{-iH(x)t}$$

This encoding method is commonly used in quantum kernel methods, quantum support vector machines (QSVMs), and quantum generative models, where quantum feature maps are applied to represent data in a high-dimensional Hilbert space. Hamiltonian encoding is particularly effective for quantum-enhanced machine learning, where classical models struggle to represent complex correlations in data.

Choosing the Right Quantum Data Encoding Method

Each encoding method has trade-offs in terms of efficiency, implementation complexity, and suitability for different QML models:

Table 18: Advantages and limitations of encoding methods.

Encoding Method	Advantages	Limitations
Basis Encoding	Simple and straightforward	Requires one qubit per bit (not scalable)

Encoding Method	Advantages	Limitations
Amplitude Encoding	Efficient (exponential compression of classical data)	Hard to implement due to complex state preparation
Angle Encoding	Easy to implement with quantum gates	Requires multiple qubits for high-dimensional data
Hamiltonian Encoding	Powerful for quantum kernels and quantum machine learning	Complex, requires quantum evolution over time

In practical Quantum Machine Learning (QML) applications, Amplitude Encoding is preferred for data-intensive tasks, while Angle Encoding is widely used for Quantum Neural Networks (QNNs) and Variational Quantum Circuits (VQCs) due to its compatibility with quantum hardware.

Quantum-enhanced data encoding plays a crucial role in improving the performance of machine learning and artificial intelligence models by allowing quantum computers to process classical data in a more efficient and expressive manner. By leveraging quantum states to represent complex data structures, Quantum Machine Learning (QML) can achieve computational advantages over traditional machine learning approaches.

One significant application of quantum data encoding is in Quantum Support Vector Machines (QSVMs), where quantum feature maps are used to transform classical data into a high-dimensional quantum space. This approach enhances the ability of support vector machines (SVMs) to find optimal decision boundaries for classification tasks. By encoding data in quantum states, QSVMs can identify intricate patterns that classical SVMs struggle to detect, potentially leading to more accurate predictions and improved machine learning performance.

Another area where quantum data encoding is making an impact is in Quantum Neural Networks (QNNs). In these models, angle encoding is often used to represent classical data as quantum states, which are then processed by variational quantum circuits. QNNs leverage quantum parallelism and entanglement to improve the efficiency of neural network computations, offering a promising approach for solving optimization problems, pattern recognition, and deep learning tasks.

Quantum Generative Models also benefit from quantum data encoding, particularly in Quantum Generative Adversarial Networks (QGANs). These models use Hamiltonian encoding to represent complex probability distributions, allowing quantum computers to generate synthetic data with higher fidelity compared to classical GANs. Quantum generative models have potential applications in drug discovery, financial modelling, and AI-driven creativity, where the generation of realistic and diverse data samples is crucial.

In Quantum Kernel Methods, classical data is mapped into a high-dimensional quantum space using quantum feature maps, enhancing the ability of machine learning algorithms to perform classification, clustering, and anomaly detection. Quantum kernels allow for efficient similarity measurements between data points, enabling quantum-enhanced versions of kernel-based learning algorithms such as support vector machines, principal component analysis (PCA), and k-means clustering. These techniques could revolutionize fields such as finance, healthcare, and cybersecurity, where complex data relationships need to be understood and analysed efficiently.

As quantum hardware continues to advance, more sophisticated quantum data encoding techniques will emerge, further enhancing the capabilities of Quantum AI systems. Future developments in quantum encoding methods will enable deeper integration between quantum computing and artificial intelligence, leading to breakthroughs in optimization, autonomous decision-making, and large-scale data analysis. The potential of QML to outperform classical machine learning in specific domains is becoming increasingly evident, making quantum data encoding a foundational component of next-generation AI models.

Practical Example of Quantum Data Encoding and Representation

To illustrate Quantum Data Encoding and Representation, let's consider a simple binary classification problem where we want to classify data points based on a two-feature dataset. In classical machine learning, data points are represented as vectors in a two-dimensional space. However, in Quantum Machine Learning (QML), these data points need to be encoded into quantum states before they can be processed by a quantum algorithm.

Problem Statement: Binary Classification with Quantum Encoding

Assume we have a dataset with two features **(x_1, x_2)** that represent two classes (Class A and Class B). The goal is to encode this data into quantum states so that a quantum classifier can process it.

Sample	Feature x_1	Feature x_2	Class
A_1	0.2	0.8	A
A_2	0.5	0.5	A
B_1	0.9	0.1	B
B_2	0.8	0.2	B

Each data point (x_1, x_2) will be mapped into a quantum state using Angle Encoding (since it is efficient for small datasets and works well on near-term quantum devices).

Step 1: Encoding the Classical Data into Quantum States

In Angle Encoding, classical features are mapped onto the rotation angles of qubit gates. The two features, x_1 and x_2, will be used to rotate qubits along the Y-axis (Ry gates) of a quantum circuit. Each data point is encoded as follows:

$$|\psi\rangle = R_y(x_1)R_y(x_2)|0\rangle$$

For our dataset, this means applying Ry(x_1) and Ry(x_2) gates to a quantum state:

Sample	Quantum State Representation
A_1	(R_y(0.2) R_y(0.8)
A_2	(R_y(0.5) R_y(0.5)
B_1	(R_y(0.9) R_y(0.1)
B_2	(R_y(0.8) R_y(0.2)

Each data point is now represented in the quantum state space, ready to be processed by a quantum classifier.

Step 2: Implementing Quantum Data Encoding in Qiskit

Below is a Python implementation using Qiskit to encode the dataset into a quantum circuit:

```
from qiskit import QuantumCircuit, Aer, transpile, execute
from qiskit.visualization import plot_bloch_multivector
import numpy as np

# Define sample data points for encoding
data_points = [
    (0.2, 0.8),   # A1
    (0.5, 0.5),   # A2
    (0.9, 0.1),   # B1
    (0.8, 0.2)    # B2
]

# Create a quantum circuit with one qubit for encoding
qc = QuantumCircuit(1)

# Encode the first data point (x1, x2) using Ry gates
x1, x2 = data_points[0]
```

```
qc.ry(x1 * np.pi, 0)  # Rotate by x1
qc.ry(x2 * np.pi, 0)  # Rotate by x2

# Display the quantum circuit
print(qc.draw())

# Simulate the statevector to visualize the encoded state
simulator = Aer.get_backend('statevector_simulator')
compiled_circuit = transpile(qc, simulator)
job = execute(compiled_circuit, simulator)
result = job.result()
statevector = result.get_statevector()
plot_bloch_multivector(statevector)
```

Step 3: Processing Encoded Data in a Quantum Classifier

Once the data points are encoded into quantum states, we can use quantum classifiers such as:

- Quantum Support Vector Machines (QSVMs) with a quantum kernel method
- Quantum Neural Networks (QNNs) that use variational quantum circuits

For instance, we can apply a simple quantum classifier that uses Hadamard gates and measurement operations to separate Class A from Class B based on interference patterns.

Step 4: Measuring and Extracting Results

After quantum encoding, the quantum states can be measured to extract feature patterns that distinguish different classes. The measurement results are then used in classical post-processing for classification.

- If the measured qubit state collapses to $|0\rangle$, it belongs to Class A.
- If the measured qubit state collapses to $|1\rangle$, it belongs to Class B.

By running quantum circuits for multiple data points and analysing the measurement distributions, we can train a quantum classifier to distinguish between Class A and Class B.

This example demonstrates Quantum Data Encoding and Representation using Angle Encoding for a binary classification problem. The classical data is converted into quantum states using Ry gate rotations, which can then be processed by quantum machine learning models such as Quantum Neural Networks (QNNs) and Quantum Support Vector Machines (QSVMs).

As quantum hardware improves, more advanced encoding techniques such as Amplitude Encoding and Quantum Kernel Encoding will enable faster and more accurate Quantum Machine Learning models, unlocking new possibilities for AI, pattern recognition, and complex decision-making.

Quantum Neural Networks

Quantum Neural Networks (QNNs) represent a groundbreaking intersection of quantum computing and artificial intelligence, encompassing paradigms of Quantum Machine Learning (QML). They fundamentally extend classical artificial neural networks (ANNs) by harnessing quantum phenomena such as superposition, entanglement, and interference, which can potentially lead to superior computational efficiency and acceleration of data processing tasks. Specifically, while classical ANNs rely heavily on matrix operations and non-linear activation functions, QNNs exploit quantum circuits and qubits for data manipulation, enhancing their capabilities in tasks such as optimization, pattern recognition, and deep learning applications [445, 446].

The theoretical roots of QNNs can be traced back to early explorations into quantum computation that posited quantum effects could influence cognitive functions, connecting with Subhash Kak's and Ron Chrisley's work in quantum mind theory [445]. However, much of the subsequent research focuses on the integration of classical neural architecture with quantum principles to improve the performance and training efficacy of such networks. A primary motivation for developing QNNs is the challenge of training classical neural networks, particularly in large datasets or complex problem domains, where features of quantum mechanics, such as quantum parallelism, could be leveraged as a computational resource [446, 447].

Currently, most QNN architectures are designed as feed-forward networks, similar to classical structures, where multiple layers of qubits process input data sequentially. Research indicates that these architectures can differ in their qubit count across layers, thus providing flexibility in their design and functionality [448, 449]. Moreover, QNNs can be categorized based on their data interaction framework: quantum computers processing classical data, classical systems processing quantum data, and true quantum systems managing quantum data [448].

Practical QNNs have emerged chiefly as a result of attempts to devise quantum analogues to traditional neural units, such as the perceptron. Challenges arise in aligning non-linear activation functions typically seen in classical models with quantum mechanics, which operates under linear transformations leading to probabilistic outputs. Proposals to resolve these discrepancies include the introduction of quantum versions of activation mechanisms and various training techniques [446, 447, 450]. The advancement of quantum neural architectures not only aids in efficient computational processes, but also allows for the development of quantum associative memory frameworks, which outperform classical counterparts in particular applications due to their unique quantum memory retrieval strategies [451, 452].

Despite their promising potential, QNNs remain largely theoretical, with their implementation in extensive physical systems still in its infancy [453, 454]. Existing research indicates keen interest in enhancing the trainability of QNNs through overcoming inherent challenges such as so-called "barren plateaus," which significantly hinder optimization and performance [447, 455, 456]. Consequently, research is directed toward achieving quantum advantages that provide meaningful distinctions over classical methods [457].

Barren Plateaus in Quantum Machine Learning and Optimization

Barren plateaus refer to a critical challenge in Quantum Machine Learning (QML) and Variational Quantum Algorithms (VQAs), where the gradient of the cost function vanishes exponentially as the number of qubits increases. This issue significantly hampers the training of Quantum Neural Networks (QNNs) and the optimization of Variational Quantum Circuits (VQCs) because gradient-based optimization methods struggle to update parameters effectively.

The problem arises when the parameter space of a quantum circuit becomes too large, leading to an optimization landscape where nearly all gradient values approach zero. As a result, the optimization process becomes stuck, preventing meaningful learning. This phenomenon is particularly problematic for variational quantum algorithms, which rely on tuning quantum gate parameters using classical optimization techniques.

Barren plateaus are a consequence of randomized quantum states and deep quantum circuits, which cause gradients to diminish. Several key factors contribute to this issue:

When a quantum circuit is initialized with random parameters, the resulting evolution of quantum states can generate an almost uniform distribution over the Hilbert space. This randomness leads to an expected gradient value that is close to zero, making parameter updates ineffective. Since optimization in quantum circuits relies on finding informative gradients, this randomness significantly slows down learning.

In variational quantum algorithms, cost functions are often defined as expectation values of observables (for example, measuring a qubit's state after computation). However, as the number of qubits increases, the Hilbert space expands exponentially, which causes gradient values to shrink at the same rate. This makes it increasingly difficult to locate optimal parameter updates, leading to a barren plateau.

Excessive circuit depth can also cause barren plateaus. When a quantum circuit has too many layers, the system tends to generate highly entangled quantum states, spreading information across multiple qubits. This results in a loss of local sensitivity in parameter updates, meaning that small changes to quantum gate parameters have minimal impact on the cost function.

Unlike classical deep learning, where neural network loss landscapes often contain sharp valleys and hills that guide optimization, quantum cost landscapes can become flat due to interference effects. This lack of structure makes standard gradient-based optimizers (such as Adam, SGD, or RMSprop) ineffective in navigating the optimization process.

The barren plateau phenomenon can be expressed mathematically through the variance of the cost function gradient. Consider a quantum cost function $C(\theta)$ that depends on a set of trainable quantum gate parameters θ. The gradient of this cost function is given by:

$$\frac{\partial C(\theta)}{\partial \theta_i}$$

For a randomly initialized quantum circuit, the variance of this gradient is known to decay exponentially with the number of qubits n:

$$\mathrm{Var}\left(\frac{\partial C(\theta)}{\partial \theta_i}\right) \propto \frac{1}{2^n}$$

This equation indicates that as n increases, the variance of the gradient approaches zero, making it nearly impossible for an optimization algorithm to locate an effective learning direction. For example, in a system with n=50 qubits, the gradient variance is so small that meaningful updates become infeasible.

Barren plateaus introduce serious challenges in training Quantum Neural Networks (QNNs) and optimizing Variational Quantum Circuits (VQCs) for real-world applications such as quantum chemistry, finance, and combinatorial optimization. One major issue is slow or stalled training, where small gradient values prevent learning progress. Since optimization algorithms rely on gradient information, they fail to make effective updates, leading to extended training times or complete convergence failure.

QNNs trained under barren plateaus may also exhibit poor model generalization. If the optimization process cannot efficiently update parameters, the trained model may struggle to recognize new data patterns, making it ineffective for real-world classification and prediction tasks.

Many hybrid quantum-classical algorithms rely on classical optimization techniques such as gradient descent. However, if the gradients in a quantum circuit are too small, classical optimizers fail to improve performance, leading to inefficient learning processes.

Researchers are actively developing methods to overcome barren plateaus and improve the trainability of quantum models. One approach is layerwise training and circuit pruning, where instead of training an entire quantum circuit at once, only a subset of layers is optimized at a time. This approach prevents excessive randomization of quantum states and maintains gradient magnitude during optimization.

Another method involves using local cost functions instead of global cost functions that depend on all qubits. By focusing on only a subset of qubits, gradients remain larger and more stable, reducing the likelihood of encountering barren plateaus.

Designing problem-specific quantum architectures is also a promising approach. Instead of using deep, randomly initialized circuits, researchers develop structured variational quantum circuits that are better suited to specific optimization problems. This helps maintain a controlled level of entanglement, preventing the exponential decay of gradients.

Quantum-inspired initialization techniques can further improve optimization. Rather than initializing circuits with random parameters, researchers use pre-trained classical models or heuristic-based quantum gate initialization strategies. This ensures that circuits begin closer to optimal configurations, avoiding the randomization effects that contribute to barren plateaus.

Some researchers explore gradient-free optimization techniques, such as genetic algorithms, evolutionary strategies, and reinforcement learning, to optimize quantum circuits without relying on traditional gradient-based methods. These techniques provide an alternative way to update parameters without suffering from gradient vanishing.

Another effective strategy is entanglement regularization, where the number of entangling operations (such as CNOT or CZ gates) is carefully controlled. By limiting excessive entanglement, circuits remain more stable, and gradients retain meaningful values for a longer time.

Barren plateaus represent a major obstacle in Quantum Machine Learning (QML) and variational quantum computing, where gradients vanish exponentially as the number of qubits increases. This effect makes training Quantum Neural Networks (QNNs) and optimizing Variational Quantum Circuits (VQCs) extremely challenging, leading to slow training, poor generalization, and inefficient hybrid optimization.

Through careful quantum circuit design, alternative cost function strategies, and improved initialization methods, researchers are finding ways to mitigate barren plateaus and enhance the trainability of quantum models. As quantum hardware and algorithms continue to advance, these solutions will be critical for unlocking scalable quantum machine learning, AI, and optimization applications.

Quantum Neural Network Functionality

Quantum Neural Networks (QNNs) function differently from classical neural networks, which rely on layers of perceptrons (neurons) connected by weighted edges. Instead of processing data using conventional mathematical operations, QNNs leverage quantum circuits where qubits encode and manipulate information. This quantum-based structure allows QNNs to exploit principles like superposition, entanglement, and interference, potentially offering computational advantages over classical machine learning models.

The first step in a QNN is Quantum Data Encoding, where classical input data is converted into quantum states. This transformation is achieved using Quantum Feature Maps or specific Quantum Data Encoding techniques, such as Angle Encoding, Amplitude Encoding, or Basis Encoding. The choice of encoding method determines how effectively the quantum model can represent complex data patterns and correlations.

Once the data is encoded into quantum states, it passes through Quantum Circuit Layers, which serve as quantum equivalents of classical neurons. These layers consist of parametric quantum gates such as RX, RY, and RZ rotations, which apply transformations to the qubits. Additionally, entanglement operations, such as controlled-NOT (CNOT) gates, allow qubits to interact, enabling the QNN to learn and process information in a fundamentally different way from classical models. The combination of these quantum gates forms the Quantum Perceptrons, which are responsible for feature extraction and representation learning.

After quantum computation is performed, the final step is Measurement and Classical Post-Processing. Since quantum mechanics dictates that qubits exist in a superposition of states until they are measured, the measurement process collapses the quantum state to extract meaningful information. The measured outcomes are then fed into classical machine learning algorithms for further analysis, classification, or optimization. This hybrid approach combines the power of quantum processing with the robustness of classical models, making QNNs a practical tool for near-term quantum devices.

A defining feature of QNNs is their use of Variational Quantum Circuits (VQCs), which are tunable quantum circuits optimized through iterative learning. These circuits contain adjustable parameters that are trained using gradient-based or gradient-free optimization methods, similar to backpropagation in classical neural networks. By fine-tuning these parameters, QNNs can learn complex patterns in data and improve their predictive accuracy. The integration of VQCs into quantum machine learning frameworks makes QNNs adaptable and suitable for various applications, including optimization problems, pattern recognition, and deep learning tasks.

Architecture of Quantum Neural Networks

A typical QNN consists of three main components:

1. Quantum Input Layer (Quantum Encoding)

Classical data must first be encoded into quantum states using an encoding scheme. Angle Encoding is commonly used for QNNs, where classical features xxx are converted into quantum states via rotation gates:

$$|\psi(x)\rangle = R_y(x)|0\rangle$$

For multi-qubit systems, multiple features are encoded across different qubits using:

$$|\psi(x)\rangle = R_y(x_1)R_y(x_2)R_y(x_3)...|00...0\rangle$$

2. Quantum Hidden Layers (Quantum Perceptrons)

Instead of classical neurons, QNNs use parametric quantum gates (quantum perceptrons) that apply unitary transformations to the quantum states. These quantum layers include:

- **Parameterized Rotation Gates (RX, RY, RZ):** These gates act as adjustable weights in quantum neural networks.
- **Entanglement Layers:** Qubits interact using CNOT (CX) gates or other multi-qubit operations to create quantum correlations.
- **Variational Quantum Circuits (VQCs):** Quantum circuits with trainable parameters optimized using classical methods (e.g., gradient descent).

3. Quantum Measurement & Output Layer

After processing, the final quantum state is measured to extract useful information. Since quantum measurements collapse the quantum state, multiple runs (shots) are required to extract statistical patterns from quantum circuits. The measured outputs are then mapped to classical probability distributions for classification or regression tasks.

For binary classification, a single qubit's measurement outcome (0 or 1) can determine the class. In more complex models, measurement statistics from multiple qubits are used for final decision-making.

Example: Implementing a Simple Quantum Neural Network in Qiskit

Below is a Python implementation of a simple QNN using Qiskit, where quantum gates act as learnable parameters in a variational quantum circuit.

```
from qiskit import QuantumCircuit, Aer, transpile, execute
from qiskit.circuit import Parameter
import numpy as np

# Define a simple 2-qubit Quantum Neural Network
theta = Parameter('θ')  # Trainable parameter

qc = QuantumCircuit(2)
qc.ry(theta, 0)  # Apply rotation on qubit 0
qc.cx(0, 1)  # Apply entanglement (CNOT gate)
qc.ry(theta, 1)  # Apply rotation on qubit 1

# Display the quantum circuit
print(qc.draw())

# Simulate the quantum circuit
backend = Aer.get_backend('statevector_simulator')
theta_value = np.pi / 4
bound_qc = qc.bind_parameters([(theta, theta_value)])
job = execute(bound_qc, backend)result = job.result()
print(result.get_statevector())  # Print quantum state
```

This quantum circuit forms a simple Quantum Neural Network layer, where θ (theta) acts as a trainable weight, and the CNOT gate introduces entanglement between qubits. By optimizing θ using classical machine learning techniques, we can train this quantum model for classification or regression tasks.

Advantages, Challenges and Applications of Quantum Neural Networks

Quantum Neural Networks (QNNs) provide several theoretical advantages over classical neural networks by leveraging the unique properties of quantum computing. One of the most significant benefits is the exponential speedup for certain problems. Unlike classical networks, QNNs exploit quantum parallelism, which allows them to evaluate multiple states simultaneously. This capability can significantly reduce computation time for complex tasks such as optimization, feature mapping, and high-dimensional data processing.

Another key advantage of QNNs is their ability to perform efficient high-dimensional feature mapping. Quantum circuits can encode and represent complex data structures in Hilbert space more efficiently than classical neural networks, leading to enhanced pattern recognition and data clustering. This property makes QNNs particularly well-suited for tasks requiring deep feature extraction, such as image processing, anomaly detection, and financial modeling.

QNNs may also help in reducing overfitting, a common issue in classical deep neural networks. Classical models often require large datasets to generalize effectively, but QNNs, due to their ability to encode richer representations with fewer parameters, have the potential to generalize better even with limited data. This characteristic is especially valuable in domains where collecting large labelled datasets is challenging, such as medical imaging, genomics, and quantum chemistry.

Another area where QNNs excel is quantum-enhanced optimization. Many machine learning tasks, including deep network training, reinforcement learning, and kernel-based methods, involve solving complex optimization problems. QNNs can leverage quantum-enhanced optimization algorithms such as Quantum Approximate Optimization Algorithm (QAOA) and Variational Quantum Eigensolver (VQE), which may outperform classical optimization techniques in specific problem domains.

Despite their promising potential, Quantum Neural Networks face several practical challenges that must be addressed before they can be widely adopted. One of the biggest challenges is hardware limitations. Current quantum computers suffer from noise, decoherence, and gate errors, making it difficult to implement deep QNN architectures. The limited number of available high-fidelity qubits restricts the scalability of QNNs, as deeper circuits require more stable and error-resistant quantum gates.

The training process for QNNs is another significant challenge. Unlike classical neural networks, which use well-established backpropagation techniques, QNNs require hybrid quantum-classical optimization methods. Computing gradients in quantum circuits is non-trivial due to the probabilistic nature of quantum measurements, making QNN training more complex and computationally expensive. Researchers are actively developing techniques such as parameter-shift rules and quantum natural gradients to improve QNN training efficiency.

Scalability is another major hurdle in the development of QNNs. While classical deep learning models can be efficiently scaled on GPUs and TPUs, quantum circuits are currently constrained by the number of available qubits and connectivity between them. Implementing large-scale QNNs requires substantial advancements in quantum hardware, qubit coherence, and error correction techniques.

Measurement overhead is another limitation. Since quantum systems must be measured to extract useful information, obtaining meaningful outputs from QNNs requires multiple circuit executions (shots). This process is computationally expensive and introduces additional uncertainty, which can affect the reliability of QNN models in real-world applications.

To overcome these challenges, researchers are focusing on developing hybrid quantum-classical algorithms, where QNNs are used for feature representation and processing, while classical deep learning models handle final classification and decision-making. This approach combines the strengths of both quantum and classical computing, allowing for practical QNN implementations on near-term quantum hardware.

Although Quantum Neural Networks are still in the early stages of research and development, they have promising applications across multiple industries. One potential application is Quantum Image Processing (QIP), where QNNs can be used for image recognition, feature extraction, and pattern detection. Quantum models can enhance classical image processing techniques by efficiently handling high-dimensional representations and improving computational efficiency.

In financial modelling, QNNs have the potential to optimize portfolio risk management, fraud detection, and high-frequency trading strategies. Quantum-enhanced learning algorithms can analyse complex financial datasets and uncover correlations that classical models might overlook.

Another key area where QNNs show promise is in drug discovery and material science. Simulating molecular interactions and predicting chemical properties requires solving complex quantum mechanics problems. QNNs, combined with quantum chemistry simulations, can accelerate the discovery of new pharmaceutical drugs, catalysts, and novel materials.

Quantum-enhanced Natural Language Processing (NLP) is another emerging field where QNNs can be used to improve text classification, sentiment analysis, and language translation. By leveraging quantum kernel methods and quantum-enhanced word embeddings, QNNs may enable more efficient semantic understanding and knowledge representation in large-scale NLP tasks.

QNNs also offer advantages in solving complex optimization problems, particularly in areas such as supply chain logistics, scheduling, and machine learning hyperparameter tuning. By utilizing Quantum Annealing and Variational Quantum Optimization techniques, QNNs can solve combinatorial optimization problems more efficiently than classical approaches.

Several companies, including IBM, Google, Xanadu, and Rigetti, are actively researching and developing QNN applications for machine learning, AI, and optimization tasks. As quantum hardware continues to advance, Quantum Neural Networks are expected to play a crucial role in the future of AI and data-driven decision-making.

Variational Quantum Algorithms for Machine Learning

Variational Quantum Algorithms (VQAs) constitute a pivotal aspect of Quantum Machine Learning (QML), particularly for their capacity to exploit the hybrid quantum-classical approach that aligns well with the limitations of Noisy Intermediate-Scale Quantum (NISQ) devices. NISQ devices, characterized by their limited qubit count and noise sensitivity, benefit from VQAs due to their effective use of parameterized quantum circuits (PQCs) that enable optimization through classical methods [458].

These algorithms primarily operate through an iterative process that includes the preparation of quantum states, execution of variational circuits, and classical optimization to minimize an objective function. The first step, quantum state preparation, involves encoding classical data into quantum states, typically using various encoding strategies such as Angle Encoding or Amplitude Encoding [436, 443]. The encoded data then undergoes transformations via a PQC, which consists of rotation gates and entangling gates that incorporate trainable parameters [459]. The output produces probabilistic results that are fundamental to capturing the intricate relationships within the training data.

The classical optimization phase is vital as it adjusts the variational parameters to enhance model efficacy. A plethora of classical optimizers, including Gradient Descent and SPSA, are employed to refine these parameters, thus driving the algorithm towards a converged solution [460]. In this hybrid framework, the quantum processor and classical computer play complementary roles, with the former addressing computationally intensive tasks while the latter manages overall optimization [461].

In terms of practical applications, VQAs stand out in their versatility for tasks such as quantum classification and quantum regression, allowing for significant advancements in real-world applications like drowsiness detection from EEG signals [462]. Moreover, the continual refinement of these algorithms is directed towards improving measurement precision and minimizing computational overhead, which is crucial for harnessing quantum advantage in machine learning scenarios [441].

A notable aspect of VQAs is their intrinsic noise resilience. Unlike purely quantum algorithms, VQAs can leverage classical computation to mitigate the impact of noise found in NISQ devices, thus potentially providing improved performance in real-world applications [463]. This highlights the dual strength of VQAs, rendering them a suitable choice for near-term quantum applications while also paving the way for future developments in quantum machine learning.

Types of Variational Quantum Algorithms for Machine Learning

Among these algorithms, several notable applications can significantly enhance classical machine learning tasks, which are critical in both artificial intelligence and quantum computing.

1. Variational Quantum Classifiers (VQC) employ parameterized quantum circuits for classification tasks, akin to classical models such as Support Vector Machines (SVMs) and Neural Networks. VQCs utilize quantum circuits trained to map input features onto different quantum states that correspond to specific classes. The output state of the quantum circuit is measured to classify input data points. Techniques in classical optimization are employed to adjust and minimize classification errors, enabling continuous improvement of the classifier [464, 465]. The innovative nature of VQCs lies in their ability to operate in a high-dimensional quantum feature space, potentially yielding better classification performance than classical counterparts [466].

2. Quantum Kernel Methods represent another significant advancement in Quantum Machine Learning (QML). These methods leverage quantum circuits to compute inner products between data in a feature space that expands exponentially with dimensionality, facilitating improved pattern recognition. Quantum Kernel Methods allow for effective separation of non-linearly separable data, defining a leading duality with classical kernel methods commonly used in algorithms like SVMs [467, 468]. The effectiveness of these methods is further amplified by the exploration of quantum feature maps and their relation to quantum computation, which promotes a competitive edge in handling complex datasets [469].

3. Quantum Neural Networks (QNNs) utilize VQCs as quantum perceptrons, essentially mimicking the architecture of classical neural networks. The fundamental advantage of QNNs resides in their integration of quantum phenomena such as entanglement and superposition to enhance information processing capabilities, potentially expediting training processes and improving representation [470, 471]. The training of QNNs mirrors classical deep learning frameworks, where parameters are optimized using classical techniques to achieve improved performance on tasks such as image recognition and natural language processing [472].

4. Quantum Generative Models (QGANs) extend the capabilities of generative models in QML by incorporating a competitive architecture similar to classical Generative Adversarial Networks (GANs). QGANs comprise generator circuits that synthesize quantum data samples and discriminator circuits that evaluate the authenticity of these samples. This iterative training method allows QGANs to outperform classical generative models by exploiting the unique properties of quantum mechanics to produce more complex data distributions [473]. The entangling nature of quantum circuits in QGANs allows them to tackle limitations faced by traditional GANs, providing an enhanced modelling framework [474].

5. Quantum Approximate Optimization Algorithm (QAOA) serves another important role in the realm of variational algorithms, particularly beneficial for solving combinatorial optimization

problems prevalent in machine learning applications like feature selection and scheduling. By training quantum circuits to minimize objective functions, QAOA holds promise for tackling NP-hard problems [475]. Its ability to traverse large search spaces more efficiently than classical alternatives is a critical advantage in applications including supply chain logistics and social network analysis [465, 469].

6. Variational Quantum Eigensolver (VQE), while primarily used for quantum chemistry simulations, is also applicable to machine learning tasks focusing on feature extraction and dimensionality reduction. VQE identifies optimal patterns within datasets by calculating the ground-state energy of quantum systems, thereby enhancing the performance of machine learning algorithms [476, 477]. With its applications extending beyond traditional realms, VQE contributes significantly to the innovation landscape in quantum machine learning.

Advantages, Challenges and Limitations of Variational Quantum Algorithms for Machine Learning

Variational Quantum Algorithms (VQAs) provide several advantages in quantum machine learning by utilizing parameterized quantum circuits optimized through hybrid quantum-classical methods. One of the primary benefits is their efficient use of noisy quantum hardware. Since VQAs require relatively short-depth quantum circuits, they are well-suited for Noisy Intermediate-Scale Quantum (NISQ) devices, which currently have limited coherence times and qubit connectivity. This makes VQAs more practical than quantum algorithms requiring full fault tolerance.

Another advantage is their quantum speedup in optimization. Many VQAs exploit quantum parallelism and entanglement, enabling them to explore multiple potential solutions simultaneously. This parallelism provides potential advantages over classical optimization methods, especially in solving combinatorial and high-dimensional problems that are computationally expensive for classical approaches.

VQAs also allow for high-dimensional feature representation, which is a crucial factor in machine learning. Quantum circuits can encode data in exponentially large Hilbert spaces, providing a richer mathematical framework for representing complex patterns. This capability enhances pattern recognition, clustering tasks, and anomaly detection, making quantum machine learning potentially more powerful than classical models in specific applications.

Another important benefit of VQAs is hybrid quantum-classical learning, where quantum systems work alongside classical machine learning models. This hybrid approach allows quantum computers to handle high-dimensional transformations, while classical processors perform gradient optimization and decision-making. The ability to integrate quantum models with existing AI frameworks makes VQAs more accessible and applicable to real-world problems.

Variational Quantum Algorithms however, face several significant challenges that must be addressed before they can be widely adopted. One of the biggest hurdles is hardware limitations. Current quantum computers have limited qubits, high gate error rates, and short coherence times, which restrict the depth of quantum circuits that can be executed reliably. Since VQAs rely on iterative quantum-classical loops, these hardware constraints can slow down convergence and limit the size of problems that can be efficiently solved.

Another major challenge is the issue of barren plateaus, where gradients vanish exponentially as the number of qubits increases. This phenomenon makes it difficult to optimize quantum circuits effectively, as the loss function landscape becomes flat, preventing meaningful updates to the model parameters. Overcoming this issue requires improved initialization strategies, adaptive optimization techniques, and alternative training methods that maintain useful gradient values throughout learning.

The complexity of hybrid optimization also presents difficulties in training VQAs. Since quantum circuits do not support standard backpropagation, classical optimizers must be used to update quantum parameters. However, classical optimization methods, such as gradient descent, Adam, and COBYLA, may not always converge efficiently when applied to quantum circuits, requiring the development of new quantum-aware optimization techniques.

Scalability remains a key concern for VQAs. While they work well for small-scale problems, their performance on large datasets remains uncertain due to the measurement overhead required to extract useful information from quantum states. Since quantum measurements collapse the quantum state, multiple circuit executions (shots) are needed to obtain meaningful results, increasing computational costs. Addressing scalability issues will require advancements in quantum error mitigation, improved measurement strategies, and efficient circuit designs.

Quantum Support Vector Machines

Quantum Support Vector Machines (QSVMs) represent a significant advancement in the domain of machine learning by incorporating the principles of quantum computing to enhance classification performance and speed. This innovation builds on classical Support Vector Machines (SVMs), known for their effectiveness in binary classification tasks where the aim is to identify the optimal hyperplane that differentiates data points belonging to distinct categories. QSVMs leverage quantum feature mapping and kernel methods to operate on data more efficiently than their classical counterparts, thereby offering promising avenues for Quantum Machine Learning (QML) applications.

The fundamental principles that differentiate QSVMs from classical SVMs lie in the application of quantum kernel methods. The quantum kernel, defined through inner products of quantum states, allows for the transformation of data into a high-dimensional feature space, enabling intricate classifications that are often infeasible with classical SVMs [478, 479]. There is empirical evidence suggesting that QSVMs can outperform classical SVMs, particularly in high-dimensional feature spaces where conventional methods struggle due to computational constraints [457, 480]. This enhancement is primarily attributed to the quantum ability to perform complex computations with high efficiency and speed, as explored in various studies [481].

Central to the efficacy of QSVMs is the implementation of quantum feature maps that facilitate kernel computations in ways that are not only faster but also capable of capturing complex decision boundaries within data [465, 482]. For instance, Bartkiewicz et al. highlight the importance of kernel-based approaches in QML, emphasizing that quantum circuits may be tailored to specific applications to achieve optimal classification results [483]. Furthermore, experimental demonstrations, such as the detection of breast cancer using QSVMs on real quantum processors, provide tangible evidence of the enhancements QSVMs offer over classical models [480, 484].

The theoretical underpinnings of QSVMs are complemented by practical implementations that underscore their potential applicability across various domains. A recent study by Saeedi et al. discusses how quantum algorithms can expedite training processes on large datasets, which is a notable challenge for traditional machine learning frameworks [485]. Additionally, a comparative performance analysis shows that QSVMs not only offer quantum speed-up in certain scenarios but also maintain competitive accuracy when benchmarked against classical SVMs [486]. These findings advocate for the continued exploration of QSVMs, particularly as quantum computing technology evolves, providing a platform for solving increasingly complex classification challenges in diverse fields.

QSVMs function similarly to classical SVMs but take advantage of quantum computing techniques to perform feature mapping and kernel evaluations in a more computationally efficient way. The process of implementing QSVMs consists of three main steps:

1. Quantum Feature Mapping

In classical SVMs, data points are transformed into a higher-dimensional space where a linear decision boundary can separate them. In QSVMs, this transformation is performed using Quantum Feature Maps, where classical input data is encoded into a quantum state.

This encoding process uses Parameterized Quantum Circuits (PQCs) to map classical data into a high-dimensional quantum Hilbert space. The idea is that quantum computing can provide a more expressive and efficient representation of data, allowing complex patterns to be distinguished more easily than in classical SVMs.

Some common methods for quantum data encoding include:

- Angle Encoding – Data points are mapped onto the rotation angles of quantum gates (e.g., RX, RY, RZ).

- Amplitude Encoding – The probability amplitudes of quantum states represent data values.
- Hamiltonian Encoding – The input data is encoded using Hamiltonian evolution in a quantum system.

2. Quantum Kernel Computation

A key advantage of QSVMs is their ability to leverage Quantum Kernels, which compute the similarity between data points in a quantum feature space. Classical kernel methods often require exponentially large computations to evaluate the inner product of high-dimensional feature vectors. However, QSVMs can perform these calculations natively within a quantum circuit, significantly reducing computational overhead.

In a Quantum Kernel Method, a quantum circuit is applied to two different input states |x⟩ and |x'⟩ to compute the kernel function:

$$K(x, x') = |\langle\psi(x)|\psi(x')\rangle|^2$$

This approach allows quantum computers to evaluate complex feature mappings more efficiently, particularly for datasets where classical kernel functions would require exponential resources.

3. Training and Classification

Once the quantum feature mapping and kernel computation are performed, QSVMs follow the standard Support Vector Machine optimization process. A classical optimizer is used to find the optimal support vectors that define the decision boundary.

The Quantum Kernel Matrix, which contains the pairwise similarities between all training samples, is computed using quantum circuits. A classical SVM solver then uses this matrix to train the model, determining the best hyperplane for classification.

For new test data points, the quantum kernel function is evaluated again, and the trained model predicts the class label based on the computed similarities.

Quantum Support Vector Machines (QSVMs) Using IBM Qiskit

IBM Qiskit is an open-source quantum computing framework developed by IBM to facilitate quantum programming, research, and application development. At its core, Qiskit allows users to design quantum algorithms, execute them on real quantum computers, and analyse their results through a high-level Python library.

Qiskit is structured into multiple components, each serving a distinct purpose in quantum computing. The `qiskit.circuit` module enables users to construct and manipulate quantum circuits, allowing researchers and developers to design complex quantum algorithms in a structured manner. This modular approach simplifies the process of working with quantum gates and quantum operations.

For simulation and testing, Qiskit provides `qiskit.aer`, a high-performance quantum simulator that runs quantum circuits on classical hardware. This allows researchers to validate quantum algorithms before deploying them on actual quantum hardware. Quantum simulation is essential for debugging and refining quantum circuits, ensuring they behave as expected before execution on a real quantum processor.

To interface with IBM Quantum's cloud-based quantum computers, Qiskit includes `qiskit.ibm`, which allows users to submit jobs to IBM's quantum processors and retrieve results. This feature provides access to IBM's extensive quantum hardware infrastructure, supporting quantum research and real-world applications.

Qiskit extends its capabilities beyond basic quantum programming through domain-specific libraries. These include `qiskit-nature` for quantum chemistry, `qiskit-optimization` for solving optimization problems, and `qiskit-machine-learning` for quantum-enhanced AI and ML applications. These specialized libraries expand the practical applications of quantum computing across multiple scientific and engineering disciplines.

Implementing QSVM with Qiskit

Qiskit provides several tools for implementing QSVMs, including quantum feature maps, quantum kernels, and quantum classifiers. Below is an example of training a QSVM using Qiskit and PennyLane.

Step 1: Installing Dependencies

```
!pip install scikit-learn==1.2.1
!pip install pennylane==0.26
!pip install qiskit==0.39.2
!pip install qiskit_machine_learning==0.5.0
```

Step 2: Loading and Preprocessing Data

```
import numpy as np
from sklearn.datasets import load_wine
from sklearn.model_selection import train_test_split
from sklearn.preprocessing import MaxAbsScaler

x, y = load_wine(return_X_y=True)
```

```
x = x[:59+71]
y = y[:59+71]

x_tr, x_test, y_tr, y_test = train_test_split(x, y,
train_size=0.9)
scaler = MaxAbsScaler()
x_tr = scaler.fit_transform(x_tr)
x_test = scaler.transform(x_test)
x_test = np.clip(x_test, 0, 1)
```

Step 3: Implementing a Quantum Kernel

```
import pennylane as qml
from sklearn.svm import SVC

nqubits = 4
dev = qml.device("lightning.qubit", wires=nqubits)

@qml.qnode(dev)
def kernel_circ(a, b):
    qml.AmplitudeEmbedding(a, wires=range(nqubits), pad_with=0,
normalize=True)
    qml.adjoint(qml.AmplitudeEmbedding(b, wires=range(nqubits),
pad_with=0, normalize=True))
    return qml.probs(wires=range(nqubits))

def qkernel(A, B):
    return np.array([[kernel_circ(a, b)[0] for b in B] for a in
A])

svm = SVC(kernel=qkernel).fit(x_tr, y_tr)
```

Step 4: Evaluating Model Performance

```
from sklearn.metrics import accuracy_score
print(accuracy_score(svm.predict(x_test), y_test))
```

IBM Qiskit provides a powerful platform for developing Quantum Support Vector Machines (QSVMs), allowing researchers to leverage quantum-enhanced feature mapping and kernel methods for improved classification tasks. By encoding classical data into high-dimensional quantum states, QSVMs can theoretically outperform classical SVMs for certain complex problems.

Quantum Computing in Big Data and Optimization

Quantum computing has the potential to revolutionize big data analytics and optimization by leveraging quantum parallelism and entanglement to process large-scale datasets and solve complex optimization problems more efficiently than classical computers. Traditional computing systems struggle with the exponential growth of data, requiring massive computational resources to analyse, optimize, and extract meaningful insights. Quantum computing offers exponential speedup for specific algorithms, making it a powerful tool for big data processing, machine learning, and combinatorial optimization.

Quantum computing represents a paradigm shift in computation that directly intersects with big data analytics, offering transformative capabilities that standard classical computing cannot match. This synthesis will explore how quantum computing enhances data analytics through three main aspects: quantum speedup for data processing, quantum machine learning for big data, and quantum-assisted data clustering.

Quantum computing allows for significant speedups in processing large datasets through advanced algorithms. The Quantum Fourier Transform (QFT) is a notable algorithm that accelerates operations such as signal processing and spectral analysis, which are essential in applications like time-series forecasting and financial modelling [487]. Furthermore, Grover's Algorithm optimizes the search process in unsorted databases, reducing the time complexity from $O(N)$ in classical computing to $O(\sqrt{N})$ in quantum computing. This enhancement vastly improves efficiencies in big data searches, enabling quicker extraction of insights from complex datasets [488].

Additionally, Quantum Principal Component Analysis (QPCA) dramatically increases the efficiency of dimensionality reduction, allowing for the swift analysis of high-dimensional datasets typical in machine learning. The implementation of QPCA can execute dimensionality reduction exponentially faster than classical PCA, thus facilitating the processing of intricate datasets more effectively [488, 489]. By transforming the dataset into a lower-dimensional space while preserving variance, users can derive meaningful patterns and trends more rapidly [489].

Quantum Machine Learning (QML) constructs a bridge between quantum technology and artificial intelligence, yielding enhanced algorithms particularly adept at handling complex and extensive datasets. For instance, Quantum Support Vector Machines (QSVMs) benefit from quantum parallelism, allowing for faster classification and better scalability in handling massive datasets. Variational Quantum Circuits (VQCs) improve model efficiency in deep learning by optimizing parameterized layers specifically designed for handling big data applications [490, 491]. Quantum Kernel Methods also play a crucial role in QML, offering enhanced capabilities for classification and anomaly detection by mapping data into higher-dimensional quantum spaces, enabling the identification of more complex relationships in the data [492, 493].

Quantum Neural Networks (QNNs) utilize quantum properties to process multiple inputs simultaneously, enhancing the efficiency and effectiveness of deep learning models [491, 494]. The convergence of quantum computing and neural networks aims to significantly elevate the performance of machine learning tasks, especially in large datasets where classical methods may struggle.

Quantum algorithms also present innovative approaches in data clustering, a fundamental aspect of big data analysis. Techniques such as Quantum k-Means Clustering harness quantum superposition and entanglement to achieve higher efficiency in data grouping tasks, which can significantly impact various applications including customer segmentation in marketing, cybersecurity for anomaly detection, and tailored healthcare solutions [461, 491]. By leveraging quantum computing's capabilities, clustering algorithms are accelerated, thus enabling quicker processing times compared to their classical counterparts that typically suffer from scalability issues with growing data sizes [461, 495].

The implications of quantum-assisted clustering extend into practical applications, where the ability to analyse and group large datasets rapidly can lead to improved decision-making strategies and enhanced service delivery.

Quantum optimization is essential for solving large-scale problems across various industries, including logistics, finance, healthcare, and engineering. Classical optimization methods struggle with the exponential growth of solution spaces, making them inefficient for real-time decision-making. Quantum computing offers a significant advantage by leveraging exponential or quadratic speedups through quantum annealing, variational quantum algorithms, and quantum-inspired techniques.

Quantum annealing is particularly effective for combinatorial optimization problems, where finding the best solution requires evaluating an immense number of possible configurations. Companies like D-Wave have developed quantum annealers that specialize in solving such problems. In logistics, quantum annealing optimizes route planning for supply chain management, such as solving the Traveling Salesperson Problem (TSP). In finance, portfolio optimization is enhanced by quickly identifying the most efficient combination of assets to maximize returns while minimizing risks. In workforce management and production scheduling, quantum annealing helps optimize job assignments and resource allocation, reducing costs and improving efficiency. Unlike classical heuristic methods, quantum annealing naturally finds low-energy solutions within complex optimization landscapes, providing faster and more efficient results.

Variational Quantum Algorithms (VQAs) offer another promising approach to quantum optimization. Hybrid quantum-classical algorithms such as the Quantum Approximate Optimization Algorithm (QAOA) are designed to solve discrete optimization problems efficiently. QAOA is particularly effective for solving graph-related problems, including Max-Cut, Boolean satisfiability, and partitioning problems, which have applications in machine learning, logistics, and AI. Quantum Gradient Descent further enhances optimization tasks by leveraging quantum superposition to process multiple variables simultaneously, making it useful for machine learning applications and advanced financial modelling. These quantum algorithms enable new levels of computational efficiency in solving highly complex problems that classical methods struggle to address.

Even before fault-tolerant quantum computers become widely available, quantum-inspired optimization techniques are already making an impact. Quantum-inspired neural networks are being used to enhance deep learning models, leading to better decision-making in AI applications. Quantum Monte Carlo methods are applied in financial risk analysis and statistical simulations to improve accuracy in predicting market behaviour. Tensor networks, which are inspired by quantum mechanics, are utilized in quantum chemistry, drug discovery, and material science to model and optimize molecular interactions. By integrating quantum-inspired approaches into classical computing frameworks, businesses can benefit from quantum acceleration while waiting for fully scalable quantum hardware.

As quantum computing continues to evolve, industries are increasingly exploring its applications in big data analytics and optimization. The financial sector is experiencing a transformation with quantum computing, particularly in portfolio management, where quantum algorithms optimize investment strategies. Fraud detection benefits from quantum-enhanced machine learning, which can identify anomalies in financial transactions more efficiently than classical methods. Quantum Monte Carlo methods enable more accurate risk modelling, improving decision-making in banking and investment management.

In healthcare and drug discovery, quantum computing is accelerating research and pharmaceutical development. Molecular simulations are optimized to identify potential drug candidates faster, improving the efficiency of drug discovery pipelines. Quantum-enhanced data analysis allows for better prediction of disease patterns, leading to personalized treatment plans and improved patient care. Protein folding analysis, a critical challenge in medicine and genetic research, benefits from quantum computing's ability to model complex biological interactions.

Supply chain and logistics are also being optimized using quantum computing. Quantum annealing enhances route planning, ensuring the most efficient delivery routes are determined in real-time. Warehouse inventory management benefits from quantum-enhanced AI, which can predict demand fluctuations and optimize storage solutions. In manufacturing, quantum algorithms improve production scheduling, minimizing costs and maximizing operational efficiency.

Artificial intelligence and machine learning are advancing rapidly with quantum computing. Natural language processing (NLP) sees improvements through quantum-enhanced learning, leading to more accurate speech recognition and chatbot interactions. Reinforcement learning is accelerated by quantum algorithms, optimizing decision-making in robotics and automation. Quantum computing also enhances image and video processing, improving applications in computer vision and AI-powered media analytics.

In cybersecurity and cryptography, quantum computing presents both a challenge and an opportunity. While quantum computers pose a threat to traditional encryption methods by breaking classical cryptographic algorithms, they also enable new security solutions. Quantum Key Distribution (QKD) allows for ultra-secure communication networks, ensuring data security in an era where classical encryption may become obsolete. Quantum-resistant cryptography is being developed to safeguard digital assets and communications from quantum attacks, providing future-proof encryption solutions.

Despite the promise of quantum computing in big data analytics and optimization, several challenges must be addressed. Current quantum hardware has limitations, including a limited number of qubits and high error rates, restricting its scalability for large-scale applications. The development of more fault-tolerant quantum systems is necessary to optimize large datasets effectively. Many quantum algorithms remain in early research stages, requiring further theoretical refinement and experimental validation. Hybrid quantum-classical computing models are essential for practical deployment, as integrating quantum computing with existing classical infrastructure will be key to its widespread adoption.

Ongoing research in quantum error correction, quantum AI, and hybrid computing is steadily improving the feasibility of quantum computing for real-world applications. As quantum technology matures, its impact on big data analytics and optimization will become increasingly transformative, offering new solutions for industries that rely on high-performance computing for decision-making and problem-solving.

Chapter 9

Applications of Quantum Computing

As we have explored in previous chapters, quantum computing harnesses the unique properties of quantum mechanics—such as superposition, entanglement, and quantum parallelism—to solve complex problems that are intractable for classical computers. We have examined the fundamental principles of quantum mechanics, the nature of qubits and quantum gates, the power of quantum algorithms, and the role of error correction and noise mitigation in making quantum computing viable. We have also delved into the different hardware approaches being pursued by leading research groups and technology companies and discussed the impact of quantum computing on security through quantum cryptography. Furthermore, we have explored how quantum computing intersects with artificial intelligence and machine learning, opening new possibilities for optimization, pattern recognition, and predictive modelling.

With this foundation in place, we now turn to the real-world applications of quantum computing—the ultimate test of its potential to transform industries and scientific research. While the technology is still in its early stages, quantum computing is already showing promise in domains such as chemistry and material science, financial modelling, healthcare and drug discovery, climate research, and logistics optimization. By leveraging quantum capabilities, researchers and businesses aim to tackle problems that would take classical supercomputers thousands or even millions of years to solve. This chapter provides an overview of how quantum computing is being applied today, what breakthroughs are on the horizon, and how industries are preparing for the quantum revolution.

Quantum Simulations in Chemistry and Material Science

Quantum simulations are a powerful application of quantum computing that can revolutionize chemistry and material science by accurately modelling molecular interactions, chemical reactions, and material properties at the quantum level. Classical computers struggle with simulating complex quantum systems because the computational cost grows exponentially with the size of the system. Quantum computers, on the other hand, naturally follow quantum mechanics, allowing them to efficiently model electronic structures, molecular dynamics, and reaction mechanisms that are otherwise intractable for classical methods.

Quantum Simulations in Chemistry

Quantum simulations in chemistry represent a pivotal frontier in computational chemistry, facilitating unprecedented insights into molecular structures, bonding, and reaction mechanisms through advanced algorithms and quantum computing architectures. This exploration encompasses three primary aspects: electronic structure calculations, reaction mechanisms and catalysis, and quantum molecular dynamics (QMD), each substantiated through an array of studies.

Quantum computers hold the capability to efficiently solve the Schrödinger equation for complex molecular systems, significantly enhancing computational efficiency compared to classical methods like Density Functional Theory (DFT) and Coupled-Cluster methods, which scale poorly with molecule size. Algorithms such as the Variational Quantum Eigensolver (VQE) and Quantum Phase Estimation (QPE) have emerged as instrumental in these calculations, enabling the determination of ground state energies and electronic wavefunctions with high accuracy. Specifically, VQE utilizes a hybrid quantum-classical approach to optimize quantum circuits for finding the lowest energy configurations of molecules [187, 196]. On the other hand, QPE is adept at estimating eigenvalues of quantum systems, facilitating a deeper understanding of molecular electronic structure and properties [496, 497]. These quantum methodologies not only predict molecular characteristics but also optimize reaction pathways and advance material design, expanding the horizons of both theoretical and applied chemistry [498].

The application of quantum simulations extends prominently into catalysis and drug discovery. By providing detailed insights into reaction mechanisms, especially in enzyme reactions and catalytic processes, quantum computing can simulate transition states and intermediate complexes, thereby elucidating energy barriers and reaction rates. Notably, Santoro et al. underscore the efficacy of quantum chemical methodologies in rationalizing experimental outcomes and directing further research into catalytic systems [499]. Quantum simulations also have the potential to transform catalyst design for processes like hydrogen production and carbon capture, thus fostering environmentally sustainable practices [500]. In drug discovery, the capability to simulate protein-ligand interactions allows for the design of pharmaceuticals with optimized binding affinities, which can minimize adverse effects [501]. This nuanced understanding accelerates advancements across the chemical and pharmaceutical landscapes, underpinning the significance of quantum computing in catalysis [498].

QMD simulations stand as a powerful tool for examining the temporal evolution of molecular systems under diverse conditions, effectively capturing quantum effects that are approximated in classical models. Quantum computers can directly model quantum wavefunctions, advancing the accuracy of predictions in various domains such as reaction kinetics, solvent effects, and quantum tunnelling phenomena, which are critical in processes like enzyme catalysis and nuclear fusion [95, 502]. The ability of quantum systems to explore these dynamics is particularly advantageous in the study of biochemical reactions, complex nanomaterials, and extreme conditions often encountered in astrophysics and high-energy physics [497, 503]. This represents a substantial step forward in understanding molecular behaviour and interactions at a fundamental level.

Practical Examples Illustrating the Application of Quantum Simulations Across Various Areas of Chemistry

The use of quantum computers for electronic structure calculations has been transformative, particularly in solving Schrödinger's equation for complex molecules. For instance, researchers at IBM implemented the Variational Quantum Eigensolver (VQE) algorithm to precisely calculate the ground state energy of small molecules such as Lithium Hydride (LiH) and Hydrogen (H_2) [504, 505]. These calculations are critical in contexts like battery materials and hydrogen storage technologies, illustrating quantum computing's potential to provide accurate electronic properties vital for practical applications [505].

In addition to simulating small molecules, advancements in Density Functional Theory (DFT) have been made by Google's Quantum AI team, which combined quantum and classical approaches to improve predictions of chemical properties for larger organic molecules [506]. This progress signifies a leap towards more effective material design in sectors such as organic solar cells and molecular electronics.

Harvard researchers also utilized Quantum Phase Estimation (QPE) to model electronic structures of water (H_2O) and methane (CH_4), enabling a deeper understanding of interactions such as hydrogen bonding—integral to many biochemical reactions and energy applications [505].

Quantum computing is redefining the modelling of reaction mechanisms essential for industrial catalysis and drug discovery. A notable example is the simulation of the Haber-Bosch process, pivotal for ammonia production, carried out by Microsoft and the Pacific

Northwest National Laboratory (PNNL) using quantum simulations. This work aims to optimize nitrogen-fixing catalysts, potentially lowering energy consumption and CO_2 emissions [507, 508].

In the realm of hydrogen fuel cells, the Toyota Research Institute employed quantum algorithms to enhance the efficiency of platinum-based catalysts, thereby minimizing dependence on scarce materials [509]. Moreover, collaborations between Pfizer and IBM are exploring quantum simulations for enzyme reactions to streamline drug development processes for diseases like Alzheimer's and cancer [510].

Similarly, ExxonMobil, in partnership with IBM Quantum, is investigating catalysts for converting CO_2 into useful chemicals, reinforcing quantum computing's role in sustainable processes to combat climate change [511].

Quantum simulations are also making significant impacts in drug discovery. For instance, Biogen and Xanadu Quantum Technologies are leveraging quantum simulations to predict protein folding, an essential factor in understanding several diseases, including Alzheimer's and COVID-19 [511]. This results in more rapid and accurate therapeutic design.

Additionally, GlaxoSmithKline (GSK) and Cambridge Quantum Computing are implementing VQE algorithms to model drug molecules, facilitating the design of new drugs that target disease-related proteins more effectively [512]. Roche and Google Quantum AI's collaborative effort to utilize quantum-enhanced machine learning is further innovating drug-target interaction predictions, paving the way for personalized medicine [513].

Quantum simulations are critical in advancing energy storage systems. For instance, IBM and Daimler have employed quantum computing to model lithium-air batteries, known for their substantial potential energy density compared to conventional lithium-ion batteries [514]. This research aims to enhance battery safety and longevity.

Volkswagen, in collaboration with Google Quantum AI, is researching solid-state batteries utilizing quantum simulations to improve ionic conductivity, thus enhancing the performance and charging speeds of electric vehicles (EVs) [505, 507]. NASA and IBM are also jointly exploring high-temperature superconductors, aiming to boost electrical transmission efficiencies using quantum simulations [515].

Research into green chemistry applications is also prominent in quantum simulations. BASF and Google Quantum AI are modelling biodegradable plastics to develop eco-friendly materials through quantum-level insights into polymer chemistry [516]. Meanwhile, Shell and Microsoft Quantum's focus on biofuels through quantum-enhanced simulations is advancing the efficiency of biomass conversion processes relevant to renewable energy [517, 518].

Quantum Simulations in Material Science

Quantum simulations are significantly shaping advancements in material science, particularly in the design and discovery of new materials with desirable properties. The ability of quantum computing to simulate materials at the atomic level allows researchers to optimize characteristics such as electronic and magnetic properties. This has considerable implications for a variety of applications, including high-performance batteries, superconductors, and topological materials.

The capacity of quantum computers to simulate materials' electronic properties is pivotal in discovering new superconductors and topological materials. Superconductors, which can conduct electricity without resistance, are sought after for efficient energy transmission. Research indicates that leveraging quantum simulations can expedite the identification of materials that exhibit superconductivity at elevated temperatures, thereby potentially transforming electronic technologies [519]. Topological materials, characterized by surface states that are robust against disorder and impurities, are also being explored through quantum simulations for applications in quantum computing and spintronics—the intersection of spin and charge in electronic devices [520].

Moreover, these simulations enable the engineering of novel magnetic materials tailored for advanced applications in data storage and quantum sensing. The precision in designing materials with specific electronic band structures is crucial for innovations in semiconductors and quantum device fabrication. Consequently, quantum simulations represent a transformative approach to material science, significantly enhancing our ability to discover and design materials with exotic properties [519].

Quantum simulations are also at the forefront of developing next-generation batteries and energy storage solutions. Significant research efforts are concentrating on improving materials for lithium-ion batteries and solid-state batteries. Studies have demonstrated that using quantum modelling accelerates the discovery of high-performance battery materials, addressing critical parameters such as efficiency and lifespan [521]. For instance, exploring novel electrode materials through quantum simulations has shown promise in enhancing the efficiency of lithium-ion batteries, which are crucial for modern electronics [521].

In the context of solid-state batteries, quantum simulations facilitate the identification of suitable solid electrolytes that could outperform liquid electrolytes in durability and safety. This aspect holds particular importance as the demand for safer and more efficient energy storage systems escalates alongside global energy needs [521]. Furthermore, simulating the interactions of hydrogen molecules with solid materials aids in advancements in hydrogen fuel cell technology, thereby contributing towards sustainable energy sources [522].

The realm of photovoltaics and light-emitting diodes (LEDs) also benefits immensely from quantum simulations, particularly in optimizing material compositions for devices like perovskite solar cells and organic LEDs (OLEDs). These simulations enhance the design of materials to improve light-matter interactions, which directly influences the efficiency of energy conversion in solar cells [521]. For example, research has shown that specific quantum simulations can facilitate the optimization of perovskite materials, which are leading candidates for next-generation solar cells due to their unique properties [521].

Moreover, advancements in quantum dot technology—semiconductor particles that exhibit quantum mechanical properties—are paving the way for groundbreaking developments in imaging and sensing technologies [523]. Overall, by harnessing the computational power of quantum simulations, significant improvements in the performance and efficiency of renewable energy solutions can be achieved.

Quantum computing is actively revolutionizing materials science by providing powerful tools for simulating complex atomic interactions, a capability that traditional classical computers struggle to match. It enables precise modelling of atomic structures, electronic properties, and energy transport mechanisms, which are crucial for understanding and designing new materials. Research highlights this transformation, emphasizing quantum simulations as a means for rapid advancements in areas such as superconductors, energy storage materials, and applications in renewable energy technologies. These methodologies are anticipated to dramatically decrease the time and cost associated with experimental testing needed for material discovery and optimization [498, 524].

In the realm of superconductivity, quantum computers facilitate the simulation of electron behaviour and magnetic properties on an atomic level. Companies such as IBM Research and Harvard University employ quantum algorithms to investigate the electronic band structures of materials like cuprates and iron-based superconductors. These innovations could lead to advancements such as lossless electricity transmission for power grids, improved magnetic levitation train technologies, and enhanced high-performance computing solutions [525, 526]. Additionally, Microsoft's StationQ Lab is exploring topological materials to identify those capable of hosting topological quantum states necessary for the development of Majorana-based topological qubits, promising more reliable and error-resistant quantum computing [527]. Furthermore, quantum magnetism simulations by firms like Samsung and Google Quantum AI aim to create new magnetic memory materials that improve data storage density [528].

Battery technology is another key area benefitting from quantum simulations. Automakers like Daimler AG and Volkswagen are leveraging these techniques to model lithium-sulphur battery chemistry and solid-state electrolytes, aiming to create batteries with increased energy density, safety, and faster charging capabilities to address the accelerating demand for electric vehicles and renewable energy storage solutions [498, 524]. The research led by the Toyota Research Institute on metal-organic frameworks seeks to enhance hydrogen storage solutions for fuel cell technologies, promoting further advancements in hydrogen-powered applications [527].

Moreover, in optics and photovoltaics, quantum computing is optimizing the development of solar cells and organic light-emitting diodes (OLEDs). Collaborations between Oxford PV and IBM Quantum are focusing on perovskite material enhancements that could potentially surpass 30% efficiency in solar panel technology. These advancements rely on quantum methods that accurately predict material performance, stability, and defect behaviours crucial for longevity in energy applications [524, 529]. Companies like LG and Samsung are applying quantum algorithms to boost the performance of organic semiconductors for OLEDs, thereby enhancing the energy efficiency and brightness of displays in modern electronics [524, 529].

The overall impact of quantum computing on material science is profound, as it accelerates the discovery of superconductors, energy storage materials, and advanced optoelectronic devices. Major players in the technology and automotive sectors, including IBM, Google, and Toyota, are leveraging quantum capabilities to innovate and streamline the development of materials, leading to safer batteries, more efficient solar panels, and increased computational power. As quantum hardware continues to improve, the implications for energy management, electronics, and the broader field of quantum computing promise significant advancements in sustainable technologies and computational efficiency [526, 530].

While quantum simulations offer tremendous potential, several challenges must be addressed to fully realize their capabilities. One of the primary limitations is the current state of quantum hardware. Existing quantum computers have a limited number of qubits, short coherence times, and high gate error rates, making it difficult to accurately simulate large molecules and complex materials. These constraints restrict the size and precision of quantum simulations, preventing researchers from modelling highly intricate molecular interactions.

Scaling quantum simulations to handle large-scale chemistry and material science problems remains another major hurdle. Many quantum chemistry problems, such as simulating biomolecules or novel materials, require millions of qubits to capture the full complexity of molecular structures and electron interactions. Given that today's quantum processors have only a few hundred noisy qubits, achieving this level of scalability requires significant advancements in quantum hardware and qubit connectivity.

Quantum error correction is essential for reliable quantum simulations, but current quantum systems are far from being fault-tolerant. Quantum noise and decoherence introduce errors in calculations, leading to inaccurate or unstable results. Without robust error correction techniques, the usefulness of quantum simulations is limited to small-scale problems that can tolerate a degree of imprecision. Developing practical fault-tolerant quantum computers remains a critical research focus for ensuring that quantum simulations produce consistently accurate and reliable results.

Another key challenge is integrating quantum simulations with classical computing methods. While quantum computers can offer exponential speedups for certain tasks, many chemical and material simulations still require classical approximations and computational techniques. Hybrid quantum-classical approaches, where quantum simulations handle the most computationally intensive calculations while classical computers manage the rest, are necessary to bridge the gap between existing methodologies and quantum advantages. Efficient hybrid algorithms and seamless integration between quantum and classical workflows are essential for practical applications in industry and research.

Despite these challenges, ongoing advancements in quantum hardware, error correction, and hybrid computing are steadily improving quantum simulations. Researchers are developing more robust quantum algorithms, refining quantum error mitigation techniques, and exploring new quantum processor architectures to enhance performance. As quantum technology progresses, these efforts will bring quantum simulations closer to practical use in chemistry, materials science, and beyond, ultimately transforming the way scientists discover new materials and design innovative molecular systems.

Financial Modelling and Risk Analysis

Financial institutions rely on complex mathematical models to assess risk, optimize portfolios, and forecast market trends. Classical computers struggle with the exponential growth of financial data and the computational intensity of risk analysis. Quantum computing offers the potential to revolutionize financial modelling by accelerating computations, improving optimization techniques, and enabling more accurate risk assessments.

Quantum computing provides significant advantages in solving mathematical problems fundamental to financial modelling. One of the key applications is Monte Carlo simulations, which are widely used in pricing derivatives, risk assessment, and portfolio management. Classical Monte Carlo methods require a large number of random samples to approximate probabilities, making them computationally expensive. Quantum Monte Carlo (QMC), leveraging quantum parallelism and amplitude amplification, can potentially speed up simulations, reducing the time complexity from $O(N)$ to $O(\sqrt{N})$, where N is the number of samples needed for an accurate estimate.

Another important application is Quantum Gradient Descent (QGD), which can enhance machine learning models used in financial forecasting. Traditional gradient descent algorithms require iterative calculations over large datasets, which can be slow and computationally expensive. Quantum-enhanced optimization can provide a speedup in finding optimal trading strategies and financial risk assessments by leveraging quantum linear algebra techniques.

Portfolio optimization is a critical aspect of financial modelling, where investors seek to maximize returns while minimizing risk. Classical portfolio optimization relies on methods such as Markowitz's Modern Portfolio Theory (MPT), which involves solving quadratic programming problems. These calculations become computationally demanding as the number of assets increases.

Quantum computing offers solutions through Quantum Approximate Optimization Algorithm (QAOA) and Variational Quantum Eigensolver (VQE), which can efficiently solve quadratic optimization problems. These quantum algorithms allow financial institutions to quickly determine optimal asset allocation, balancing risk and return under complex constraints. Companies such as Goldman Sachs and JPMorgan Chase are actively exploring quantum optimization techniques for portfolio risk management.

Risk analysis involves calculating exposure to market fluctuations, credit defaults, and financial shocks. Classical methods use Value-at-Risk (VaR) and Conditional Value-at-Risk (CVaR) models, which require heavy computation to simulate various market scenarios.

Risk management is a critical component of financial modelling, allowing institutions to assess their exposure to market fluctuations, credit defaults, and economic downturns. Two of the most widely used risk assessment metrics are Value-at-Risk (VaR) and Conditional Value-at-Risk (CVaR). These measures help financial institutions, hedge funds, and portfolio managers quantify potential losses and make informed investment decisions.

Classical financial models for calculating VaR and CVaR rely on statistical methods, Monte Carlo simulations, and historical data analysis. However, these approaches become computationally expensive, particularly when dealing with high-dimensional financial portfolios. Quantum computing introduces a new approach to risk assessment, accelerating simulations, optimizing portfolio allocations, and improving risk quantification.

VaR is a statistical measure that estimates the worst possible loss of an investment portfolio within a given confidence level (e.g., 95% or 99%) over a specific time period. Mathematically, it is expressed as:

$$P(L > VaR) \leq 1 - \alpha$$

where L represents the loss distribution and α\alphaα is the confidence level.

For example, if a portfolio has a 1-day 95% VaR of $10 million, it means there is a 95% probability that the portfolio will not lose more than $10 million in a single day.

VaR can be computed using different methods, including historical simulation, variance-covariance analysis, and Monte Carlo simulations. The Monte Carlo method is the most flexible but computationally expensive. As the number of financial assets increases, the required number of simulations grows exponentially, making real-time risk assessment difficult.

Quantum computing enhances VaR calculations through Quantum Monte Carlo (QMC), which provides a quadratic speedup compared to classical Monte Carlo methods. Instead of requiring O(N) simulations, quantum parallelism allows QMC to estimate VaR in O(√N) time.

Quantum algorithms such as Quantum Amplitude Estimation (QAE) reduce the number of required Monte Carlo samples, enabling faster probability estimation. Quantum Principal Component Analysis (QPCA) improves risk factor modelling by analysing asset correlations more efficiently than classical PCA. Additionally, Quantum Gradient Descent (QGD) helps optimize risk functions in complex financial models. By leveraging QAE, financial institutions can compute VaR with fewer simulations while maintaining accuracy, enabling faster risk analysis in high-frequency trading and portfolio management.

CVaR, also known as Expected Shortfall (ES), is a more conservative risk measure than VaR. It estimates the expected loss beyond the VaR threshold, capturing tail risk in extreme market conditions. Mathematically, it is defined as:

$$CVaR_\alpha = \mathbb{E}[L|L > VaR_\alpha]$$

For example, if a portfolio has a 1-day 95% CVaR of $15 million, it means that in the worst 5% of cases, the expected loss will be at least $15 million.

Since CVaR considers tail risk, it requires additional computations beyond VaR, such as Monte Carlo simulations to evaluate losses in extreme scenarios, optimization techniques to manage downside risk, and heavy-tailed distribution models to capture financial market anomalies. These calculations significantly increase the computational burden, making them impractical for large portfolios using classical methods.

Quantum algorithms improve CVaR calculations by enhancing both probability estimation and optimization. Quantum Amplitude Amplification (QAA) increases sampling efficiency, while Quantum Approximate Optimization Algorithm (QAOA) helps optimize portfolio allocation to minimize expected shortfall. Variational Quantum Eigensolver (VQE) is used to find the lowest energy state of an optimization problem, improving risk-adjusted portfolio management. By leveraging these quantum techniques, CVaR can be computed and optimized more efficiently, allowing traders and risk managers to react faster to financial downturns.

Quantum computing enhances risk analysis through Quantum Monte Carlo simulations and Quantum Principal Component Analysis (QPCA). Quantum Monte Carlo accelerates the evaluation of risk exposure by reducing the computational complexity of simulations. QPCA improves risk management by analysing correlations in financial data, helping banks and hedge funds detect hidden market risks and price complex financial instruments more accurately.

Derivatives pricing is another area where quantum computing can provide a competitive edge. Pricing complex derivatives, such as options and swaps, involves solving high-dimensional partial differential equations (PDEs) or running thousands of simulations. Quantum Machine Learning (QML) can improve the accuracy of pricing models by recognizing non-linear patterns in financial markets, allowing for better predictions of asset movements.

Financial institutions rely on machine learning models to detect fraud, identify anomalous transactions, and improve regulatory compliance. Traditional fraud detection models struggle with massive datasets and the complexity of modern financial transactions.

Quantum Machine Learning (QML) offers a new approach by enabling quantum-enhanced pattern recognition and anomaly detection. Quantum algorithms can process large datasets more efficiently, uncovering fraudulent transactions with greater accuracy. Companies such as Mastercard and Wells Fargo are researching quantum AI to improve fraud detection, cybersecurity, and anti-money laundering (AML) techniques.

Despite its potential, quantum computing in finance faces several challenges. Quantum hardware limitations, including qubit decoherence and gate errors, restrict the size and complexity of quantum models that can be executed today. Algorithm development is still in its early stages, and practical quantum speedups for financial applications have yet to be demonstrated at a large scale. Additionally, integration with classical financial systems requires hybrid quantum-classical approaches, as current quantum processors are not yet capable of fully replacing traditional computing infrastructure.

Financial institutions are increasingly investing in quantum research to maintain a competitive edge in the finance sector. Companies such as JPMorgan Chase, Citigroup, HSBC, and Barclays are actively collaborating with leading quantum technology firms, including IBM Quantum, Google Quantum AI, and D-Wave, to explore practical applications of quantum computing within financial contexts. This strategic partnership is primarily focused on the development of quantum algorithms that are expected to outperform traditional methods while ensuring effective integration into existing financial models.

JPMorgan Chase and other financial entities recognize that quantum computing has the potential to transform financial operations, particularly in complex tasks like derivative pricing, risk analysis, and portfolio optimization. Recent research highlights the significance of quantum algorithms designed for these tasks, showcasing advancements such as improved Monte Carlo simulations for option pricing, which offer substantial speed-ups compared to classical techniques and provide pathways to achieving quantum advantage in finance [531-533]. The evolution of these algorithms underscores the necessity of robust benchmarking practices in quantum software engineering, ensuring that newly developed methods can be integrated with traditional financial models [198, 534].

Moreover, the intersection of artificial intelligence (AI) and quantum computing is identified as a critical area of innovation. Quantum-enhanced machine learning techniques hold promise for solving challenging problems in finance, such as accurate forecasting of financial crashes and optimizing asset allocations [535, 536]. As financial firms explore these technologies, there is a focus on creating algorithms that leverage quantum computational advantages without disrupting established frameworks in financial modelling [537].

The development of hybrid algorithms, which integrate insights from classical and quantum computing, represents another key area of research necessary for achieving practical implementations [424, 538]. As quantum technology matures, these developments are essential for financial institutions aiming to harness quantum computing for strategic advantages. The future landscape of financial computing may be significantly influenced by these efforts as institutions seek to navigate the complexities of quantum algorithm integration and performance validation [539, 540].

As quantum hardware improves and more efficient quantum algorithms are developed, financial institutions will gain access to powerful tools for portfolio optimization, risk assessment, fraud detection, and derivative pricing. Quantum computing has the potential to revolutionize financial modelling by providing faster, more accurate, and data-driven decision-making, ultimately transforming the future of the finance industry.

Healthcare and Drug Discovery

Quantum computing represents a significant advancement in the domain of drug discovery and molecular modelling, providing new methodologies for simulating molecular interactions and enhancing the design of therapeutics. This exploration can be encapsulated in four key areas: quantum simulations for drug discovery, accelerating protein folding and biomolecular interactions, precision medicine and genomic analysis, and the broader implications for healthcare systems.

Quantum computing holds immense capabilities for simulating molecular structures with higher precision than classical computers. Techniques such as the Variational Quantum Eigensolver (VQE), Quantum Phase Estimation (QPE), and Quantum Monte Carlo (QMC) enable more accurate modelling of molecular behaviours and interactions, critical in drug design. For instance, VQE is adept at determining the ground-state energy of molecular systems, thereby facilitating simulations of drug-protein complexes, while QPE offers robust electronic structure predictions [541-543]. Companies such as Boehringer Ingelheim are leveraging these quantum techniques in collaboration with IBM Quantum to create novel small-molecule drugs targeting complex diseases such as Alzheimer's [544].

The challenge of understanding protein folding—a complex process influenced by myriad potential conformations—presents a significant hurdle in biomedicine. Quantum computing enhances our ability to simulate folding mechanisms, which is crucial for predicting drug interactions with specific targets [545]. Furthermore, by optimizing molecular docking approaches, quantum algorithms can accurately identify optimal binding sites and improve the efficacy of new therapeutics [546, 547]. Notable efforts by organizations such as Google

Quantum AI and Pfizer aim to model multifaceted proteins and expedite vaccine development through these methodologies [544, 548].

In the realm of genomic sequencing, quantum computing can drastically improve data analysis, allowing for personalized medicine tailored to individual genetic profiles. Utilizing Quantum Machine Learning (QML) techniques can enhance capabilities in Genome-Wide Association Studies (GWAS) to correlate genetic mutations with diseases, and in discovering biological markers essential for early disease detection [549, 550]. Furthermore, companies like Menten AI are pioneering quantum-inspired algorithms to refine protein design and drug discovery processes [551]. By integrating genomic data analysis with quantum computing, researchers can significantly improve the predictive accuracy of drug responses based on genetic information.

Beyond direct applications in drug discovery, quantum computing can optimize clinical trials and enhance healthcare efficiency. By employing quantum algorithms to select patient groups based on extensive datasets, optimize scheduling, and improve drug dosage calculations, healthcare practices can be streamlined [544, 545]. For example, collaborations between Amgen and IBM Quantum are investigating how these methods can be applied to refine clinical trial logistics [545, 546]. Furthermore, advancements in quantum-enhanced imaging techniques could revolutionize diagnostics by improving image resolution and reducing noise, ultimately leading to better disease detection capabilities [546].

Climate Modelling and Complex Systems

Quantum computing is rapidly emerging as an innovative element in tackling complex scientific challenges, especially in the realms of climate modelling and environmental sustainability. Traditional supercomputing methods, despite their importance as crucial computational tools, encounter significant limitations when simulating climate dynamics characterized by complexity and vast datasets. This is largely due to the non-linear behaviours and chaotic dynamics evident in these systems, which quantum algorithms are uniquely positioned to handle efficiently.

Climate models fundamentally depend on resolving partial differential equations (PDEs) and conducting extensive statistical analyses. Classical techniques for modelling weather and climate, such as finite element analysis (FEA) and numerical weather prediction (NWP), generally require immense computational power, demanding substantial time investments, often stretching into days or weeks [552, 553]. Quantum computing, however, has the potential to simplify and expedite this process. For instance, the application of quantum differential equation solvers can address the Navier-Stokes equations, crucial for modelling fluid dynamics within climate systems [554]. Moreover, implementing quantum Monte Carlo methods enhances the quantification of uncertainties prevalent in long-term climate forecasts, which is vital for robust climate science [554].

In the specific context of atmospheric and oceanic modelling, quantum computing holds promise for accurately simulating intricate systems such as atmospheric circulation and ocean currents. Advanced quantum algorithms can lead to improvements in predicting processes like cloud formation and precipitation [555]. For example, leveraging quantum simulations enables the analysis of interaction dynamics between different atmospheric layers and ocean movements, offering a level of detail beyond traditional models [554]. Notably, collaborative efforts between organizations such as NASA and IBM Quantum aim to create quantum-assisted climate models that address pivotal issues such as global warming and polar melt, indicating a strategic alignment towards using quantum technologies for impactful climate research [554].

Moreover, quantum computing applications extend to the analysis of extreme weather events, including hurricanes, droughts, and heat waves, which necessitate real-time data processing from sensor and satellite sources. Quantum machine learning (QML) holds the potential to enhance these predictive capabilities, optimizing models for forecasting tornado paths and improving flood prediction accuracy [554, 556]. Collaborations like that between Google's Quantum AI team and NOAA exemplify the potential of quantum technology in refining early warning systems and bolstering disaster response frameworks, thus fortifying climate resilience [554].

Furthermore, the role of quantum computing extends to carbon capture and climate mitigation strategies. Quantum simulations can accelerate the design of catalysts necessary for efficient carbon capture technologies and optimize chemical reactions essential for direct air capture of CO_2, potentially revolutionizing strategies for carbon sequestration [556]. Additionally, corporations such as ExxonMobil and Microsoft Quantum are engaging in utilizing quantum assistance to enhance their capabilities in renewable energy optimization and effective resource allocation, indicative of a paradigm shift in corporate responsibility towards environmental sustainability [554, 557].

Nonetheless, while the prospects of quantum computing in climate science are promising, several challenges persist. Scalability of quantum hardware remains a significant barrier, as high-precision climate models necessitate millions of qubits, a current limitation within existing quantum processors [553]. Additionally, quantum noise and error rates impede the accuracy of simulations, emphasizing the need for advanced error correction techniques and the development of hybrid systems integrating classical and quantum computational methods [554].

Looking to the future, initiatives such as IBM and NASA's Quantum Climate Initiative illustrate a commitment to merging quantum technology with climate science, aiming for broader applications in sustainability and complex systems analysis [558]. As research progresses, the integration of quantum computing into climate modelling appears set to enhance our understanding of climate dynamics and lead to innovative solutions to mitigate environmental risks [553].

Logistics and Supply Chain Optimization

Quantum computing presents a transformative approach to optimizing logistics and supply chain management, addressing complex problems that exceed the capabilities of classical computational methods. By utilizing quantum algorithms, businesses can enhance route optimization, inventory management, and demand forecasting—streamlining operations across global supply networks and potentially yielding substantial economic benefits. Quantum parallelism allows organizations to process extensive datasets, enable superior resource allocation, and adapt rapidly to supply chain disruptions, a pivotal capacity in today's volatile markets [23, 559].

Challenges in classical supply chain optimization predominantly arise from the limitations of traditional methods, including linear programming and heuristic algorithms. These classical approaches often fail to deliver efficient solutions when applied to large-scale logistics problems involving thousands of variables. Core operational challenges such as route optimization—where efficient delivery routes must be determined amid dynamic variables like traffic and fuel costs—and inventory management—balancing supply with demand to avoid stockouts—highlight the inadequacies of conventional frameworks. Moreover, demand forecasting is complicated by the inherent unpredictability of consumer behaviour, further emphasizing the need for advanced computational solutions. Quantum computing offers heightened speed and efficiency by providing exponential solutions to these problems, positioning it as a compelling alternative to classical optimization techniques [559, 560].

One promising technique within quantum computing is quantum annealing, which is highly effective for combinatorial optimization problems found in logistics. Companies like D-Wave have demonstrated its application in the Traveling Salesperson Problem (TSP) and the Vehicle Routing Problem (VRP), key areas for minimizing delivery routes and optimizing fleet operations. Case studies, such as Volkswagen's implementation of quantum-based routing for taxis in Beijing, illustrate tangible improvements in operational efficiency and travel times by leveraging real-time data processing [561]. Furthermore, Variational Quantum Algorithms (VQAs), including the Quantum Approximate Optimization Algorithm (QAOA), enable detailed analysis of resource allocation and workflow efficiencies, critical for warehouse logistics and production scheduling [560].

Quantum Machine Learning (QML) adds another layer of sophistication for demand forecasting by employing quantum neural networks and quantum kernel methods for more precise analysis of vast datasets. Initiatives from Google Quantum AI are exploring these QML techniques to enhance forecasting accuracy, which is crucial for maintaining appropriate inventory levels and managing supply chain risks effectively. However, no specific reference supporting this was cited in the response, so this statement remains unverified.

Despite these advances, the path to fully integrating quantum computing into logistics and supply chain systems is fraught with challenges. Hardware limitations, such as the restricted number of qubits, pose significant hurdles to scalability. Additionally, issues related to quantum noise and error rates can undermine the reliability of optimization outcomes. Hybrid models that integrate classical computing with quantum solutions are necessary to create a cohesive framework for practical applications [561]. Nevertheless, leading technology firms aim to overcome these barriers by improving quantum hardware and refining error correction methods, which bode well for the future of quantum-enhanced supply chain management [23, 559].

Chapter 10

Quantum Computing in the Cloud

As quantum computing advances, accessibility remains one of its biggest challenges. Until recently, quantum computers were only available in highly specialized research facilities, requiring extensive infrastructure and expertise to operate. However, with the emergence of cloud-based quantum computing, more people than ever before—researchers, developers, businesses, and even students—can experiment with quantum technology without needing direct access to expensive hardware.

This chapter explores the growing field of Quantum Computing in the Cloud, a paradigm that enables users to access quantum computing resources remotely through platforms such as IBM Quantum Experience, Google Quantum AI, and Amazon Braket. The ability to run quantum algorithms online has dramatically changed the way quantum computing is researched and applied. Cloud-based quantum computing offers a Quantum as a Service (QaaS) model, allowing users to develop and test quantum programs without the need for physical quantum machines.

Building on previous chapters, which have covered the foundations of quantum mechanics, quantum algorithms, hardware approaches, and real-world applications, this chapter provides a practical perspective on how quantum computing can be accessed and utilized today. It discusses the technical aspects of cloud-based quantum computing, including how users can access and run quantum algorithms online, the types of quantum hardware available through cloud providers, and the limitations of current cloud quantum computing services.

Additionally, this chapter examines the challenges of scaling quantum cloud services, including network latency, computational constraints, and security concerns. As the industry moves toward commercializing quantum computing, cloud platforms play a crucial role in democratizing access to quantum technology. By leveraging the power of the cloud, researchers and businesses can begin integrating quantum solutions into practical applications, accelerating the transition from theoretical research to real-world impact.

Introduction to Cloud-Based Quantum Computing

Cloud-based quantum computing represents a significant advancement in the accessibility of quantum technologies, allowing users from various backgrounds to engage with quantum processors through internet platforms. This approach alleviates the necessity for organizations to invest in the complex and costly infrastructure traditionally associated with quantum computing, allowing researchers, developers, and businesses to experiment with quantum algorithms and applications on demand [562, 563].

Major technology companies, such as IBM, Google, Amazon, Microsoft, and Rigetti, have been pioneers in the establishment of cloud-based quantum computing platforms. These platforms are designed to democratize access to quantum hardware and simulators, facilitating user experimentation with quantum algorithms while integrating seamlessly with classical computing workflows [562]. For instance, users can execute quantum circuits and run applications in a hybrid quantum-classical environment using tools like IBM's Qiskit and Google's Cirq, which are vital for developing and testing quantum algorithms in real-world scenarios [562-564].

The advent of Quantum-as-a-Service (QaaS) models has profoundly impacted the landscape of quantum computing research. By removing the requirement for dedicated infrastructures, these models have accelerated the development of innovative quantum applications and methodologies. Users can now conduct experiments involving quantum gates, variational quantum algorithms (VQAs), and quantum simulations without needing extensive specialized knowledge in quantum hardware [563, 564]. Furthermore, platforms such as Rigetti's Forest and IBM's services have illustrated the practical applications of quantum techniques in areas like machine learning and quantum optimization through accessible cloud resources [564, 565].

The collaborative nature of cloud-based quantum platforms encourages rapid advancement in quantum computing capabilities. For example, experiments conducted on public quantum computers, such as those from IBM and Rigetti, validate the feasibility of complex quantum tasks such as quantum adder implementations and autoencoders, showcasing how cloud infrastructure supports robust experimental endeavours within the quantum realm [566-568]. The integration of classical and quantum computing resources via cloud services further enhances the potential for innovative solutions to computational problems that have historically challenged classical computing capabilities, thereby heralding a new era of scientific and technological exploration [569, 570].

Cloud-based quantum computing operates through a hybrid cloud model, where users interact with quantum processors remotely while leveraging classical cloud infrastructure for preprocessing and post-processing tasks. This approach enables researchers, developers, and businesses to experiment with quantum algorithms without requiring direct access to specialized quantum hardware. The cloud-based system consists of multiple components that facilitate quantum computation, from hardware access to software frameworks and simulators.

Cloud platforms provide access to real quantum processors, allowing users to run quantum experiments on advanced qubit architectures. These include superconducting qubits used by IBM Quantum, Google Quantum AI, and Rigetti; trapped-ion qubits utilized by IonQ and Quantinuum; photonic quantum processors developed by Xanadu; and quantum annealers offered by D-Wave. By accessing these quantum devices remotely, users can execute quantum computations without having to manage cryogenic cooling systems or perform qubit calibration, making quantum computing more accessible and scalable.

To develop and execute quantum programs, cloud-based quantum computing platforms support various quantum programming languages and software development kits (SDKs). IBM Qiskit provides a Python-based framework for quantum circuit design and execution, while Google Cirq is tailored for near-term quantum algorithms on Google's Sycamore quantum processor. Microsoft Q# and Azure Quantum SDK offer quantum programming integrated with Microsoft's cloud ecosystem, and Amazon Braket serves as a cloud-based quantum development environment supporting multiple quantum hardware backends. These frameworks allow researchers to create quantum circuits, optimize algorithms, and integrate quantum computations into classical workflows, facilitating experimentation and innovation in quantum computing.

Before deploying a quantum algorithm on physical quantum hardware, cloud platforms provide quantum simulators that replicate quantum behaviour on classical computers. IBM Aer is a high-performance quantum circuit simulator designed for noise-aware simulations, while Google qsim utilizes tensor network-based quantum emulation for large-scale circuit simulations. Microsoft's Quantum Development Kit (QDK) offers hybrid quantum simulators for algorithm testing and debugging, and Amazon Braket Simulators provide classical emulation of quantum circuits to refine algorithm performance. These simulators help users develop and test quantum applications, optimize quantum circuit designs, and mitigate quantum noise before running computations on real quantum processors.

Once a quantum circuit is developed, it is submitted as a job request to a cloud-based quantum computer for execution. The process begins with preprocessing, where the quantum algorithm is compiled and optimized for execution on specific quantum hardware. The compiled quantum circuit is then executed on real quantum hardware or a simulator, depending on the user's choice and hardware availability. After execution, the system performs measurement and post-processing, where quantum measurement results are collected, analysed, and returned to the user.

Most cloud-based quantum platforms utilize a job queue system, where users submit their quantum tasks and receive results asynchronously. Due to the high demand and limited availability of quantum processors, tasks are often scheduled in a queue, prioritizing access based on the user's subscription level or computing needs. By providing managed quantum execution environments, cloud-based quantum computing platforms streamline the process of running quantum experiments, making quantum resources more accessible to researchers and businesses worldwide.

Cloud-based quantum computing removes the barrier of expensive infrastructure, making quantum technology accessible to a wider range of users, including students, researchers, startups, and enterprises. Instead of requiring specialized quantum hardware, users can remotely access quantum processors and scale their quantum workloads as needed. This democratization of quantum computing allows for greater participation in quantum research and innovation, without the prohibitive costs of setting up dedicated quantum labs.

Building and maintaining a quantum computing lab requires substantial financial investment in quantum processors, dilution refrigerators, and quantum control electronics. Cloud computing offers a cost-efficient alternative by operating on a pay-as-you-go model, where users only pay for the quantum resources they consume. This model enables businesses and researchers to experiment with quantum computing without committing to large capital expenditures, making it more practical for commercial applications and academic research.

Cloud-based quantum platforms accelerate quantum algorithm development by allowing software developers to prototype, test, and optimize quantum programs across different quantum architectures. This capability is essential for advancing quantum machine learning, cryptography, materials science, and complex optimization problems. By leveraging cloud access, developers can experiment with various quantum processors, compare performance, and refine their algorithms without hardware constraints, significantly reducing the development cycle for quantum applications.

Many cloud platforms offer hybrid quantum-classical computing environments, where quantum algorithms can interact with classical computing resources to enhance performance. This hybrid approach is critical for near-term quantum applications, as current quantum computers are still in their early stages and require classical preprocessing and post-processing for effective computation. By integrating quantum and classical computing, cloud-based platforms enable researchers to harness quantum advantages while mitigating the limitations of today's quantum hardware.

One of the primary challenges of cloud-based quantum computing is limited hardware availability. Due to the high demand for quantum resources and the relatively small number of quantum processors, users often experience long wait times when submitting jobs on public quantum cloud platforms. Some vendors offer priority access through enterprise partnerships or paid subscription plans, but availability remains a bottleneck for widespread adoption.

Current quantum processors suffer from quantum noise, decoherence, and gate errors, which impact computation accuracy. Quantum systems are highly sensitive to external disturbances, leading to errors in quantum operations. Cloud users must implement error mitigation techniques such as zero-noise extrapolation (ZNE) and quantum error correction (QEC) to improve computational results. Until large-scale, fault-tolerant quantum computers become available, noise and error rates will remain significant challenges for cloud-based quantum computing.

Bandwidth and latency issues can arise when running quantum computations over the cloud, especially when transferring large datasets for hybrid quantum-classical applications. Since quantum circuits must be executed remotely, the time required for data transmission and result retrieval can introduce delays. Future advancements in quantum networking and cloud infrastructure will be necessary to reduce these latency concerns and enable more seamless quantum computing experiences.

Security and data privacy are also critical concerns in cloud-based quantum computing. Running quantum computations on cloud platforms raises questions about data confidentiality, secure quantum execution, and potential vulnerabilities in quantum workloads. While existing encryption methods help protect data during transmission, future advancements in homomorphic encryption for quantum computing and secure multi-party quantum computation may be required to ensure full security in cloud-based quantum applications.

Despite these challenges, cloud-based quantum computing continues to advance, offering a scalable and cost-effective solution for accessing quantum resources. As quantum hardware improves, along with error correction techniques and quantum networking capabilities, cloud-based quantum computing will play a central role in accelerating quantum research and real-world applications.

IBM Quantum Platform, Google Quantum AI, and Amazon Braket

Cloud-based quantum computing platforms have revolutionized access to quantum resources, enabling researchers, developers, and businesses to experiment with quantum algorithms without requiring dedicated hardware. Among the leading providers, IBM Quantum Platform, Google Quantum AI, and Amazon Braket offer distinct quantum computing environments, supporting various quantum hardware and software frameworks. Each platform provides access to quantum processors, simulators, and programming tools, enabling advancements in quantum research, machine learning, optimization, and cryptography.

IBM Quantum Platform

IBM Quantum Platform, formerly known as IBM Quantum Experience (IBM Q), is one of the most widely used cloud-based quantum computing platforms. It provides access to real quantum processors and high-performance simulators, enabling users to run quantum algorithms, experiment with quantum circuits, and integrate quantum computing into scientific research and enterprise solutions. Developed by IBM Quantum, the platform supports a broad range of applications, from fundamental research to commercial deployment.

IBM Q utilizes superconducting qubit-based quantum processors, built on transmon qubits arranged in a lattice structure. The platform offers a variety of backend devices, ranging from 5-qubit systems designed for educational purposes to 127-qubit processors, such as Eagle, intended for more advanced research. IBM has outlined an ambitious roadmap for scaling up its quantum systems, including the Osprey processor with 433 qubits and the Condor processor with 1,121 qubits, aimed at achieving high-performance quantum computing.

To facilitate quantum programming and algorithm development, IBM provides Qiskit, an open-source Python-based quantum computing framework. Qiskit offers a modular approach to designing and simulating quantum circuits. Qiskit Terra enables circuit design and optimization, while Qiskit Aer provides high-performance quantum simulators. Qiskit Ignis includes tools for error mitigation and noise modelling, helping researchers improve quantum circuit fidelity. Additionally, domain-specific modules such as Qiskit Machine Learning and Qiskit Finance enable quantum applications in artificial intelligence and financial modelling.

IBM Quantum Platform is widely used across various fields, including quantum chemistry, where it is applied to simulate molecular structures for drug discovery. In optimization problems, the platform helps solve complex combinatorial challenges in logistics and finance. Quantum-enhanced machine learning models, such as support vector machines and generative models, benefit from IBM's quantum capabilities. In cryptography, researchers use IBM Quantum for studying quantum-safe encryption methods and quantum key distribution (QKD), ensuring future security in the era of quantum computing.

With its extensive adoption in academia, research, and industry, IBM Quantum Platform has established itself as a leading cloud-based quantum computing solution. The continuous advancements in hardware scalability, quantum algorithms, and practical applications make it a key player in the quantum computing ecosystem.

Google Quantum AI

Google Quantum AI is a research initiative dedicated to advancing quantum computing through the development of scalable superconducting qubit processors, innovative quantum algorithms, and experiments demonstrating quantum supremacy. The platform is designed to support near-term quantum applications, hybrid quantum-classical computing models, and deep learning integration through TensorFlow Quantum (TFQ). By focusing on both hardware and software advancements, Google Quantum AI is working toward practical quantum computing solutions.

Google's quantum processors are based on superconducting qubits using transmon technology, similar to IBM's approach but with specific optimizations to enhance coherence times, gate fidelities, and qubit connectivity. One of the most significant achievements of Google Quantum AI is the development of Sycamore, a 53-qubit processor that demonstrated quantum supremacy by solving a specific random sampling problem faster than the world's most powerful classical supercomputers. Google is now focusing on building larger, fault-tolerant quantum processors, aiming to scale beyond 1,000 physical qubits in the coming years.

To facilitate quantum programming and algorithm development, Google provides Cirq, an open-source quantum computing framework optimized for near-term quantum devices. Cirq is designed to help researchers develop variational quantum algorithms (VQAs), implement quantum error correction techniques, and explore quantum-enhanced machine learning applications. Additionally, TensorFlow Quantum (TFQ) extends Google's AI ecosystem by integrating quantum circuits into classical deep learning workflows, enabling the study of quantum neural networks and hybrid artificial intelligence models.

Google Quantum AI's research focuses on multiple key areas, including quantum supremacy experiments that demonstrate computational advantages over classical supercomputers. The team is actively working on quantum error correction techniques, particularly surface codes and logical qubits, to improve the reliability of quantum computations. In quantum chemistry, Google's quantum processors are being used to simulate complex molecular structures, advancing materials science and pharmaceutical research. The platform also explores quantum optimization, leveraging hybrid quantum-classical algorithms to solve combinatorial problems more efficiently.

With a strong emphasis on achieving practical, fault-tolerant quantum computing, Google Quantum AI continues to push the boundaries of quantum information science. By improving qubit stability, developing scalable quantum processors, and integrating quantum computing with artificial intelligence, Google is positioning itself as a leader in the future of quantum technology.

Amazon Braket

Amazon Braket is Amazon Web Services' (AWS) cloud-based quantum computing platform, designed to provide businesses, researchers, and developers with access to multiple quantum hardware providers. Unlike IBM and Google, which primarily focus on proprietary quantum hardware, Amazon Braket operates as a multi-vendor platform, supporting a range of quantum computing architectures. This approach allows users to experiment with different quantum technologies without being restricted to a single ecosystem.

Amazon Braket provides access to quantum processors from various hardware partners, including superconducting qubits from Rigetti Computing for universal gate-based quantum computing, trapped-ion qubits from IonQ and Quantinuum, which offer long coherence times and high-fidelity operations, and quantum annealers from D-Wave, specialized for solving optimization problems. By integrating multiple quantum hardware platforms, Amazon Braket allows users to explore different quantum computing approaches and select the most suitable architecture for their specific applications.

Amazon Braket includes a fully managed quantum development environment within AWS, providing a range of software tools and programming frameworks. The Braket SDK, a Python-based toolkit, enables users to design, test, and execute quantum algorithms with ease. The platform also supports hybrid quantum-classical workflows, integrating with AWS services such as SageMaker for machine learning applications and AWS Lambda for automated cloud processing. Additionally, Braket Simulators offer high-performance classical simulations, allowing users to test and refine their quantum circuits before running them on real quantum hardware.

Amazon Braket supports a wide range of quantum applications across multiple industries. In finance, it enables quantum-enhanced risk analysis and portfolio optimization using quantum Monte Carlo methods. In logistics and supply chain management, it helps optimize route planning and warehouse operations with quantum algorithms. Quantum AI and machine learning applications benefit from improved feature selection and quantum kernel methods, enhancing large-scale data processing. In drug discovery and material science, superconducting and trapped-ion qubits facilitate molecular simulations, accelerating the search for new pharmaceuticals and advanced materials.

By providing access to multiple quantum computing architectures within a unified cloud platform, Amazon Braket offers a flexible and scalable solution for businesses and researchers exploring quantum computing. Its multi-vendor approach ensures that users can experiment with various quantum technologies without long-term commitment to a single hardware provider, making it an attractive option for those looking to integrate quantum computing into their workflows.

Platform Comparison and Access

IBM Quantum Experience, Google Quantum AI, and Amazon Braket each offer unique approaches to cloud-based quantum computing. IBM focuses on scalability and commercial applications, Google is pioneering quantum supremacy and deep learning integration, while Amazon Braket provides a multi-hardware ecosystem for broad industry adoption.

Table 19: Comparison: IBM Quantum Experience vs. Google Quantum AI vs. Amazon Braket.

Feature	IBM Quantum Experience	Google Quantum AI	Amazon Braket
Quantum Hardware	Superconducting Qubits	Superconducting Qubits	Superconducting, Trapped-Ion, Quantum Annealers
Programming Framework	Qiskit	Cirq, TensorFlow Quantum	Braket SDK (Python)
Simulators	IBM Aer	qsim, Tensor Network Simulators	Braket Simulators
Access Model	Public Cloud, Enterprise Plans	Research Access, Cloud Integration	Pay-as-You-Go Multi-Vendor Model
Special Features	Quantum Volume Scaling, Cloud-Based Job Queue	Quantum Supremacy, Deep Learning Integration	Hybrid Quantum-Classical Workflows
Target Audience	Researchers, Businesses, Developers	AI & ML Researchers, Quantum Scientists	Industry, Financial Firms, R&D

As quantum technology advances, these platforms will continue to expand, offering improved quantum processors, better error correction, and more robust quantum-classical integration. Whether for academic research, industrial applications, or enterprise solutions, cloud-based quantum computing is making quantum technology more accessible and practical for real-world use cases.

Accessing cloud-based quantum computing platforms enables researchers, developers, and businesses to experiment with quantum algorithms and applications without the need for specialized hardware. Here's how to access each platform, along with associated costs and examples.

IBM Quantum Platform: To begin using the IBM Quantum Platform, visit the IBM Quantum Computing website. Users can sign up for an account to access IBM's quantum processors and simulators. IBM offers different access plans:

- **Open Plan**: Provides free access to certain quantum systems and simulators, suitable for educational purposes and initial experimentation.
- **Pay-As-You-Go Plan**: Offers access to premium quantum systems with higher qubit counts for a fee, catering to more advanced research and development needs.

Costs: The Open Plan allows users to run small-scale experiments at no cost. For more extensive usage, the Pay-As-You-Go Plan involves costs based on the quantum resources utilized. Specific pricing details can be found on IBM's Quantum Computing Pricing page.

Example: A researcher aiming to simulate molecular structures for drug discovery can use the Open Plan to access IBM's 5-qubit systems for preliminary tests. For more complex simulations requiring higher qubit counts, the Pay-As-You-Go Plan provides access to advanced processors like the 127-qubit 'Eagle'.

Google Quantum AI: Google Quantum AI focuses on research collaborations and does not offer direct public access to its quantum processors. Interested parties can explore Google's quantum computing initiatives and resources through the Google Quantum AI website. Developers can utilize Cirq, an open-source Python framework for designing, simulating, and executing quantum circuits on Google's quantum processors.

Costs: Since direct access to Google's quantum hardware is limited to research collaborations, there is no publicly available pricing information at the time of writting. However, using the Cirq framework for simulation purposes is free and accessible to anyone.

Example: An academic institution collaborating with Google Quantum AI can conduct experiments on Google's quantum processors to explore quantum error correction techniques. Independent developers can use Cirq to design and simulate quantum circuits locally without incurring costs.

Amazon Bracket: Amazon Braket is AWS's quantum computing service that provides access to various quantum hardware providers. To start, users need an AWS account, which can be created on the AWS Management Console. Once logged in, navigate to the Amazon Braket service to begin designing and running quantum algorithms.

Costs: Amazon Braket offers a pay-as-you-go pricing model, with costs varying based on the chosen quantum hardware and usage duration. For instance [571]:

- **Simulators**: Using on-demand simulators like SV1 costs $0.075 per minute. AWS Free Tier provides one hour of free simulation time per month for the first twelve months.
- **Quantum Processing Units (QPUs)**: Costs vary by provider. For example, running a task on Rigetti's quantum computer incurs a per-task charge of $0.30, plus $0.00090 per shot. Detailed pricing information is available on the Amazon Braket Pricing page.

Example: A data scientist developing a quantum algorithm for financial risk analysis can use Amazon Braket's simulators to test the algorithm at a rate of $0.075 per minute. After refining the algorithm, they can execute it on a QPU, such as Rigetti's, with costs calculated based on the number of tasks and shots performed.

Quantum as a Service (QaaS)

Quantum as a Service (QaaS) is a cloud-based model that provides remote access to quantum computing resources, allowing businesses, researchers, and developers to experiment with quantum algorithms without requiring direct access to physical quantum hardware. Similar to traditional Software as a Service (SaaS) or Infrastructure as a Service (IaaS) models, QaaS enables users to run quantum programs via the cloud, eliminating the need for specialized quantum infrastructure.

Quantum as a Service (QaaS) allows users to access quantum computing hardware and software remotely through cloud-based platforms. This eliminates the need for direct access to quantum processors and specialized infrastructure, making quantum computing more widely available. QaaS providers host quantum hardware and integrate software tools that enable users to develop, test, and execute quantum algorithms without requiring in-house quantum technology.

One of the key components of QaaS is quantum hardware access, which allows users to run quantum circuits on real quantum processors. These processors are built on different quantum computing architectures, including superconducting qubits, trapped-ion qubits, photonic quantum processors, and quantum annealers. Leading QaaS providers such as IBM Quantum, Google Quantum AI, and Rigetti offer superconducting qubit-based processors. IonQ and Quantinuum specialize in trapped-ion qubits, known for their long coherence times and high-fidelity operations. Xanadu provides photonic quantum processors that leverage quantum light for computational tasks. D-Wave, on the other hand, focuses on quantum annealers, which are particularly suited for solving optimization problems.

Another essential aspect of QaaS is cloud-based quantum programming environments, which provide the software tools needed to write and execute quantum programs. Various quantum programming frameworks support quantum circuit development and algorithm implementation. IBM Qiskit is a widely used Python-based framework that allows users to design and simulate quantum circuits. Google Cirq is optimized for near-term quantum devices, enabling researchers to develop quantum variational algorithms and error correction techniques. Amazon Braket SDK provides tools for accessing multiple quantum hardware providers through a unified interface. PennyLane, developed by Xanadu, specializes in quantum machine learning, enabling hybrid quantum-classical deep learning models. D-Wave's Ocean SDK is designed specifically for quantum annealing, helping businesses solve optimization problems such as logistics and scheduling.

Before deploying quantum programs on actual quantum processors, QaaS platforms offer quantum simulators and hybrid integration to allow users to test and refine their algorithms on classical hardware. Quantum simulators help in debugging quantum circuits and optimizing algorithm performance. Many QaaS providers also support hybrid quantum-classical computing, where quantum algorithms work alongside classical systems to improve efficiency. Hybrid integration is crucial for near-term applications, as quantum computers today are still limited by hardware constraints, requiring classical resources to preprocess and interpret quantum results.

The final stage in the QaaS workflow is cloud-based job submission and execution. Users submit their quantum computing tasks through cloud-based platforms, where the provider schedules and executes them on available quantum hardware. The job queue system manages multiple user requests, prioritizing access based on resource availability and usage plans. Once the quantum computation is complete, the provider returns the results for further analysis. This seamless cloud-based execution allows businesses, researchers, and developers to access cutting-edge quantum technology without the challenges of maintaining and managing physical quantum hardware.

Quantum as a Service (QaaS) offers several advantages, making quantum computing more accessible and practical for businesses, researchers, and developers. One of the key benefits is that it eliminates the need for organizations to invest in specialized quantum hardware. Quantum computers require highly controlled environments, including millikelvin temperatures and advanced quantum control electronics, which make them expensive and difficult to maintain. With QaaS, users can access quantum processors remotely, removing the financial and logistical burden of owning quantum hardware.

Another advantage of QaaS is its pay-as-you-go pricing model, which makes quantum computing more affordable. Instead of requiring a significant upfront investment, QaaS platforms allow users to pay for quantum computing resources based on usage. Some providers offer subscription plans or tiered pricing, ensuring that businesses and researchers can experiment with quantum algorithms at a manageable cost before committing to larger-scale applications.

Scalability is another important feature of QaaS. Users can start with quantum simulators to test and optimize their algorithms before running them on real quantum processors. As quantum hardware continues to improve, businesses can scale their quantum workloads as needed, seamlessly transitioning from simulations to actual quantum computations without requiring additional infrastructure investments. This flexibility makes it easier for organizations to explore quantum applications at their own pace.

QaaS also enables hybrid quantum-classical computing, which is essential for solving complex problems that require both classical and quantum resources. Many QaaS platforms integrate quantum computing with traditional cloud infrastructure, allowing users to leverage quantum algorithms for optimization, artificial intelligence, and scientific computing. Hybrid models are particularly useful for near-term quantum applications, as classical computing resources help compensate for the current limitations of quantum hardware.

One of the most significant impacts of QaaS is its ability to accelerate research and development across multiple industries. Universities, startups, and enterprises can experiment with state-of-the-art quantum algorithms for applications such as drug discovery, materials science, financial modelling, and logistics optimization. By providing remote access to quantum processors and powerful simulation tools, QaaS lowers the barriers to entry for quantum computing, enabling more organizations to explore its transformative potential.

Leading QaaS Providers

Several major technology companies and research institutions offer QaaS solutions, each providing unique quantum hardware and software capabilities.

IBM Quantum: IBM Quantum is one of the earliest providers of cloud-based quantum computing. It allows users to access superconducting qubit processors via IBM Cloud.

- **Framework**: Qiskit
- **Quantum Processors**: 5-qubit to 127-qubit (Eagle)
- **Key Offerings**: Free and premium access to quantum hardware, enterprise-grade quantum computing, quantum education initiatives

Google Quantum AI: Google Quantum AI provides Sycamore superconducting qubit processors through the cloud, focusing on research and industry applications.

- **Framework**: Cirq
- **Quantum Processors**: Sycamore (53 qubits), larger models in development
- **Key Offerings**: Hybrid AI-quantum integration via TensorFlow Quantum (TFQ), research-driven access

Amazon Bracket: Amazon Braket is a multi-vendor QaaS platform, offering access to various quantum processors.

- **Framework**: Braket SDK
- **Quantum Processors**: Rigetti (superconducting qubits), IonQ & Quantinuum (trapped-ion), D-Wave (quantum annealing)
- **Key Offerings**: AWS-integrated quantum development, support for hybrid computing and machine learning via AWS SageMaker

Microsoft Azure Quantum: Microsoft provides access to multiple quantum hardware providers through Azure Quantum.

- **Framework**: Q# (Quantum Development Kit)
- **Quantum Processors**: IonQ (trapped-ion), Quantinuum, Rigetti (superconducting)
- **Key Offerings**: Cloud-native quantum computing, hybrid quantum-classical algorithms

D-Wave Leap: D-Wave Leap offers quantum annealing as a service, optimized for optimization and logistics problems.

- **Framework**: Ocean SDK
- **Quantum Processors**: Quantum annealers up to 5000+ qubits
- **Key Offerings**: Optimization algorithms for logistics, finance, AI, and supply chain management

Microsoft Azure Quantum - Example of Quantum as a Service (QaaS)

Microsoft Azure Quantum is a cloud-based quantum computing platform that provides businesses, researchers, and developers access to multiple quantum computing hardware providers and quantum development tools. It operates under the Quantum as a Service (QaaS) model, allowing users to run quantum algorithms, test quantum simulations, and develop hybrid quantum-classical applications without needing to own or maintain quantum hardware.

Azure Quantum is part of Microsoft's broader Azure cloud ecosystem, which means it integrates seamlessly with classical cloud computing resources. It offers a flexible and scalable approach to quantum computing, allowing users to choose from various quantum hardware providers and software development frameworks. Microsoft is also developing its own topological qubits as part of its long-term quantum computing vision.

Azure Quantum is designed to be hardware-agnostic, meaning that users can experiment with different quantum computing architectures within the same platform. It provides access to a range of quantum hardware backends, including:

- Superconducting qubits from Rigetti Computing (gate-based quantum computing)
- Trapped-ion qubits from IonQ and Quantinuum (high-fidelity quantum operations)
- Quantum annealers from D-Wave (optimization-focused quantum computing)
- Quantum simulators from Microsoft for testing quantum circuits before execution on real hardware

One of the major advantages of Azure Quantum is its multi-vendor approach, allowing users to access different types of quantum hardware without being locked into a single technology. Each quantum hardware provider within Azure Quantum specializes in different quantum computing methodologies:

- **IonQ**: Provides trapped-ion qubits with long coherence times and high-fidelity quantum operations, making it suitable for quantum chemistry and error-sensitive computations.
- **Quantinuum (formerly Honeywell)**: Offers high-precision trapped-ion quantum processors designed for scalable quantum error correction.
- **Rigetti Computing**: Delivers superconducting qubit-based gate-model quantum processors optimized for executing variational quantum algorithms (VQAs).
- **D-Wave**: Provides quantum annealing processors tailored for solving combinatorial optimization problems such as logistics, portfolio optimization, and scheduling.

Users can select the best quantum hardware for their specific use case, whether it's optimization, simulation, or machine learning.

Azure Quantum supports multiple quantum programming languages and development tools, allowing users to write quantum algorithms, optimize circuits, and execute computations efficiently. The primary tools include:

1. Q# and the Quantum Development Kit (QDK): Microsoft's proprietary Q# (Quantum Sharp) programming language is specifically designed for developing quantum applications. It includes:

- High-level abstractions for quantum circuits and algorithms
- Integration with classical programming languages like Python and C#
- Tools for quantum error correction and resource estimation

- Libraries for quantum chemistry, machine learning, and cryptography

The Quantum Development Kit (QDK) includes a quantum simulator that allows users to test quantum algorithms before running them on real quantum hardware.

2. Python and OpenQASM Support: Azure Quantum also supports Python-based quantum frameworks such as Qiskit (IBM) and Cirq (Google), allowing researchers to port their quantum code from other platforms. Additionally, OpenQASM (Quantum Assembly Language) is supported, enabling interoperability with other quantum computing environments.

3. Hybrid Quantum-Classical Computing: Azure Quantum allows for hybrid computing, where quantum algorithms work alongside classical computing resources. This is particularly useful for variational quantum algorithms (VQAs) that require frequent interaction between quantum and classical systems. Microsoft provides hybrid quantum computing tools through Azure Machine Learning (Azure ML) and classical HPC (High-Performance Computing) integration.

Before executing quantum programs on real quantum processors, users can test and optimize their circuits using Azure's quantum simulators. These simulators help debug quantum programs, optimize quantum gate sequences, and estimate required quantum resources.

Azure Quantum provides three main simulation environments:

1. **Full-State Quantum Simulator**: Allows users to simulate quantum circuits on classical cloud infrastructure with high fidelity.
2. **Resource Estimation Tool**: Predicts the quantum resources (qubits, gates, and coherence time) needed for executing a quantum algorithm.
3. **Noise-Aware Simulators**: Models real-world quantum noise to test error mitigation techniques before execution on quantum hardware.

Azure Quantum is widely used across various industries, including finance, logistics, healthcare, and energy. The platform enables businesses and researchers to leverage quantum computing for solving complex problems that are difficult for classical computers.

In logistics and supply chain management, companies use Azure Quantum to optimize delivery routes, manage warehouse inventory, and improve production scheduling. By leveraging quantum annealing technology from D-Wave, available through Azure Quantum, businesses can minimize downtime and maximize efficiency in their supply chain operations. Quantum algorithms help identify the most efficient routes for delivery networks, reducing costs and improving overall logistics management.

Financial institutions utilize Azure Quantum for portfolio optimization, risk analysis, and fraud detection. Quantum Monte Carlo simulations assist in optimizing investment portfolios, while quantum-enhanced machine learning models improve fraud detection by identifying anomalies in financial transactions. Quantum algorithms also enhance option pricing and derivative calculations, enabling more accurate financial forecasting.

In the pharmaceutical industry and quantum chemistry research, Azure Quantum plays a crucial role in accelerating drug discovery and materials science. Researchers simulate molecular interactions to develop new drugs, model quantum chemistry to discover novel catalysts, and improve the design of energy-efficient batteries. Quantum simulations provide more accurate predictions for chemical properties, reducing the time and cost of developing new materials.

Machine learning and artificial intelligence benefit from the integration of Azure Quantum with quantum-enhanced algorithms. Quantum machine learning techniques improve feature selection for big data analysis, optimize kernel methods for classification tasks, and enhance deep learning architectures. Microsoft's Quantum Machine Learning (QML) libraries enable seamless integration between quantum computing and classical AI models, helping researchers explore hybrid quantum-classical approaches.

Azure Quantum follows a pay-as-you-go pricing model, allowing users to pay only for the quantum resources they consume. The cost structure varies based on the quantum hardware provider, such as IonQ, Rigetti, Quantinuum, or D-Wave, as well as the number of shots (executions) performed on a quantum processor. Users can choose between running simulations on classical quantum emulators or executing computations on real quantum hardware, with different pricing tiers depending on the selected approach.

To encourage adoption and research, Azure Quantum offers free trials and cloud credits for academic institutions. Students and researchers can explore quantum computing without incurring costs, gaining hands-on experience with real quantum processors and simulators. This initiative supports the growth of the quantum computing ecosystem by making advanced quantum tools accessible to a broader audience.

Users can access Azure Quantum through:

1. **Microsoft Azure Portal**: A cloud-based dashboard for submitting quantum jobs and managing quantum resources.
2. **Azure CLI (Command Line Interface)**: Allows developers to interact with quantum hardware through scripts.
3. **Jupyter Notebooks**: Integrated within Azure Quantum, enabling interactive quantum programming in Q# and Python.

Microsoft Azure Quantum is a leading Quantum as a Service (QaaS) platform that provides cloud-based access to multiple quantum hardware providers, quantum programming tools, and hybrid quantum-classical integration. It enables businesses, researchers, and developers to explore quantum computing without requiring specialized hardware, making quantum technology accessible at scale.

D-Wave Leap - Cloud-Based Quantum Computing Platform for Optimization Example

D-Wave Leap is D-Wave's Quantum as a Service (QaaS) platform, providing cloud-based access to D-Wave's quantum annealers. It allows researchers, businesses, and developers to solve real-world optimization problems using quantum annealing, a specialized form of quantum computing designed for solving combinatorial optimization challenges.

Unlike gate-based quantum computers from IBM, Google, and Rigetti, D-Wave's quantum processors are designed to find low-energy solutions to optimization problems by leveraging quantum annealing. This makes them particularly useful for logistics, scheduling, financial modelling, drug discovery, and machine learning.

D-Wave Leap provides remote access to D-Wave's quantum hardware, enabling users to run quantum algorithms, develop applications, and integrate quantum computing into classical workflows. The platform allows researchers, developers, and businesses to leverage quantum annealing for solving optimization problems without requiring specialized quantum hardware infrastructure.

Leap provides access to D-Wave's latest quantum annealers, including the Advantage Quantum System, which features over 5000 qubits with 15-way connectivity between qubits. This architecture enables efficient solutions for large-scale optimization problems. The platform also supports the D-Wave 2000Q, a previous-generation quantum processor with 2000 qubits, which remains available for experimentation. Looking ahead, the upcoming Advantage2 system is expected to further increase qubit count and connectivity, enhancing problem-solving capabilities.

Since quantum annealers do not provide a universal quantum computing model, D-Wave Leap integrates classical algorithms with quantum annealing through hybrid solvers. These solvers combine quantum and classical computing resources to improve optimization outcomes. D-Wave offers hybrid solvers designed for discrete optimization, which is useful for scheduling and logistics problems, continuous optimization, which is applied in finance and engineering, graph-based problems that analyse relationships in networked data, and quantum-enhanced machine learning.

Leap provides a comprehensive quantum development environment, allowing users to write, test, and deploy quantum algorithms through multiple software tools. The Ocean SDK, a Python-based software development kit, enables quantum programming, while D-Wave Hybrid provides a framework for building hybrid quantum-classical algorithms. The platform also includes a cloud-based integrated development environment (IDE), where users can write and execute quantum code without needing local quantum hardware.

Before running computations on real quantum processors, users can refine their algorithms using quantum simulators. Leap includes simulated annealing, a classical algorithm that mimics quantum annealing, allowing users to test optimization problems before executing them on quantum hardware. Additionally, QPU emulators provide classical simulations of quantum processors, enabling debugging and performance benchmarking. These tools help users analyse problem formulations and adjust parameters before submitting jobs to physical quantum processors.

D-Wave Leap follows a cloud-based job execution model, ensuring seamless access to quantum resources. Users submit quantum tasks through the Leap platform, where they are processed by D-Wave's quantum hardware or hybrid solvers. Once computations are completed, results are returned to users for analysis. By providing real-time access to quantum computing, Leap eliminates the need for dedicated quantum hardware, making quantum resources accessible to a broader audience across multiple industries.

D-Wave Leap, a platform for quantum computing, is utilized in various fields for solving real-world problems where combinatorial optimization is essential. Here, we discuss specific applications across multiple domains, supported by existing literature.

D-Wave's quantum annealers have demonstrated effectiveness in optimizing logistics and supply chain operations. A prime example is Volkswagen's initiative in Beijing, where D-Wave was applied to optimize taxi fleet routing, resulting in a significant reduction in traffic congestion. Similarly, Save-On-Foods, a Canadian grocery retailer, utilized D-Wave's quantum capabilities to enhance their delivery route optimization, which led to lowered travel costs and improved efficiency in their supply chain operations [572]. The efficiency of such optimizations is supported by computational studies demonstrating the application of quantum annealing in vehicle routing problems [573].

The financial sector has leveraged D-Wave Leap for various applications, particularly in portfolio optimization and risk assessment. BBVA, a global banking institution, successfully employed D-Wave to enhance investment portfolios by improving returns while managing risk factors effectively [573, 574]. Furthermore, Deloitte explored quantum annealing algorithms for fraud detection, utilizing the platform to analyse complex transaction patterns that traditional methods struggled with [575]. The robustness of quantum computing in financial optimizations is further highlighted by scalability studies showing the potential of quantum methods in portfolio management [573]

D-Wave has also made strides in drug discovery, where it is utilized to accelerate protein design and facilitate molecular simulations. For instance, Menten AI implemented D-Wave technology to refine protein designs crucial for drug development [576]. Additionally, collaborations between D-Wave and pharmaceutical companies are focusing on employing quantum simulations of molecular chemistry to discover new drug formulations [577]. This integration of quantum principles in drug design underscores the shift towards computational methods that can potentially shorten the timeframe and lower costs associated with traditional R&D methodologies [578].

D-Wave's Leap is instrumental in advancing machine learning tasks such as feature selection, clustering, and optimizing neural networks. A notable collaboration with TensorFlow Quantum aims to integrate quantum computing into deep learning frameworks, showcasing the potential for quantum-enhanced models in handling complex datasets more efficiently than classical alternatives [579]. These advancements demonstrate how D-Wave technologies can significantly expedite processing tasks that involve vast amounts of data and intricate relationships [573].

The field of cybersecurity is evolving with the incorporation of quantum computing capabilities. D-Wave Leap is employed for cryptographic applications, enabling researchers to explore quantum annealing techniques for key generation and cybersecurity analysis [575]. Governments and organizations are investigating quantum-safe encryption strategies to safeguard data against potential quantum attacks in the future [580]. Given the projected advancements in quantum technology, D-Wave's contributions may play a pivotal role in developing robust systems to counteract emerging security threats [575].

D-Wave Leap is available as a cloud-based service, accessible through Leap's web interface, API, and Python SDK (Ocean SDK).

1. Free Tier for Quantum Exploration: D-Wave offers free access to its quantum computers through Leap's Free Tier, which includes:

- One minute of free QPU time per month for running quantum computations.
- Access to hybrid solvers to experiment with real-world optimization problems.
- Online learning resources and tutorials to help users get started.

2. Pay-As-You-Go and Enterprise Plans: For commercial and research users, Leap offers pay-as-you-go pricing based on:

- The number of quantum processing unit (QPU) accesses.
- The complexity of optimization problems.
- The use of hybrid solvers for large-scale computations.

Enterprise users can also purchase priority access and dedicated quantum computing resources.

D-Wave Leap supports multiple quantum programming tools:

- **Python** (Ocean SDK, D-Wave Hybrid)

- **Jupyter Notebooks** for interactive quantum computing experiments
- **C++ and Java APIs** for enterprise-level quantum integration

Developers can write and execute quantum annealing programs using D-Wave's Ocean SDK, which includes:

- **dimod**: A low-level library for defining optimization problems.
- **dwave-system**: Tools for connecting to D-Wave's quantum processors.
- **dwave-hybrid**: A framework for hybrid quantum-classical computing.

D-Wave Leap is a leading QaaS platform for quantum annealing, enabling users to solve complex optimization problems in logistics, finance, AI, and scientific research. With cloud-based access, hybrid solvers, and free-tier quantum computing, Leap democratizes access to quantum resources.

Accessing and Running Quantum Algorithms Online

Programming Languages and Software Frameworks

Each of the major cloud-based quantum computing platforms supports specific programming languages and software frameworks to enable users to write and execute quantum algorithms. Below is an overview of the programming languages used for processing on each platform.

IBM Quantum Platform: Qiskit (Python-Based)

IBM Quantum Platform uses Qiskit, an open-source quantum computing framework written in Python. Qiskit provides a suite of tools for designing, optimizing, and executing quantum circuits on IBM's superconducting qubit-based quantum processors.

- **Primary Language**: Python
- **Programming Framework**: Qiskit
- **Key Components**:
    - **Qiskit Terra** – Enables quantum circuit creation, optimization, and compilation.
    - **Qiskit Aer** – Provides high-performance quantum circuit simulation.
    - **Qiskit Ignis** – Includes tools for quantum error mitigation and noise modelling.

  - **Qiskit Machine Learning, Finance, and Optimization** – Specialized modules for AI, financial modelling, and optimization tasks.

Qiskit allows users to define quantum circuits using gate-based programming, execute them on IBM's quantum hardware or simulators, and retrieve the results for analysis. The Python-based API makes it accessible for researchers, educators, and industry professionals integrating quantum computing into various applications.

Example: Writing a Quantum Circuit in Qiskit

```
from qiskit import QuantumCircuit, Aer, transpile, assemble,
execute
from qiskit.visualization import plot_histogram

# Create a simple quantum circuit
qc = QuantumCircuit(2)  # 2-qubit circuit
qc.h(0)  # Apply Hadamard gate to the first qubit
qc.cx(0, 1)  # Apply CNOT gate (entangling both qubits)
qc.measure_all()

# Execute the circuit using a simulator
simulator = Aer.get_backend('aer_simulator')
result = execute(qc, simulator).result()
plot_histogram(result.get_counts())
```

IBM Quantum Platform primarily supports Python through Qiskit, making it a widely used framework for quantum research and development.

Google Quantum AI: Cirq (Python-Based)

Google Quantum AI relies on Cirq, an open-source Python framework optimized for near-term quantum devices. Cirq is designed to work efficiently with Google's Sycamore superconducting quantum processors and supports quantum applications such as error correction and variational quantum algorithms.

- **Primary Language**: Python
- **Programming Framework**: Cirq
- **Key Features**:
  - Designed for noisy intermediate-scale quantum (NISQ) devices.
  - Provides tools for quantum circuit design, noise modelling, and execution.

- Supports quantum variational algorithms for machine learning and optimization.
- Integrates with TensorFlow Quantum (TFQ) to enable hybrid quantum-classical deep learning models.

Example: Creating and Running a Quantum Circuit in Cirq

```
import cirq

# Define two qubits
qubit_1, qubit_2 = cirq.LineQubit.range(2)

# Create a simple quantum circuit
circuit = cirq.Circuit(
    cirq.H(qubit_1),  # Hadamard gate
    cirq.CNOT(qubit_1, qubit_2),  # CNOT gate
    cirq.measure(qubit_1, qubit_2)  # Measurement
)

# Simulate the quantum circuit
simulator = cirq.Simulator()
result = simulator.run(circuit, repetitions=1000)
print(result)
```

Google's Cirq framework is designed to be lightweight and optimized for research and experimental work in quantum computing, particularly for hardware-specific optimizations.

Amazon Braket: Braket SDK (Python-Based)

Amazon Braket is hardware-agnostic and supports multiple quantum processors, including superconducting qubits (Rigetti), trapped-ion qubits (IonQ, Quantinuum), and quantum annealers (D-Wave). It provides a Python-based SDK for designing and running quantum circuits across multiple hardware platforms.

- **Primary Language**: Python
- **Programming Framework**: Amazon Braket SDK
- **Key Features**:
    - Provides access to multiple quantum architectures within a single cloud platform.
    - Supports both gate-based and annealing quantum computing models.

- Allows integration with AWS cloud services (e.g., AWS Lambda, SageMaker).
- Offers high-performance classical simulators for testing circuits before deployment.

Example: Writing and Running a Quantum Circuit in Amazon Braket

```
from braket.circuits import Circuit
from braket.aws import AwsQuantumTask
from braket.devices import LocalSimulator

# Define a quantum circuit
circuit = Circuit().h(0).cnot(0, 1).measure(0, 1)

# Run on a local simulator
device = LocalSimulator()
task = device.run(circuit, shots=1000)
result = task.result()
print(result.measurement_counts)
```

Amazon Braket enables users to experiment with different quantum hardware platforms without being restricted to a single vendor, making it ideal for businesses, research institutions, and startups exploring quantum computing.

Table 20: Comparison of Programming Frameworks.

Platform	**Programming Language**	**Quantum Framework**	**Use Cases**
IBM Quantum Platform	Python	Qiskit	Gate-based quantum computing, quantum chemistry, optimization, AI
Google Quantum AI	Python	Cirq	Variational quantum algorithms, error correction, hybrid AI models
Amazon Braket	Python	Braket SDK	Multi-vendor quantum computing, annealing, optimization

While all three platforms primarily use Python, the specific quantum frameworks vary:

- Qiskit is designed for gate-based quantum computing with an emphasis on modularity.

- Cirq is optimized for Google's quantum processors and NISQ applications.
- Amazon Braket SDK allows users to switch between different quantum hardware providers.

Each platform provides simulators, enabling users to test quantum circuits on classical machines before running them on real quantum hardware. Hybrid computing is a common theme across all platforms, allowing users to integrate quantum algorithms with classical computing resources.

Challenges of Scaling Quantum Cloud Services

The scaling of quantum cloud services presents several intricate challenges stemming from hardware limitations, networking infrastructure requirements, software integration, security concerns, and cost-efficiency issues. These challenges are critical as the demand for quantum computing resources continues to rise among researchers, enterprises, and developers.

A significant barrier to scaling quantum cloud services is the current limitations in quantum hardware, particularly regarding the quality and availability of qubits. Current quantum processors exhibit short coherence times, high gate error rates, and noise interference, which directly impacts their ability to execute large-scale quantum computations [166, 581]. Various qubit technologies, such as superconducting qubits, trapped-ion qubits, and photonic qubits, exhibit distinct technological constraints that impede efficient scaling of quantum hardware [582, 583]. Furthermore, implementing effective quantum error correction (QEC) presents formidable challenges. Achieving fault tolerance necessitates a multitude of physical qubits to accurately encode a single logical qubit, making it difficult for current quantum systems to support the deep quantum circuits required for complex algorithms without extensive QEC methods [584, 585].

Quantum cloud services necessitate robust networking infrastructure to facilitate remote access to quantum computing resources. High latency in transferring quantum circuit definitions and measurement results can significantly affect the performance of quantum computations, especially within quantum-classical hybrid workflows [586]. Moreover, the development of a Quantum Internet incorporating quantum repeaters and entanglement-based communication systems is still largely experimental, creating additional constraints on the scalability of distributed quantum computing infrastructures [582, 586]. The limited access to high-fidelity quantum hardware compounds these issues since maintaining such sophisticated processors involves considerable cost, resulting in extended queue times for users [587].

With current quantum computers not yet capable of exceeding the performance of classical supercomputers for many applications, the effective integration of quantum and classical computing resources is essential. Optimizing hybrid workflows, which require significant pre-processing and post-processing by classical systems, poses noteworthy integration challenges [588]. Additionally, users often rely on classical simulators for quantum circuits, which can be computationally expensive when dealing with larger circuits, thereby escalating operational costs [589, 590]. The diversity of programming languages and frameworks across different quantum cloud services further complicates cross-compatibility, making the development of universal applications more difficult [588].

As quantum cloud services expand, the importance of security and data privacy becomes paramount. Quantum computations involving sensitive data, such as cryptographic keys and proprietary information, contend with unique vulnerabilities different from those in classical cloud systems. Protecting quantum data requires advancing beyond conventional encryption methods to safeguard quantum states during processing and transmission [591, 592]. The susceptibility of quantum hardware to noise and environmental interference can result in erroneous computations and facilitate eavesdropping, thereby compromising data integrity [591, 593]. The transition to post-quantum cryptography remains critical for protecting data from the capabilities of future quantum attacks [592].

The cost associated with establishing and maintaining quantum cloud services remains one of the most pervasive challenges. High operational expenses arise from the sophisticated requirements for maintaining quantum computing systems, such as dilution refrigerators for superconducting qubits and precision lasers for trapped-ion systems [587, 594]. To encourage wider adoption, providers must establish cost-effective pricing models and create incentives for research and enterprise partnerships that balance accessibility with financial sustainability [588, 595]. Furthermore, demonstrating a clear return on investment (ROI) for businesses remains essential; the lack of compelling quantifiable outcomes may deter enterprises from investing in quantum solutions at scale [591].

Chapter 11

The Business of Quantum Computing

As quantum computing transitions from theoretical research to practical implementation, the potential for commercial applications and industry-wide impact is becoming increasingly evident. The foundational concepts explored in earlier chapters—ranging from quantum mechanics, qubits, quantum gates, and algorithms to error correction, hardware advancements, and cryptographic applications—have set the stage for understanding the broader implications of quantum computing. Now, the focus shifts to the business landscape, where quantum technology is evolving into a commercially viable field with significant investment and strategic interest from corporations, governments, and startups.

This chapter delves into the rapidly expanding quantum computing industry, examining key players, investment trends, and the growing ecosystem of quantum startups. It explores how major technology companies, including IBM, Google, Microsoft, and Amazon, are spearheading developments, while governments and research institutions are fostering innovation through funding and policy initiatives. Additionally, the chapter highlights the concept of *Quantum as a Service (QaaS)*—a model that allows businesses and researchers to access quantum computing resources via cloud-based platforms, lowering entry barriers to this transformative technology.

The discussion also extends to the road toward achieving *quantum advantage*, where quantum computers consistently outperform classical systems for practical, real-world applications. As quantum computing moves closer to commercialization, ethical and economic considerations emerge, including workforce implications, regulatory challenges, and the potential for quantum-driven economic disruption. This chapter provides insights into these factors, offering a comprehensive view of how quantum computing is shaping the future of technology, business, and society.

The Quantum Computing Industry Landscape

The quantum computing industry is evolving rapidly, with technology giants, academic institutions, and startups competing to develop scalable quantum hardware, sophisticated algorithms, and real-world applications. The landscape of this industry is divided into several key areas, including hardware development, quantum software ecosystems, cloud-based quantum computing, research institutions, and commercial applications.

The race to build practical quantum computers is being led by major technology companies and specialized startups. Different hardware approaches are being explored, including superconducting qubits, trapped-ion qubits, photonic quantum computing, and topological qubits.

Superconducting qubits have emerged as a prominent technology in the field of quantum computing, primarily due to their high-speed gate operations and scalability potential. This category of qubits is predominantly based on Josephson junctions, which enable rapid manipulation of quantum states. Companies like IBM Quantum and Google Quantum AI have made significant strides with their superconducting qubit technologies. IBM has developed transmon-based quantum processors, such as Eagle (127 qubits) and Osprey (433 qubits), with ambitious plans for future architectures like Condor (1,121 qubits) [596]. Similarly, Google's Sycamore processor, which features 53 qubits, has been pivotal in demonstrating quantum supremacy, showcasing the potential of superconducting qubits for practical applications in quantum computing [597]. Rigetti Computing continues to innovate with their superconducting processors, providing cloud-based access through their Rigetti Quantum Cloud Services (QCS), which significantly enhances accessibility for researchers and developers worldwide [562].

Trapped-ion qubits represent a complementary technology in quantum computing, utilizing electromagnetic fields for the manipulation of ions via laser pulses. This method is lauded for its high fidelity and extended coherence times, which are critical metrics in quantum information processing. IonQ specializes in commercial applications of trapped-ion quantum computers, enabling access through major cloud services such as AWS, Google Cloud, and Microsoft Azure, effectively democratizing quantum computing capabilities [598]. Additionally, Quantinuum, a merger between Honeywell Quantum Solutions and Cambridge Quantum Computing, is also focused on advancing trapped-ion technologies, further establishing this method as a significant contender in the quantum computing landscape [598].

Photonic quantum computing employs photons as qubits and is uniquely positioned for applications in quantum communication and networking. Xanadu is a leading entity in this field, providing cloud access to their Borealis processor, which is designed for Gaussian boson sampling, a task central to quantum communication [599]. PsiQuantum's efforts in developing scalable photonic quantum computers aim to enhance their capabilities for commercial and industrial applications, indicating a robust market push toward photonic technologies [599]. The integration of photonic systems with existing quantum architectures provides an expansive avenue for further research and applications [600].

Quantum annealing is specialized for solving optimization problems rather than serving as a universal quantum computing framework. D-Wave Systems has pioneered this approach and offers access to their quantum annealer via the D-Wave Leap platform, making it a practical tool for businesses addressing complex optimization challenges [601]. This focus on optimization aligns with the needs of industries where such problems are prevalent, demonstrating the applicability of quantum computing beyond traditional computing paradigms.

Lastly, topological qubits represent an exciting frontier in the quest for stable and fault-tolerant quantum computing. By relying on Majorana fermions, topological quantum computing seeks to create qubits that are inherently resistant to errors, thus promising significant improvements over current technologies. Microsoft's StationQ initiative is at the forefront of this research, illustrating the potential for this technology to revolutionize quantum computing and enhance error resilience [601]. With continued investments and advancements in these various technologies, the landscape of quantum computing is rapidly evolving, characterized by diverse approaches to overcome the challenges associated with noise and error correction in quantum systems.

While hardware development is crucial, quantum computing also requires robust software frameworks and quantum algorithms to enable practical applications. IBM Quantum provides Qiskit, a Python-based open-source quantum programming framework for circuit design, simulation, and execution. Google Quantum AI offers Cirq, optimized for near-term quantum devices, particularly Google's Sycamore processor. Amazon Braket provides a cloud-based framework for developing quantum algorithms across multiple quantum hardware platforms. Xanadu's PennyLane specializes in quantum machine learning and variational quantum algorithms.

Several quantum algorithms demonstrate potential quantum advantage. Shor's Algorithm is used for integer factorization, posing a threat to classical cryptographic methods like RSA and ECC. Grover's Algorithm provides a quadratic speedup for unstructured search problems. Variational Quantum Eigensolver (VQE) is widely used for quantum chemistry simulations. The Quantum Approximate Optimization Algorithm (QAOA) helps solve combinatorial optimization problems in logistics and finance.

Startups and research groups are actively developing quantum applications in areas such as drug discovery, materials science, financial modelling, artificial intelligence, and cybersecurity.

Quantum computing is increasingly being delivered as a cloud service, allowing enterprises, researchers, and developers to access quantum resources remotely. IBM Quantum offers quantum cloud access through IBM Cloud and Qiskit Runtime. Google Quantum AI provides access to quantum hardware via Google Cloud and Cirq. Amazon Braket is a multi-vendor quantum cloud service that supports Rigetti, IonQ, and D-Wave hardware. Microsoft Azure Quantum provides cloud access to IonQ, Quantinuum, and Rigetti, with an emphasis on hybrid quantum-classical computing. D-Wave Leap is a cloud platform dedicated to quantum annealing and optimization problems.

These cloud services enable users to experiment with quantum algorithms without needing their own quantum hardware, accelerating quantum research and enterprise adoption.

Governments and academic institutions play a significant role in advancing quantum computing through funding and research collaborations.

Leading universities conducting cutting-edge quantum research include MIT, Harvard, and Stanford, focusing on quantum algorithms, quantum information theory, and quantum cryptography. The University of Waterloo's Perimeter Institute is a leading centre for quantum information and quantum foundations. QuTech, a collaboration between Delft University and TNO in the Netherlands, researches quantum networking and scalable quantum computing.

The U.S. National Quantum Initiative (NQI) provides funding for quantum research and development in academia and industry. The European Union's Quantum Flagship is a €1 billion initiative to advance European quantum computing and networking. China's Quantum Program leads advancements in quantum communication and satellite-based quantum key distribution (QKD). Australia's Quantum Technology Roadmap focuses on quantum cybersecurity and quantum materials.

Governments are investing in quantum-safe encryption, quantum networks, and national security applications to stay ahead in the global quantum race.

Quantum computing is beginning to impact various industries, from finance to healthcare and supply chain optimization. JPMorgan Chase, Goldman Sachs, and BBVA are exploring quantum Monte Carlo methods for portfolio optimization and risk management. NASDAQ and Mastercard are investigating quantum-enhanced fraud detection and transaction security.

Companies such as BASF, Pfizer, and Roche use quantum simulations for molecular modelling and drug development. Quantum chemistry startups like Zapata and ProteinQure are developing quantum tools for bioinformatics and materials science. Automotive and logistics companies such as Volkswagen and Daimler are using quantum optimization for traffic management and autonomous vehicle routing. FedEx and DHL are exploring quantum logistics to improve global shipping and delivery optimization.

IBM and Microsoft are working on post-quantum cryptography (PQC) to secure encryption against future quantum attacks. China's Quantum Satellite (Micius) has demonstrated quantum key distribution (QKD) for ultra-secure communication networks.

The quantum computing industry is expanding rapidly, with significant advancements in hardware development, software ecosystems, cloud services, research, and commercial applications. While challenges remain in scaling quantum hardware, achieving quantum error correction, and identifying practical use cases, the industry is progressing toward achieving real-world quantum advantage.

With increasing investments in quantum research, hybrid quantum-classical computing, and artificial intelligence integration, quantum computing is poised to revolutionize industries ranging from finance and drug discovery to logistics and cybersecurity. The future of the industry will be defined by how quickly quantum hardware scales and how efficiently businesses can harness quantum algorithms for practical applications.

Investments and Startups in Quantum Computing

Quantum computing is currently at the forefront of technological innovation, garnering significant interest and investment from various stakeholders including venture capital firms, government entities, and large corporations. The momentum in this sector is primarily driven by the promising applications of quantum technologies, which could fundamentally disrupt several industries, including finance, healthcare, cybersecurity, and logistics [602, 603].

In recent years, startups in the quantum computing space have attracted billions of dollars in venture capital, signalling strong confidence among investors in the potential of this technology [604]. Notable firms like Bessemer Venture Partners and Sequoia Capital have made significant investments in leading companies such as Rigetti Computing and PsiQuantum, respectively. These firms recognize the potential long-term returns associated with quantum technologies as they emerge from their nascent stages. For instance, companies like IonQ and Rigetti have each raised substantial funding, exceeding $100 million before going public [604].

Governments worldwide are increasing their investments in quantum computing as a strategy to maintain technological supremacy and national security. In the United States, the National Quantum Initiative has allocated over $1.2 billion for quantum research, while the European Union has committed €1 billion to its Quantum Flagship Program. Furthermore, China's investments in quantum technology are substantial, estimated to exceed $10 billion, reinforcing its position as a key player in the field [605]. In Canada, the National Quantum

Strategy has earmarked $360 million for related research and startups, and Australia is funding various applications including quantum cybersecurity [606, 607].

Corporate entities are heavily involved in the quantum computing landscape, advancing their own initiatives through direct acquisitions and partnerships with startups. Companies like IBM, Google, and Microsoft have invested in quantum research and development, facilitating the transition of quantum technologies into practical applications [604, 608]. For instance, IBM Quantum and Google Quantum AI are at the forefront of superconducting qubit technology innovations through strategic collaborations [606]. Amazon and Honeywell also play significant roles in enhancing quantum computing infrastructure and developing commercial applications, thereby accelerating innovation in the field [604, 609].

Startups focusing on quantum hardware are integral to the advancement of the field. Rigetti Computing stands out as a leader in superconducting quantum processors, while IonQ has gained recognition as the first publicly traded quantum company, specializing in trapped-ion technology [605]. D-Wave Systems has pioneered quantum annealing techniques that provide cloud access to its solutions, and PsiQuantum is making progress in the development of photonic quantum computers, successfully raising substantial venture capital to fuel its ambitions [605].

On the software front, startups like Zapata Computing and Cambridge Quantum (now part of Quantinuum) are developing critical algorithms and applications for various industries, capitalizing on the unique capabilities of quantum computing to enhance machine learning and optimization processes [603]. Companies such as QC Ware focus on providing enterprise solutions by integrating quantum computing with traditional machine learning techniques [604].

With the growing capabilities of quantum computing, cybersecurity concerns have emerged. Startups like ID Quantique are pioneering quantum key distribution methods to secure communication channels against potential threats posed by quantum computers [606]. This sector is witnessing innovations aimed at creating quantum-safe encryption solutions to fend against future cyberattacks, pushing for a secure quantum-enabled communication framework [607].

The quantum computing sector has seen several startups transition to public companies via SPAC mergers and IPOs, providing investors opportunities to engage with the burgeoning market. IonQ, as the first publicly traded quantum computing enterprise, exemplifies this trend alongside other firms like Rigetti and D-Wave [604]. Such public offerings enable direct investments into the expected growth trajectory of quantum technologies, albeit with market uncertainties surrounding scalability and the commercialization of quantum products [603].

Despite the significant influx of investment, quantum startups face critical challenges related to technology scalability and commercialization. The complexities of building fault-tolerant quantum systems remain a significant hurdle, with error correction technologies still under development [606, 610]. Additionally, the market for quantum applications is still maturing, complicating revenue generation and investment returns for startups navigating this pioneering phase [606, 607].

Projections indicate that the quantum computing industry could reach a valuation of $65 billion by 2030, driven largely by growing demands for quantum cloud computing and quantum cybersecurity solutions [602, 603]. Continued government funding and corporate collaborations are anticipated to play vital roles in this trajectory, with the integration of quantum computing into existing industries likely to spur broader adoption and innovation [608, 609].

Governments and Corporate Involvement (DARPA, NIST, Google, IBM, Microsoft)

Quantum computing has become a strategic priority for both governments and corporations due to its potential to revolutionize industries, enhance national security, and unlock new scientific discoveries. Governments are heavily investing in quantum research and development, while major tech companies are leading advancements in quantum hardware, software, and cloud-based quantum services. Organizations such as DARPA, NIST, Google, IBM, and Microsoft are at the forefront of this technological revolution.

Government Involvement in Quantum Computing

DARPA, the U.S. Department of Defense's research agency, has played a major role in advancing quantum technologies, particularly in military and defence applications. The agency has invested heavily in quantum research, focusing on areas such as quantum sensing, secure communication, and quantum computing to enhance national security and strategic defence capabilities.

One of DARPA's key initiatives is the Optimization with Noisy Intermediate-Scale Quantum Devices (ONISQ) program, which investigates how near-term quantum processors can address complex optimization problems relevant to defence operations. This program aims to leverage quantum computational power to improve logistical planning, resource allocation, and mission-critical decision-making.

Another significant program is the Quantum-Inspired Classical Computing (QuICC) program, which seeks to develop quantum-inspired algorithms that enhance the performance of classical computers. By integrating principles of quantum mechanics into classical computation, this initiative enables more efficient processing in optimization, machine learning, and cryptographic applications.

DARPA actively collaborates with universities, national laboratories, and industry leaders to develop quantum-resistant cryptographic solutions. Given the potential threat of quantum computers breaking classical encryption methods, DARPA's research is focused on ensuring secure military communications, encrypted data transmission, and cyber defence strategies. These efforts are essential for protecting sensitive information and maintaining secure communication networks in an era of emerging quantum threats.

The National Institute of Standards and Technology (NIST) plays a crucial role in advancing quantum computing by establishing standards for quantum cryptography, quantum-resistant encryption, and quantum measurement techniques. As quantum technology progresses, NIST is at the forefront of ensuring secure and reliable quantum-based applications.

One of NIST's most significant initiatives is the Post-Quantum Cryptography Standardization Initiative, which is focused on developing encryption methods that will remain secure even in the presence of large-scale quantum computers. Since quantum algorithms like Shor's algorithm threaten traditional cryptographic protocols such as RSA and ECC, NIST is working to standardize quantum-resistant encryption algorithms to ensure long-term data security.

Through its Quantum Information Program, NIST conducts fundamental research in quantum metrology, quantum networks, and quantum mechanics. These efforts contribute to advancements in precise quantum measurements, secure quantum communication, and the development of quantum computing frameworks that meet rigorous performance standards.

NIST is also actively working on Quantum Random Number Generation (QRNG) to enhance cryptographic security. Unlike classical random number generators, which rely on algorithms and can be predicted, QRNGs leverage quantum mechanics to produce truly unpredictable and high-entropy randomness. This research is expected to improve encryption protocols, secure communications, and cybersecurity frameworks in both governmental and commercial sectors.

As a leading institution in quantum research and standardization, NIST continues to shape the future of quantum technologies by ensuring their security, reliability, and integration into real-world applications.

Governments around the world are investing heavily in quantum computing to secure technological leadership and drive innovation across industries. These initiatives aim to advance quantum computing, quantum communication, and quantum sensing while ensuring national security and economic competitiveness.

In the United States, the National Quantum Initiative (NQI) has allocated $1.2 billion in federal funding to accelerate quantum research and foster industry partnerships. The initiative supports quantum computing development, quantum networking, and quantum cryptography through collaboration between national laboratories, universities, and private sector leaders.

The European Quantum Flagship Program, a €1 billion initiative, is dedicated to advancing quantum technologies across multiple fields. This program focuses on quantum computing, quantum communication, and quantum sensing, bringing together European academic institutions, startups, and technology companies to strengthen Europe's position in the global quantum race.

China is making significant strides in quantum technology through its Quantum Computing Program, which leads in quantum communication and quantum key distribution (QKD). With state-backed research and large-scale investments, China has achieved breakthroughs in satellite-based QKD, quantum-secure communication networks, and superconducting quantum computing.

Canada's National Quantum Strategy is investing in quantum startups and academic research to build a strong quantum ecosystem. The country is home to leading quantum computing firms, including D-Wave, Xanadu, and 1QBit, which specialize in quantum annealing, photonic quantum computing, and quantum software development.

Australia is actively shaping its quantum future through its Quantum Technology Roadmap, which focuses on quantum security, quantum materials science, and quantum-enhanced AI. The government is supporting both academic research and private sector collaborations to position Australia as a leader in quantum innovation.

These government-led initiatives ensure the development of secure, quantum-resistant systems and foster collaboration between academia and industry. By funding quantum research, providing infrastructure support, and encouraging international partnerships, governments are driving the global race toward practical quantum computing applications.

Corporate Involvement in Quantum Computing

Google Quantum AI is a leader in quantum computing, focusing on superconducting qubits and pioneering quantum supremacy experiments. One of its most notable achievements was the development of the Sycamore processor, a 53-qubit superconducting quantum processor that demonstrated quantum supremacy in 2019 by solving a random sampling problem exponentially faster than classical supercomputers. This marked a significant milestone in quantum computing, proving that quantum processors could outperform classical machines in specific tasks.

Beyond quantum supremacy, Google is actively advancing quantum error correction research to develop logical qubits and move toward fault-tolerant quantum computing. This research aims to scale quantum processors beyond current limitations and create practical quantum computing systems. To support developers and researchers, Google created Cirq, an open-source quantum programming framework optimized for near-term quantum algorithms and error correction techniques. Additionally, Google is integrating quantum computing with artificial intelligence through TensorFlow Quantum (TFQ), which enables the development of hybrid quantum-classical AI models for advanced machine learning applications.

IBM Quantum is one of the most established players in quantum computing, offering cloud-based quantum computing and enterprise quantum solutions through its IBM Quantum Experience platform. This cloud-based service provides both public and enterprise users with access to IBM's superconducting qubit quantum processors, allowing them to experiment with quantum algorithms and applications. IBM supports quantum software development through Qiskit, a Python-based open-source quantum computing framework for designing and optimizing quantum circuits.

IBM has an ambitious quantum roadmap, aiming to develop a 1,121-qubit quantum processor, Condor, by 2025. The company has long-term plans to scale quantum processors beyond 1,000,000 qubits, making fault-tolerant quantum computing a reality. IBM is also a leader in enterprise adoption of quantum computing, collaborating with Fortune 500 companies to develop quantum solutions for industries such as finance, drug discovery, and logistics.

Microsoft Azure Quantum is developing a hybrid quantum computing ecosystem that integrates classical cloud computing with quantum hardware. Unlike competitors focusing on superconducting and trapped-ion qubits, Microsoft is pursuing topological qubits, which theoretically offer greater error resistance and stability. Although topological qubits are still under development, they have the potential to revolutionize quantum computing by reducing the need for complex error correction.

Azure Quantum is a cloud-based quantum computing platform that provides access to multiple quantum hardware providers, including IonQ, Quantinuum, Rigetti, and D-Wave. Microsoft also offers the Quantum Development Kit (QDK), featuring Q#, a dedicated quantum programming language optimized for quantum algorithm development. Microsoft is heavily investing in quantum machine learning and optimization, working with industries such as finance and energy to apply quantum algorithms for solving complex optimization problems.

Other major industry players are also making significant contributions to quantum computing. Amazon Braket, AWS's quantum cloud service, provides access to multiple quantum hardware platforms, including IonQ, Rigetti, and D-Wave, making it a versatile choice for developers and enterprises. Intel Quantum Computing is leveraging its expertise in semiconductor manufacturing to research spin qubits and superconducting qubits, aiming for scalable quantum hardware solutions.

Honeywell/Quantinuum is advancing trapped-ion quantum computing, focusing on high-precision quantum operations and enterprise applications. D-Wave Systems, a pioneer in quantum annealing, specializes in solving optimization problems in logistics, finance, and machine learning, providing practical quantum computing solutions through its D-Wave Leap cloud platform.

The quantum computing industry is rapidly evolving, with major corporations, research institutions, and startups racing to develop scalable quantum hardware, software frameworks, and commercial applications. Each company brings a unique approach to quantum computing, driving innovation toward achieving practical quantum advantage.

Governments and corporations are playing a critical role in advancing quantum computing by funding research, developing quantum hardware, and creating practical applications for industries. DARPA and NIST are ensuring national security and cryptographic advancements, while companies like Google, IBM, Microsoft, and Amazon are competing to build the most scalable quantum processors.

As investments continue to rise, collaborations between governments, universities, and industry leaders will accelerate the path toward practical quantum advantage. While challenges such as error correction, hardware scalability, and algorithm development remain, the combined efforts of public and private institutions are driving quantum computing closer to real-world applications across multiple sectors.

The Road to Commercial Quantum Advantage

Achieving commercial quantum advantage—the point at which quantum computers consistently outperform classical computers in solving real-world business problems—requires advancements across hardware, software, and applications. While quantum computers have already demonstrated quantum supremacy in specific artificial benchmarks, practical commercial advantage remains the next major milestone. This journey involves overcoming technical challenges, improving error correction, expanding use cases, and ensuring scalability to meet industry demands.

One of the primary challenges in attaining commercial quantum advantage is the scalability of quantum hardware. Current systems utilize superconducting qubits, which are favoured for their relatively high fidelity and speed. For instance, companies such as IBM and Google are making strides in developing quantum processors, with some proposals suggesting the possibility of systems with over 1 million qubits as essential for practical applications beyond the capabilities of classical systems [88, 584]. In contrast, trapped-ion qubits provide high-fidelity operations and long coherence times, yet they struggle with scalability [88, 611]. Photonic quantum computing presents an innovative alternative, with firms like PsiQuantum targeting a scalable model using photons [612]. Meanwhile, Microsoft is exploring topological qubits, which, if successful, could dramatically reduce error rates and enhance robustness [584].

This race towards scalable quantum computing emphasizes the necessity for large-scale fault-tolerant systems. Achieving such systems will demand sophisticated quantum error correction techniques and improvements in qubit coherence times [88, 543].

The current era of quantum computing, known as the Noisy Intermediate-Scale Quantum (NISQ) era, is characterized by systems that are susceptible to errors due to decoherence and operational noise [88, 543]. To transition towards commercial viability, it's crucial to implement logical qubits constructed from multiple physical qubits via error correction mechanisms. Techniques such as surface codes and topological error correction are pivotal in mitigating errors, which enables quantum processors to maintain operational integrity [21, 88].

Additionally, more effective noise mitigation strategies, including zero-noise extrapolation, are being developed to enhance the computational accuracy of quantum systems [88]. Enhanced qubit connectivity is equally essential to facilitate optimal entanglement and gate fidelity, which are crucial for efficient quantum operations [75].

Advancing quantum software is equally crucial for achieving commercial quantum advantage. Companies are developing quantum algorithms tailored to a variety of sectors, including finance, logistics, and pharmaceuticals. For example, the Quantum Approximate Optimization Algorithm (QAOA) addresses optimization challenges in supply chains [543], while the Variational Quantum Eigensolver (VQE) has significant applications in drug discovery [21, 612]. Quantum Monte Carlo methods improve financial risk modelling capabilities [613].

Moreover, hybrid quantum-classical computing models are emerging as a vital strategy. Given the current limitations of NISQ devices, integrating quantum algorithms alongside classical computing resources enables businesses to tackle larger-scale problems more effectively [614, 615].

The increasing accessibility and capability of quantum computing are opening up numerous applications across industries. In finance, for instance, quantum algorithms enhance portfolio optimization and fraud detection processes [613, 616]. In healthcare, quantum simulations play a crucial role in drug discovery and genetic analysis, while logistics companies benefit from quantum optimization in route planning and demand forecasting [11, 613].

These advancements illustrate a growing enterprise interest in quantum solutions, with companies such as JPMorgan and Pfizer leading the charge [613]. As these technologies mature, their ability to solve previously intractable problems is expected to reshape various fields.

Cloud-based quantum computing platforms are instrumental in democratizing access to quantum technologies, thus facilitating research and development. IBM Quantum, Google Quantum AI, and Microsoft Azure Quantum have established platforms that allow researchers and industries to experiment with quantum processors without the need for substantial hardware investments [613, 617]. These cloud services enable organizations to explore innovative applications of quantum technology effectively, expediting the journey toward practical quantum advantage.

Despite rapid advancements, various challenges must be addressed to achieve widespread commercial utility. These include overcoming scalability issues, advancing quantum error correction methods, developing efficient quantum algorithms, enhancing hardware performance, and ensuring seamless integration into existing IT infrastructures [618, 619]. The collaborative efforts of governments, corporations, and research institutions are essential to address these hurdles and pave the way for future quantum innovations [620].

The road to commercial quantum advantage is expected to unfold in several phases. The years 2025 to 2030 will likely see improvements in NISQ devices and hybrid quantum-classical algorithms, while the 2030s may witness the advent of fault-tolerant quantum computers and more scalable solutions [619]. Looking beyond 2040, the full integration of quantum computing across industries could lead to unprecedented innovations, highlighting the urgency for early investment in this emerging technology [620].

The journey to commercial quantum advantage is marked by significant challenges and opportunities across various domains, from hardware scalability to algorithm efficiency. Continued investment and research in these areas will usher in an era of quantum computing capable of transforming industries and enhancing computational capabilities. As the infrastructure and applications evolve, businesses that strategically adopt quantum technologies stand to gain competitive advantages, ultimately redefining the computational landscape.

Ethical and Economic Implications

Quantum computing holds the potential to revolutionize industries, from finance and healthcare to cybersecurity and artificial intelligence. However, alongside its benefits, quantum technology raises significant ethical and economic concerns that must be addressed to ensure responsible development and equitable distribution of its advantages. These challenges span privacy, security, economic impact, job displacement, and global power dynamics.

Ethical Implications of Quantum Computing

Quantum computing presents an array of ethical and operational challenges in the domain of cybersecurity and privacy, primarily due to its potential to disrupt classical encryption methods fundamental to current digital security protocols. The emergence of powerful quantum algorithms, particularly Shor's algorithm, threatens to render widely used cryptographic techniques such as RSA (Rivest-Shamir-Adleman) and ECC (Elliptic Curve Cryptography) vulnerable to decryption [621]. This vulnerability poses significant risks, particularly in sectors that rely on secure communications, such as finance, healthcare, and government, where breaches could result in catastrophic data exposure and significant privacy violations [622]. Furthermore, the potential for mass surveillance enabled by quantum decryption capabilities creates an environment ripe for ethical concerns surrounding personal privacy rights and cybersecurity integrity [622].

To address these imminent threats, significant efforts are underway in the realm of Post-Quantum Cryptography (PQC). Governments and organizations are developing quantum-resistant algorithms aimed at securing data against the impending capabilities of quantum computers [622, 623]. However, this transition is fraught with urgency as there exists a critical window wherein quantum decryption techniques might outpace the adoption of secure alternatives. The scenario suggests a risk where state and non-state actors could exploit quantum technologies for surveillance or cyberattacks, raising profound ethical dilemmas for policymakers concerning these emerging realities [624].

Compounding these issues is the concept of the "quantum divide," which highlights the disparity in technological capabilities associated with quantum computing across different nations and corporations. Currently, quantum computing infrastructure is concentrated within a select group of technologically advanced nations and corporations, such as IBM, Microsoft, and Google [625]. This oligopolistic access creates a risk of monopolization, resulting in significant advantages in fields like scientific research, economic dominance, and military applications. To mitigate such inequalities, a push for open-source quantum solutions, cloud-based Quantum-as-a-Service (QaaS) platforms, and international collaborations is paramount to democratizing access and preventing technological elitism [623].

The interface of artificial intelligence (AI) and quantum computing introduces further complexities regarding surveillance and data ethics. Organizations with access to advanced quantum AI technologies possess the capability to enhance facial recognition, predictive analytics, and autonomous decision-making systems, potentially escalating mass surveillance initiatives [622]. Ethical concerns arise regarding biases inherent in these systems; if quantum machine learning models are trained on biased datasets, they may perpetuate discrimination across various sectors, such as hiring practices and law enforcement [626]. Thus, establishing transparent governance frameworks amidst the development of quantum AI becomes essential to address potential misconduct and ensure equitable outcomes.

From a national security perspective, quantum computing's impact is profound, reshaping traditional paradigms of cyberwarfare and cryptographic security. The advent of technologies like Quantum Key Distribution (QKD) promises secure methods of communication [624]. However, the exclusive use of these advancements by military or governmental entities could leave civilian infrastructures vulnerable, ultimately elevating the stakes in global cybersecurity debates [627]. Thus, international cooperation and regulatory frameworks are necessary to establish norms for responsible use of quantum technologies in combating cyber threats and maintaining global cybersecurity.

Economic Implications of Quantum Computing

Quantum computing is regarded as a transformative technology that can reshape various industries by introducing groundbreaking opportunities and disruptions. It holds the potential to optimize operations in sectors such as finance, healthcare, logistics, and many more, offering vast benefits in efficiency, precision, and capability. The underlying principles of quantum algorithms enable significant advancements in complex problem-solving, from financial modelling and risk assessment to accelerated drug discovery and optimal logistics management [533, 536].

In finance, for instance, quantum computing is anticipated to enhance risk analysis, fraud detection, and investment modelling significantly. Quantum algorithms, including Monte Carlo simulations, can refine strategies for derivatives pricing and asset management, paving the way for institutions to tackle computationally intensive tasks more effectively [533, 536]. However, the introduction of quantum computing into high-frequency trading may also precipitate instability within financial markets, particularly if early adopters gain an overwhelming advantage over traditional methods [535, 537].

The implications for workforce dynamics are profound. As quantum computing automates complex tasks, the demand for human analysts in targeted sectors may decline, thus leading to job displacement. Conversely, this technological evolution will also yield new career paths in areas like quantum programming, cybersecurity, and machine learning [535, 628]. Experts in quantum error correction and hardware engineering will become increasingly valuable as industries adapt to this disruptive technology. Governments and educational institutions must prioritize investments in quantum education programs to ensure that the workforce is adequately equipped for these changes [629, 630].

Healthcare is another sector poised for substantial transformation through quantum computing. Leading pharmaceutical companies are investing in quantum chemistry to expedite drug discovery processes, which could facilitate personalized medicine and advanced material development [37, 631]. However, challenges arise concerning the pricing and accessibility of these innovations. If quantum-driven advancements are monopolized by pharmaceutical giants, they could lead to inflated costs that disproportionately impact low-income populations. Collaborative efforts among governments, healthcare providers, and research entities are crucial to ensuring equitable access to these benefits [629, 631].

Logistics and supply chains stand to gain significantly as well, leveraging quantum computing for the optimization of transportation routes and operational efficiency. Although quantum algorithms can yield cost reductions, they may also automate roles traditionally filled by human workers, necessitating proactive strategies to retrain affected individuals and support a smooth transition [628, 632]. The ability of quantum systems to enhance energy distribution and promote sustainable practices further underscores the technology's transformative potential in facilitating a green economy [632, 633].

However, the rapid ascent of quantum research has prompted intense competition for patents, raising concerns regarding monopolization in this emerging field. Major tech companies are striving for dominance through aggressive patent filing, sparking debates about the balance between scientific collaboration and commercialization. To foster an environment where innovation can thrive, ethical licensing norms must be established to deter anti-competitive behaviours within the quantum landscape [630, 634].

Chapter 12

Challenges and the Future of Quantum Computing

As we have explored throughout this book, quantum computing represents one of the most promising yet complex technological advancements of our time. From the foundational principles of quantum mechanics to the development of quantum algorithms, hardware innovations, and real-world applications, each chapter has demonstrated how quantum systems are poised to revolutionize industries ranging from cryptography and artificial intelligence to materials science and logistics. However, despite the remarkable progress made in recent years, significant challenges remain that must be addressed before quantum computing can achieve widespread adoption and deliver on its full potential.

This chapter examines the key obstacles that researchers and engineers face in scaling quantum systems, including decoherence, error correction, and the difficulty of maintaining stable qubits. It also explores the theoretical and physical constraints that could limit the scalability of quantum computers, such as energy efficiency, quantum control, and fault tolerance. Beyond technical hurdles, the chapter considers the broader implications of quantum technology, including its potential to disrupt cybersecurity, reshape industries, and introduce new ethical and economic challenges.

Looking ahead, the future of quantum computing is filled with both promise and uncertainty. Predictions for the next decade range from advancements in fault-tolerant quantum computing and the development of a quantum internet to potential breakthroughs in distributed quantum systems. As classical computing approaches the limits of Moore's Law, quantum computing is emerging as a powerful alternative that could redefine the boundaries of what is computationally possible. While it remains in its early stages, continued investment, interdisciplinary collaboration, and theoretical advancements will determine how soon quantum computing can move from research labs into practical, large-scale deployment.

This final chapter serves as a roadmap for what lies ahead, providing insights into how quantum computing might evolve in the coming years and what researchers, businesses, and policymakers need to consider as we transition into the quantum era.

Overcoming Technical Barriers (Decoherence, Scalability, Control)

Quantum computing faces several technical barriers that must be addressed before it can achieve large-scale, fault-tolerant computation. The three major challenges are decoherence, scalability, and control. Scientists and engineers are actively developing solutions to overcome these obstacles and bring quantum computing closer to practical applications.

Decoherence is one of the most significant challenges in quantum computing. Unlike classical bits, which remain stable as 0s and 1s, quantum bits (qubits) exist in superposition states that are highly sensitive to their environment. Any interaction with external noise, temperature fluctuations, or electromagnetic radiation can cause a qubit to lose its quantum state, leading to errors in computations. This fragility makes quantum systems difficult to scale and maintain for extended periods.

To combat decoherence, researchers are developing quantum error correction (QEC) techniques. Unlike classical error correction, which can rely on simple redundancy by adding extra bits for parity checks, quantum error correction requires a more sophisticated approach due to the no-cloning theorem, which prevents direct copying of quantum states. Instead of duplicating information, QEC distributes quantum information across multiple physical qubits to encode a more stable logical qubit.

One of the most widely studied quantum error correction methods is surface codes, which use multiple physical qubits to encode a logical qubit. This allows for error detection and correction without directly measuring the qubit's state, which would otherwise collapse its quantum properties. Google and IBM are actively working on implementing surface codes to improve the reliability of their quantum processors. Another promising approach is topological qubits, being developed by Microsoft. These qubits, based on Majorana fermions, theoretically offer greater resistance to noise due to their unique quantum properties. Although topological qubits have not yet been demonstrated experimentally, they hold promise for robust quantum computing.

Researchers are also exploring bosonic codes, a technique used in continuous-variable quantum computing. In this approach, quantum information is encoded in superconducting cavities, allowing for more efficient error correction than traditional qubit-based methods. While still in the early stages of research, bosonic codes could provide a pathway toward improved error resilience.

Despite advancements in error correction, current quantum computers still have high error rates, and QEC techniques require a large number of physical qubits to create a single fault-tolerant logical qubit. This means that practical large-scale quantum computation will require millions of qubits to implement robust error correction. As qubit coherence times improve and error correction methods become more efficient, quantum computing will move closer to achieving fault-tolerant systems capable of solving real-world problems.

Scalability remains one of the biggest challenges in quantum computing, as increasing the number of qubits while maintaining stability and connectivity is a complex task. Today's most advanced quantum processors contain only tens to a few hundred qubits, but large-scale quantum applications will require millions of qubits to implement robust error correction and achieve practical quantum advantage. The difficulty lies not only in adding more qubits but also in ensuring that they remain coherent, properly interconnected, and resistant to errors as the system expands.

Efforts to scale quantum systems are being explored through different quantum architectures. Superconducting qubit architectures, used by IBM, Google, and Rigetti, focus on increasing the number of superconducting qubits on a single chip while improving coherence times and minimizing crosstalk between qubits. IBM's roadmap includes developing a 1,121-qubit processor (Condor) by 2025, with plans to scale beyond that in the coming years. These efforts involve advances in fabrication techniques, qubit arrangement, and cryogenic cooling systems.

Trapped-ion quantum computers, pursued by companies like IonQ and Quantinuum, use charged atoms trapped in electromagnetic fields and manipulated with lasers. Trapped-ion qubits offer long coherence times and high-fidelity operations, making them a promising candidate for scalable quantum computing. However, scaling these systems requires precise laser control over individual ions, and increasing the number of trapped ions without losing operational accuracy remains a major challenge.

Photonic quantum computing, being developed by Xanadu and PsiQuantum, uses light-based qubits instead of matter-based qubits. This approach could theoretically scale more easily by leveraging existing fiber-optic infrastructure, which is already widely used in telecommunications. However, photonic quantum computing still requires breakthroughs in quantum error correction to maintain stability and enable large-scale computation.

Another potential solution for scalability is quantum interconnects, where multiple smaller quantum processors are linked together via quantum networks to form a larger distributed quantum system. This modular quantum computing approach could help overcome the physical limitations of single-chip scaling, making it possible to build large-scale quantum computers using smaller, interconnected quantum nodes.

Scaling quantum systems is not just about increasing the number of qubits; it also requires improving qubit connectivity, reducing noise, and ensuring that error rates remain low as systems grow. Hybrid approaches that combine different quantum technologies—such as integrating superconducting qubits with photonic networks—may provide a more effective way to scale quantum computing and accelerate the path toward fault-tolerant quantum systems.

Precise control of qubits is essential for reliable quantum computation. Unlike classical bits, qubits require highly sensitive control mechanisms such as microwave pulses, laser beams, or electromagnetic fields to manipulate quantum states. Small errors in these control operations can lead to gate errors, reducing computation accuracy and increasing the likelihood of decoherence. Developing robust qubit control techniques is a key focus of quantum computing research.

One critical area of advancement is high-fidelity quantum gates. The accuracy of quantum operations depends on gate fidelity, which measures how precisely qubits can be manipulated. Companies like Google and IBM have achieved two-qubit gate fidelities above 99%, a significant milestone toward practical quantum computation. However, fault-tolerant quantum computing requires even lower error rates, necessitating further improvements in quantum hardware and control precision.

Another challenge in qubit control is the need for cryogenic control electronics. Superconducting qubits operate at extremely low temperatures (below -273°C), meaning traditional electronic control systems cannot function at these conditions. IBM and Intel are developing cryogenic control chips that can operate at ultra-low temperatures, reducing the need for bulky external control systems. These chips integrate qubit control circuits closer to the quantum processor, improving efficiency and scalability.

AI-assisted quantum control is emerging as a powerful tool to enhance qubit stability. Machine learning techniques are being used to optimize pulse sequences for quantum gates, reducing errors and improving system performance. AI-driven control systems can dynamically fine-tune quantum operations based on real-time feedback, helping to mitigate fluctuations and improve coherence times.

Another key area is scalable qubit readout methods. Measuring quantum states without disturbing them, known as quantum non-demolition measurement, is a major challenge in quantum computing. Advanced techniques, such as dispersive readout for superconducting qubits and fluorescence detection for trapped-ion qubits, allow researchers to extract information without collapsing the quantum system prematurely. These advancements are critical for enabling error correction and improving computational stability.

Efforts to improve qubit control mechanisms are crucial for making quantum computers more stable, reliable, and commercially viable. With ongoing progress in quantum gate fidelity, cryogenic electronics, AI-driven control, and scalable qubit readout, the industry is moving closer to achieving fault-tolerant quantum computation and unlocking the full potential of quantum technology.

Future Directions and Breakthroughs

Overcoming the technical barriers associated with decoherence, scalability, and control in quantum computing is a multi-faceted challenge that necessitates the integration of hardware innovations, software optimizations, and advancements in materials science. The convergence of these areas is vital for realizing practical quantum processors, as each element significantly influences the overall performance of quantum architectures.

The development of Hybrid Quantum-Classical Architectures is one promising approach. By leveraging high-performance classical computing alongside near-term quantum processors, this hybrid model can effectively optimize workloads, thereby capitalizing on the strengths of both quantum and classical computing paradigms. Research indicates that such modular architectures, which facilitate communication between quantum and classical components, are essential for reducing overhead associated with qubit operations and managing non-local communications [635, 636]. Companies like IBM, Google, and Microsoft are actively exploring these architectures to enhance computational efficiency and effectiveness [28, 637]. Integration strategies such as these not only aim to address the shortcomings of current quantum systems but also pave the way for more scalable and robust quantum applications.

In parallel, progress in Quantum Networking and Distributed Quantum Computing is essential for establishing a future quantum internet. Quantum networks can enable the interconnection of multiple quantum processors using quantum teleportation and entanglement-based protocols, facilitating distributed quantum computing across large distances [83, 638]. These technologies promise enhanced scalability by mitigating the challenges posed by decoherence and communication overhead, allowing multiple devices to function synergistically [639, 640]. The development of a structured quantum internet will revolutionize data exchange procedures, enabling secure communications and novel applications in quantum sensing and distributed processing [641].

Another significant area is the exploration of New Qubit Materials. Various alternative materials, including silicon-based spin qubits, neutral atom qubits, and diamond-based systems, are under investigation as potential means to enhance the stability and scalability of quantum processors [30, 642, 643]. Innovations in materials science will be critical in addressing decoherence—a leading contributor to the loss of quantum information due to environmental interactions. For instance, improved coherence times and lower error rates in qubit implementations can directly impact the overall fidelity of quantum computing systems, making them more viable for complex computations [644, 645].

Lastly, the innovation of Error-Resilient Quantum Algorithms is crucial for making quantum computing more practical in the near term. Researchers are developing quantum algorithms that can operate under conditions of imperfect hardware, allowing some tolerance for errors without sacrificing performance [646, 647]. These advancements in algorithm design will be pivotal in utilizing near-term quantum systems to deliver meaningful results, even as the technology matures. The shift toward error-tolerant strategies represents not merely a technical goal but a foundational change in how quantum computational tasks can be realized in real-world applications.

Quantum computing is still in its early stages, but progress in error correction, qubit scalability, and precision control is accelerating its development. While current quantum processors face significant limitations, ongoing advancements in hardware, software, and system architecture are gradually overcoming these barriers. As these technical challenges are addressed, quantum computing will transition from experimental research to practical, large-scale applications, unlocking its full potential across industries such as finance, healthcare, materials science, and artificial intelligence.

Theoretical Limits and Physical Constraints

The exploration of quantum computing reveals numerous theoretical limits and physical constraints that are pivotal in understanding its feasibility and future. These limitations significantly influence the scalability, reliability, and practical implementation of quantum processors, underscoring the necessity for innovative solutions to achieve commercially viable quantum computing.

A critical theoretical constraint is the no-cloning theorem, which prohibits the duplication of arbitrary unknown quantum states. This principle restricts the adoption of classical error-correction techniques commonly employed in classical computing and introduces unique challenges in quantum communication systems, where maintaining information integrity is paramount [648]. Furthermore, while quantum algorithms such as Grover's algorithm enhance computational efficiency with a quadratic speedup, this advantage is not uniform across all problems [649]. Problems that fall into NP-complete categories remain unsolved regarding their efficient resolution through quantum methods, leading to uncertainty about the ultimate capabilities of quantum computing [650]. As such, the relationship between BQP (Bounded-Error Quantum Polynomial Time) and classical complexity classes such as P and NP remains a focal point of ongoing research to determine the computational boundaries of quantum systems [651].

Additionally, the overhead associated with quantum error correction (QEC) poses a formidable challenge. Implementing QEC requires the encoding of multiple physical qubits into a single logical qubit to mitigate the effects of decoherence and noise. This exponential qubit overhead complicates the scaling of fault-tolerant quantum computers, rendering them resource-intensive and difficult to manage with current technological capabilities [652, 653]. Thus, it becomes vital to alter traditional computational models to accommodate the unique paradigms of quantum systems [654].

Physical limitations are equally significant, most notably quantum decoherence, where the interaction between qubits and their external environment leads to loss of quantum information. To counteract decoherence, quantum processors necessitate operation in ultra-low temperature environments, typically near absolute zero. Yet even in such controlled settings, factors such as thermal fluctuations and electromagnetic interference can introduce noise that compromises computational fidelity [655, 656].

Achieving high qubit fidelity represents another substantial hurdle. Gate operations can incur errors, reducing overall computational accuracy. The challenge becomes exacerbated as operations on qubits may result in accumulating fidelity errors over extended periods [657]. Therefore, enhancing gate performance to fault-tolerant levels is imperative for reliable quantum operations [652]. Moreover, the architecture of quantum systems, particularly superconducting qubits, is constrained by connectivity limits; as the complexity of interconnected qubits increases, so do the unintended error rates [648].

The pursuit of solutions to these challenges is underway. Ongoing research in QEC seeks to streamline the resource requirements, facilitating large-scale quantum computations with reduced overhead [658, 659]. The investigation of topological qubits, which inherently resist environmental noise, is also gaining traction as a promising avenue to improve the robustness of qubits and their associated operations [655].

Furthermore, the development of hybrid quantum-classical models is emerging as a method to leverage the strengths of both computational systems, wherein classical components handle elements of computation that do not benefit from quantum properties [660]. The synergy of quantum and classical systems could alleviate some of the burdens associated with scaling and resource allocation.

Quantum Internet and Distributed Quantum Computing

Quantum internet and distributed quantum computing are emerging fields that aim to interconnect quantum computers across networks, enabling secure communication, distributed processing, and large-scale quantum information exchange. These advancements will revolutionize fields such as cryptography, data security, high-performance computing, and cloud-based quantum services.

Quantum Internet

The quantum internet represents a transformative advance in communication technology, predicated on the principles of quantum mechanics rather than classical theories. Unlike traditional networks that utilize bits, the quantum internet operates with qubits, which can exist simultaneously in multiple states (superposition) and be entangled regardless of distance. This unique feature allows for exceptionally secure communication protocols, such as Quantum Key Distribution (QKD), which utilizes the peculiarities of quantum states to detect eavesdropping attempts effectively, thus ensuring the integrity of the shared keys between parties [661, 662].

Quantum Key Distribution stands as a cornerstone of quantum internet communication, leveraging quantum mechanics to provide security guarantees unattainable by classical systems. The most notable protocols, BB84 and E91, operationalize this concept by revealing any eavesdropping through the collapse of quantum states [661, 662]. China's Micius satellite has exemplified the practical application of QKD, successfully demonstrating key exchange over long distances—up to 1,200 kilometres—thus underlining the feasibility of a global quantum communication infrastructure [661]. Noteworthy developments by companies like Toshiba and ID Quantique in commercial QKD systems further illustrate the growing accessibility of quantum-secure communication for various sectors, including government and enterprise [662].

An additional technological requirement for the successful deployment of the quantum internet is the development of quantum repeaters. These devices are essential for mitigating signal degradation across extensive distances, which is a significant limitation within quantum communication systems. Quantum repeaters establish entanglement links and correct errors without direct measurement, thereby maintaining the quantum properties of the signals over long ranges. Institutions such as Harvard and MIT are at the forefront of this research, experimenting with nitrogen-vacancy (NV) centres in diamonds as well as trapped ions, both promising candidates for reliable quantum memory with prolonged coherence times necessary for repeaters [639, 663]. Furthermore, initiatives by Delft University's QuTech aim to innovate multi-node entanglement networks, which are critical for enabling robust long-distance quantum communications [639, 663].

The architecture of the quantum internet heavily relies on entanglement-based networking. This approach facilitates instantaneous correlations between qubits through techniques such as entanglement swapping. The Quantum Network Explorer (QNE) project in the Netherlands is actively testing these concepts in real-world scenarios, seeking to establish a scalable quantum communication network [639, 663]. Major tech leaders like Google, IBM, and Microsoft are investing significantly in research regarding entanglement-based communication protocols, aiming to interconnect quantum processors for distributed computing, thereby enhancing computational capabilities surpassing classical limitations [661, 663, 664].

Furthermore, the quantum internet enables secure cloud-based quantum computing, permitting users to execute computations remotely on quantum processors while ensuring their data remains confidential through mechanisms like blind quantum computing. This capability is essential for sectors requiring stringent data protection measures, including banking, healthcare, and governmental operations [662]. As the advancement of quantum communication technologies progresses, the emergence of a fully operational quantum internet stands to revolutionize global data security and elevate computational capacities to unprecedented levels [639, 662, 663].

Distributed Quantum Computing

Distributed quantum computing represents a paradigm shift in quantum information processing, wherein multiple quantum processors are interconnected through quantum networks to harness a more scalable computational framework. Unlike traditional approaches that rely on a singular, large quantum processor, distributed quantum computing facilitates partitioned computations across various quantum nodes, thereby amplifying computational capability while addressing the intrinsic limitations associated with single-chip configurations [663, 665]. This architectural innovation enhances the overall power of quantum computations by allowing disparate quantum systems to collaborate synergistically [666]. The ability to consolidate processing power across multiple interconnected nodes not only fosters scalability but also potentially mitigates issues such as qubit error rates due to the distribution of computational loads [640].

At the core of this distributed model are quantum networking and interconnects, which are vital for maintaining coherence and entanglement between quantum processors as they exchange quantum information. Leading technology companies, including IBM, Google, and Intel, are actively refining quantum interconnect methodologies utilizing microwave links, optical fibre, or superconducting cables [663]. Such quantum interconnects are essential for maintaining entangled states over spatial separations, which ensures the integrity of quantum information as it traverses between processing nodes [114]. Additionally, advancements in quantum teleportation—where quantum states are transmitted between nodes without the need for direct physical transfer—are pivotal for enhancing efficiency in distributed quantum systems [667]. The implications of such technologies could significantly transform how quantum computations are performed, especially in noisy environments typically encountered in distributed setups [668].

Modular and hybrid quantum architectures are being explored as promising frameworks that offer scalable alternatives in contrast to traditional large, monolithic quantum processors. Modular quantum computing, by linking multiple small-scale processors, addresses scaling challenges presented by single-chip systems, thus maintaining performance levels while minimizing qubit errors [640]. Hybrid quantum-classical approaches also allow quantum processors to complement classical computing capabilities by executing specialized computations while conventional systems manage problem areas unsuitable for quantum speedup [663, 669].

Furthermore, the emergence of cloud-based quantum computing services and Quantum-as-a-Service (QaaS) platforms have opened extensive avenues for distributed quantum computing. Platforms such as IBM Quantum, Amazon Braket, and Microsoft Azure Quantum provide users access to remote quantum processors, enabling computational resources to be utilized without the prohibitive costs associated with hardware investments [661]. Google's Quantum AI division is also developing multi-node networking solutions that interconnect superconducting processors, bolstering the accessibility and scalability of quantum computing resources [114]. Such infrastructure supports not only greater computational power but also democratizes access to quantum technologies for researchers and businesses alike.

Finally, addressing the challenges of fault tolerance in quantum computation remains a priority within distributed quantum contexts. When computational loads are distributed across multiple entangled qubits in different locations, systems can exhibit enhanced fault tolerance, effectively reducing the error rates of individual qubits [666]. Innovations in topological qubits, which boast inherent stability against decoherence, are being pursued to strengthen the resilience of quantum computations [667]. Moreover, the utilization of quantum error correction codes is crucial for maintaining stability in multi-node systems, thus enabling longer computations with less susceptibility to error occurrences [640]. As networking technologies continue to evolve, the potential for distributed quantum computing to achieve practical quantum advantage grows increasingly tangible.

Practical Applications of Quantum Internet and Distributed Quantum Computing

Practical applications of the quantum internet and distributed quantum computing will revolutionize multiple industries by enabling ultra-secure communication, scalable cloud-based quantum computing, advanced AI optimization, and global scientific collaboration. These advancements will enhance cybersecurity, improve computational efficiency, and facilitate breakthroughs in science and technology.

Ultra-secure communication and cybersecurity will be one of the most immediate benefits of the quantum internet. Governments and enterprises will use quantum-secure encryption to protect sensitive communications from cyber threats. Quantum key distribution (QKD) will ensure that any interception attempt collapses the quantum state, making eavesdropping impossible. This will be particularly crucial for financial transactions, healthcare records, and military communications. Quantum-secure voting systems will prevent electoral fraud, while financial institutions will rely on quantum encryption to safeguard digital transactions from hacking.

Scalable quantum cloud computing will enable researchers and businesses to access distributed quantum processing power through cloud-based platforms. By linking quantum processors via the quantum internet, quantum computing resources will become more accessible for solving complex problems in quantum machine learning, drug discovery, and financial modelling. Cloud providers such as IBM Quantum, Google Quantum AI, Amazon Braket, and Microsoft Azure Quantum are already building infrastructure to support distributed quantum computing, allowing users to harness quantum algorithms without needing their own quantum hardware.

Quantum-enhanced AI and optimization will benefit from distributed quantum computing by enabling more efficient problem-solving in logistics, traffic control, and scientific simulations. Quantum AI models will leverage the quantum internet to process vast datasets more efficiently, optimizing global supply chains, reducing transportation costs, and improving smart city planning. Quantum-enhanced AI will also accelerate scientific simulations, allowing researchers to model complex molecular interactions, design new materials, and optimize energy grid distribution with unprecedented accuracy.

Global quantum networks for scientific collaboration will connect quantum laboratories worldwide, enabling researchers to conduct joint experiments in materials science, high-energy physics, and chemistry. Distributed quantum computing will allow international research teams to share entangled qubits and perform quantum simulations across different geographic locations in real time. This will lead to groundbreaking discoveries in quantum chemistry, condensed matter physics, and fundamental quantum mechanics. By creating a globally interconnected quantum research environment, the quantum internet will foster innovation and accelerate progress in quantum science and technology.

The development of a full-scale quantum internet and distributed quantum computing will take years, requiring improvements in quantum hardware, networking infrastructure, and error correction methods. Governments, academia, and technology companies are investing heavily in building prototypes and testing real-world applications.

The U.S., China, and the EU have launched multi-billion-dollar quantum networking programs, and private companies like IBM, Google, Microsoft, and Amazon are leading commercial development. Quantum internet testbeds are already operational in the Netherlands, China, and the U.S., demonstrating early-stage quantum network functionality.

With continued breakthroughs in quantum networking, entanglement distribution, and multi-node quantum computing, the vision of a global quantum internet and fully distributed quantum processing is becoming more achievable, promising a future of secure communication, scalable quantum computing, and revolutionary scientific discovery.

Predictions for the Next 10-20 Years

The landscape of quantum computing is poised for significant transformation over the next 10 to 20 years, with advancements in hardware, algorithms, and applications anticipated to shape various industries. This transformation is categorized into near-term applications within the next 5 to 10 years and the long-term transition towards fault-tolerant quantum computing in the subsequent decade.

Next 5-10 Years: Near-Term Quantum Applications and Hybrid Computing

In the immediate future, the focus of quantum computing is set to narrow on developing noisy intermediate-scale quantum (NISQ) devices and integrating quantum processing with classical computing frameworks. Companies such as IBM, Google, and Microsoft are at the forefront, aiming to scale quantum hardware significantly. IBM's target to deliver a 1,121-qubit processor named Condor by 2025 exemplifies these efforts [670]. Furthermore, improvements in quantum error correction (QEC) techniques are expected to enhance coherence times and reduce error rates, paving the way for the emergence of logical qubits [169].

The commercialization of Quantum as a Service (QaaS) platforms will further democratize access to quantum computing resources, empowering industries like finance, logistics, and pharmaceuticals to leverage quantum optimization algorithms and achieve substantial efficiencies [35, 671]. For instance, early adopters in supply chain management can utilize quantum computing for logistics optimization and complex financial modelling. Additionally, the integration of quantum-enhanced machine learning models is projected to outperform classical algorithms in specific sectors, such as drug discovery and fraud detection [672].

Moreover, the evolution of quantum cryptography will take centre stage, with growing adoption of post-quantum cryptography (PQC) frameworks to safeguard data against future quantum threats [671]. Concurrently, advancements in quantum key distribution (QKD) networks will ensure secure communication channels, which are crucial for sectors such as finance and national security [673].

10-20 Years: Fault-Tolerant Quantum Computing and Widespread Adoption

Looking further ahead, there is a consensus that by the 2040s, operational large-scale fault-tolerant quantum computers featuring millions of logical qubits will become a reality, surpassing the capabilities of classical supercomputers in solving complex computational problems [670]. Such advancements will lead to breakthroughs in materials science and drug discovery, as companies will increasingly rely on quantum simulations to identify new pharmaceuticals and materials more efficiently [674].

Furthermore, as quantum neural networks (QNNs) develop, they are expected to revolutionize artificial intelligence, enabling unprecedented advancements in deep learning and natural language processing [672]. This technological evolution will be complemented by the establishment of a quantum internet, integrating quantum repeaters and entanglement-based communication networks, thereby greatly enhancing global secure data transmission [670, 675].

The implications for cybersecurity are profound, as traditional encryption methods will likely become obsolete, necessitating widespread adoption of quantum-safe encryption standards [676]. The global infrastructure will require adaptations to accommodate quantum-secure cloud storage solutions, ensuring protection against future cyber threats.

Challenges and Roadblocks

Notwithstanding the promising trajectory, several challenges persist in the realm of quantum computing. Issues related to scalability and the manufacturing of quantum processors capable of maintaining coherence at scale represent significant hurdles [169]. Compounded by the complexity of developing effective quantum error correction methods, these challenges must be addressed to realize fault-tolerant quantum computing [674, 675]. Additionally, concerns regarding energy consumption and operational costs continue to be prominent, particularly due to the extreme cooling requirements of many quantum systems [35].

Moreover, ethical concerns have emerged regarding the potential widening of the technological divide, placing nations and corporations with advanced quantum capabilities at an advantage [670]. Subsequently, a framework addressing the ethical implications of quantum computing, particularly in cybersecurity and AI bias, is urgently needed.

As quantum computing transitions from an experimental to a mainstream technology over the next 10 to 20 years, its integration across numerous industries will enable unprecedented innovation and problem resolution capability. The near-term focus on hybrid systems and optimization will pave the way for longer-term achievements in fault-tolerant quantum systems, revolutionizing fields such as computational science, secure communications, and artificial intelligence. Despite existing technical and ethical challenges, continuous investment and research in quantum computing are expected to propel forward the next wave of technological advancements.

Quantum Computing's Role in a Post-Moore's Law Era

Moore's Law, which predicts that the number of transistors on a microchip doubles approximately every two years, has been the driving force behind classical computing advancements for decades. However, as transistors approach atomic scales, physical limitations such as heat dissipation, quantum tunnelling, and fabrication complexity are slowing progress. The post-Moore's Law era presents a significant challenge for traditional computing architectures, requiring new paradigms to sustain computational advancements. Quantum computing, with its fundamentally different approach to information processing, offers a potential solution to break past these limits.

Moore's Law, originally articulated by Gordon Moore in his 1965 paper, predicts that the number of transistors on a microchip would double approximately every 18 to 24 months, while the cost per transistor would decline simultaneously. This observation has significant implications for computing power, as it suggests that performance will essentially double every couple of years at a constant price [677, 678]. Since its inception, Moore's Law has served as a foundational guideline for semiconductor development and has been observed continually throughout the decades, influencing technology industries, economic models, and consumer electronics profoundly [679].

The empirical nature of Moore's Law points to its reliance on ongoing advancements in semiconductor manufacturing technologies, such as improvements in photolithography and materials science. For instance, innovations in excimer laser lithography and extreme ultraviolet (EUV) lithography have enabled manufacturers to achieve the necessary reductions in feature size to maintain the pace envisaged by Moore [680, 681]. Such technological advancements highlight the intricate relationship between physical capability and economic incentives in chip manufacturing, where increased production volumes foster continual improvements and significant investment in next-generation technologies [682].

Despite its long-standing influence, there has been considerable discourse regarding the sustainability of Moore's Law in light of emerging physical and economic limits. As transistors approach atomic scales, quantum effects may increasingly hinder further advancements [683]. Researchers have forecasted that the continuity of Moore's Law could be jeopardized and that potential transitions to alternative computing architectures, such as quantum computing or neuromorphic systems, may be necessary as explored in futuristic discourses on semiconductor technology [684]. Nonetheless, the historical trajectory of CPU development closely aligns with Moore's Law, suggesting that while physical limits might eventually challenge it, the law itself has accurately framed the narrative of technological progress in microelectronics for over half a century [678, 679].

As classical computing approaches its physical constraints, several challenges emerge that hinder the continued scaling of computational power. The first major limitation is transistor miniaturization. Traditional silicon transistors are reaching atomic scales, where quantum effects such as electron leakage due to quantum tunnelling make further miniaturization impractical. As transistors become smaller, it becomes increasingly difficult to control electron behaviour, leading to unintended current flows and reduced reliability of electronic circuits.

Another significant challenge is increasing power consumption. As transistor density grows, managing heat dissipation becomes a critical issue. The more transistors packed onto a chip, the greater the power requirements, and with that comes excess heat generation. This heat buildup limits performance gains, as overheating can degrade processor efficiency and reliability, requiring more advanced cooling solutions that add to system complexity and cost.

The breakdown of Dennard scaling further complicates the situation. Dennard scaling states that as transistors shrink, their power density remains constant, meaning that smaller transistors should not require significantly more power. However, as chips have become denser, power leakage and energy dissipation have increased, leading to diminishing returns in performance per watt. This shift has forced the industry to explore alternative computing architectures and new materials to sustain improvements in processing power.

Finally, there are limits to classical parallelism. Modern computing relies on multi-core processors and parallel processing to enhance performance. However, there is a fundamental ceiling to how much computational power can be extracted through parallelization. Many problems cannot be efficiently parallelized, and increasing the number of cores does not necessarily translate to faster execution due to communication overhead and memory bottlenecks.

These challenges collectively signal the end of Moore's Law as a reliable predictor of computing advancements. As classical systems reach their physical and architectural limits, alternative computing paradigms such as quantum computing, neuromorphic computing, and optical computing are being explored to sustain the next wave of technological progress.

Quantum computing provides an alternative approach to computation by leveraging the principles of quantum mechanics. Unlike classical computers, which rely on binary bits to process information sequentially or in parallel, quantum computers utilize qubits, which exist in superposition and can represent multiple states simultaneously. This fundamental difference enables quantum systems to perform computations beyond the limits of classical machines, offering new pathways for solving problems that become intractable as Moore's Law slows down.

One of the key advantages of quantum computing is its exponential speedup for certain types of problems. Classical algorithms struggle with problems that require massive amounts of computational resources, such as integer factorization and unstructured search. Quantum algorithms like Shor's Algorithm can factor large numbers exponentially faster than classical algorithms, which has profound implications for cryptography. Similarly, Grover's Algorithm provides a quadratic speedup for searching unsorted databases. In addition to these fundamental problems, quantum computing is particularly effective in quantum simulation, where it can model quantum systems that classical computers cannot efficiently simulate. This is critical for advancements in material science, drug discovery, and chemistry, where researchers need to understand molecular interactions at the quantum level. Quantum computing also excels at optimization problems, using algorithms such as the Quantum Approximate Optimization Algorithm (QAOA) and the Variational Quantum Eigensolver (VQE) to solve complex combinatorial problems in logistics, finance, and artificial intelligence.

As full-scale quantum computers are still in development, hybrid quantum-classical computing is emerging as a practical solution for bridging the gap between current quantum capabilities and real-world applications. Instead of replacing classical computers entirely, quantum accelerators act as specialized co-processors that handle specific tasks more efficiently than classical systems. Quantum Processing Units (QPUs) can assist in optimization, cryptography, and AI model training while working alongside traditional CPUs and GPUs. The cloud-based quantum computing model is also gaining traction, with companies like IBM, Google, and Amazon providing Quantum-as-a-Service (QaaS) solutions. These platforms allow businesses, researchers, and developers to access quantum resources remotely, making quantum computing more widely available without requiring dedicated quantum hardware.

Quantum computing is also expected to revolutionize artificial intelligence and machine learning by enhancing computational efficiency in ways that classical systems cannot match. Quantum Machine Learning (QML) introduces new methods for improving pattern recognition, data classification, and feature selection. Quantum neural networks (QNNs) and quantum kernel methods offer advantages in training deep learning models, enabling AI systems to process large datasets more efficiently. One of the biggest bottlenecks in AI development is the time required for training deep learning models, and quantum computing has the potential to significantly accelerate this process, making AI systems more powerful and scalable.

Beyond computation, quantum computing is expected to drive breakthroughs in post-silicon materials and technologies. As the limitations of traditional semiconductors become more apparent, researchers are turning to quantum simulations to develop new materials that could replace silicon-based processors. Quantum computers can help identify high-temperature superconductors, which could lead to ultra-efficient computing systems with minimal energy loss. Additionally, quantum simulations of nanotechnology and quantum materials could enable innovations in next-generation computing architectures, leading to faster, more efficient, and more sustainable computing solutions.

As Moore's Law slows, quantum computing presents a path forward for maintaining exponential growth in computational power. Whether through quantum simulation, hybrid computing, or AI acceleration, quantum technology is poised to redefine the future of computing by overcoming classical limitations and unlocking new possibilities in science, engineering, and industry.

Conclusion

Recap of Key Takeaways

Quantum computing represents a fundamental shift in computational power and problem-solving capabilities. One of its primary advantages is the exponential speedup it offers for certain types of problems that classical computers struggle with. Traditional algorithms often require immense computational resources to tackle problems such as integer factorization and unstructured search. Quantum algorithms like Shor's Algorithm can factor large numbers exponentially faster than classical methods, presenting significant implications for cryptography. Similarly, Grover's Algorithm enables quadratic speedup for searching unsorted databases. Beyond these foundational applications, quantum computing is particularly effective for quantum simulations, which allow for modelling complex quantum systems that classical computers cannot efficiently process. This is crucial for advancements in material science, drug discovery, and chemistry, where understanding molecular interactions at the quantum level is essential.

Despite the promise of quantum computing, full-scale quantum systems are still under development. Hybrid quantum-classical computing has emerged as a practical bridge between existing quantum capabilities and real-world applications. Rather than replacing classical computers, quantum processors act as specialized co-processors that accelerate specific tasks more efficiently than classical systems. Quantum Processing Units (QPUs) can assist with optimization, cryptography, and artificial intelligence model training while working alongside traditional CPUs and GPUs. Cloud-based quantum computing is also gaining momentum, with companies like IBM, Google, and Amazon offering Quantum-as-a-Service (QaaS) platforms. These services allow researchers, businesses, and developers to remotely access quantum resources, eliminating the need for dedicated hardware.

Quantum computing is poised to revolutionize artificial intelligence and machine learning by significantly enhancing computational efficiency. Quantum Machine Learning (QML) introduces new methods for pattern recognition, data classification, and feature selection. Quantum neural networks (QNNs) and quantum kernel methods hold promise for training deep learning models more efficiently. One of the most pressing challenges in AI development is the lengthy training time for deep learning models. Quantum computing has the potential to dramatically accelerate this process, enabling more powerful and scalable AI systems.

Beyond computation, quantum computing is expected to drive major breakthroughs in post-silicon materials and computing architectures. As classical semiconductor technology reaches its physical limits, researchers are turning to quantum simulations to develop alternative materials. Quantum computers could help discover high-temperature superconductors, which could lead to ultra-efficient computing systems with minimal energy loss. Additionally, quantum simulations of nanotechnology and advanced quantum materials could facilitate innovations in next-generation computing hardware, leading to faster, more efficient, and more sustainable systems.

With the slowing of Moore’s Law, quantum computing offers a pathway for continued exponential growth in computational power. Whether through quantum simulation, hybrid computing, or AI acceleration, quantum technology is expected to redefine the future of computing by overcoming classical limitations and unlocking new possibilities across various scientific, engineering, and industrial domains.

The Path Forward for Researchers and Developers

As quantum computing transitions from theoretical exploration to practical implementation, researchers and developers face a unique set of challenges and opportunities. Advancements in quantum hardware, software, and applications are critical to unlocking quantum advantage and making quantum systems viable for real-world problems. The path forward requires collaboration between academia, industry, and government initiatives, focusing on scalability, error correction, software development, and quantum applications.

The advancement of quantum hardware is a prominent research challenge, particularly in developing scalable and reliable quantum systems. Current quantum processors, such as superconducting qubits and trapped-ion qubits, face several challenges, including limited qubit counts and significant noise and decoherence impacting computational power. Efforts by leading companies focus on mitigating these issues. IBM's 1,121-qubit Condor chip and Google's strategy for error-corrected quantum computing represent significant milestones toward scalability in superconducting qubit technology [219, 685]. Trapped-ion quantum systems, such as those being developed by IonQ and Quantinuum, show promise due to their long coherence times and high fidelity operations, albeit with the caveat of requiring precise laser control for effective scalability [686].

Furthermore, the exploration of photonic quantum computing, as pursued by companies like Xanadu and PsiQuantum, presents an innovative approach leveraging existing fibre-optic networks for potential large-scale implementations [564]. Microsoft's work on topological qubits aims at achieving enhanced error resistance and scalability, positioning this technology as a frontrunner if successfully realized [687]. In summary, the evolution in qubit technology, through advancements in superconducting, trapped-ion, photonic, and topological qubits, is pivotal for addressing the limitations of current quantum processors. However, for quantum hardware to transition into widely usable systems, improvements in qubit connectivity, noise reduction, and the creation of modular designs are essential [688].

In parallel with hardware developments, effective quantum error correction (QEC) methodologies must be advanced to attain fault tolerance in quantum computing. Modern quantum processors generally exhibit high error rates, constraining the quantum circuit depth that can be executed reliably. Surface codes are the most extensively studied QEC approach, permitting the encoding of logical qubits across multiple physical qubits to enable error detection and correction without direct measurement, with entities like IBM and Google actively implementing this strategy [689]. Other innovative methods under exploration include bosonic codes and topological quantum computing, which focus on utilizing continuous-variable quantum states and Majorana fermions, respectively, to enhance error resistance [585].

Additionally, leveraging machine learning for adaptive error correction dynamically adjusts quantum gate operations in real-time, which could be crucial in mitigating errors as they occur [478, 690]. Therefore, developing robust QEC techniques is crucial for catalysing the transition from current noisy intermediate-scale quantum (NISQ) devices toward fully functional fault-tolerant quantum systems.

The development of quantum software is equally crucial for practical quantum computing applications. Open-source quantum programming frameworks such as IBM's Qiskit, Google's Cirq, and Microsoft's Q# equip researchers and developers with the necessary tools to create algorithms suited for both near-term and future quantum hardware [691]. Hybrid quantum-classical computing strategies are being increasingly investigated, combining quantum computations with classical resources to enhance overall solving capabilities across various domains, including logistics and materials science [692]. The exploration of variational quantum algorithms (VQAs), like the Variational Quantum Eigensolver (VQE) and Quantum Approximate Optimization Algorithm (QAOA), showcases potential applications in chemistry simulations and combinatorial optimization [693].

Moreover, integrating quantum machine learning (QML) methodologies with classical algorithm frameworks holds promise for improving data analysis and recognition tasks [690].

Consequently, the ongoing progress in quantum programming and algorithmic approaches is pivotal in realizing more accessible and practical quantum computing solutions.

Lastly, quantum computing's influence across diverse industries, including healthcare, finance, logistics, and cybersecurity, underlines the necessity for expanding quantum applications. Companies like Pfizer and Merck harness quantum simulations to expedite drug discovery processes, while financial institutions explore quantum algorithms for enhanced risk management and trading strategies [694]. Quantum algorithms are also proving beneficial in optimizing logistical operations, thereby improving operational efficiency and reducing costs [686]. The urgency for post-quantum cryptography emerges from the potential risks posed by quantum systems to classical encryption mechanisms, prompting active development in this area [695].

Finally, the emergence of Quantum-as-a-Service (QaaS) platforms significantly democratizes access to quantum computing resources, enabling wider participation in quantum research without the burden of substantial infrastructure investments. Organizations like IBM, Google, and Microsoft provide remote access to diverse quantum processors through their cloud services, fostering an ecosystem conducive to innovation and collaboration [562, 564]. The support from government initiatives worldwide, such as the U.S. National Quantum Initiative and the European Quantum Flagship, is integral to advancing these technologies further [685].

Achieving quantum advantage, where quantum computers outperform classical systems for meaningful real-world tasks, remains a key goal for researchers. To reach this milestone, several critical advancements are necessary. One of the primary objectives is improving qubit coherence times and gate fidelities. Quantum systems are highly sensitive to environmental disturbances, and increasing the stability of qubits while minimizing errors is essential for reliable quantum computation. Enhancing gate fidelities will allow quantum operations to be performed with greater accuracy, reducing the impact of noise and improving overall system performance.

Another major challenge is advancing quantum error correction to achieve fault tolerance. Current quantum computers are prone to errors due to decoherence and noise, which limits their computational depth. Developing efficient quantum error correction codes will enable deeper quantum circuits, allowing quantum processors to perform complex computations without excessive errors. This requires a significant increase in the number of physical qubits dedicated to error correction, making scalability an equally important focus.

Scaling up quantum hardware while maintaining low noise and high connectivity between qubits is essential for practical quantum computing. As researchers work to build quantum processors with thousands or even millions of qubits, ensuring that these qubits interact efficiently without introducing unwanted interference is a significant technical hurdle. Approaches such as modular quantum computing and quantum networking are being explored to address these scalability challenges.

Beyond hardware improvements, the development of practical quantum applications that demonstrate real-world value is crucial. Quantum computing must move beyond theoretical algorithms to solving tangible problems in business, science, and engineering. Fields such as quantum chemistry, financial modelling, logistics optimization, and quantum artificial intelligence offer promising areas where quantum advantage can be realized.

Ensuring widespread access to quantum computing through cloud-based platforms and educational initiatives is another key step in fostering the quantum revolution. Quantum cloud services from IBM, Google, Amazon, and Microsoft provide researchers and businesses with remote access to quantum processors, enabling experimentation without requiring physical quantum hardware. Expanding quantum education programs will also be vital in preparing the next generation of scientists and engineers to develop and implement quantum technologies.

The next decade will be critical in shaping the future of quantum computing, as researchers and developers continue pushing the boundaries of hardware, algorithms, and applications. As the field progresses, quantum computing will transition from experimental research to commercial adoption, driving innovation across multiple industries. With continued advancements in scalability, error correction, and quantum software, the long-term vision of practical, fault-tolerant quantum computing is steadily becoming a reality.

How to Get Involved in Quantum Computing

Quantum computing is a rapidly evolving field that presents numerous opportunities for students, researchers, developers, and industry professionals to contribute to its advancement. Whether individuals are new to the field or have experience in computing, mathematics, or physics, multiple pathways exist for getting involved in quantum computing.

The first step in engaging with quantum computing is to develop a strong foundation in quantum mechanics, linear algebra, and computer science. Since quantum computing operates on principles different from classical computing, understanding concepts such as superposition, entanglement, and quantum gates is essential. Mathematics and physics are key areas of study, particularly linear algebra, probability theory, and quantum mechanics. A background in computer science, including knowledge of classical computing, algorithms, and programming fundamentals, is also beneficial. Many platforms, including edX, Coursera, Udacity, and MIT OpenCourseWare, offer quantum computing courses taught by leading experts.

Gaining hands-on experience with quantum programming is another crucial step. Several open-source quantum computing frameworks provide access to real quantum hardware and simulators. IBM offers Qiskit, a Python-based framework for developing quantum applications on IBM Quantum processors. Google's Cirq is optimized for near-term quantum applications, while Microsoft's Q# and Quantum Development Kit are integrated with Azure

Quantum. Amazon Braket SDK allows users to develop quantum programs across multiple quantum hardware providers, and PennyLane supports quantum machine learning and hybrid quantum-classical computing. Each of these platforms provides extensive documentation and tutorials to help beginners write and execute quantum algorithms.

Many quantum computing companies provide cloud-based access to real quantum hardware, allowing users to experiment with quantum programming without requiring access to physical quantum processors. IBM Quantum Experience offers free cloud access to IBM's quantum processors, while Google Quantum AI provides cloud-based quantum computing through Google Cloud. Microsoft Azure Quantum supports multiple quantum hardware backends, and Amazon Braket offers access to quantum processors from IonQ, Rigetti, and D-Wave. D-Wave Leap provides quantum annealing access for optimization problems. Most of these platforms offer free access with limited resources, enabling users to test small-scale quantum circuits and gain practical experience.

Participating in quantum hackathons and competitions is an excellent way to collaborate with other quantum enthusiasts, solve real-world problems, and gain hands-on experience. IBM hosts the Quantum Challenge, which periodically provides tasks that can be solved using Qiskit. MIT iQuHACK is an annual hackathon focused on quantum computing, while QHack by Xanadu is a quantum programming competition emphasizing machine learning. The Quantum Open Source Foundation (QOSF) Mentorship Program provides structured learning opportunities for beginners to contribute to open-source quantum projects. These events often offer mentorship, educational workshops, and networking opportunities with quantum professionals.

Engaging with the quantum computing community can accelerate learning and provide valuable networking opportunities. The IBM Quantum Community includes forums, Slack channels, and local meetups for quantum computing discussions. Quantum Computing Stack Exchange is a Q&A platform dedicated to quantum topics. The Qiskit Advocates Program brings together a global community of quantum computing enthusiasts. Organizations like the Quantum Open Source Foundation (QOSF) support open-source quantum computing projects. Additionally, LinkedIn and Discord host various industry and academic quantum computing groups. By joining these communities, participants can collaborate on research, stay updated on industry trends, and explore potential career opportunities.

Quantum computing offers career paths in academia, research, and industry. Depending on their background and expertise, individuals can explore different pathways. In academia and research, students can pursue undergraduate, master's, or Ph.D. programs in quantum computing, physics, or computer science. Many universities, including MIT, Stanford, the University of Waterloo, and Delft University, offer specialized quantum research programs. In the industry, companies such as IBM, Google, Microsoft, Amazon, and startups like Rigetti, Xanadu, and IonQ are actively hiring quantum researchers, software developers, and engineers. Government institutions such as NASA, NIST, and DARPA are also investing in quantum computing research, particularly for defense, cybersecurity, and national security applications. Many companies offer internships and fellowships for students and early-career professionals looking to enter the field.

Contributing to open-source quantum computing projects provides a practical way to gain experience and collaborate with experts. Qiskit (IBM Quantum), Cirq (Google Quantum AI), PennyLane (Xanadu), ProjectQ (ETH Zurich), and D-Wave Ocean SDK are some of the leading open-source projects in quantum computing. Contributing to these projects on GitHub can help individuals build a portfolio and connect with experienced quantum developers.

Staying updated on quantum computing advancements is crucial, given the rapid pace of innovation in the field. Following research papers on arXiv.org, which has a dedicated quantum physics section, can provide insight into the latest breakthroughs. Subscribing to newsletters and blogs such as *Quantum Zeitgeist*, *Quantum Computing Report*, and the *IBM Quantum Blog* helps keep track of industry news and developments. Attending quantum computing conferences, such as the IEEE Quantum Computing Conference, Quantum Tech Congress, and APS March Meeting, provides opportunities to learn about cutting-edge research, network with professionals, and explore industry trends.

Quantum computing presents exciting opportunities for students, researchers, and professionals to contribute to the next era of computing. Whether through self-learning, academic programs, open-source contributions, or industry involvement, multiple avenues exist to get started. As quantum hardware, software, and applications continue to evolve, those who invest time and effort in learning today will be well-positioned to shape the future of quantum technology.

Final Thoughts on the Future of Quantum

Quantum computing represents one of the most transformative technological advancements of the 21st century. While it is still in its early stages, rapid progress in quantum hardware, software, and applications suggests that we are moving closer to realizing its full potential. The transition from experimental research to practical implementation will require continued breakthroughs in quantum error correction, qubit scalability, and hybrid quantum-classical computing.

The future of quantum computing will likely unfold in phases. In the near term, quantum devices will remain noisy and primarily serve as accelerators for specific tasks such as optimization, cryptography, and quantum simulations. Quantum cloud computing will continue to democratize access, enabling researchers, startups, and enterprises to experiment with quantum algorithms without the need for specialized hardware. The integration of quantum computing with artificial intelligence, cybersecurity, and materials science will lead to practical use cases that provide a competitive advantage in these industries.

As we move toward the long-term vision of fault-tolerant, universal quantum computing, industries will need to prepare for both the opportunities and challenges that come with this technology. Financial institutions must adopt quantum-resistant encryption to safeguard digital transactions. Pharmaceutical companies will leverage quantum simulations to develop new drugs and treatments more efficiently. Governments and national security agencies must address the geopolitical implications of quantum computing, particularly in the fields of cybersecurity and quantum communications.

One of the most significant challenges will be ensuring equitable access to quantum computing. Without international collaboration, there is a risk that only a few technologically advanced nations and corporations will dominate the field, exacerbating global inequalities in science and technology. Open-source quantum initiatives, educational programs, and cloud-based quantum services will play a crucial role in making quantum technology more inclusive and widely available.

Despite the obstacles ahead, the trajectory of quantum computing is clear. As research institutions, technology companies, and governments invest in this field, quantum computing will gradually move from theory to application. The coming decades will witness the emergence of quantum-enhanced industries, unlocking new frontiers in computing power that were previously unimaginable.

For those entering the field, now is the time to learn, experiment, and contribute to the growing quantum ecosystem. Whether through academic research, software development, or engineering advancements, the individuals and organizations shaping quantum computing today will define the future of computing for generations to come.

Appendices

Appendix A - Mathematical Foundations of Quantum Computing (Linear Algebra, Probability, Hilbert Spaces)

Quantum computing relies heavily on mathematical concepts from linear algebra, probability theory, and Hilbert spaces. Understanding these foundations is essential for working with quantum states, quantum gates, and quantum algorithms.

1. Linear Algebra for Quantum Computing

Linear algebra is the backbone of quantum mechanics and quantum computing. Qubits, quantum gates, and quantum circuits are represented using vectors and matrices.

Linear algebra is a fundamental branch of mathematics that deals with vectors, matrices, vector spaces, and linear transformations. It provides essential tools for solving systems of linear equations, understanding geometric transformations, and analysing mathematical models in various scientific and engineering fields.

1.1 What is Linear Algebra?

Linear algebra is the study of linear equations, vector spaces, linear transformations, and matrices. It allows us to manipulate multi-dimensional data efficiently and is widely used in computer science, physics, machine learning, cryptography, and quantum mechanics.

1.1.1. What is a Linear Equation?

A linear equation is an equation where each term is either a constant or a product of a constant and a variable, and there are no exponents or products of variables.

The general form of a linear equation is:

$$u_1x_1 + u_2x_2 + \cdots + u_nx_n = v$$

where:

- $u_1, u_2, \ldots, u_n$ are coefficients,
- $x_1, x_2, \ldots, x_n$ are unknown variables,
- v is a constant.

For example, the equation:

$$2x + 3y = 6$$

is a linear equation because it represents a straight line when plotted on a graph.

A system of linear equations consists of multiple linear equations with the same set of variables. The goal is to find the values of the variables that satisfy all equations simultaneously.

For example:

$$2x + 3y = 6$$

$$4x + 6y = 12$$

These two equations form a system that can be solved to find *x* and *y*.

1.1.2. Key Concepts in Linear Algebra

1.1.2.1. Vectors and Vector Spaces

A vector is a mathematical object that represents both a magnitude and a direction. In two dimensions, a vector can be written as:

$$\mathbf{v} = \begin{bmatrix} x \\ y \end{bmatrix}$$

A vector space is a set of vectors that can be added together and multiplied by scalars (real numbers). A vector space must satisfy properties like closure under addition and scalar multiplication.

Example of vector addition:

$$\begin{bmatrix} 2 \\ 3 \end{bmatrix} + \begin{bmatrix} 1 \\ 4 \end{bmatrix} = \begin{bmatrix} 3 \\ 7 \end{bmatrix}$$

1.1.2.2. Matrices and Matrix Operations

A matrix is a rectangular array of numbers, arranged in rows and columns. Matrices are used to represent linear transformations and systems of equations.

A general matrix is written as:

$$A = \begin{bmatrix} a_{11} & a_{12} \\ a_{21} & a_{22} \end{bmatrix}$$

Matrix Addition and Subtraction

Matrices of the same size can be added or subtracted by adding or subtracting corresponding elements.

Example:

$$A = \begin{bmatrix} 1 & 2 \\ 3 & 4 \end{bmatrix}, \quad B = \begin{bmatrix} 5 & 6 \\ 7 & 8 \end{bmatrix}$$

$$A + B = \begin{bmatrix} 1+5 & 2+6 \\ 3+7 & 4+8 \end{bmatrix} = \begin{bmatrix} 6 & 8 \\ 10 & 12 \end{bmatrix}$$

Matrix Multiplication

The product of two matrices *A* and *B* is found by multiplying the rows of *A* by the columns of *B*.

Example:

$$A = \begin{bmatrix} 1 & 2 \\ 3 & 4 \end{bmatrix}, \quad B = \begin{bmatrix} 5 & 6 \\ 7 & 8 \end{bmatrix}$$

$$AB = \begin{bmatrix} (1 \cdot 5 + 2 \cdot 7) & (1 \cdot 6 + 2 \cdot 8) \\ (3 \cdot 5 + 4 \cdot 7) & (3 \cdot 6 + 4 \cdot 8) \end{bmatrix}$$

$$= \begin{bmatrix} 19 & 22 \\ 43 & 50 \end{bmatrix}$$

1.1.3. Eigenvalues and Eigenvectors

An eigenvector of a square matrix is a vector that only gets stretched or shrunk when the matrix is applied to it, rather than changing direction. The factor by which it is stretched is called the eigenvalue.

Mathematically, if *A* is a square matrix and **v** is an eigenvector with eigenvalue λ, then:

$$A\mathbf{v} = \lambda\mathbf{v}$$

Example: Find the Eigenvalues of a Matrix

Given:

$$A = \begin{bmatrix} 3 & 0 \\ 8 & 6 \end{bmatrix}$$

To find the eigenvalues, solve:

$$\det(A - \lambda I) = 0$$

$$\begin{vmatrix} 3-\lambda & 0 \\ 8 & 6-\lambda \end{vmatrix} = 0$$

Expanding the determinant:

$$(3-\lambda)(6-\lambda) - (8 \cdot 0) = 0$$

$$(3-\lambda)(6-\lambda) = 0$$

Solving for λ:

$$\lambda = 3, 6$$

Thus, the eigenvalues of *A* are 3 and 6.

1.1.4. Solving Systems of Linear Equations

A system of equations can be solved using Gaussian elimination, matrix inverses, or Cramer's rule.

Example: Solve a System of Equations

$$2x + 3y = 5$$

$$4x - y = 11$$

Using the matrix form:

$$\begin{bmatrix} 2 & 3 \\ 4 & -1 \end{bmatrix} \begin{bmatrix} x \\ y \end{bmatrix} = \begin{bmatrix} 5 \\ 11 \end{bmatrix}$$

Solving using elimination:

Multiply the first equation by **2**:

$$4x + 6y = 10$$

Now subtract from the second equation:

$$(4x - y) - (4x + 6y) = 11 - 10$$

$$-7y = 1 \quad \Rightarrow \quad y = -\frac{1}{7}$$

Substituting into the first equation:

$$2x + 3(-1/7) = 5$$

$$2x - 3/7 = 5$$

$$2x = 5 + 3/7 = 38/7$$

$$x = 19/7$$

Thus, the solution is:

$$x = \frac{19}{7}, \quad y = -\frac{1}{7}$$

Linear algebra is a powerful tool used in numerous applications, from solving equations to designing AI algorithms. By understanding vectors, matrices, eigenvalues, and transformations, we can unlock a wide range of mathematical and computational techniques that power modern technology.

1.2. Vectors and State Representations

A qubit (quantum bit) is represented as a two-dimensional complex vector in a Hilbert space. In the Dirac notation (bra-ket notation), a qubit state is written as:

$$|\psi\rangle = \alpha|0\rangle + \beta|1\rangle$$

where:

- $|0\rangle = \begin{bmatrix} 1 \\ 0 \end{bmatrix}$,
- $|1\rangle = \begin{bmatrix} 0 \\ 1 \end{bmatrix}$,
- α, β are complex numbers satisfying $|\alpha|^2 + |\beta|^2 = 1$ (normalization condition).

For example, the Hadamard state (equal superposition of |0⟩ and |1⟩ is:

$$|+\rangle = \frac{1}{\sqrt{2}} \begin{bmatrix} 1 \\ 1 \end{bmatrix}$$

In quantum computing, state representations describe the state of a quantum system, and these states are mathematically represented as vectors in a complex vector space called Hilbert space. Understanding vectors and their role in quantum computing is crucial for working with qubits, quantum gates, and quantum circuits.

A vector is a mathematical object that has both magnitude and direction. In quantum mechanics, a vector represents the state of a quantum system. For example, the state of a qubit (the basic unit of quantum information) is a vector in a two-dimensional complex space.

The most fundamental quantum states are called basis states, typically denoted as:

$$|0\rangle = \begin{bmatrix} 1 \\ 0 \end{bmatrix}, \quad |1\rangle = \begin{bmatrix} 0 \\ 1 \end{bmatrix}$$

These vectors represent the classical states 0 and 1, similar to classical bits.

A qubit can exist in a superposition of both |0⟩ and |1⟩, meaning it is represented as:

$$|\psi\rangle = \alpha|0\rangle + \beta|1\rangle$$

where:

- α and β are complex numbers that determine the probability amplitudes of measuring the qubit in state |0⟩ or |1⟩|.
- These amplitudes must satisfy the normalization condition:

$$|\alpha|^2 + |\beta|^2 = 1$$

This ensures that the total probability of measuring the qubit in any state is 100%.

Example: Superposition State

If a qubit is in the equal superposition state:

$$|\psi\rangle = \frac{1}{\sqrt{2}}|0\rangle + \frac{1}{\sqrt{2}}|1\rangle$$

Then the probabilities of measuring the qubit are:

- $P(0) = \left|\frac{1}{\sqrt{2}}\right|^2 = \frac{1}{2}$ (50%)
- $P(1) = \left|\frac{1}{\sqrt{2}}\right|^2 = \frac{1}{2}$ (50%)

This means if we measure the qubit many times, we expect to see |0⟩ and |1⟩| with equal frequency.

In quantum mechanics, we use Dirac notation to represent quantum states:

- A column vector (state vector) is written as a ket: |ψ⟩.
- The conjugate transpose (row vector) is written as a bra: ⟨ψ|

For example:

$$|0\rangle = \begin{bmatrix}1\\0\end{bmatrix}, \quad \langle 0| = \begin{bmatrix}1 & 0\end{bmatrix}$$

$$|1\rangle = \begin{bmatrix}0\\1\end{bmatrix}, \quad \langle 1| = \begin{bmatrix}0 & 1\end{bmatrix}$$

This notation is useful when computing inner products and expectation values.

Inner Products and Norms

Inner Product

The inner product between two quantum states $|\psi\rangle$ and $|\phi\rangle$ is given by:

$$\langle\psi|\phi\rangle = \sum_i \psi_i^* \phi_i$$

where ψ_i^* is the complex conjugate of ψ_i .

Example: Inner Product Calculation

Let's compute the inner product of $|0\rangle$ and $|1\rangle$:

$$\langle 0|1\rangle = \begin{bmatrix}1 & 0\end{bmatrix} \cdot \begin{bmatrix}0\\1\end{bmatrix} = (1 \times 0) + (0 \times 1) = 0$$

Since the inner product is zero, this means $|0\rangle$ and $|1\rangle$ are orthogonal.

Norm (Length of a Vector)

The norm (or length) of a quantum state $|\psi\rangle$ is:

$$||\psi|| = \sqrt{\langle\psi|\psi\rangle}$$

For example, the norm of $|0\rangle$ is:

$$||0|| = \sqrt{\langle 0|0\rangle} = \sqrt{1} = 1$$

Quantum states must always be **normalized,** meaning their norm must be **1.**

Multi-Qubit Systems and Tensor Products

For systems with multiple qubits, we use the tensor product to describe their states.

For two qubits:

$$|00\rangle = |0\rangle \otimes |0\rangle = \begin{bmatrix}1\\0\end{bmatrix} \otimes \begin{bmatrix}1\\0\end{bmatrix} = \begin{bmatrix}1\\0\\0\\0\end{bmatrix}$$

Similarly:

$$|01\rangle = \begin{bmatrix}0\\1\end{bmatrix} \otimes \begin{bmatrix}1\\0\end{bmatrix} = \begin{bmatrix}0\\1\\0\\0\end{bmatrix}$$

For a general two-qubit state:

$$|\psi\rangle = \alpha|00\rangle + \beta|01\rangle + \gamma|10\rangle + \delta|11\rangle$$

where $\alpha, \beta, \gamma, \delta$ are complex numbers satisfying:

$$|\alpha|^2 + |\beta|^2 + |\gamma|^2 + |\delta|^2 = 1$$

This is essential for representing entangled states, such as:

$$|\Phi^+\rangle = \frac{1}{\sqrt{2}}(|00\rangle + |11\rangle)$$

which is an example of a Bell state, a fundamental entangled quantum state.

1.3. Inner Product and Norms

The inner product (dot product in complex space) between two vectors $|\psi\rangle$ and $|\phi\rangle$ is:

$$\langle\psi|\phi\rangle = \sum_i a_i^* b_i$$

where a_i^* is the complex conjugate of a_i.

The norm (or length) of a quantum state is given by:

$$||\psi|| = \sqrt{\langle\psi|\psi\rangle}$$

This ensures that quantum states remain normalized.

Inner products and norms are fundamental mathematical concepts used in linear algebra, which forms the basis of quantum mechanics and quantum computing. These concepts help us understand quantum state normalization, probabilities, and measurement outcomes.

The inner product measures the similarity between two vectors. In quantum computing, it helps determine how much one quantum state overlaps with another.

In Dirac notation (bra-ket notation), the inner product between two quantum states $|\psi\rangle$ and $|\phi\rangle$ is written as:

$$\langle\psi|\phi\rangle$$

Mathematically, if we represent quantum states as column vectors, the inner product is computed as:

$$\langle\psi|\phi\rangle = \sum_i \psi_i^* \phi_i$$

where:

- ψ_i^* is the complex conjugate of ψ_i .
- The sum runs over all components of the vectors.

The inner product is useful for:

- Checking orthogonality: If $\langle\psi|\phi\rangle = 0$, the states are orthogonal (completely different).
- Computing probabilities: The probability of measuring one state in another is given by the square of the absolute value of their inner product.

Example 1: Inner Product of Basis States

Let's compute the inner product of the two computational basis states $|0\rangle$ and $|1\rangle$.

The standard basis states are:

$$|0\rangle = \begin{bmatrix} 1 \\ 0 \end{bmatrix}, \quad |1\rangle = \begin{bmatrix} 0 \\ 1 \end{bmatrix}$$

The inner product:

$$\langle 0|1\rangle = \begin{bmatrix} 1 & 0 \end{bmatrix} \cdot \begin{bmatrix} 0 \\ 1 \end{bmatrix}$$

$$= (1 \times 0) + (0 \times 1) = 0$$

Since the inner product is zero, the states are orthogonal, meaning they are completely distinct.

Example 2: Inner Product of Superposition States

Consider the Hadamard state, which is an equal superposition of |0⟩ and |1⟩:

$$|+\rangle = \frac{1}{\sqrt{2}}(|0\rangle + |1\rangle)$$

We want to compute the inner product between |+⟩ and |0⟩.

$$\langle +|0\rangle = \left(\frac{1}{\sqrt{2}}\langle 0| + \frac{1}{\sqrt{2}}\langle 1|\right)|0\rangle$$

Using ⟨0|0⟩=1 and ⟨1|0⟩=0, we get:

$$\langle +|0\rangle = \frac{1}{\sqrt{2}} \times 1 + \frac{1}{\sqrt{2}} \times 0 = \frac{1}{\sqrt{2}}$$

This means |+⟩ has a nonzero overlap with |0⟩, so they are not orthogonal.

The norm (or length) of a quantum state vector is a measure of its magnitude. In quantum mechanics, all valid quantum states must have a norm of 1 (they must be normalized).

The norm of a quantum state $|\psi\rangle$ is given by:

$$\||\psi\|| = \sqrt{\langle\psi|\psi\rangle}$$

Since quantum states must be normalized, we require:

$$\langle\psi|\psi\rangle = 1$$

This ensures that the total probability of measuring any possible outcome is 100%.

Example 3: Checking if a State is Normalized

Consider the quantum state:

$$|\psi\rangle = \frac{1}{\sqrt{5}}\begin{bmatrix}2\\1\end{bmatrix}$$

To check if it is normalized, compute:

$$\langle\psi|\psi\rangle = \left(\frac{1}{\sqrt{5}}\begin{bmatrix}2 & 1\end{bmatrix}\right)\cdot\left(\frac{1}{\sqrt{5}}\begin{bmatrix}2\\1\end{bmatrix}\right)$$
$$= \frac{1}{5}\left(2^2+1^2\right) = \frac{1}{5}\times(4+1) = \frac{5}{5} = 1$$

Since $\langle\psi|\psi\rangle = 1$, this state is properly normalized.

Example 4: Normalizing a State

Suppose we have an unnormalized quantum state:

$$|\phi\rangle = \begin{bmatrix}3\\4\end{bmatrix}$$

The norm is:

$$\|\phi\| = \sqrt{3^2+4^2} = \sqrt{9+16} = \sqrt{25} = 5$$

To normalize it, we divide by its norm:

$$|\phi_{\text{normalized}}\rangle = \frac{1}{5}\begin{bmatrix}3\\4\end{bmatrix} = \begin{bmatrix}\frac{3}{5}\\\frac{4}{5}\end{bmatrix}$$

Now:

$$\left(\frac{3}{5}\right)^2 + \left(\frac{4}{5}\right)^2 = \frac{9}{25} + \frac{16}{25} = \frac{25}{25} = 1$$

This confirms the state is properly normalized.

The inner product determines how much one quantum state overlaps with another. This is critical for understanding probability measurements in quantum mechanics.

For example, the probability of measuring state $|\phi\rangle$ when the system is in state $|\psi\rangle$ is:

$$P = |\langle\phi|\psi\rangle|^2$$

If the states are orthogonal (inner product = 0), the probability is zero.

Since quantum states represent probabilities, they must always be normalized $(\|\psi\| = 1)$. If a state is not normalized, measurement probabilities will not sum to 100%, violating fundamental rules of quantum mechanics.

Quantum operations (such as quantum gates) are represented by unitary matrices, which preserve inner products and norms. This ensures that quantum operations do not change the total probability of outcomes.

A matrix *U* is unitary if:

$$U^\dagger U = I$$

where $U^\dagger$ is the Hermitian conjugate (complex conjugate transpose) of *U*.

1.4. Matrices and Operators

Quantum gates and transformations are represented by unitary matrices *U*, which satisfy:

$$U^\dagger U = I$$

where $U^\dagger$ is the Hermitian conjugate (complex conjugate transpose of *U*). Examples of common quantum gates include:

- **Pauli Matrices**:

$$X = \begin{bmatrix} 0 & 1 \\ 1 & 0 \end{bmatrix}, \quad Y = \begin{bmatrix} 0 & -i \\ i & 0 \end{bmatrix}, \quad Z = \begin{bmatrix} 1 & 0 \\ 0 & -1 \end{bmatrix}$$

- **Hadamard Gate**:

$$H = \frac{1}{\sqrt{2}} \begin{bmatrix} 1 & 1 \\ 1 & -1 \end{bmatrix}$$

Applying the Hadamard gate to |0⟩:

$$H|0\rangle = \frac{1}{\sqrt{2}} \begin{bmatrix} 1 & 1 \\ 1 & -1 \end{bmatrix} \begin{bmatrix} 1 \\ 0 \end{bmatrix} = \frac{1}{\sqrt{2}} \begin{bmatrix} 1 \\ 1 \end{bmatrix} = |+\rangle$$

In quantum computing, matrices and operators play a fundamental role in describing quantum states, quantum gates, and quantum measurements. Since quantum mechanics is a linear algebra-based theory, quantum operations are represented using matrices, and quantum states are expressed as vectors.

A matrix is a rectangular array of numbers arranged in rows and columns. Matrices are used in quantum computing to represent quantum gates and transformations that act on quantum states.

For example, a 2×2 matrix:

$$A = \begin{bmatrix} a & b \\ c & d \end{bmatrix}$$

can act on a 2D vector (which represents a single qubit):

$$|\psi\rangle = \begin{bmatrix} x \\ y \end{bmatrix}$$

When we multiply a matrix with a vector, we get a new vector:

$$A|\psi\rangle = \begin{bmatrix} a & b \\ c & d \end{bmatrix} \begin{bmatrix} x \\ y \end{bmatrix} = \begin{bmatrix} ax + by \\ cx + dy \end{bmatrix}$$

This operation transforms the quantum state $|\psi\rangle$, just like a quantum gate transforms a qubit.

An operator is a mathematical function that acts on quantum states, represented using matrices. Operators describe how quantum systems evolve, interact, and are measured.

Types of Quantum Operators:

1. Unitary Operators – Represent quantum gates (e.g., Hadamard, Pauli-X).
2. Hermitian Operators – Represent observable quantities (e.g., energy, spin).
3. Projection Operators – Describe measurements (probabilities of outcomes).

In quantum mechanics, all valid quantum operations must be unitary, meaning they preserve the norm of the quantum state.

A matrix *U* is unitary if:

$$U^\dagger U = I$$

where:

- $U^\dagger$ is the Hermitian conjugate (complex conjugate transpose).
- *I* is the identity matrix.

Example: Hadamard Gate (H)

The Hadamard gate creates superposition and is represented by the matrix:

$$H = \frac{1}{\sqrt{2}} \begin{bmatrix} 1 & 1 \\ 1 & -1 \end{bmatrix}$$

Applying *H* to $|0\rangle$:

$$H|0\rangle = \frac{1}{\sqrt{2}} \begin{bmatrix} 1 & 1 \\ 1 & -1 \end{bmatrix} \begin{bmatrix} 1 \\ 0 \end{bmatrix} = \frac{1}{\sqrt{2}} \begin{bmatrix} 1 \\ 1 \end{bmatrix} = |+\rangle$$

Now the qubit is in **equal superposition.**

Common Quantum Operators

1. Pauli Matrices (Quantum Bit Flips)

The Pauli matrices are fundamental operators in quantum mechanics.

1. **Pauli-X (Bit Flip)**

$$X = \begin{bmatrix} 0 & 1 \\ 1 & 0 \end{bmatrix}$$

Acts like a **classical NOT gate**, flipping |0⟩ to |1⟩ and vice versa.

2. **Pauli-Y (Phase & Bit Flip)**

$$Y = \begin{bmatrix} 0 & -i \\ i & 0 \end{bmatrix}$$

Combines bit-flipping and phase-flipping.

3. **Pauli-Z (Phase Flip)**

$$Z = \begin{bmatrix} 1 & 0 \\ 0 & -1 \end{bmatrix}$$

Flips the **phase** of |1⟩| while leaving |0⟩ unchanged.

Example: Applying the Pauli-X Operator

If we apply *X* to |0⟩:

$$X|0\rangle = \begin{bmatrix} 0 & 1 \\ 1 & 0 \end{bmatrix} \begin{bmatrix} 1 \\ 0 \end{bmatrix} = \begin{bmatrix} 0 \\ 1 \end{bmatrix} = |1\rangle$$

Applying it again:

$$X|1\rangle = \begin{bmatrix} 0 & 1 \\ 1 & 0 \end{bmatrix} \begin{bmatrix} 0 \\ 1 \end{bmatrix} = \begin{bmatrix} 1 \\ 0 \end{bmatrix} = |0\rangle$$

The Pauli-X gate acts like a NOT gate, swapping |0⟩ and |1⟩.

2. Measurement Operators (Observables)

In quantum mechanics, measuring a state collapses it to one of its possible outcomes. Measurement is represented by projection operators.

For a single qubit, measurement in the computational basis (|0⟩,|1⟩) is represented by the matrices:

$$M_0 = |0\rangle\langle 0| = \begin{bmatrix} 1 & 0 \\ 0 & 0 \end{bmatrix}, \quad M_1 = |1\rangle\langle 1| = \begin{bmatrix} 0 & 0 \\ 0 & 1 \end{bmatrix}$$

If the quantum state is:

$$|\psi\rangle = \alpha|0\rangle + \beta|1\rangle$$

Then the probability of measuring |0⟩ is:

$$P(0) = \langle\psi|M_0|\psi\rangle = |\alpha|^2$$

and the probability of measuring |1⟩ is:

$$P(1) = \langle\psi|M_1|\psi\rangle = |\beta|^2$$

Since total probability must be 1:

$$|\alpha|^2 + |\beta|^2 = 1$$

2. Probability Theory in Quantum Computing

Probability is the mathematical way of measuring how likely something is to happen. It helps us predict outcomes when there is uncertainty. For example, if you flip a coin, you don't know whether it will land on heads or tails, but probability can tell you that each outcome has a 50% chance.

How is Probability Measured?

Probability is always a number between 0 and 1:

- 0 means the event is impossible (e.g., rolling a 7 on a standard six-sided die).
- 1 means the event is certain (e.g., the sun rising tomorrow).
- 0.5 (or 50%) means the event has an equal chance of happening or not happening (e.g., flipping heads on a fair coin).

The formula for probability is:

$$P(A) = \frac{\text{Number of ways A can happen}}{\text{Total possible outcomes}}$$

Where:

- *P(A)* is the probability of event *A* happening.
- The number of ways A can happen refers to the favourable outcomes.
- The total possible outcomes are all the possible things that could happen.

Example 1: Coin Flip

When flipping a fair coin, there are two possible outcomes: heads or tails. Since both are equally likely, the probability of flipping heads is:

$$P(\text{Heads}) = \frac{1}{2} = 0.5 = 50\%$$

Similarly, the probability of flipping tails is also 0.5 (or 50%).

Example 2: Rolling a Die

If you roll a standard six-sided die, there are six possible outcomes: 1, 2, 3, 4, 5, or 6.

The probability of rolling a 3 is:

$$P(3) = \frac{1}{6} = 0.1667 = 16.67\%$$

The probability of rolling an even number (2, 4, or 6) is:

$$P(\text{even}) = \frac{3}{6} = 0.5 = 50\%$$

Basic Probability Rules

1. Rule of Complementary Events

The probability of something not happening is:

$$P(\text{Not A}) = 1 - P(A)$$

Example

If the probability of rain tomorrow is 0.3 (30%), then the probability of no rain is:

1 - 0.3 = 0.7 (70%)

2. The Addition Rule (OR Probability)

If you want to know the probability of either event *A* or event *B* happening, you use:

$$P(A \text{ or } B) = P(A) + P(B) - P(A \text{ and } B)$$

Example

In a deck of 52 playing cards, what is the probability of drawing a heart or a king?

- There are 13 hearts.
- There are 4 kings.
- One of those kings is a heart, so we don't count it twice.

3. The Multiplication Rule (AND Probability)

If you want to find the probability of two events happening together, you multiply their probabilities.

For independent events:

$$P(A \text{ and } B) = P(A) \times P(B)$$

Example

What is the probability of flipping two heads in a row?

$$P(\text{Heads on first flip}) = \frac{1}{2}$$

$$P(\text{Heads on second flip}) = \frac{1}{2}$$

$$P(\text{Both heads}) = \frac{1}{2} \times \frac{1}{2} = \frac{1}{4} = 25\%$$

For dependent events (where one outcome affects the other), the probabilities change.

Example: Drawing Two Aces from a Deck

The probability of drawing an Ace first from a deck of 52 cards is:

$$P(\text{First Ace}) = \frac{4}{52} = \frac{1}{13}$$

After drawing one ace, only 3 aces remain in 51 cards, so:

$$P(\text{Second Ace}) = \frac{3}{51}$$

Now multiply:

$$P(\text{Two Aces}) = \frac{4}{52} \times \frac{3}{51} = \frac{12}{2652} = \frac{1}{221} = 0.45\%$$

Expected Value (Average Outcome)

The expected value (EV) is the average result over many trials.

$$EV = \sum(\text{Value} \times \text{Probability})$$

Example: Expected Value of a Dice Roll

Each number 1, 2, 3, 4, 5, 6 on a die has a probability of $\frac{1}{6}$.

The expected value of rolling the die is:

$$EV = (1 \times \frac{1}{6}) + (2 \times \frac{1}{6}) + (3 \times \frac{1}{6}) + (4 \times \frac{1}{6}) + (5 \times \frac{1}{6}) + (6 \times \frac{1}{6})$$

$$EV = \frac{1+2+3+4+5+6}{6} = \frac{21}{6} = 3.5$$

So, on average, the result of rolling a die is 3.5, even though you can never actually roll a 3.5.

2.1. Born's Rule

Born's Rule is one of the most fundamental principles in quantum mechanics. It tells us how to calculate the probability of measuring a particular outcome when observing a quantum system. Unlike classical physics, where outcomes are deterministic (meaning they can be predicted exactly), quantum mechanics deals with probabilities.

When measuring a qubit in state $|\psi\rangle = \alpha|0\rangle + \beta|1\rangle$, the probability of obtaining |0⟩ or |1⟩ is given by Born's Rule:

$$P(0) = |\alpha|^2, \quad P(1) = |\beta|^2$$

For example, measuring |+⟩ gives:

$$P(0) = \left|\frac{1}{\sqrt{2}}\right|^2 = \frac{1}{2}, \quad P(1) = \left|\frac{1}{\sqrt{2}}\right|^2 = \frac{1}{2}$$

Born's Rule states that the probability of obtaining a specific measurement outcome in a quantum system is given by the square of the absolute value of the wavefunction's amplitude.

Mathematically, if a quantum system is in a state $|\psi\rangle$, and we measure an observable (like position or spin) associated with an eigenstate $|\phi\rangle$, the probability of getting that outcome is:

$$P(\phi) = |\langle\phi|\psi\rangle|^2$$

where:

- $P(\phi)$ is the probability of measuring the system in state $|\phi\rangle$.
- $\langle\phi|\psi\rangle$ is the inner product (overlap) between the quantum state $|\psi\rangle$ and the eigenstate $|\phi\rangle$.
- The absolute value squared $|\cdot|^2$ ensures the probability is a real, non-negative number.

In simple terms, Born's Rule tells us that quantum systems do not behave like classical objects. Instead of a definite outcome, a measurement gives different possible results with certain probabilities.

For example, if we measure the spin of an electron, it could be spin-up or spin-down, but we cannot predict with certainty which one we will get. Born's Rule tells us how to compute the probability of each outcome.

Worked Example: Single Qubit in Superposition

Consider a qubit in the quantum state:

$$|\psi\rangle = \frac{3}{5}|0\rangle + \frac{4}{5}|1\rangle$$

Here:

- The coefficient 3/5 is the amplitude for state $|0\rangle$.
- The coefficient 4/5 is the amplitude for state $|1\rangle$.

To find the probability of measuring $|0\rangle$, we apply Born's Rule:

$$P(0) = \left|\frac{3}{5}\right|^2 = \frac{9}{25} = 0.36$$

To find the probability of measuring $|1\rangle$:

$$P(1) = \left|\frac{4}{5}\right|^2 = \frac{16}{25} = 0.64$$

Since probabilities must sum to 1, we check:

$$0.36 + 0.64 = 1$$

This confirms that the quantum state is correctly normalized.

Another Example: The Hadamard Gate and Superposition

The Hadamard Gate creates a superposition state from |0⟩. If we apply the Hadamard gate to |0⟩, we get:

$$H|0\rangle = \frac{1}{\sqrt{2}}|0\rangle + \frac{1}{\sqrt{2}}|1\rangle$$

Applying Born's Rule:

$$P(0) = \left|\frac{1}{\sqrt{2}}\right|^2 = \frac{1}{2} = 0.5$$

$$P(1) = \left|\frac{1}{\sqrt{2}}\right|^2 = \frac{1}{2} = 0.5$$

This means if we measure the qubit, we will get 0 or 1 with equal probability (50%).

Born's Rule is one of the fundamental postulates of quantum mechanics. It states that the probability of measuring a quantum system in a particular state is given by the squared magnitude of the wavefunction's amplitude. It explains why quantum mechanics is probabilistic rather than deterministic.

Understanding Born's Rule is essential for working with quantum mechanics, quantum computing, and quantum cryptography. It provides the mathematical framework for how measurements work and how we can extract useful information from quantum systems.

2.2. Expectation Values

In quantum mechanics, the expectation value of an observable *A* in state |ψ⟩ is:

$$\langle A\rangle = \langle\psi|A|\psi\rangle$$

For example, measuring the expectation value of the Pauli-Z operator in |+⟩:

$$\langle +|Z|+\rangle = \frac{1}{\sqrt{2}}\begin{bmatrix}1 & 1\end{bmatrix}\begin{bmatrix}1 & 0\\0 & -1\end{bmatrix}\frac{1}{\sqrt{2}}\begin{bmatrix}1\\1\end{bmatrix} = 0$$

which means $|+\rangle$ has an equal probability of measuring 0 or 1.

In quantum mechanics, the expectation value represents the average value of a physical observable (such as energy, position, or momentum) when measured repeatedly on an ensemble of identical quantum systems. It provides insight into the most likely outcome of a measurement but does not necessarily correspond to any single measurement result.

Mathematically, the expectation value $\langle A\rangle$ of an observable A (which is represented by a Hermitian operator $\hat{A}$) in a quantum state $|\psi\rangle$ is given by:

$$\langle A\rangle = \langle\psi|\hat{A}|\psi\rangle$$

This formula means:

- $|\psi\rangle$ is the quantum state.
- $\hat{A}$ is the operator corresponding to the observable.
- $\langle\psi|$ is the conjugate transpose (bra) of $|\psi\rangle$.
- The inner product $\langle\psi|\hat{A}|\psi\rangle$ gives a weighted sum of all possible outcomes.

The expectation value represents the average result we expect if we measure the system many times.

Example 1: Expectation Value of a Qubit in the $|+\rangle$ State

Consider a qubit in the superposition state:

$$|+\rangle = \frac{1}{\sqrt{2}}|0\rangle + \frac{1}{\sqrt{2}}|1\rangle$$

We want to calculate the expectation value of the Pauli Z operator, which is given by:

$$\hat{Z} = \begin{bmatrix}1 & 0\\0 & -1\end{bmatrix}$$

Applying the expectation value formula:

$$\langle Z\rangle = \langle +|\hat{Z}|+\rangle$$

First, write $|+\rangle$ as a column vector:

$$|+\rangle = \frac{1}{\sqrt{2}}\begin{bmatrix}1\\1\end{bmatrix}$$

Compute $\hat{Z}|+\rangle$:

$$\hat{Z}|+\rangle = \begin{bmatrix}1 & 0\\0 & -1\end{bmatrix} \cdot \frac{1}{\sqrt{2}}\begin{bmatrix}1\\1\end{bmatrix} = \frac{1}{\sqrt{2}}\begin{bmatrix}1\\-1\end{bmatrix}$$

Now compute the inner product:

$$\langle +|\hat{Z}|+\rangle = \left(\frac{1}{\sqrt{2}}\begin{bmatrix}1 & 1\end{bmatrix}\right) \cdot \left(\frac{1}{\sqrt{2}}\begin{bmatrix}1\\-1\end{bmatrix}\right)$$
$$= \frac{1}{2}(1 \cdot 1 + 1 \cdot (-1)) = \frac{1}{2}(1 - 1) = 0$$

So, the expectation value of $\hat{Z}$ in the |+⟩ state is 0.

Interpretation: If we measure the *Z*-component of this qubit's spin many times, the average result will be 0, meaning the qubit is equally likely to be found in state |0⟩ (spin up) or |1⟩ (spin down).

Example 2: Expectation Value of Position in a Continuous System

For a continuous system (such as a quantum particle in space), the expectation value of position $\hat{x}$ is given by:

$$\langle x\rangle = \int_{-\infty}^{\infty} \psi^*(x) x \psi(x) dx$$

where:

- $\psi(x)$ is the wavefunction of the particle.
- x is the position variable.
- $\psi^*(x)$ is the complex conjugate of the wavefunction.

If $\psi(x)$ is centred around $x = 2$, then $\langle x\rangle \approx 2$, meaning the most probable location of the particle is around *x*=2.

3. Hilbert Spaces in Quantum Computing

Quantum states exist in Hilbert spaces, which are vector spaces with inner products, allowing for the description of superposition and entanglement.

3.1. Tensor Product for Multi-Qubit States

For two qubits, their combined state is represented using the tensor product:

$$|\psi\rangle \otimes |\phi\rangle = \begin{bmatrix} a_1 \\ a_2 \end{bmatrix} \otimes \begin{bmatrix} b_1 \\ b_2 \end{bmatrix} = \begin{bmatrix} a_1 b_1 \\ a_1 b_2 \\ a_2 b_1 \\ a_2 b_2 \end{bmatrix}$$

For example, the two-qubit state:

$$|00\rangle = |0\rangle \otimes |0\rangle = \begin{bmatrix} 1 \\ 0 \end{bmatrix} \otimes \begin{bmatrix} 1 \\ 0 \end{bmatrix} = \begin{bmatrix} 1 \\ 0 \\ 0 \\ 0 \end{bmatrix}$$

In quantum computing, when dealing with multiple qubits, we need a way to describe their combined state mathematically. This is done using the tensor product, which allows us to construct multi-qubit systems from individual qubits.

A single qubit is represented as a two-dimensional vector:

$$|q\rangle = \alpha|0\rangle + \beta|1\rangle = \begin{bmatrix} \alpha \\ \beta \end{bmatrix}$$

where α\alphaα and β\betaβ are complex numbers satisfying $|\alpha|^2 + |\beta|^2 = 1.$

If we have two qubits, we cannot just add them like classical bits. Instead, we use the tensor product (also called the Kronecker product) to combine their state representations.

For example, if one qubit is in the state $|q_1\rangle$ and another qubit is in the state $|q_2\rangle$, the combined state is given by:

$$|q_1\rangle \otimes |q_2\rangle$$

which creates a four-dimensional vector.

Tensor Product of Two Qubits

If we have two qubits in the states:

$$|q_1\rangle = \begin{bmatrix}\alpha_1 \\ \beta_1\end{bmatrix}, \quad |q_2\rangle = \begin{bmatrix}\alpha_2 \\ \beta_2\end{bmatrix}$$

Their combined state is found using the tensor product:

$$|q_1\rangle \otimes |q_2\rangle = \begin{bmatrix}\alpha_1 \\ \beta_1\end{bmatrix} \otimes \begin{bmatrix}\alpha_2 \\ \beta_2\end{bmatrix}$$

which results in:

$$|q_1 q_2\rangle = \begin{bmatrix}\alpha_1 \begin{bmatrix}\alpha_2 \\ \beta_2\end{bmatrix} \\ \beta_1 \begin{bmatrix}\alpha_2 \\ \beta_2\end{bmatrix}\end{bmatrix} = \begin{bmatrix}\alpha_1\alpha_2 \\ \alpha_1\beta_2 \\ \beta_1\alpha_2 \\ \beta_1\beta_2\end{bmatrix}$$

Thus, the two-qubit system is now represented as a 4D vector.

Example: Tensor Product of Basis States

Let's calculate the tensor product of two **basis states**:

1. **Computational Basis State |00⟩**

$$|0\rangle \otimes |0\rangle = \begin{bmatrix}1 \\ 0\end{bmatrix} \otimes \begin{bmatrix}1 \\ 0\end{bmatrix} = \begin{bmatrix}1 \times 1 \\ 1 \times 0 \\ 0 \times 1 \\ 0 \times 0\end{bmatrix} = \begin{bmatrix}1 \\ 0 \\ 0 \\ 0\end{bmatrix}$$

which represents the |00⟩ state.

2. **Computational Basis State |01⟩**

$$|0\rangle \otimes |1\rangle = \begin{bmatrix}1 \\ 0\end{bmatrix} \otimes \begin{bmatrix}0 \\ 1\end{bmatrix} = \begin{bmatrix}1 \times 0 \\ 1 \times 1 \\ 0 \times 0 \\ 0 \times 1\end{bmatrix} = \begin{bmatrix}0 \\ 1 \\ 0 \\ 0\end{bmatrix}$$

which represents the |01⟩ state.

3. **Computational Basis State |10⟩|10\rangle|10⟩**

$$|1\rangle \otimes |0\rangle = \begin{bmatrix}0 \\ 1\end{bmatrix} \otimes \begin{bmatrix}1 \\ 0\end{bmatrix} = \begin{bmatrix}0 \times 1 \\ 0 \times 0 \\ 1 \times 1 \\ 1 \times 0\end{bmatrix} = \begin{bmatrix}0 \\ 0 \\ 1 \\ 0\end{bmatrix}$$

which represents the **|10⟩ state.**

4. **Computational Basis State |11⟩|11\rangle|11⟩**

$$|1\rangle \otimes |1\rangle = \begin{bmatrix} 0 \\ 1 \end{bmatrix} \otimes \begin{bmatrix} 0 \\ 1 \end{bmatrix} = \begin{bmatrix} 0 \times 0 \\ 0 \times 1 \\ 1 \times 0 \\ 1 \times 1 \end{bmatrix} = \begin{bmatrix} 0 \\ 0 \\ 0 \\ 1 \end{bmatrix}$$

which represents the **|11⟩|11\rangle|11⟩ state.**

Entanglement and the Tensor Product

The tensor product also allows us to describe entanglement, a uniquely quantum phenomenon where qubits become correlated in a way that classical bits cannot.

For example, the Bell state (an entangled state) is:

$$|\Phi^+\rangle = \frac{1}{\sqrt{2}}(|00\rangle + |11\rangle)$$

This state cannot be written as a simple tensor product of two separate qubits, meaning that measurement of one qubit instantly affects the other, regardless of the distance between them.

Tensor Product of Matrices (Multi-Qubit Gates)

The tensor product is also used for multi-qubit quantum gates.

For example, if we apply two Hadamard gates to a two-qubit system:

$$\begin{aligned} H \otimes H &= \frac{1}{\sqrt{2}} \begin{bmatrix} 1 & 1 \\ 1 & -1 \end{bmatrix} \otimes \frac{1}{\sqrt{2}} \begin{bmatrix} 1 & 1 \\ 1 & -1 \end{bmatrix} \\ &= \frac{1}{2} \begin{bmatrix} 1 & 1 \\ 1 & -1 \end{bmatrix} \otimes \begin{bmatrix} 1 & 1 \\ 1 & -1 \end{bmatrix} \\ &= \frac{1}{2} \begin{bmatrix} 1 \cdot 1 & 1 \cdot 1 & 1 \cdot 1 & 1 \cdot -1 \\ 1 \cdot 1 & 1 \cdot -1 & 1 \cdot 1 & 1 \cdot -1 \\ 1 \cdot 1 & 1 \cdot 1 & -1 \cdot 1 & -1 \cdot -1 \\ 1 \cdot 1 & 1 \cdot -1 & -1 \cdot 1 & -1 \cdot -1 \end{bmatrix} \\ &= \frac{1}{2} \begin{bmatrix} 1 & 1 & 1 & -1 \\ 1 & -1 & 1 & -1 \\ 1 & 1 & -1 & 1 \\ 1 & -1 & -1 & -1 \end{bmatrix} \end{aligned}$$

which is the Hadamard transformation for a two-qubit system.

3.2. Entanglement and Bell States

A key quantum feature is entanglement, where qubits share a state that cannot be factored into individual qubits. The Bell states are maximally entangled two-qubit states:

$$|\Phi^+\rangle = \frac{1}{\sqrt{2}}(|00\rangle + |11\rangle)$$

$$|\Phi^-\rangle = \frac{1}{\sqrt{2}}(|00\rangle - |11\rangle)$$

$$|\Psi^+\rangle = \frac{1}{\sqrt{2}}(|01\rangle + |10\rangle)$$

$$|\Psi^-\rangle = \frac{1}{\sqrt{2}}(|01\rangle - |10\rangle)$$

If we measure one qubit in an entangled pair, the other qubit's state is instantly determined, no matter how far apart they are.

3.3. Density Matrices for Mixed States

For a pure quantum state $|\psi\rangle$, its density matrix is:

$$\rho = |\psi\rangle\langle\psi|$$

For example, for $|+\rangle$:

$$\rho = |+\rangle\langle+| = \frac{1}{2}\begin{bmatrix} 1 & 1 \\ 1 & 1 \end{bmatrix}$$

For mixed states, the density matrix is:

$$\rho = \sum_i p_i |\psi_i\rangle\langle\psi_i|$$

which is used in quantum noise models and open quantum systems.

In quantum mechanics and quantum computing, density matrices provide a way to describe both pure states and mixed states. While pure states represent an exact quantum state of a system, mixed states describe statistical ensembles of different quantum states, capturing uncertainty or decoherence effects.

A density matrix (or density operator) is a mathematical representation of a quantum system that accounts for both pure states and mixed states. It generalizes the traditional state vector $|\psi\rangle$ approach.

For a quantum state |ψ⟩, the corresponding density matrix is:

$$\rho = |\psi\rangle\langle\psi|$$

This matrix provides a complete description of the quantum state, including probabilities of measurements and quantum coherences.

Pure State Example

If a system is in a pure state:

$$|\psi\rangle = \alpha|0\rangle + \beta|1\rangle$$

The density matrix is:

$$\rho = |\psi\rangle\langle\psi| = \begin{bmatrix}\alpha \\ \beta\end{bmatrix}\begin{bmatrix}\alpha^* & \beta^*\end{bmatrix} = \begin{bmatrix}|\alpha|^2 & \alpha\beta^* \\ \alpha^*\beta & |\beta|^2\end{bmatrix}$$

This fully describes the quantum system when we are certain about its state.

Mixed States: When We Have Uncertainty

A mixed state occurs when we have a probabilistic mixture of multiple quantum states. Instead of knowing the system is in one specific quantum state, we only have a statistical distribution of possible states.

For example, if a system is in state |ψ1⟩ with probability p_1 and state |ψ2⟩ with probability p_2, the density matrix is:

$$\rho = p_1|\psi_1\rangle\langle\psi_1| + p_2|\psi_2\rangle\langle\psi_2|$$

Unlike pure states, mixed states arise in situations where:

- The system is part of a larger system and we ignore part of it.
- The system undergoes decoherence due to interaction with the environment.
- We lack full information about the quantum state.

Example of a Mixed State

Suppose we have a system that is 50% in |0⟩ and 50% in |1⟩. The density matrix is:

$$\rho = \frac{1}{2}|0\rangle\langle 0| + \frac{1}{2}|1\rangle\langle 1|$$

Expanding:

$$\rho = \frac{1}{2}\begin{bmatrix}1 & 0\\0 & 0\end{bmatrix} + \frac{1}{2}\begin{bmatrix}0 & 0\\0 & 1\end{bmatrix} = \begin{bmatrix}0.5 & 0\\0 & 0.5\end{bmatrix}$$

This is different from the pure state superposition $\frac{1}{\sqrt{2}}(|0\rangle + |1\rangle)$, which has nonzero off-diagonal elements representing quantum coherence.

Measurement and Expectation Values Using Density Matrices

The probability of measuring an observable *A* (such as energy or spin) is given by the expectation value:

$$\langle A \rangle = \mathrm{Tr}(\rho A)$$

This equation generalizes Born's rule and applies even to mixed states.

Example: Measuring the Pauli *Z*-Operator

For the Pauli matrix $Z = \begin{bmatrix}1 & 0\\0 & -1\end{bmatrix}$,

If our quantum system is in the mixed state:

$$\rho = \begin{bmatrix}0.5 & 0\\0 & 0.5\end{bmatrix}$$

Then,

$$\begin{aligned}\langle Z \rangle &= \mathrm{Tr}\left(\begin{bmatrix}0.5 & 0\\0 & 0.5\end{bmatrix} \cdot \begin{bmatrix}1 & 0\\0 & -1\end{bmatrix}\right)\\ &= \mathrm{Tr}\left(\begin{bmatrix}0.5 & 0\\0 & -0.5\end{bmatrix}\right)\\ &= 0.5 - 0.5 = 0\end{aligned}$$

This means that measuring *Z* in this mixed state has an expected value of zero, indicating a completely random mixture of spin-up and spin-down.

Appendix B - Programming Quantum Computers (Qiskit, Cirq, Braket, PennyLane)

Quantum computing programming involves using specialized frameworks to create and manipulate quantum circuits. In this tutorial, we will explore four of the most popular quantum programming frameworks:

1. **Qiskit** (IBM) – Open-source framework for IBM Quantum hardware and simulators.
2. **Cirq** (Google) – Optimized for near-term quantum devices, especially Google's Sycamore processor.
3. **Amazon Braket** – A cloud-based quantum computing service supporting multiple hardware backends.
4. **PennyLane** – A quantum machine learning library that integrates with multiple quantum platforms.

Each section includes a basic installation guide, how to create a quantum circuit, and running quantum programs on simulators or real quantum hardware.

1. Qiskit: Programming IBM Quantum Computers

Qiskit is an open-source Python framework for quantum computing developed by IBM. It allows users to create quantum circuits, run them on IBM Quantum devices, and analyse results.

Qiskit provides tools for building and running quantum programs on both simulators and real IBM Quantum hardware. The framework is continuously evolving to accommodate the latest advancements in quantum technology, with new features and enhancements released regularly.

Qiskit consists of multiple components, including the Qiskit SDK, which allows users to build and execute quantum circuits, and Qiskit Runtime, a cloud-based execution environment for running quantum workloads efficiently. Additionally, several independent open-source projects integrate with Qiskit, extending its functionality beyond IBM's ecosystem.

1.1 Installing Qiskit

First, install Qiskit using pip:

```
pip install qiskit
```

If you want to install visualization and additional features, use:

```
pip install qiskit[visualization]
```

1.2 Creating a Basic Quantum Circuit

A quantum circuit is a series of quantum operations (gates) applied to qubits. Below, we create a simple quantum circuit using the Hadamard gate.

```
from qiskit import QuantumCircuit

# Create a quantum circuit with 1 qubit and 1 classical bit
qc = QuantumCircuit(1, 1)

# Apply a Hadamard gate (H) to create superposition
qc.h(0)

# Measure the qubit
qc.measure(0, 0)

# Draw the circuit
print(qc.draw())
```

1.3 Running on a Quantum Simulator

Qiskit allows users to simulate quantum circuits before running them on real quantum computers.

```
from qiskit import Aer, execute

# Select the quantum simulator
simulator = Aer.get_backend('qasm_simulator')

# Execute the circuit
job = execute(qc, simulator, shots=1024)

# Get results
result = job.result()
counts = result.get_counts(qc)
print("Measurement Results:", counts)
```

1.4 Running on IBM Quantum Hardware

To run your program on a real quantum device:

1. Create an IBM Quantum account at IBM Quantum
2. Get your API Token and save it.

Then, authenticate and run:

```
from qiskit import IBMQ

# Load IBM account
IBMQ.save_account('YOUR_API_TOKEN')
IBMQ.load_account()

# Get a real quantum computer
provider = IBMQ.get_provider(hub='ibm-q')
backend = provider.get_backend('ibmq_lima')  # Choose a quantum
device

# Execute the circuit on real quantum hardware
job = execute(qc, backend, shots=1024)
result = job.result()
counts = result.get_counts(qc)
print("Quantum Computer Results:", counts)
```

1.5 Detail

The Core Components of Qiskit

Qiskit SDK

The Qiskit SDK is the main package that provides the fundamental tools for programming quantum computers. It supports the creation, manipulation, and execution of quantum circuits and includes various libraries for quantum information processing.

Key features of the Qiskit SDK include:

- **Circuit-building tools (`qiskit.circuit`)** – Enables the creation and manipulation of quantum circuits, including registers, instructions, gates, and control flow.
- **Circuit library (`qiskit.circuit.library`)** – Provides pre-built quantum circuits and gates that can be used as building blocks for quantum computations.
- **Quantum information library (`qiskit.quantum_info`)** – Contains tools for working with quantum states, operators, and channels, helping users analyze quantum systems.
- **Transpiler (`qiskit.transpiler`)** – Optimizes quantum circuits for specific hardware by adapting them to device topology and reducing noise.
- **Primitives (`qiskit.primitives`)** – Provides the foundational definitions for the Sampler and Estimator primitives, which are essential for executing quantum programs on different quantum hardware backends.

To install the Qiskit SDK, simply run:

```
pip install qiskit
```

Qiskit Runtime

Qiskit Runtime is IBM's cloud-based quantum computing service, designed to streamline quantum execution and optimize workloads. It allows users to run quantum programs efficiently by leveraging both classical and quantum computing resources.

Features of Qiskit Runtime:

- **Error Mitigation & Suppression** – Uses techniques like dynamical decoupling and readout mitigation to improve the accuracy of quantum computations.
- **Execution Modes** – Supports three different execution modes:
    - **Job** – A single execution of a quantum task.

  - **Session** – Runs multiple jobs iteratively, optimizing execution.
  - **Batch** – Submits multiple jobs simultaneously for parallel processing.
- **Integration with IBM Cloud** – Users can access Qiskit Runtime via the IBM Quantum Platform or IBM Cloud.

To install the Qiskit Runtime package:

```
pip install qiskit-ibm-runtime
```

While the Qiskit SDK is fully open-source, parts of Qiskit Runtime (such as the backend execution software) are proprietary. However, the client-side interface remains open-source, allowing users to develop their own integrations.

The Qiskit Development Workflow

The Four Steps of Quantum Programming in Qiskit

Qiskit follows a structured workflow for quantum computing applications:

1. **Map the problem** – Define the problem and translate it into a quantum circuit using the Qiskit SDK.
2. **Optimize for hardware** – Use the Qiskit Transpiler to adapt circuits for specific quantum devices, minimizing errors and execution time.
3. **Execute on hardware** – Run the quantum program on either a simulator or a real IBM quantum processor using Qiskit Runtime.
4. **Post-process results** – Analyse and interpret the output of quantum computations to extract meaningful insights.

All steps except "Execute on hardware" use the Qiskit SDK, while the execution phase is managed through Qiskit Runtime Service.

Qiskit Transpiler as a Service

The Qiskit Transpiler Service is an experimental cloud-based transpilation service that enhances circuit optimization by leveraging IBM Quantum's cloud infrastructure and AI-powered transpilation methods. It allows for more efficient compilation and error mitigation when running on real hardware.

To install the transpiler service:

```
pip install qiskit-ibm-transpiler
```

This service is currently available to IBM Quantum Premium Plan users.

Advanced Features and Extensions

Qiskit Serverless

For large-scale quantum computing applications, Qiskit Serverless enables users to manage workloads across quantum and classical resources efficiently.

Features of Qiskit Serverless:

- **Parallel task execution** – Enables efficient use of quantum-classical hybrid computations.
- **Cloud-based workload management** – Runs long-duration tasks remotely.
- **Multi-cloud compatibility** – Supports deployment on various cloud platforms.

To install Qiskit Serverless:

```
pip install qiskit-serverless
```

Qiskit Functions

Qiskit Functions accelerate the discovery of quantum algorithms by providing pre-built computational functions. These include:

- **Circuit Functions** – Handle error mitigation, suppression, and optimization.
- **Application Functions** – Enable domain-specific quantum workflows in industries like finance, chemistry, and machine learning.

Qiskit Functions can be installed via:

```
pip install qiskit-ibm-catalog
```

Qiskit Addons

Qiskit addons extend the capabilities of the Qiskit SDK, allowing researchers to explore advanced quantum algorithms.

Some key addons include:

- **Sample-based Quantum Diagonalization (SQD):**

```
pip install qiskit-addon-sqd
```

- **Approximate Quantum Compilation (AQC):**

```
pip install qiskit-addon-aqc-tensor[quimb-jax]
```

- **Operator Backpropagation (OBP):**

```
pip install qiskit-addon-obp
```

The Qiskit Ecosystem

Beyond Qiskit, there are many open-source projects that integrate with it to expand its capabilities. Some notable ones include:

- **Qiskit Aer (`qiskit-aer`)** – A quantum simulator with realistic noise models.

```
pip install qiskit-aer
```

- **mthree (`mthree`)** – A library for measurement error mitigation.

```
pip install mthree
```

- **qBraid SDK (`qbraid`)** – A cloud-based quantum runtime framework for managing quantum jobs across multiple providers.

```
pip install qbraid
```

Simulating Quantum Gates in Qiskit: A Step-by-Step Guide

Quantum computing operates on qubits, which are manipulated using quantum gates to perform computations. Qiskit, IBM's open-source quantum computing framework, allows users to design and simulate quantum circuits, visualize quantum states, and run quantum programs on both simulators and real quantum processors.

In this guide, we will explore how to simulate simple quantum gates in Qiskit, visualize quantum states using Bloch spheres, measure qubits, and analyse quantum circuit outputs statistically.

1. Setting Up a Quantum Circuit in Qiskit

A quantum circuit consists of qubits (quantum bits) and classical bits (used for measurement). To create a simple circuit with one qubit and one classical bit, we use the QuantumCircuit class from Qiskit:

```
# Load Qiskit functions
```

```
from qiskit import QuantumCircuit

# Create a quantum circuit with 1 qubit and 1 classical bit
qc = QuantumCircuit(1, 1)

# Draw the circuit ('mpl' asks to render using Matplotlib)
qc.draw(output='mpl')
```

Understanding the Initial Qubit State

Before applying any quantum gates, let's check the initial state of the qubit using state vectors and the Bloch sphere representation:

```
# Import state vector and visualization tools
from qiskit.quantum_info import Statevector
from qiskit.visualization import plot_bloch_multivector

# Get the initial state vector of the qubit
init_state = Statevector(qc)
print(init_state)

# Visualize the qubit state on a Bloch sphere
plot_bloch_multivector(init_state)
```

Output:

```
Statevector([1.+0.j, 0.+0.j], dims=(2,))
```

This means the qubit is in the state |0⟩|0\rangle|0⟩, which is the default initial state of all qubits.

Applying Quantum Gates

Hadamard Gate (H Gate)

The Hadamard gate creates a superposition state, meaning the qubit will be in an equal probability of measuring |0⟩ or |1⟩.

```
qc.h(0)   # Apply a Hadamard gate to qubit 0
qc.draw(output='mpl')
```

Now, let’s check the updated state vector and visualize the qubit again:

```
h_state = Statevector(qc)
print(h_state)

plot_bloch_multivector(h_state)
```

Output:

```
Statevector([0.70710678+0.j, 0.70710678+0.j], dims=(2,))
```

This means the qubit is now in a superposition state:

$$|+\rangle = \frac{|0\rangle + |1\rangle}{\sqrt{2}}$$

Measuring the Qubit

To observe the quantum state, we must measure the qubit and store the result in the classical bit:

```
qc.measure(0, 0)  # Measure qubit 0 and store the result in
classical bit 0
qc.draw(output='mpl')
```

Running the Circuit on a Simulator

We now execute the circuit multiple times (2000 shots) on a quantum simulator and analyze the results:

```
from qiskit.providers.basic_provider import BasicSimulator
from qiskit.visualization import plot_histogram

backend = BasicSimulator()

# Run the circuit on a simulator with 2000 shots
result = backend.run(qc, shots=2000).result()

# Extract the counts and visualize
counts = result.get_counts()
plot_histogram(counts)
```

Expected Output:

Since the Hadamard gate creates an equal probability of measuring |0⟩ and |1⟩, we expect the histogram to show approximately 50% of each outcome.

Multi-Qubit Gates: The CNOT Gate

Multi-qubit gates such as the CNOT (Controlled-NOT) gate enable interactions between qubits. The CNOT gate flips the target qubit if the control qubit is in state |1⟩|1\rangle|1⟩.

Creating a 2-Qubit Circuit

```
qc = QuantumCircuit(2)    # Create a circuit with 2 qubits
qc.x(0)                   # Apply an X gate to prepare q0 in
state |1⟩
print(Statevector(qc))
qc.draw(output='mpl', initial_state=True)
```

Applying the CNOT Gate

```
qc.cx(0, 1)  # Apply a CNOT with q0 as control and q1 as target
print(Statevector(qc))
qc.draw(output='mpl')
```

Expected Outcome:

If q_0=|1⟩, the CNOT gate will flip q_1 from |0⟩ to |1⟩.

Visualizing the Qubits on Bloch Spheres

```
plot_bloch_multivector(qc)
```

Implementing a Half-Adder Using Quantum Gates

A half-adder adds two bits and produces a sum and a carry bit. The quantum circuit for a half-adder is built using CNOT and Toffoli gates.

```
qc_ha = QuantumCircuit(4,2)

# Encode inputs (q1 + q0) = (1 + 1)
qc_ha.x(0)
qc_ha.x(1)

qc_ha.barrier()

# Quantum algorithm
```

```
qc_ha.cx(0,2)        # CNOT with q0 as control and q2 as target
qc_ha.cx(1,2)        # CNOT with q1 as control and q2 as target
qc_ha.ccx(0,1,3)     # CCNOT (Toffoli) with q0, q1 as controls
and q3 as target

qc_ha.barrier()

# Measure outputs
qc_ha.measure(2,0)  # Measure qubit 2 into classical bit 0
qc_ha.measure(3,1)  # Measure qubit 3 into classical bit 1

qc_ha.draw(output = 'mpl', initial_state=True)
```

Running on a Quantum Simulator

```
result = backend.run(qc_ha).result()
counts = result.get_counts()
plot_histogram(counts)
```

The results show that for an input of (1+1), the output is "10", meaning:

- The sum is stored in classical bit 0.
- The carry is stored in classical bit 1.

Running on a Fake Quantum Backend

A fake backend mimics a real quantum processor and allows testing of circuits under realistic noise conditions.

```
from qiskit.providers.fake_provider import GenericBackendV2

# Generate a 4-qubit fake backend
backend = GenericBackendV2(num_qubits=4)

# Transpile and execute
transpiled_circuit = transpile(qc_ha, backend)
result = backend.run(transpiled_circuit).result()

# Visualize results
counts = result.get_counts()
plot_histogram(counts)
```

This simulation allows users to test error mitigation techniques and optimize quantum circuits before running them on real hardware.

Implementing Grover's Algorithm (2-Qubit Example)

Grover's algorithm amplifies the probability of measuring a specific "winning state." Here's an example where the target state is $|11\rangle$:

```
gc = QuantumCircuit(2,2)

# Put qubits into superposition
gc.h([0,1])

gc.barrier()

# Oracle function for |11>
gc.cz(0,1)

gc.barrier()

# Diffusion step (Grover iteration)
gc.h([0,1])
gc.z([0,1])
gc.cz(0,1)
gc.h([0,1])

gc.barrier()

# Measurement
gc.measure([0,1], [0,1])

gc.draw(output = 'mpl')
```

Running Grover's Algorithm

```
result = backend.run(gc, shots=5000).result()
counts = result.get_counts()
plot_histogram(counts)
```

The algorithm amplifies the probability of measuring |11⟩|11\rangle|11⟩, showing the power of quantum search.

2. Cirq: Programming Google Quantum AI Processors

Cirq is an open-source framework developed by Google, optimized for near-term quantum applications.

2.1 Installing Cirq

To install Cirq:

```
pip install cirq
```

2.2 Creating a Quantum Circuit in Cirq

Cirq uses qubits and gates to build quantum circuits.

```
import cirq

# Define a qubit
qubit = cirq.LineQubit(0)

# Create a quantum circuit
circuit = cirq.Circuit(
    cirq.H(qubit),  # Apply Hadamard gate
    cirq.measure(qubit)  # Measure the qubit
)

print("Quantum Circuit:\n", circuit)
```

2.3 Running on a Quantum Simulator

Cirq provides quantum simulators to test circuits before running them on real hardware.

```
simulator = cirq.Simulator()
result = simulator.run(circuit, repetitions=1000)
print("Measurement Results:", result.histogram(key='0'))
```

2.4 Running on Google Quantum Hardware

Google provides cloud access to Sycamore and other quantum processors via Quantum Engine.

1. Set up Google Cloud Quantum API at Google Cloud.
2. Authenticate with your Google Cloud account.

3. Submit jobs using:

```
import cirq_google

# Create an Engine object
engine = cirq_google.Engine(project_id='your-google-cloud-
project')

# Run on Sycamore processor
job = engine.run(program=circuit, repetitions=1000)
print("Quantum Computer Results:", job.results())
```

2.5 Detail

Installing Cirq

Before using Cirq, you need to install it. The following Python code checks if Cirq is installed and installs it if necessary.

```
try:
    import cirq
except ImportError:
    print("Installing Cirq...")
    !pip install --quiet cirq
    print("Installed Cirq.")
    import cirq
```

Additionally, `cirq_google` can be imported to access Google's quantum devices:

```
import cirq_google
```

Defining Qubits in Cirq

In Cirq, qubits are the fundamental units of quantum circuits. They can be defined in different ways:

- `cirq.NamedQubit`: Labels qubits with arbitrary names.
- `cirq.LineQubit`: Labels qubits with numbers in a linear sequence.
- `cirq.GridQubit`: Labels qubits using grid coordinates.

Examples of Qubit Definitions

```
# Named Qubits
q0 = cirq.NamedQubit('source')
```

```
q1 = cirq.NamedQubit('target')

# Line Qubits (1D Array)
q3 = cirq.LineQubit(3)
q0, q1, q2 = cirq.LineQubit.range(3)  # Creates LineQubit(0),
LineQubit(1), LineQubit(2)

# Grid Qubits (2D Lattice)
q4_5 = cirq.GridQubit(4, 5)
qubits = cirq.GridQubit.square(4)  # Creates a 4x4 grid of
qubits
```

GridQubits are particularly useful when working with Google's quantum processors, as they are structured in a grid layout.

Quantum Gates and Operations

Cirq differentiates between:

- Gates: Abstract operations that can be applied to qubits.
- Operations: Gates applied to specific qubits.

Example of Gates

```
# Standard quantum gates
cnot_gate = cirq.CNOT
pauli_z = cirq.Z
sqrt_x_gate = cirq.X**0.5  # Square root of X gate
sqrt_sqrt_y = cirq.YPowGate(exponent=0.25)  # Y gate raised to
1/4 power
```

Applying Gates to Qubits

```
q0, q1 = cirq.LineQubit.range(2)

# Example of operations (applying gates to qubits)
z_op = cirq.Z(q0)
not_op = cirq.CNOT(q0, q1)
sqrt_iswap_op = cirq.SQRT_ISWAP(q0, q1)
```

Cirq supports a wide range of quantum gates, including Hadamard (H), Pauli (X, Y, Z), CNOT, CZ, SWAP, and Toffoli (CCX).

Constructing Quantum Circuits

A Quantum Circuit in Cirq is a collection of Moments, where each moment consists of a set of operations that act simultaneously on different qubits.

Creating a Simple Quantum Circuit

```
# Create a quantum circuit
circuit = cirq.Circuit()
qubits = cirq.LineQubit.range(3)

# Add Hadamard gates to all qubits
circuit.append(cirq.H(q) for q in qubits)

print(circuit)
```

Output:

```
0: ───H───
1: ───H───
2: ───H───
```

Appending Multi-Qubit Gates

```
# Create a circuit with multi-qubit gates
circuit = cirq.Circuit()
circuit.append(cirq.CNOT(q0, q1))
circuit.append(cirq.SWAP(q1, q2))

print(circuit)
```

Output:

```
0: ───@───
      │
1: ───X───×───
          │
2: ───────×───
```

Key Takeaways:

- Cirq automatically optimizes the placement of gates.

- Multiple single-qubit gates can be merged into a single moment.
- Multi-qubit operations are placed sequentially if they act on overlapping qubits.

Running Quantum Simulations

Cirq provides a Simulator to execute quantum circuits. There are two methods:

1. simulate() - Gives access to the quantum wave function (useful for debugging).
2. run() - Mimics actual quantum hardware by measuring bit-string outputs.

Simulating a Bell State Circuit

The Bell state is an entangled quantum state:

$$|\Psi\rangle = \frac{1}{\sqrt{2}}(|00\rangle + |11\rangle)$$

```
# Create a Bell state circuit
bell_circuit = cirq.Circuit()
q0, q1 = cirq.LineQubit.range(2)

# Apply Hadamard and CNOT
bell_circuit.append([cirq.H(q0), cirq.CNOT(q0, q1)])

# Simulate the circuit
s = cirq.Simulator()
results = s.simulate(bell_circuit)
print(results)
```

Expected Output:

```
output vector: 0.707|00⟩ + 0.707|11⟩
```

Measuring the Qubits

To obtain measurement results, we add a measurement operation:

```
# Add measurement to the circuit
bell_circuit.append(cirq.measure(q0, q1, key='result'))

# Run the circuit multiple times
samples = s.run(bell_circuit, repetitions=1000)
```

Visualizing the Results

```
import matplotlib.pyplot as plt
cirq.plot_state_histogram(samples, plt.subplot())
plt.show()
```

The histogram should show **~50% probability for |00⟩ and |11⟩**, confirming quantum entanglement.

Advanced Features

Parameterized Gates & Sweeps

Cirq allows defining parameterized quantum circuits, useful for variational algorithms.

```
import sympy

# Create a parameterized X gate
q = cirq.GridQubit(1, 1)
circuit = cirq.Circuit(cirq.X(q) ** sympy.Symbol('t'),
cirq.measure(q, key='m'))

# Define a parameter sweep from 0 to 2
param_sweep = cirq.Linspace('t', start=0, stop=2, length=50)

# Run the simulation
trials = s.run_sweep(circuit, param_sweep, repetitions=100)

# Plot results
x_data = [trial.params['t'] for trial in trials]
y_data = [trial.histogram(key='m')[1] / 100.0 for trial in
trials]
plt.scatter(x_data, y_data)
plt.xlabel("t (X gate exponent)")
plt.ylabel("Probability of measuring |1⟩")
plt.show()
```

This simulates an X rotation sweep, showing how the probability of measuring |1⟩ varies.

Using Google's Quantum Devices

Cirq provides access to Google's Sycamore processor, which uses a 54-qubit grid.

```
print(cirq_google.Sycamore)
```

This prints the qubit connectivity of Sycamore.

Validating a Circuit for Sycamore

```
q0 = cirq.GridQubit(5, 6)
q1 = cirq.GridQubit(5, 5)
q2 = cirq.GridQubit(4, 5)

# Valid operation (adjacent qubits)
adjacent_op = cirq_google.SYC(q0, q1)

# Invalid operation (non-adjacent qubits)
nonadjacent_op = cirq_google.SYC(q0, q2)

# Validate a working circuit
working_circuit = cirq.Circuit(adjacent_op)
cirq_google.Sycamore.validate_circuit(working_circuit)

# Try validating an invalid circuit
bad_circuit = cirq.Circuit(nonadjacent_op)
try:
    cirq_google.Sycamore.validate_circuit(bad_circuit)
except ValueError as e:
    print(e)
```

3. Amazon Braket: Programming Quantum Computers in AWS

Amazon Braket is a cloud-based quantum computing platform that allows access to multiple quantum hardware providers.

Amazon Braket is a fully managed quantum computing service that allows users to experiment with different quantum hardware, including gate-based quantum processors (QPU) from IonQ and Rigetti, as well as quantum annealers from D-Wave. This tutorial will guide you through setting up Amazon Braket, creating your first quantum circuit, and running it on a simulator or real quantum hardware.

3.1 Installing Amazon Braket

To install the Braket SDK:

```
pip install amazon-braket-sdk boto3
```

3.2 Creating a Quantum Circuit in Braket

Braket uses the PennyLane-style circuit approach to define quantum programs.

```
from braket.circuits import Circuit

# Create a simple quantum circuit
circuit = Circuit().h(0).cnot(0, 1).measure(0, 1)

# Print the circuit
print(circuit)
```

3.3 Running on a Quantum Simulator

Amazon Braket provides simulators to test quantum circuits before running on hardware.

```
from braket.devices import LocalSimulator

# Define a local simulator
simulator = LocalSimulator()

# Run the circuit
task = simulator.run(circuit, shots=1000)
result = task.result()
print("Simulation Results:", result.measurement_counts)
```

3.4 Running on Real Quantum Hardware

Amazon Braket provides access to IonQ, Rigetti, and D-Wave quantum computers.

1. Set up AWS IAM credentials in your AWS account.
2. Run on a real device:

```
from braket.aws import AwsDevice

# Choose a quantum device
device = AwsDevice("arn:aws:braket:::device/qpu/ionq/Aria-1")

# Run the quantum program
task = device.run(circuit, shots=1000)
result = task.result()
print("Quantum Computer Results:", result.measurement_counts)
```

3.5 Detail

Setting Up Amazon Braket

Before writing your first quantum program, you need to set up Amazon Braket in your AWS account. Follow these steps:

1. Sign in to the AWS Management Console and navigate to Amazon Braket.
2. Amazon Braket is available in select regions:
    - US East (N. Virginia)
    - US West (N. California)
    - US West (Oregon)
3. Choose a storage location: Braket uses Amazon S3 to store quantum job results. You can use an existing S3 bucket or create a new one.
4. Accept the terms and conditions: Since Braket runs on third-party quantum hardware, data may be processed outside AWS facilities.
5. Once Amazon Braket is enabled, you'll see available quantum processors (QPU) and AWS-managed simulators.
6. Choose a quantum device:
    - Run on a quantum simulator for testing and debugging.
    - Run on a real quantum computer (QPU) for real execution.

Note: Amazon Braket notebooks are fully managed Jupyter notebooks using Amazon SageMaker. You pay for the usage, so turn them off when not needed to avoid unnecessary charges.

Writing Your First Quantum Circuit: The Bell State

A Bell State is the "Hello, World" program of quantum computing. It creates two qubits that become entangled, meaning they are correlated regardless of distance.

Understanding the Bell State Circuit

- Start with two qubits both in state |0>.
- Apply a Hadamard gate (H) to the first qubit. This puts it in superposition (equal probability of being 0 or 1).
- Apply a CNOT gate with:
    - The first qubit as the control.
    - The second qubit as the target.
- This operation entangles the qubits, producing the Bell State.

Bell State Logic

Input	Output
**	00⟩ →
**	01⟩ →
**	10⟩ →
**	11⟩ →

Final state: The qubits are entangled, with a 50% probability of measuring |00⟩ or |11⟩.

Implementing the Bell State Circuit in Amazon Braket

Running the Circuit on a Local Simulator

Using Amazon Braket's Python SDK, we define and simulate the Bell State:

```
# Import Braket SDK
from braket.circuits import Circuit
from braket.devices import LocalSimulator
```

```
# Create the Bell State Circuit
bell = Circuit().h(0).cnot(control=0, target=1)

# Create a local simulator device
device = LocalSimulator()

# Run the circuit 1000 times
task = device.run(bell, shots=1000)

# Print the results
print(task.result().measurement_counts)
```

Expected Output (approximately):

```
Counter({11': 503, '00': 497)
```

Since the Bell State creates an equal superposition, measuring the qubits should yield 50% |00⟩ and 50% |11⟩.

Running the Circuit on a Managed Simulator

Instead of using a local simulator, we can run the same circuit on a managed simulator (Amazon SV1) using AWS cloud resources.

Using an AWS Managed Simulator

Replace the local simulator with `AwsDevice` and specify an S3 bucket for storing results.

```
# Import AwsDevice
from braket.aws import AwsDevice

# Choose the managed simulator
device = AwsDevice("arn:aws:braket:::device/quantum-
simulator/amazon/sv1")

# Specify an S3 bucket for storing results
s3_output_folder = ("amazon-braket-Your-Bucket-Name", "folder-
name")

# Run the circuit on the managed simulator
task = device.run(bell, shots=1000)

# Print results
print(task.result().measurement_counts)
```

Available AWS Simulators

Amazon Braket provides different simulator ARNs for different use cases:

- **State Vector Simulator (SV1)**
  `arn:aws:braket:::device/quantum-simulator/amazon/sv1`
- **Tensor Network Simulator (TN)**
  `arn:aws:braket:::device/quantum-simulator/amazon/tn`
- **Density Matrix Simulator (DM1)**
  `arn:aws:braket:::device/quantum-simulator/amazon/dm1`

Running on a Real Quantum Device

To execute on a real quantum processor, replace the simulator ARN with a Quantum Processing Unit (QPU) ARN.

Example:

```
# Run on an actual QPU (example: Rigetti Aspen-M)
device = AwsDevice("arn:aws:braket:::device/qpu/rigetti/Aspen-
M")
```

Available QPUs include:

- IonQ trapped ion QPU
- Rigetti superconducting QPU
- D-Wave quantum annealer (for optimization problems)

Note: Real quantum hardware has queue times and potential errors, unlike simulators.

Building More Advanced Circuits

You can build circuits with more qubits and gates using Amazon Braket's SDK.

Example: Creating a Four-Qubit Entangled State

```
from braket.circuits import Circuit

# Define a 4-qubit quantum circuit
```

```
my_circuit = Circuit().h(range(4))  # Apply Hadamard to all
qubits

# Add controlled NOT gates to create entanglement
my_circuit.cnot(control=0, target=2)
my_circuit.cnot(control=1, target=3)

print(my_circuit)
```

Listing Available Gates in Amazon Braket

Amazon Braket supports a wide range of quantum gates:

```
from braket.circuits import Gate
import string

gate_set = [attr for attr in dir(Gate) if attr[0] in
string.ascii_uppercase]
print(gate_set)
```

Output:

```
['CCNot', 'CNot', 'H', 'I', 'S', 'Swap', 'T', 'X', 'Y', 'Z',
'Rx', 'Ry', 'Rz']
```

These gates allow implementing quantum logic, entanglement, and transformations.

Using D-Wave Quantum Annealers

Amazon Braket also supports D-Wave quantum annealers, which solve optimization problems.

Formulating an Optimization Problem

```
from braket.ocean_plugin import BraketDWaveSampler
from dimod import BinaryQuadraticModel

# Define a Binary Quadratic Model (BQM)
linear = dict(x1=-1, x2=-1)
quadratic = dict([(('x1', 'x2'), 2)])
bqm = BinaryQuadraticModel(linear, quadratic, 0, dimod.BINARY)

# Choose D-Wave QPU
```

```
sampler = BraketDWaveSampler(s3_folder,
'arn:aws:braket:::device/qpu/d-wave/DW_2000Q_6')

# Solve the problem
sampleset = sampler.sample(bqm, num_reads=1000)
print(sampleset.aggregate())
```

Quantum annealers find global minima of optimization problems.

4. PennyLane: Quantum Machine Learning with Hybrid Quantum-Classical Computing

PennyLane is a Python framework designed for quantum machine learning (QML). PennyLane is an open-source quantum machine learning (QML) library that allows you to create and train quantum neural networks (QNNs) and hybrid quantum-classical models. It integrates seamlessly with popular machine learning libraries such as TensorFlow, PyTorch, and Keras, enabling users to train quantum models using automatic differentiation techniques.

Key Features of PennyLane:

- **Flexibility**:
    - Design custom quantum circuits.
    - Use for tasks like classification, regression, generative modeling, and reinforcement learning.
    - Combine quantum and classical components.
- **Compatibility**:
    - Works with TensorFlow, PyTorch, and Keras.
    - Supports gradient-based optimization and model serialization.
    - Provides built-in optimizers (SGD, Adam, Nesterov Momentum, etc.).
- **Portability**:
    - Supports multiple quantum backends like IBM Q, Rigetti Forest, Google Cirq, Microsoft Q#, Strawberry Fields, and ProjectQ.
    - Easily switch between different quantum hardware without modifying the code.
- **Accessibility**:
    - Offers tutorials and examples for quantum computing beginners.
    - Encourages community contributions and collaboration.

4.1 Installing PennyLane

To install:

```
pip install pennylane
```

4.2 Creating a Hybrid Quantum-Classical Neural Network

PennyLane integrates with PyTorch and TensorFlow for quantum ML.

```
import pennylane as qml
import numpy as np

# Define a quantum device
dev = qml.device("default.qubit", wires=1)

# Define a quantum node (QNode)
@qml.qnode(dev)
def quantum_circuit(theta):
    qml.RX(theta, wires=0)
    return qml.expval(qml.PauliZ(0))

# Compute the expectation value for θ = π/4
theta = np.pi / 4
print("Quantum Expectation Value:", quantum_circuit(theta))
```

4.3 Running PennyLane on Real Hardware

PennyLane supports multiple quantum platforms, including IBM, Google, and Amazon Braket.

To use PennyLane with Braket:

```
dev = qml.device("braket.aws.qubit", wires=2,
device_arn="arn:aws:braket:::device/qpu/ionq/Aria-1")

@qml.qnode(dev)
def quantum_model(theta):
    qml.RX(theta, wires=0)
    qml.CNOT(wires=[0, 1])
    return qml.expval(qml.PauliZ(1))

print("Quantum Computer Result:", quantum_model(theta))
```

4.4 Detail

Installing and Setting Up PennyLane

PennyLane requires Python 3.6+ and supports Windows, Linux, and macOS.

Installing PennyLane

The recommended installation method is via pip:

```
pip install pennylane
```

This installs:

- PennyLane core library
- Default built-in simulators (`default.qubit, default.mixed`)
- IBM Q compatibility

You can also install additional plugins for different quantum hardware:

```
pip install pennylane-qiskit   # For IBM Q devices
pip install pennylane-cirq     # For Google Cirq devices
pip install pennylane-forest   # For Rigetti Forest devices
```

Importing PennyLane

Once installed, you can start using PennyLane in Python:

```
import pennylane as qml
```

Creating Quantum Circuits in PennyLane

A quantum node (QNode) is a core component of PennyLane. It represents a trainable quantum circuit that can be executed on a quantum device or simulator.

Defining a Quantum Circuit

Below is an example of a simple quantum function:

```
import pennylane as qml

def quantum_function(x, y):
    qml.RZ(x, wires=0)  # Rotation around Z-axis
    qml.CNOT(wires=[0,1])  # Controlled-NOT gate
    qml.RY(y, wires=1)  # Rotation around Y-axis
    return qml.expval(qml.PauliZ(1))  # Expectation value of
PauliZ
```

Instead of calling them qubits, PennyLane refers to them as wires.

Choosing a Quantum Device

Before executing a quantum function, we must define a quantum device:

```
dev = qml.device('default.qubit', wires=2, shots=1000)
```

Here:

- default.qubit → Uses a classical simulator.
- wires=2 → Uses a 2-qubit quantum system.
- shots=1000 → Runs the circuit 1000 times to obtain results.

Creating a QNode

A QNode encapsulates the quantum function and device:

```
@qml.qnode(dev)
def circuit(x, y):
    return quantum_function(x, y)
```

Now, we can run the quantum circuit:

```
import numpy as np
print(circuit(np.pi/3, 0.5))
```

To visualize the quantum circuit:

```
print(circuit.draw())
```

Building a Variational Classifier

A variational classifier is a machine learning model that learns to classify data using trainable quantum circuits.

Creating a Quantum Device

```
dev = qml.device("default.qubit", wires=2)
```

Defining a Variational Quantum Circuit

We define a layered variational circuit:

```
def variational_layer(W):
    qml.Rot(W[0, 0], W[0, 1], W[0, 2], wires=0)
    qml.Rot(W[1, 0], W[1, 1], W[1, 2], wires=1)
    qml.CNOT(wires=[0, 1])
```

Encoding Data into Quantum States

Data inputs must be encoded into quantum states:

```
def quantum_state(x):
    qml.RY(x[0], wires=0)
    qml.CNOT(wires=[0, 1])
    qml.RY(x[1], wires=1)
```

Defining the Quantum Node

```
@qml.qnode(dev)
def circuit(weights, angles):
    quantum_state(angles)
    for W in weights:
        variational_layer(W)
    return qml.expval(qml.PauliZ(0))
```

Defining the Variational Classifier

```
def variational_classifier(var, x):
    weights = var[0]
    bias = var[1]
    return circuit(weights, x) + bias
```

Training the Variational Classifier

To train the classifier, we need:

1. A loss function.
2. An optimizer.
3. A dataset (e.g., Iris dataset).

Defining the Loss Function

```
def square_loss(labels, predictions):
    loss = 0
```

```
    for l, p in zip(labels, predictions):
        loss = loss + (l - p) ** 2
    return loss / len(labels)
```

Defining the Accuracy Function

```
def accuracy(labels, predictions):
    correct = sum(abs(l - p) < 1e-5 for l, p in zip(labels,
predictions))
    return correct / len(labels)
```

Preparing the Dataset

```
data = np.loadtxt("data/data.txt")
X = data[:, 0:2]

# Normalize Data
padding = 0.3 * np.ones((len(X), 1))
X_pad = np.c_[X, padding]
normalization = np.sqrt(np.sum(X_pad ** 2, axis=-1))
X_norm = (X_pad.T / normalization).T

# Convert to quantum state encoding
features = np.array([angles(x) for x in X_norm])
Y = data[:, -1]

# Split into training and validation sets
num_train = int(0.75 * len(Y))
feats_train = features[:num_train]
Y_train = Y[:num_train]
feats_val = features[num_train:]
Y_val = Y[num_train:]
```

Optimizing the Model

```
from pennylane.optimize import NesterovMomentumOptimizer

opt = NesterovMomentumOptimizer(0.01)
batch_size = 5
var = (0.01 * np.random.randn(6, 2, 3), 0.0)  # Initialize
weights and bias

for it in range(60):
    batch_index = np.random.randint(0, len(feats_train),
(batch_size,))
```

```
    var = opt.step(lambda v: cost(v, feats_train[batch_index],
Y_train[batch_index]), var)

    acc_train = accuracy(Y_train, [variational_classifier(var,
f) for f in feats_train])
    acc_val = accuracy(Y_val, [variational_classifier(var, f)
for f in feats_val])
    print("Iteration " + str(it + 1) + " | Accuracy Train: " +
str(acc_train) + " | Accuracy Val: " + str(acc_val))
```

Visualizing the Decision Boundary

```
import matplotlib.pyplot as plt

xx, yy = np.meshgrid(np.linspace(0.0, 1.5, 20), np.linspace(0.0,
1.5, 20))
X_grid = np.array([[x, y] for x, y in zip(xx.flatten(),
yy.flatten())])
predictions = [variational_classifier(var, angles(x)) for x in
X_grid]
Z = np.reshape(predictions, xx.shape)

plt.contourf(xx, yy, Z, cmap="RdBu", alpha=0.8)
plt.colorbar()
plt.show()
```

Appendix C - Glossary of Key Terms

A

Amplitude

A complex number representing the probability of a quantum state being measured in a particular state. The probability is given by the square of the absolute value of the amplitude.

Ancilla Qubit

An auxiliary qubit used temporarily during quantum computations, often employed for error correction and complex quantum operations.

Ansatz

A trial quantum circuit structure, commonly used in variational quantum algorithms such as the Variational Quantum Eigensolver (VQE) and Quantum Approximate Optimization Algorithm (QAOA). It serves as an initial framework for optimization.

Atom Trap

A technique that uses electromagnetic fields to hold individual atoms in place. Atom traps are frequently used in neutral atom quantum computing.

B

Bacon–Shor Code

A type of quantum error-correcting code that encodes quantum information into a subsystem of a Hilbert space. It allows simplified error correction procedures.

Bell State

A maximally entangled two-qubit quantum state, fundamental in quantum teleportation, superdense coding, and quantum networking.

Bloch Sphere

A graphical representation of a qubit's state on a 3D sphere, with the north and south poles corresponding to the classical states $|0\rangle$ and $|1\rangle$.

Boson Sampling

A quantum computational model that focuses on simulating boson (photon) interactions in an optical network. It is believed to be infeasible for classical computers.

Bra-Ket Notation (Dirac Notation)

A mathematical notation used to represent quantum states, where **⟨ψ|** (bra) represents a row vector and **|ψ⟩** (ket) represents a column vector.

C

Circuit Depth

The number of sequential quantum gate operations in a quantum circuit, directly affecting execution time and error rates.

Classical Bit

A standard unit of classical information, which can be either **0** or **1**.

Clifford Gates

A set of quantum gates, including Hadamard (H), Phase (S), and Controlled-NOT (CNOT), that form the foundation for quantum error correction.

Coherence Time

The duration for which a qubit retains its quantum state before decoherence disrupts its superposition or entanglement.

Controlled Gate

A quantum gate that applies an operation based on the state of a control qubit, such as the **CNOT** gate.

Cryogenic Cooling

The process of cooling quantum hardware to extremely low temperatures (millikelvin range) to minimize noise and decoherence.

Crosstalk

Unwanted interference between qubits in a quantum processor, leading to computational errors.

Cloud-Based Quantum Computing

Access to quantum computing resources over the internet, allowing users to execute quantum algorithms on real quantum hardware or simulators.

D

Decoherence

The loss of quantum properties due to environmental interactions, causing a quantum system to behave classically.

Deutsch-Jozsa Algorithm

An early quantum algorithm demonstrating quantum advantage by determining whether a function is constant or balanced with fewer function calls than classical algorithms.

Diagonalization

A mathematical process that expresses a matrix in terms of its eigenvalues and eigenvectors, crucial in quantum mechanics.

E

Eigenvalue & Eigenvector

In quantum mechanics, an eigenvector $|\psi\rangle$ of an operator **A** satisfies $A|\psi\rangle = \lambda|\psi\rangle$, where λ is the eigenvalue. Measurement of an observable yields eigenvalues.

Entanglement

A quantum phenomenon in which the state of one qubit is dependent on another, regardless of the distance between them.

Error Correction (Quantum Error Correction - QEC)

Techniques such as Shor's Code and Surface Codes that protect quantum information from noise and decoherence.

F

Fidelity

A measure of how close a quantum state is to another, often used in benchmarking quantum gates and error correction performance.

Fourier Transform (Quantum Fourier Transform, QFT)

A quantum transformation that maps a state to its frequency representation, essential for algorithms like Shor's Algorithm.

G

Gate Fidelity

A measure of how accurately a quantum gate performs its intended operation.

GHZ State (Greenberger-Horne-Zeilinger State)

A multi-qubit entangled state used in quantum cryptography and communication.

Grover's Algorithm

A quantum search algorithm providing a quadratic speedup over classical brute-force search.

H

Hadamard Gate (H Gate)

A fundamental quantum gate that places a qubit into an equal superposition of **|0⟩** and **|1⟩**.

Hamiltonian

An operator representing the total energy of a quantum system, governing time evolution.

I

Interference

The phenomenon where quantum probability amplitudes combine constructively or destructively, enabling quantum algorithms.

Ion Trap

A technology using electromagnetic fields to trap and manipulate charged ions for quantum computation.

J

Josephson Junction

A superconducting circuit element enabling quantum superposition and entanglement in superconducting qubits.

K

Ket (|ψ⟩)

A mathematical representation of a quantum state in Dirac notation.

L

Logical Qubit

A qubit encoded using multiple physical qubits to protect against errors.

M

Measurement

The process that collapses a quantum state into a classical outcome, producing either **0** or **1**.

Mixed State

A probabilistic mixture of quantum states, represented by a **density matrix**.

N

No-Cloning Theorem

A principle stating that an arbitrary quantum state cannot be copied perfectly.

Noise

Any disturbance causing errors and decoherence in quantum systems.

NISQ (Noisy Intermediate-Scale Quantum)

A term for near-term quantum computers with limited error correction, typically with **50-100 qubits**.

O

Oracle

A black-box function used in quantum algorithms like **Grover's search**.

P

Pauli Matrices (X, Y, Z Gates)

Fundamental quantum gates:

- **X (NOT gate)** flips **|0⟩** ↔ **|1⟩**
- **Y** introduces a phase and bit flip
- **Z** applies a phase flip

Q

Quantum Volume

A metric evaluating the performance of a quantum computer based on qubit count, gate fidelity, and connectivity.

Qubit (Quantum Bit)

The fundamental unit of quantum information, existing in a superposition of **|0⟩** and **|1⟩**.

S

Schrödinger's Cat

A thought experiment illustrating quantum superposition and measurement collapse.

Shor's Algorithm

A quantum algorithm for integer factorization, significantly faster than classical methods.

T

Teleportation (Quantum Teleportation)

A protocol for transferring quantum information between distant locations using entanglement.

Topological Qubit

A fault-tolerant qubit based on exotic **anyons**, reducing error rates.

V

Variational Quantum Eigensolver (VQE)

A hybrid quantum-classical algorithm for estimating ground-state energies in chemistry and physics.

Z

Zero Noise Extrapolation (ZNE)

An error mitigation technique reducing noise in quantum computations.

References

1. Juárez-Ramírez, R., et al., *The Foundations of Quantum Computing and Their Relation to Software Engineering.* Proceedings of the Institute for System Programming of Ras, 2024. **36**(1): p. 73-104.
2. Nielsen, M.A. and I.L. Chuang, *Quantum Computation and Quantum Information.* 2012.
3. Schuld, M., I. Sinayskiy, and F. Petruccione, *An Introduction to Quantum Machine Learning.* Contemporary Physics, 2014. **56**(2): p. 172-185.
4. Swan, M., F.M.C. Witte, and R.P.d. Santos, *Quantum Information Science.* Ieee Internet Computing, 2022. **26**(1): p. 7-14.
5. Kalai, G., *How Quantum Computers Fail: Quantum Codes, Correlations in Physical Systems, and Noise Accumulation.* 2011.
6. Venegas-Andraca, S.E., *Introductory Words: Special Issue on Quantum Image Processing Published by Quantum Information Processing.* Quantum Information Processing, 2015. **14**(5): p. 1535-1537.
7. Wang, Y., *When Quantum Computation Meets Data Science: Making Data Science Quantum.* 2022.
8. Priya, N., et al., *Quantum Computing Using Superconducting Qubits: A Review.* Evergreen, 2023. **10**(1): p. 340-347.
9. Dasgupta, S. and T.S. Humble, *Assessing the Stability of Noisy Quantum Computation.* 2022.
10. Steijl, R., *Quantum Algorithms for Fluid Simulations.* 2020.
11. Jerbi, D., *Revolutionizing Computing: A Comprehensive Introduction to Quantum Computing.* 2023.
12. Chen, J., *The Future of Quantum Computer Advantage.* American Journal of Computational Mathematics, 2023. **13**(04): p. 619-631.
13. Olorunsogo, T., B.S. Jacks, and O.A. Ajala, *Leveraging Quantum Computing for Inclusive and Responsible Ai Development: A Conceptual and Review Framework.* Computer Science & It Research Journal, 2024. **5**(3): p. 671-680.
14. Dharmawati, T., et al., *Adoption of Quantum Computing in Economic Analysis: Potential and Challenges in Distributed Information Systems.* Icst Transactions on Scalable Information Systems, 2023. **11**(1).
15. Ngoenriang, N., et al., *Optimal stochastic resource allocation for distributed quantum computing.* arXiv preprint arXiv:2210.02886, 2022.
16. Meyer, J.J., J. Borregaard, and J. Eisert, *A variational toolbox for quantum multi-parameter estimation.* npj Quantum Information, 2021. **7**(1): p. 89.
17. Zhu, D., et al., *Training of quantum circuits on a hybrid quantum computer.* Science advances, 2019. **5**(10): p. eaaw9918.
18. Gonçalves, C.P.d.S., *Quantum stochastic neural maps and quantum neural networks.* Available at SSRN 3502121, 2019.

19. Mocz, P. and A. Szasz, *Toward cosmological simulations of dark matter on quantum computers*. The Astrophysical Journal, 2021. **910**(1): p. 29.
20. Franke, S., et al., *Quantization of quasinormal modes for open cavities and plasmonic cavity quantum electrodynamics*. Physical review letters, 2019. **122**(21): p. 213901.
21. Liu, Q., *Comparisons of Conventional Computing and Quantum Computing Approaches*. Highlights in Science Engineering and Technology, 2023. **38**: p. 502-507.
22. Streif, M., F. Neukart, and M. Leib, *Solving Quantum Chemistry Problems With a D-Wave Quantum Annealer*. 2019: p. 111-122.
23. Gachnang, P., et al., *Quantum Computing in Supply Chain Management State of the Art and Research Directions*. Asian Journal of Logistics Management, 2022. **1**(1): p. 57-73.
24. Pai, A., K.M. Buddhiraju, and S.S. Durbha, *Multiclass Classification of Hyperspectral Remote Sensed Data Using QSVC*. 2022: p. 48.
25. Britt, K.A. and T.S. Humble, *High-Performance Computing With Quantum Processing Units*. Acm Journal on Emerging Technologies in Computing Systems, 2017. **13**(3): p. 1-13.
26. Yarkoni, S., et al., *Solving the Shipment Rerouting Problem With Quantum Optimization Techniques*. 2021: p. 502-517.
27. Koshka, Y. and M.A. Novotny, *Comparison of D-Wave Quantum Annealing and Classical Simulated Annealing for Local Minima Determination*. Ieee Journal on Selected Areas in Information Theory, 2020. **1**(2): p. 515-525.
28. Hernández, A.R., A.L.E. Maldonado, and J.H.D. Bermejo, *Architecture of Quantum Computing*. Journal of Engineering Research, 2024. **4**(15): p. 2-10.
29. Meter, R.V., *Quantum Computing's Classical Problem, Classical Computing's Quantum Problem*. Foundations of Physics, 2014. **44**(8): p. 819-828.
30. Ellenberger, K., et al., *Quantum Annealing Task Mapping for Heterogeneous Computing Systems*. 2024: p. 68.
31. Rojo, J., et al., *Trials and Tribulations of Developing Hybrid Quantum-Classical Microservices Systems*. 2021.
32. Khalate, P., et al., *An LLVM-based C++ Compiler Toolchain for Variational Hybrid Quantum-Classical Algorithms and Quantum Accelerators*. 2022.
33. Karamlou, A.H., et al., *Analyzing the Performance of Variational Quantum Factoring on a Superconducting Quantum Processor*. 2020.
34. McCaskey, A., et al., *Hybrid Programming for Near-Term Quantum Computing Systems*. 2018: p. 1-12.
35. McCaskey, A., et al., *XACC: A System-Level Software Infrastructure for Heterogeneous Quantum–classical Computing*. Quantum Science and Technology, 2020. **5**(2): p. 024002.
36. Barz, S., et al., *Experimental Verification of Quantum Computation*. Nature Physics, 2013. **9**(11): p. 727-731.
37. Outeiral, C., et al., *The Prospects of Quantum Computing in Computational Molecular Biology*. Wiley Interdisciplinary Reviews Computational Molecular Science, 2020. **11**(1).

38. Zhang, K., et al., *Carbon Nanotube Electron Blackbody and Its Radiation Spectra.* Proceedings of the National Academy of Sciences, 2023. **120**(6).
39. Chakravarty, S. and P. Kraus, *Can a Quantum Critical State Represent a Blackbody?* Annals of Physics, 2018. **388**: p. 135-146.
40. Ostilli, M. and C. Presilla, *Thermalization of Noninteracting Quantum Systems Coupled to Blackbody Radiation: A Lindblad-Based Analysis.* Physical Review A, 2017. **95**(6).
41. Kinross, A.W., et al., *Evolution of Quantum Fluctuations Near the Quantum Critical Point of the Transverse Field Ising Chain System.* Physical Review X, 2014. **4**(3).
42. Bolotin, A., *Justifying the Classical-Quantum Divide of the Copenhagen Interpretation.* 2014.
43. Norrgard, E.B., et al., *Quantum Blackbody Thermometry.* 2020.
44. Bell, J.S., *On the Einstein Podolsky Rosen Paradox.* 1995: p. 701-706.
45. Guo, X., et al., *The Energetics of Quantum Vacuum Friction: Field Fluctuations.* 2021.
46. Kaur, M. and M. Singh, *Quantum Double-Double-Slit Experiment With Momentum Entangled Photons.* Scientific Reports, 2020. **10**(1).
47. Zhou, H., et al., *Quantum Mechanical Double Slit for Molecular Scattering.* Science, 2021. **374**(6570): p. 960-964.
48. Andersen, A., et al., *Double-Slit Experiment With Single Wave-Driven Particles and Its Relation to Quantum Mechanics.* Physical Review E, 2015. **92**(1).
49. Liu, X., et al., *Einstein–Bohr Recoiling Double-Slit Gedanken Experiment Performed at the Molecular Level.* Nature Photonics, 2014. **9**(2): p. 120-125.
50. Harada, K., et al., *Interference Experiment With Asymmetric Double Slit by Using 1.2-MV Field Emission Transmission Electron Microscope.* Scientific Reports, 2018. **8**(1).
51. Bach, R., et al., *Controlled Double-Slit Electron Diffraction.* New Journal of Physics, 2013. **15**(3): p. 033018.
52. Peng, H., *Comprehensive Double Slit Experiments-Exploring Experimentally Mystery of Double Slit.* 2021.
53. Parrado-Rodríguez, P., et al., *Crosstalk Suppression for Fault-Tolerant Quantum Error Correction With Trapped Ions.* Quantum, 2021. **5**: p. 487.
54. Zhang, Y., et al., *Optimizing Quantum Programs Against Decoherence: Delaying Qubits Into Quantum Superposition.* 2019: p. 184-191.
55. Gisin, N., *Entanglement 25 Years After Quantum Teleportation: Testing Joint Measurements in Quantum Networks.* Entropy, 2019. **21**(3): p. 325.
56. Lee, J. and M.S. Kim, *Entanglement Teleportation via Werner States.* Physical Review Letters, 2000. **84**(18): p. 4236-4239.
57. Glancy, S. and H.M. Vasconcelos, *Methods for Producing Optical Coherent State Superpositions.* Journal of the Optical Society of America B, 2008. **25**(5): p. 712.
58. Wang, Z.-Y., et al., *Probabilistic Resumable Quantum Teleportation of a Two-Qubit Entangled State.* Entropy, 2019. **21**(4): p. 352.
59. Coles, P.J., et al., *Entropic Uncertainty Relations and Their Applications.* Reviews of Modern Physics, 2017. **89**(1).
60. Tawfik, A.N. and A.M. Diab, *A Review of the Generalized Uncertainty Principle.* Reports on Progress in Physics, 2015. **78**(12): p. 126001.

61. Gureyev, T.E., et al., *Noise-Resolution Uncertainty Principle in Classical and Quantum Systems*. Scientific Reports, 2020. **10**(1).
62. Rozema, L.A., et al., *Violation of Heisenberg's Measurement-Disturbance Relationship by Weak Measurements*. Physical Review Letters, 2012. **109**(10).
63. Tawfik, A.N. and A.M. Diab, *Generalized Uncertainty Principle: Approaches and Applications*. International Journal of Modern Physics D, 2014. **23**(12): p. 1430025.
64. Weststeijn, N., *Wigner's Friend and Relational Quantum Mechanics: A Reply to Laudisa*. Foundations of Physics, 2021. **51**(4).
65. Proietti, M., et al., *Experimental Test of Local Observer Independence*. Science Advances, 2019. **5**(9).
66. Li, Q., *Wigner's Friend Scenario and a New Interpretation of Quantum Mechanics V2*. 2023.
67. Frauchiger, D. and R. Renner, *Quantum Theory Cannot Consistently Describe the Use of Itself*. 2016.
68. Müller, M.M., S. Gherardini, and F. Caruso, *Quantum Zeno Dynamics Through Stochastic Protocols*. Annalen Der Physik, 2017. **529**(9).
69. Chaudhry, A.Z., *A General Framework for the Quantum Zeno and Anti-Zeno Effects*. Scientific Reports, 2016. **6**(1).
70. Majeed, M. and A.Z. Chaudhry, *The Quantum Zeno and Anti-Zeno Effects With Driving Fields in the Weak and Strong Coupling Regimes*. Scientific Reports, 2021. **11**(1).
71. Khalid, B. and A.Z. Chaudhry, *The Quantum Zeno and Anti-Zeno Effects: From Weak to Strong System-Environment Coupling*. The European Physical Journal D, 2019. **73**(7).
72. Zhang, X., et al., *Effect of Different Filling Tendencies on the Spatial Quantum Zeno Effect*. Scientific Reports, 2018. **8**(1).
73. Schneider, J. and I. Smalley, *What is a qubit?* 2025, IBM.
74. Grurl, T., J. Fuß, and R. Wille, *Considering Decoherence Errors in the Simulation of Quantum Circuits Using Decision Diagrams*. 2020.
75. Tamrakar, A. and R. Sharma, *Quantum Computing: A Comprehensive Review*. Turkish Journal of Computer and Mathematics Education (Turcomat), 2019. **10**(3): p. 1634-1642.
76. Martinez, J.E., et al., *Approximating Decoherence Processes for the Design and Simulation of Quantum Error Correction Codes on Classical Computers*. Ieee Access, 2020. **8**: p. 172623-172643.
77. Li, K., *The Qubit Fidelity Under Different Error Mechanisms Based on Error Correction Threshold*. Frontiers in Physics, 2022. **10**.
78. Gottesman, D., *An Introduction to Quantum Error Correction and Fault-Tolerant Quantum Computation*. 2010: p. 13-58.
79. Pushin, D.A., et al., *Experimental Realization of Decoherence-Free Subspace in Neutron Interferometry*. Physical Review Letters, 2011. **107**(15).
80. Habib, H.B., W.A. Hussein, and A.K. Abdul-Rahman, *A Hybrid Cryptosystem Based on Latin Square and the Modified BB84 Quantum Key Distribution*. Tikrit Journal of Pure Science, 2022. **27**(4): p. 100-103.

81. Munro, W.J., et al., *From Quantum Multiplexing to High-Performance Quantum Networking.* Nature Photonics, 2010. **4**(11): p. 792-796.
82. Bruzewicz, C., et al., *Trapped-Ion Quantum Computing: Progress and Challenges.* Applied Physics Reviews, 2019. **6**(2).
83. Kjærgaard, M., et al., *Superconducting Qubits: Current State of Play.* Annual Review of Condensed Matter Physics, 2020. **11**(1): p. 369-395.
84. Scappucci, G., et al., *The Germanium Quantum Information Route.* Nature Reviews Materials, 2020. **6**(10): p. 926-943.
85. Endo, S., S.C. Benjamin, and Y. Li, *Practical Quantum Error Mitigation for Near-Future Applications.* Physical Review X, 2018. **8**(3).
86. Brandhofer, S., et al., *Special Session: Noisy Intermediate-Scale Quantum (NISQ) Computers—How They Work, How They Fail, How to Test Them?* 2021.
87. Lao, L., et al., *Mapping of Lattice Surgery-Based Quantum Circuits on Surface Code Architectures.* Quantum Science and Technology, 2018. **4**(1): p. 015005.
88. Preskill, J., *Quantum Computing in the NISQ Era and Beyond.* Quantum, 2018. **2**: p. 79.
89. Molina, P.G., et al., *Noise in Digital and Digital-Analog Quantum Computation.* 2023.
90. Luca, G.D., *Survey of NISQ Era Hybrid Quantum-Classical Machine Learning Research.* Journal of Artificial Intelligence and Technology, 2021.
91. Ferraro, E., M. Fanciulli, and M.D. Michielis, *Non-Ideal X-Gate and Z-Gate in Semiconducting Spin Qubit Implementations.* 2019: p. 53.
92. Riggelen, F.v., et al., *Coherent Spin Qubit Shuttling Through Germanium Quantum Dots.* Nature Communications, 2024. **15**(1).
93. Miller, N.E. and S. Mukhopadhyay, *A Quantum Hopfield Associative Memory Implemented on an Actual Quantum Processor.* Scientific Reports, 2021. **11**(1).
94. El-Fattah, S.A., *Quantum Logic Gate From Silicon Quantum Dots.* 2023. **0**(0): p. 11-15.
95. Bauer, B., et al., *Quantum Algorithms for Quantum Chemistry and Quantum Materials Science.* Chemical Reviews, 2020. **120**(22): p. 12685-12717.
96. Huang, H.-L., et al., *Superconducting Quantum Computing: A Review.* 2020.
97. Hazra, S., et al., *Engineering Cross Resonance Interaction in Multi-Modal Quantum Circuits.* Applied Physics Letters, 2020. **116**(15).
98. He, K., et al., *Quantum Computation and Simulation With Superconducting Qubits*.* Chinese Physics B, 2021. **30**(8): p. 080304.
99. Rosenthal, E.I., et al., *Efficient and Low-Backaction Quantum Measurement Using a Chip-Scale Detector.* 2020.
100. Ishiyama, Y., et al., *Noise-Robust Optimization of Quantum Machine Learning Models for Polymer Properties Using a Simulator and Validated on the IonQ Quantum Computer.* Scientific Reports, 2022. **12**(1).
101. Saki, A.A., R.O. Topaloglu, and S. Ghosh, *Shuttle-Exploiting Attacks and Their Defenses in Trapped-Ion Quantum Computers.* Ieee Access, 2022. **10**: p. 2686-2699.
102. Suzuki, T., T. Hasebe, and T. Miyazaki, *Quantum Support Vector Machines for Classification and Regression on a Trapped-Ion Quantum Computer.* 2023.

103. Gale, E.P.G., et al., *Optimized Fast Gates for Quantum Computing With Trapped Ions.* Physical Review A, 2020. **101**(5).
104. Bushati, R., et al., *Trapped-Ion Quantum Computing With Integrated Photonics.* 2023.
105. Brown, K.R., et al., *Materials Challenges for Trapped-Ion Quantum Computers.* 2020.
106. Ikeri, H.I., A.I. Onyia, and O.J. Vwavware, *The Dependence of Confinement Energy on the Size of Quantum Dots.* International Journal of Scientific Research in Physics and Applied Sciences, 2019. **7**(2): p. 27-30.
107. Zwanenburg, F.A., et al., *Silicon Quantum Electronics.* Reviews of Modern Physics, 2013. **85**(3): p. 961-1019.
108. Kim, D., et al., *Ultrafast Optical Control of Entanglement Between Two Quantum-Dot Spins.* Nature Physics, 2010. **7**(3): p. 223-229.
109. Weperen, I.v., et al., *Charge-State Conditional Operation of a Spin Qubit.* Physical Review Letters, 2011. **107**(3).
110. Vitelli, C., et al., *Quantum State Fusion in Photons.* 2012.
111. Laccotripes, P., et al., *Spin-Photon Entanglement From a Solid-State System at Telecom Wavelengths.* 2024.
112. Jin, J., et al., *Telecom-Wavelength Atomic Quantum Memory in Optical Fiber for Heralded Polarization Qubits.* Physical Review Letters, 2015. **115**(14).
113. Duan, L. and C. Monroe, *<i>Colloquium</I>: Quantum Networks With Trapped Ions.* Reviews of Modern Physics, 2010. **82**(2): p. 1209-1224.
114. Pompili, M., et al., *Realization of a Multinode Quantum Network of Remote Solid-State Qubits.* Science, 2021. **372**(6539): p. 259-264.
115. Stolk, A., et al., *Metropolitan-Scale Heralded Entanglement of Solid-State Qubits.* Science Advances, 2024. **10**(44).
116. Gao, W., et al., *Teleportation-Based Realization of an Optical Quantum Two-Qubit Entangling Gate.* Proceedings of the National Academy of Sciences, 2010. **107**(49): p. 20869-20874.
117. Clausen, C., et al., *Quantum Storage of Heralded Polarization Qubits in Birefringent and Anisotropically Absorbing Materials.* Physical Review Letters, 2012. **108**(19).
118. Wang, H., C.-P. Yang, and F. Nori, *Robust and Scalable Optical One-Way Quantum Computation.* Physical Review A, 2010. **81**(5).
119. Feng, L., et al., *Realization of a Crosstalk-Avoided Quantum Network Node Using Dual-Type Qubits of the Same Ion Species.* Nature Communications, 2024. **15**(1).
120. Monroe, C., et al., *Large-Scale Modular Quantum-Computer Architecture With Atomic Memory and Photonic Interconnects.* Physical Review A, 2014. **89**(2).
121. Henriet, L., et al., *Quantum Computing With Neutral Atoms.* Quantum, 2020. **4**: p. 327.
122. Olmschenk, S., et al., *Randomized Benchmarking of Atomic Qubits in an Optical Lattice.* New Journal of Physics, 2010. **12**(11): p. 113007.
123. Xia, T., et al., *Randomized Benchmarking of Single-Qubit Gates in a 2D Array of Neutral-Atom Qubits.* Physical Review Letters, 2015. **114**(10).
124. Petrosyan, D., et al., *High-Fidelity Rydberg Quantum Gate via a Two-Atom Dark State.* Physical Review A, 2017. **96**(4).

125. Saffman, M., T. Walker, and K. Mølmer, *Quantum Information With Rydberg Atoms.* Reviews of Modern Physics, 2010. **82**(3): p. 2313-2363.
126. Weitenberg, C., et al., *Single-Spin Addressing in an Atomic Mott Insulator.* Nature, 2011. **471**(7338): p. 319-324.
127. Levine, H., et al., *Parallel Implementation of High-Fidelity Multiqubit Gates With Neutral Atoms.* Physical Review Letters, 2019. **123**(17).
128. Barredo, D., et al., *Synthetic Three-Dimensional Atomic Structures Assembled Atom by Atom.* Nature, 2018. **561**(7721): p. 79-82.
129. Ebert, M., et al., *Coherence and Rydberg Blockade of Atomic Ensemble Qubits.* Physical Review Letters, 2015. **115**(9).
130. Theis, L., et al., *High-Fidelity Rydberg-Blockade Entangling Gate Using Shaped, Analytic Pulses.* Physical Review A, 2016. **94**(3).
131. Shi, X., *Deutsch, Toffoli, and Cnot Gates via Rydberg Blockade of Neutral Atoms.* Physical Review Applied, 2018. **9**(5).
132. Wade, A.C.J., M. Mattioli, and K. Mølmer, *Single-Atom Single-Photon Coupling Facilitated by Atomic-Ensemble Dark-State Mechanisms.* Physical Review A, 2016. **94**(5).
133. Han, R., H.K. Ng, and B.G. Englert, *Implementing a Neutral-Atom Controlled-Phase Gate With a Single Rydberg Pulse.* Epl (Europhysics Letters), 2016. **113**(4): p. 40001.
134. Negretti, A., P. Treutlein, and T. Calarco, *Quantum Computing Implementations With Neutral Particles.* Quantum Information Processing, 2011. **10**(6): p. 721-753.
135. Lidar, D.A. and K.B. Whaley, *Decoherence-Free Subspaces and Subsystems.* 2003: p. 83-120.
136. Jeong, H., *Quantum Computation Using Weak Nonlinearities: Robustness Against Decoherence.* Physical Review A, 2006. **73**(5).
137. Zurek, W.H., *Decoherence, Einselection, and the Quantum Origins of the Classical.* Reviews of Modern Physics, 2003. **75**(3): p. 715-775.
138. Cormick, C. and J.P. Paz, *Decoherence Induced by a Dynamic Spin Environment: The Universal Regime.* Physical Review A, 2008. **77**(2).
139. Im, D.-G. and Y.-H. Kim, *Decoherence-Induced Sudden Death of Entanglement and Bell Nonlocality.* Photonics, 2022. **9**(2): p. 58.
140. Dattagupta, S., *Coherence Versus Decoherence - A Few Illustrative Examples.* Current Science, 2015. **109**(11): p. 1951.
141. Myatt, C.J., et al., *Decoherence of Quantum Superpositions Through Coupling to Engineered Reservoirs.* Nature, 2000. **403**(6767): p. 269-273.
142. Facchi, P., et al., *Control of Decoherence: Analysis and Comparison of Three Different Strategies.* Physical Review A, 2005. **71**(2).
143. Aharony, A., et al., *Partial Decoherence in Mesoscopic Systems.* Physica Scripta, 2012. **T151**: p. 014018.
144. Peng, X., D. Suter, and D.A. Lidar, *High Fidelity Quantum Memory via Dynamical Decoupling: Theory and Experiment.* Journal of Physics B Atomic Molecular and Optical Physics, 2011. **44**(15): p. 154003.
145. Annabestani, M., S.J. Akhtarshenas, and M.R. Abolhassani, *Decoherence in a One-Dimensional Quantum Walk.* Physical Review A, 2010. **81**(3).

146. Heo, J., et al., *Preparation of Quantum Information Encoded on Three-Photon Decoherence-Free States via Cross-Kerr Nonlinearities*. Scientific Reports, 2018. **8**(1).
147. Gong, J. and P. Brumer, *Intrinsic Decoherence Dynamics in Smooth Hamiltonian Systems: Quantum-Classical Correspondence*. Physical Review A, 2003. **68**(2).
148. Chandrashekar, C.M. and T. Busch, *Decoherence in Two-Dimensional Quantum Walks Using Four- And Two-State Particles*. Journal of Physics a Mathematical and Theoretical, 2013. **46**(10): p. 105306.
149. Bouchard, F., et al., *Observation of Quantum Recoherence of Photons by Spatial Propagation*. Scientific Reports, 2015. **5**(1).
150. Zhang, G., et al., *Suppression of Photon Shot Noise Dephasing in a Tunable Coupling Superconducting Qubit*. NPJ Quantum Information, 2017. **3**(1).
151. Bushev, P., et al., *Multiphoton Spectroscopy of a Hybrid Quantum System*. Physical Review B, 2010. **82**(13).
152. Eto, Y., et al., *Dissipation-Assisted Coherence Formation in a Spinor Quantum Gas*. Physical Review Letters, 2019. **122**(24).
153. Álvarez, G.A. and D. Suter, *Measuring the Spectrum of Colored Noise by Dynamical Decoupling*. Physical Review Letters, 2011. **107**(23).
154. Polonyi, J., *Instantaneous and Dynamical Decoherence*. Journal of Physics a Mathematical and Theoretical, 2018. **51**(14): p. 145302.
155. Luna, G., et al., *A Combinatorial Approach to Quantum Error Correcting Codes*. Discrete Mathematics Algorithms and Applications, 2014. **06**(04): p. 1450054.
156. Pudenz, K.L., T. Albash, and D.A. Lidar, *Error-Corrected Quantum Annealing With Hundreds of Qubits*. Nature Communications, 2014. **5**(1).
157. Young, K., M. Sarovar, and R. Blume-Kohout, *Error Suppression and Error Correction in Adiabatic Quantum Computation: Techniques and Challenges*. Physical Review X, 2013. **3**(4).
158. Paik, H., et al., *Observation of High Coherence in Josephson Junction Qubits Measured in a Three-Dimensional Circuit QED Architecture*. Physical Review Letters, 2011. **107**(24).
159. Samsonov, E.O., *Error Analysis in Circuits Building at the Quantum Computing Platform IBM Quantum Experience*. Nanosystems Physics Chemistry Mathematics, 2017: p. 272-276.
160. Carvalho, A.R.R., et al., *Error-Robust Quantum Logic Optimization Using a Cloud Quantum Computer Interface*. Physical Review Applied, 2021. **15**(6).
161. Quiroz, G. and D.A. Lidar, *High-Fidelity Adiabatic Quantum Computation via Dynamical Decoupling*. Physical Review A, 2012. **86**(4).
162. Dong, D., I.R. Petersen, and H. Rabitz, *Sampled-Data Design for Robust Decoherence Control of a Single Qubit*. 2012.
163. Wu, A., et al., *A Synthesis Framework for Stitching Surface Code With Superconducting Quantum Devices*. 2022: p. 337-350.
164. Zhao, Y., et al., *Realization of an Error-Correcting Surface Code With Superconducting Qubits*. 2021.

165. Sharma, S., S. Basnet, and R. Khanal, *Implementation of Error Correction on IBM Quantum Computing Devices.* Journal of Nepal Physical Society, 2022. **8**(1): p. 7-15.
166. Jayashankar, A. and P. Mandayam, *Quantum Error Correction: Noise-Adapted Techniques and Applications.* Journal of the Indian Institute of Science, 2022. **103**(2): p. 497-512.
167. Li, Z., et al., *Autonomous Error Correction of a Single Logical Qubit Using Two Transmons.* Nature Communications, 2024. **15**(1).
168. Pang, S., H. Xu, and M. Chen, *Construction of Binary Quantum Error-Correcting Codes From Orthogonal Array.* Entropy, 2022. **24**(7): p. 1000.
169. Day, C., *Fault-Tolerant Quantum Error Correction Without Measurements.* Physics, 2024. **17**.
170. Yao, Y., *Qubit: The Leap Into Quantum Computation.* Theoretical and Natural Science, 2024. **51**(1): p. 43-50.
171. Nguyen, N., E. Behrman, and J.E. Steck, *Quantum Learning With Noise and Decoherence: A Robust Quantum Neural Network.* Quantum Machine Intelligence, 2020. **2**(1).
172. Madhok, V. and A. Datta, *Quantum Discord as a Resource in Quantum Communication.* International Journal of Modern Physics B, 2012. **27**(01n03): p. 1345041.
173. Abdurakhimov, L.V., et al., *A Long-Lived Capacitively Shunted Flux Qubit Embedded in a 3D Cavity.* Applied Physics Letters, 2019. **115**(26).
174. Ho, S.-H., et al., *Decoherence Patterns of Topological Qubits From Majorana Modes.* New Journal of Physics, 2014. **16**(11): p. 113062.
175. Zwerger, M., et al., *Quantum Repeaters Based on Trapped Ions With Decoherence-Free Subspace Encoding.* Quantum Science and Technology, 2017. **2**(4): p. 044001.
176. Hua, M., X. Xiao, and Y. Gao, *Screening Effect in Charge Qubit.* Communications in Theoretical Physics, 2011. **56**(1): p. 74-78.
177. Liu, J., et al., *Universal Non-Adiabatic Holonomic Quantum Computation in Decoherence-Free Subspaces With Quantum Dots Inside a Cavity.* Laser Physics Letters, 2017. **14**(5): p. 055202.
178. Tosi, G., et al., *Silicon Quantum Processor With Robust Long-Distance Qubit Couplings.* Nature Communications, 2017. **8**(1).
179. Lekitsch, B., et al., *Blueprint for a Microwave Trapped Ion Quantum Computer.* Science Advances, 2017. **3**(2).
180. Crespi, A., et al., *Integrated Photonic Quantum Gates for Polarization Qubits.* Nature Communications, 2011. **2**(1).
181. Young, R., P. Birch, and C. Chatwin, *A Simplification of the Shor Quantum Factorization Algorithm Employing a Quantum Hadamard Transform.* 2018.
182. Fu, X.-Q., W.-S. Bao, and C. Zhou, *Speeding Up Implementation for Shor's Factorization Quantum Algorithm.* Chinese Science Bulletin, 2010. **55**(32): p. 3648-3653.
183. Wilmott, C. and P. Wild, *On a Generalized Quantum Swap Gate.* International Journal of Quantum Information, 2012. **10**(03): p. 1250034.

184. Tănăsescu, A., D. Constantinescu, and P.G. Popescu, *Distribution of Controlled Unitary Quantum Gates Towards Factoring Large Numbers on Today's Small-Register Devices*. Scientific Reports, 2022. **12**(1).
185. Zhou, X., et al., *Adding Control to Arbitrary Unknown Quantum Operations*. Nature Communications, 2011. **2**(1).
186. Fu, J. and X. Wu, *Effective Simulation of Quantum Entanglement Using Classical Fields Modulated With Pseudorandom Phase Sequences*. Scienceopen Research, 2015. **0**(0).
187. Peruzzo, A., et al., *A Variational Eigenvalue Solver on a Photonic Quantum Processor*. Nature Communications, 2014. **5**(1).
188. Cadney, J., *Achievability of Two Qubit Gates Using Linear Optical Elements and Post-Selection*. 2014.
189. Antipov, A.V., E.O. Kiktenko, and A.K. Fedorov, *Efficient Realization of Quantum Primitives for Shor's Algorithm Using PennyLane Library*. 2022.
190. Hwang, C.C., C.-Y. Tseng, and C.-F. Su, *Quantum Circuit Design for Computer-Assisted Shor's Algorithm*. 2022.
191. Ha, J., J. Lee, and J. Heo, *Resource Analysis of Quantum Computing With Noisy Qubits for Shor's Factoring Algorithms*. Quantum Information Processing, 2022. **21**(2).
192. Wicaksono, A.W. and A. Wicaksana, *Implementation of Shor's Quantum Factoring Algorithm Using ProjectQ Framework*. International Journal of Engineering and Advanced Technology, 2019. **8**(6s3): p. 52-56.
193. Hen, I., *Period Finding With Adiabatic Quantum Computation*. Epl (Europhysics Letters), 2014. **105**(5): p. 50005.
194. Gyöngyösi, L., *Decoherence Dynamics Estimation for Superconducting Gate-Model Quantum Computers*. Quantum Information Processing, 2020. **19**(10).
195. Gidney, C., *Halving the Cost of Quantum Addition*. Quantum, 2018. **2**: p. 74.
196. Hempel, C., et al., *Quantum Chemistry Calculations on a Trapped-Ion Quantum Simulator*. Physical Review X, 2018. **8**(3).
197. Mendelson, S.S., et al., *Quantum-Assisted Clustering Algorithms for NISQ-Era Devices*. 2019.
198. Donkers, H., et al., *QPack Scores: Quantitative Performance Metrics for Application-Oriented Quantum Computer Benchmarking*. 2022.
199. Gold, A., et al., *Entanglement Across Separate Silicon Dies in a Modular Superconducting Qubit Device*. NPJ Quantum Information, 2021. **7**(1).
200. Ramette, J., et al., *Any-to-Any Connected Cavity-Mediated Architecture for Quantum Computing With Trapped Ions or Rydberg Arrays*. 2021.
201. Landsman, K.A., et al., *Two-Qubit Entangling Gates Within Arbitrarily Long Chains of Trapped Ions*. Physical Review A, 2019. **100**(2).
202. Kandala, A., et al., *Error Mitigation Extends the Computational Reach of a Noisy Quantum Processor*. Nature, 2019. **567**(7749): p. 491-495.
203. Li, B., et al., *Pulse-Level Noisy Quantum Circuits With QuTiP*. Quantum, 2022. **6**: p. 630.
204. Mouradian, S., *Optical Engineering for Trapped Ion Quantum Computing*. 2023: p. 51.

205. Pino, J.M., et al., *Demonstration of the Trapped-Ion Quantum-CCD Computer Architecture*. 2020.
206. Zhou, X., Y. Feng, and S. Li, *Quantum Circuit Transformation: A Monte Carlo Tree Search Framework*. Acm Transactions on Design Automation of Electronic Systems, 2022. **27**(6): p. 1-27.
207. Huang, C., et al., *Classical Simulation of Quantum Supremacy Circuits*. 2020.
208. Davarzani, Z., M. Zomorodi-Moghadam, and M. Houshmand, *A Hierarchical Approach for Building Distributed Quantum Systems*. Scientific Reports, 2022. **12**(1).
209. Kumar, A., et al., *Experimental Realization of Controlled Quantum Teleportation of Arbitrary Qubit States via Cluster States*. Scientific Reports, 2020. **10**(1).
210. Taufiqi, M., et al., *Quantum Controlled Teleportation With OR-logic-gate-like Controllers in Noisy Environment*. Physica Scripta, 2023. **98**(12): p. 125120.
211. Esfahani, S.S., Z. Liao, and M.S. Zubairy, *Robust Quantum State Recovery From Amplitude Damping Within a Mixed States Framework*. Journal of Physics B Atomic Molecular and Optical Physics, 2016. **49**(15): p. 155501.
212. Zhang, Y. and K. Zhang, *GHZ Transform (I): Bell Transform and Quantum Teleportation*. 2014.
213. Zaman, F., Y. Jeong, and H. Shin, *Dual Quantum Zeno Superdense Coding*. Scientific Reports, 2019. **9**(1).
214. Henderson, J.M., et al., *Automated Quantum Oracle Synthesis With a Minimal Number of Qubits*. 2023: p. 13.
215. Daskin, A., et al., *Universal Programmable Quantum Circuit Schemes to Emulate an Operator*. The Journal of Chemical Physics, 2012. **137**(23).
216. Park, B. and D. Ahn, *Reducing CNOT Count in Quantum Fourier Transform With Linear Nearest-Neighbor Architecture*. 2023.
217. Evenbly, G. and G. Vidal, *Algorithms for Entanglement Renormalization*. Physical Review B, 2009. **79**(14).
218. Wang, B.-X., et al., *Efficient Quantum Simulation of Photosynthetic Light Harvesting*. NPJ Quantum Information, 2018. **4**(1).
219. Bravyi, S., G. Smith, and J.A. Smolin, *Trading Classical and Quantum Computational Resources*. Physical Review X, 2016. **6**(2).
220. Perez-Delgado, C.A. and P. Kok, *What Is a Quantum Computer, and How Do We Build One?* 2009.
221. Brennen, G.K., D. Song, and C.J. Williams, *Quantum-Computer Architecture Using Nonlocal Interactions*. Physical Review A, 2003. **67**(5).
222. Meter, R.V. and D. Horsman, *A Blueprint for Building a Quantum Computer*. Communications of the Acm, 2013. **56**(10): p. 84-93.
223. Wang, Z., et al., *Quantum Fourier Transform and Its Application in Shor's Algorithm*. 2020.
224. Singleton, R.L., *Shor's Factoring Algorithm and Modular Exponentiation Operators*. Quanta, 2023. **12**(1): p. 41-130.
225. Vandersypen, L.M.K., et al., *Experimental Realization of Shor's Quantum Factoring Algorithm Using Nuclear Magnetic Resonance*. Nature, 2001. **414**(6866): p. 883-887.

226. Nam, Y., Y. Su, and D. Maslov, *Approximate Quantum Fourier Transform With O(n Log(n)) T Gates.* NPJ Quantum Information, 2020. **6**(1).
227. Abbott, A.A., *De-Quantisation of the Quantum Fourier Transform.* Applied Mathematics and Computation, 2012. **219**(1): p. 3-13.
228. Lomonaco, S.J. and L.H. Kauffman, *Quantum Hidden Subgroup Algorithms.* 2007: p. 3-46.
229. Beauregard, S., *Circuit for Shor's Algorithm Using 2n+3 Qubits.* Quantum Information and Computation, 2003. **3**(2): p. 175-185.
230. Maitra, A. and S. Sarkar, *On Admissible States of Quantum Fourier Transform.* 2011.
231. Bayro-Corrachono, E. and Z. Vazquez-Flores, *Image Processing Using the Quantum Quaternion Fourier Transform.* Mathematical Methods in the Applied Sciences, 2023. **47**(3): p. 1305-1317.
232. Gogioso, S. and W.J. Zeng, *Fourier Transforms From Strongly Complementary Observables.* 2015.
233. Vorobyov, V., et al., *Quantum Fourier Transform for Quantum Sensing.* 2020.
234. Bagherimehrab, M. and A. Aspuru-Guzik, *Efficient Quantum Algorithm for All Quantum Wavelet Transforms.* Quantum Science and Technology, 2024. **9**(3): p. 035010.
235. Oshima, K., *More on Quantum Fast Fourier Transform by Linear Optics.* The Open Optics Journal, 2014. **8**(1): p. 1-2.
236. Semenenko, H. and T. Byrnes, *Implementing the Deutsch-Jozsa Algorithm With Macroscopic Ensembles.* Physical Review A, 2016. **93**(5).
237. Su, Q.-P. and C.P. Yang, *Circuit QED: Implementation of the Three-Qubit Refined Deutsch-Jozsa Quantum Algorithm.* 2014.
238. Cotler, J., H.-Y. Huang, and J.R. McClean, *Revisiting Dequantization and Quantum Advantage in Learning Tasks.* 2021.
239. Collins, D., K.W. Kim, and W.C. Holton, *Deutsch-Jozsa Algorithm as a Test of Quantum Computation.* Physical Review A, 1998. **58**(3): p. R1633-R1636.
240. Shi, F., et al., *Room-Temperature Implementation of the Deutsch-Jozsa Algorithm With a Single Electronic Spin in Diamond.* Physical Review Letters, 2010. **105**(4).
241. Choy, K., et al., *The Dynamics of Entanglement in the Adiabatic Search and Deutsch Algorithms.* Canadian Journal of Physics, 2007. **85**(10): p. 995-1021.
242. Harrow, A.W., A. Hassidim, and S. Lloyd, *Quantum Algorithm for Linear Systems of Equations.* Physical Review Letters, 2009. **103**(15).
243. Zhan, Z., *Implementation of Two-Qubit Deutsch–jozsa Algorithm With Trapped Ions.* Modern Physics Letters B, 2009. **23**(12): p. 1539-1546.
244. Combarro, E.F., et al., *On a Poset of Quantum Exact Promise Problems.* Quantum Information Processing, 2021. **20**(6).
245. Linden, N., H. Barjat, and R. Freeman, *An Implementation of the Deutsch–Jozsa Algorithm on a Three-Qubit NMR Quantum Computer.* Chemical Physics Letters, 1998. **296**(1-2): p. 61-67.
246. Pati, A.K. and S.L. Braunstein, *Deutsch-Jozsa Algorithm for Continuous Variables.* 2003: p. 31-36.

247. Ying, M., *Quantum Computation, Quantum Theory and AI.* Artificial Intelligence, 2010. **174**(2): p. 162-176.
248. Li, F.G., et al., *Transitionless Driving on Local Adiabatic Quantum Search Algorithm.* Chinese Physics B, 2018. **27**(1): p. 010308.
249. Brassard, G., P. Høyer, and A. Tapp, *Quantum Counting.* 1998: p. 820-831.
250. Arıkan, E., *An Information-Theoretic Analysis of Grover's Algorithm.* 2003: p. 339-347.
251. Shenvi, N., J. Kempe, and K.B. Whaley, *Quantum Random-Walk Search Algorithm.* Physical Review A, 2003. **67**(5).
252. Roland, J. and N.J. Cerf, *Quantum Search by Local Adiabatic Evolution.* Physical Review A, 2002. **65**(4).
253. Shapira, D., S. Mozes, and O. Biham, *Effect of Unitary Noise on Grover's Quantum Search Algorithm.* Physical Review A, 2003. **67**(4).
254. Jiayu, Z., et al., *Analysis and Simulation of Grover's Search Algorithm.* International Journal of Machine Learning and Computing, 2014: p. 21-23.
255. Ekerå, M. and J. Håstad, *Quantum Algorithms for Computing Short Discrete Logarithms and Factoring RSA Integers.* 2017: p. 347-363.
256. Lu, C.Y., et al., *Demonstration of a Compiled Version of Shor's Quantum Factoring Algorithm Using Photonic Qubits.* Physical Review Letters, 2007. **99**(25).
257. Nene, M.J. and G. Upadhyay, *Shor's Algorithm for Quantum Factoring.* 2016: p. 325-331.
258. Politi, A., J.C.F. Matthews, and J.L. O'Brien, *Shor's Quantum Factoring Algorithm on a Photonic Chip.* Science, 2009. **325**(5945): p. 1221-1221.
259. Amico, M., Z.H. Saleem, and M. Kumph, *Experimental Study of Shor's Factoring Algorithm Using the IBM Q Experience.* Physical Review A, 2019. **100**(1).
260. Guo, S., et al., *Three-Party Password Authentication and Key Exchange Protocol Based on MLWE.* Symmetry, 2023. **15**(9): p. 1750.
261. Saida, D., et al., *Factorization by Quantum Annealing Using Superconducting Flux Qubits Implementing a Multiplier Hamiltonian.* Scientific Reports, 2022. **12**(1).
262. Guerreschi, G.G. and M. Smelyanskiy, *Practical Optimization for Hybrid Quantum-Classical Algorithms.* 2017.
263. Romero, J. and A. Aspuru-Guzik, *Variational Quantum Generators: Generative Adversarial Quantum Machine Learning for Continuous Distributions.* Advanced Quantum Technologies, 2020. **4**(1).
264. Moussa, C., H. Calandra, and V. Dunjko, *To Quantum or Not to Quantum: Towards Algorithm Selection in Near-Term Quantum Optimization.* Quantum Science and Technology, 2020. **5**(4): p. 044009.
265. Morales, M.E.S., J. Biamonte, and Z. Zimborás, *On the Universality of the Quantum Approximate Optimization Algorithm.* Quantum Information Processing, 2020. **19**(9).
266. Pira, L. and C. Ferrie, *An Invitation to Distributed Quantum Neural Networks.* Quantum Machine Intelligence, 2023. **5**(2).
267. Ozaeta, A., W.v. Dam, and P.L. McMahon, *Expectation Values From the Single-Layer Quantum Approximate Optimization Algorithm on Ising Problems.* Quantum Science and Technology, 2022. **7**(4): p. 045036.

268. Farhi, E., D. Gamarnik, and S. Gutmann, *The Quantum Approximate Optimization Algorithm Needs to See the Whole Graph: Worst Case Examples*. 2020.
269. Herrman, R., et al., *Impact of Graph Structures for QAOA on MaxCut*. 2021.
270. Deng, D.-L., X. Li, and S.D. Sarma, *Quantum Entanglement in Neural Network States*. Physical Review X, 2017. **7**(2).
271. He, Z., et al., *Quantum Architecture Search With Meta-Learning*. Advanced Quantum Technologies, 2022. **5**(8).
272. Zhang, Q. and S.-M. Fei, *Coherence-Mixedness Trade-Offs*. Journal of Physics a Mathematical and Theoretical, 2024. **57**(23): p. 235301.
273. Bacon, D., et al., *Universal Fault-Tolerant Quantum Computation on Decoherence-Free Subspaces*. Physical Review Letters, 2000. **85**(8): p. 1758-1761.
274. Singh, U., et al., *Maximally Coherent Mixed States: Complementarity Between Maximal Coherence and Mixedness*. Physical Review A, 2015. **91**(5).
275. Su, Y., et al., *Coherence as Resource in Scattering Quantum Walk Search on Complete Graph*. Scientific Reports, 2018. **8**(1).
276. Zhou, Z.-W., et al., *Scalable Fault-Tolerant Quantum Computation in Decoherence-Free Subspaces*. Physical Review Letters, 2004. **93**(1).
277. Kempe, J., et al., *Theory of Decoherence-Free Fault-Tolerant Universal Quantum Computation*. Physical Review A, 2001. **63**(4).
278. Alicea, J., et al., *Non-Abelian Statistics and Topological Quantum Information Processing in 1D Wire Networks*. Nature Physics, 2011. **7**(5): p. 412-417.
279. Sarma, S.D., M. Freedman, and C. Nayak, *Majorana Zero Modes and Topological Quantum Computation*. NPJ Quantum Information, 2015. **1**(1).
280. Karzig, T., et al., *Scalable Designs for Quasiparticle-Poisoning-Protected Topological Quantum Computation With Majorana Zero Modes*. Physical Review B, 2017. **95**(23).
281. Thomas, C., et al., *Toward Durable Al-InSb Hybrid Heterostructures via Epitaxy of 2ML Interfacial InAs Screening Layers*. Physical Review Materials, 2019. **3**(12).
282. He, J.J., et al., *Correlated Spin Currents Generated by Resonant-Crossed Andreev Reflections in Topological Superconductors*. Nature Communications, 2014. **5**(1).
283. Xiao, X., J.K. Freericks, and A.F. Kemper, *Determining Quantum Phase Diagrams of Topological Kitaev-Inspired Models on NISQ Quantum Hardware*. Quantum, 2021. **5**: p. 553.
284. Qi, X.-L. and S.-C. Zhang, *Topological Insulators and Superconductors*. Reviews of Modern Physics, 2011. **83**(4): p. 1057-1110.
285. Laroche, D., et al., *Observation of the 4π-Periodic Josephson Effect in Indium Arsenide Nanowires*. Nature Communications, 2019. **10**(1).
286. Stanescu, T.D. and S. Tewari, *Majorana Fermions in Semiconductor Nanowires: Fundamentals, Modeling, and Experiment*. Journal of Physics Condensed Matter, 2013. **25**(23): p. 233201.
287. Chai, J.H. and H.K. Ng, *On the Fault-Tolerance Threshold for Surface Codes With General Noise*. 2022.
288. Paler, A. and S.J. Devitt, *An Introduction to Fault-Tolerant Quantum Computing*. 2015.
289. Bermúdez, A., et al., *Fault-Tolerant Protection of Near-Term Trapped-Ion Topological Qubits Under Realistic Noise Sources*. Physical Review A, 2019. **100**(6).

290. Menicucci, N.C., *Fault-Tolerant Measurement-Based Quantum Computing With Continuous-Variable Cluster States.* Physical Review Letters, 2014. **112**(12).
291. Campbell, E.T., H. Anwar, and D.E. Browne, *Magic-State Distillation in All Prime Dimensions Using Quantum Reed-Muller Codes.* Physical Review X, 2012. **2**(4).
292. Takita, M., et al., *Experimental Demonstration of Fault-Tolerant State Preparation With Superconducting Qubits.* Physical Review Letters, 2017. **119**(18).
293. Weinstein, Y.S., *Syndrome Measurement Strategies for the [[7,1,3]] Code.* Quantum Information Processing, 2015. **14**(6): p. 1841-1854.
294. Liu, Y., et al., *Magic State Distillation and Cost Analysis in Fault-Tolerant Universal Quantum Computation.* Quantum Science and Technology, 2023. **8**(4): p. 043001.
295. Cramer, J., et al., *Repeated Quantum Error Correction on a Continuously Encoded Qubit by Real-Time Feedback.* Nature Communications, 2016. **7**(1).
296. Cohn, I., et al., *Error Range for Encoding in Imperfect Quantum Error Correction.* 2024.
297. Litinski, D., *A Game of Surface Codes: Large-Scale Quantum Computing With Lattice Surgery.* Quantum, 2019. **3**: p. 128.
298. Paler, A. and S.J. Devitt, *An Introduction Into Fault-Tolerant Quantum Computing.* 2015.
299. Webster, P., et al., *Universal Fault-Tolerant Quantum Computing With Stabiliser Codes.* 2020.
300. Cane, R., et al., *Mitigation of Decoherence-Induced Quantum-Bit Errors and Quantum-Gate Errors Using Steane's Code.* Ieee Access, 2020. **8**: p. 83693-83709.
301. Sanders, Y.R., J.J. Wallman, and B.C. Sanders, *Bounding Quantum Gate Error Rate Based on Reported Average Fidelity.* New Journal of Physics, 2015. **18**(1): p. 012002.
302. Luo, L. and Z. Ma, *Universal Fault-Tolerant Quantum Computation Using Fault-Tolerant Conversion Schemes.* New Journal of Physics, 2019. **21**(8): p. 083032.
303. Glaudell, A.N., N.J. Ross, and J.M. Taylor, *Optimal Two-Qubit Circuits for Universal Fault-Tolerant Quantum Computation.* NPJ Quantum Information, 2021. **7**(1).
304. Kissinger, A. and J.v.d. Wetering, *Reducing the Number of Non-Clifford Gates in Quantum Circuits.* Physical Review A, 2020. **102**(2).
305. Duclos-Cianci, G. and D. Poulin, *Reducing the Quantum-Computing Overhead With Complex Gate Distillation.* Physical Review A, 2015. **91**(4).
306. Harper, R. and S.T. Flammia, *Fault-Tolerant Logical Gates in the IBM Quantum Experience.* Physical Review Letters, 2019. **122**(8).
307. Chandra, D., et al., *Quantum Topological Error Correction Codes Are Capable of Improving the Performance of Clifford Gates.* Ieee Access, 2019. **7**: p. 121501-121529.
308. Sun, K., et al., *Optical Demonstration of Quantum Fault-Tolerant Threshold.* Light Science & Applications, 2022. **11**(1).
309. Paler, A., et al., *A Fully Fault-Tolerant Representation of Quantum Circuits.* 2015: p. 139-154.
310. Brooks, P. and J. Preskill, *Fault-Tolerant Quantum Computation With Asymmetric Bacon-Shor Codes.* Physical Review A, 2013. **87**(3).

311. Fawzi, O., A. Grospellier, and A. Leverrier, *Constant Overhead Quantum Fault Tolerance With Quantum Expander Codes.* Communications of the Acm, 2020. **64**(1): p. 106-114.
312. Kueng, R., et al., *Comparing Experiments to the Fault-Tolerance Threshold.* Physical Review Letters, 2016. **117**(17).
313. Stephens, A.M., *Fault-Tolerant Thresholds for Quantum Error Correction With the Surface Code.* Physical Review A, 2014. **89**(2).
314. Dyakonov, M.I., *Revisiting the Hopes for Scalable Quantum Computation.* Jetp Letters, 2013. **98**(8): p. 514-518.
315. Krantz, P., et al., *A Quantum Engineer's Guide to Superconducting Qubits.* Applied Physics Reviews, 2019. **6**(2).
316. Bao, Z., et al., *A Cryogenic on-Chip Microwave Pulse Generator for Large-Scale Superconducting Quantum Computing.* Nature Communications, 2024. **15**(1).
317. Ahmad, M., et al., *Scalable Cryoelectronics for Superconducting Qubit Control and Readout.* Advanced Intelligent Systems, 2022. **4**(9).
318. Mariantoni, M. and A.V. Bardysheva, *High-Density Qubit Wiring: Pin-Chip Bonding for Fully Vertical Interconnects.* 2018.
319. Mallek, J., et al., *Fabrication of Superconducting Through-Silicon Vias.* 2021.
320. Ristè, D., et al., *Detecting Bit-Flip Errors in a Logical Qubit Using Stabilizer Measurements.* Nature Communications, 2015. **6**(1).
321. Kelly, J., et al., *State Preservation by Repetitive Error Detection in a Superconducting Quantum Circuit.* Nature, 2015. **519**(7541): p. 66-69.
322. Manovitz, T., et al., *A Trapped Ion Quantum Computer With Robust Entangling Gates and Quantum Coherent Feedback.* 2021.
323. Atiya, A.H. and M. Al-Temimi, *Review of Recent Laser Technology of Development Multi Qubit Gates Using Ion Trap Method.* Applied Mechanics and Materials, 2023. **915**: p. 33-42.
324. Ballance, C.J., et al., *High-Fidelity Quantum Logic Gates Using Trapped-Ion Hyperfine Qubits.* Physical Review Letters, 2016. **117**(6).
325. Shapira, Y., et al., *Robust Entanglement Gates for Trapped-Ion Qubits.* Physical Review Letters, 2018. **121**(18).
326. Aikyo, Y., et al., *Vacuum Characterization of a Compact Room-Temperature Trapped Ion System.* 2020.
327. Mount, E., et al., *Scalable Digital Hardware for a Trapped Ion Quantum Computer.* Quantum Information Processing, 2015. **15**(12): p. 5281-5298.
328. Ainsworth, R. and J.K. Slingerland, *Topological Qubit Design and Leakage.* New Journal of Physics, 2011. **13**(6): p. 065030.
329. Xue, Z.Y., et al., *Tunable Interfaces for Realizing Universal Quantum Computation With Topological Qubits.* Physical Review A, 2013. **88**(2).
330. Bonderson, P. and R.M. Lutchyn, *Topological Quantum Buses: Coherent Quantum Information Transfer Between Topological and Conventional Qubits.* Physical Review Letters, 2011. **106**(13).

331. Guan, S.Y., P.J. Chen, and T.-M. Chuang, *Topological Surface States and Superconductivity in Non-Centrosymmetric PbTaSe₂.* Japanese Journal of Applied Physics, 2021. **60**(SE): p. SE0803.
332. Wang, C., *Majorana Fermions and Its Application on Topological Quantum Computer.* Theoretical and Natural Science, 2024. **30**(1): p. 154-158.
333. Ayukaryana, N.R., M.H. Fauzi, and E.H. Hasdeo, *The Quest and Hope of Majorana Zero Modes in Topological Superconductor for Fault-Tolerant Quantum Computing: An Introductory Overview.* 2021. **2382**: p. 020007.
334. Kraus, C.V., P. Zoller, and M.A. Баранов, *Braiding of Atomic Majorana Fermions in Wire Networks and Implementation of the Deutsch-Jozsa Algorithm.* Physical Review Letters, 2013. **111**(20).
335. Li, Y., *Noise Threshold and Resource Cost of Fault-Tolerant Quantum Computing With Majorana Fermions in Hybrid Systems.* Physical Review Letters, 2016. **117**(12).
336. Deng, D.L. and L.M. Duan, *Fault-Tolerant Quantum Random-Number Generator Certified by Majorana Fermions.* Physical Review A, 2013. **88**(1).
337. Lian, B., et al., *Topological Quantum Computation Based on Chiral Majorana Fermions.* Proceedings of the National Academy of Sciences, 2018. **115**(43): p. 10938-10942.
338. Mezzacapo, A., et al., *Topological Qubits With Majorana Fermions in Trapped Ions.* New Journal of Physics, 2013. **15**(3): p. 033005.
339. Hassler, F., A. Akhmerov, and C.W.J. Beenakker, *The Top-Transmon: A Hybrid Superconducting Qubit for Parity-Protected Quantum Computation.* New Journal of Physics, 2011. **13**(9): p. 095004.
340. Vijay, S., T.H. Hsieh, and L. Fu, *Majorana Fermion Surface Code for Universal Quantum Computation.* Physical Review X, 2015. **5**(4).
341. Leijnse, M. and K. Flensberg, *Quantum Information Transfer Between Topological and Spin Qubit Systems.* Physical Review Letters, 2011. **107**(21).
342. Lutchyn, R.M., et al., *Majorana Zero Modes in Superconductor–semiconductor Heterostructures.* Nature Reviews Materials, 2018. **3**(5): p. 52-68.
343. Rokhinson, L.P., X. Liu, and J.K. Furdyna, *The Fractional a.c. Josephson Effect in a Semiconductor–superconductor Nanowire as a Signature of Majorana Particles.* Nature Physics, 2012. **8**(11): p. 795-799.
344. Stanescu, T.D., R.M. Lutchyn, and S.D. Sarma, *Majorana Fermions in Semiconductor Nanowires.* Physical Review B, 2011. **84**(14).
345. Zhang, S.-B., et al., *Topological and Holonomic Quantum Computation Based on Second-Order Topological Superconductors.* 2020.
346. Albrecht, S.M., et al., *Exponential Protection of Zero Modes in Majorana Islands.* Nature, 2016. **531**(7593): p. 206-209.
347. Tewari, S. and T.D. Stanescu, *Spin Polarization of Majorana Zero Modes and Topological Quantum Phase Transition in Semiconductor Majorana Nanowires.* 2016.
348. Chiu, C.K., et al., *Scalable Majorana Vortex Modes in Iron-Based Superconductors.* Science Advances, 2020. **6**(9).

349. Lai, H.-L., et al., *Exact Master Equation and Non-Markovian Decoherence Dynamics of Majorana Zero Modes Under Gate-Induced Charge Fluctuations*. Physical Review B, 2018. **97**(5).
350. Moore, C.H., T.D. Stanescu, and S. Tewari, *Two-Terminal Charge Tunneling: Disentangling Majorana Zero Modes From Partially Separated Andreev Bound States in Semiconductor-Superconductor Heterostructures*. Physical Review B, 2018. **97**(16).
351. Taballione, C., et al., *20-Mode Universal Quantum Photonic Processor*. 2022.
352. Takeda, S. and A. Furusawa, *Toward Large-Scale Fault-Tolerant Universal Photonic Quantum Computing*. Apl Photonics, 2019. **4**(6).
353. Hollenbach, M., et al., *A Photonic Platform Hosting Telecom Photon Emitters in Silicon*. 2021.
354. Bombin, H., et al., *Interleaving: Modular Architectures for Fault-Tolerant Photonic Quantum Computing*. 2021.
355. Lo, H.-P., et al., *Entanglement Generation Using a Controlled-Phase Gate for Time-Bin Qubits*. Applied Physics Express, 2018. **11**(9): p. 092801.
356. Ren, B.C., H.R. Wei, and F.G. Deng, *Deterministic Photonic Spatial-Polarization Hyper-Controlled-Not Gate Assisted by a Quantum Dot Inside a One-Side Optical Microcavity*. Laser Physics Letters, 2013. **10**(9): p. 095202.
357. Rajak, A., et al., *Quantum Annealing: An Overview*. Philosophical Transactions of the Royal Society a Mathematical Physical and Engineering Sciences, 2022. **381**(2241).
358. Śmierzchalski, T., et al., *Post-Error Correction for Quantum Annealing Processor Using Reinforcement Learning*. 2022.
359. Wang, Y., S. Wu, and H. Liu, *Statistical Analysis of Quantum Annealing*. Statistica Sinica, 2023.
360. Hu, F., et al., *Quantum Machine Learning With D-wave Quantum Computer*. Quantum Engineering, 2019. **1**(2).
361. Aguilera, E., et al., *Multi-Objective Portfolio Optimization Using a Quantum Annealer*. Mathematics, 2024. **12**(9): p. 1291.
362. Neukart, F., et al., *Traffic Flow Optimization Using a Quantum Annealer*. Frontiers in Ict, 2017. **4**.
363. Feld, S., et al., *A Hybrid Solution Method for the Capacitated Vehicle Routing Problem Using a Quantum Annealer*. Frontiers in Ict, 2019. **6**.
364. O'Malley, D., et al., *Nonnegative/Binary Matrix Factorization With a D-Wave Quantum Annealer*. Plos One, 2018. **13**(12): p. e0206653.
365. Wilson, B., et al., *Machine Learning Framework for Quantum Sampling of Highly Constrained, Continuous Optimization Problems*. Applied Physics Reviews, 2021. **8**(4).
366. Pourabdollah, A., G. Acampora, and R. Schiattarella, *Fuzzy Logic on Quantum Annealers*. Ieee Transactions on Fuzzy Systems, 2022. **30**(8): p. 3389-3394.
367. Haug, T. and M.S. Kim, *Scalable Measures of Magic Resource for Quantum Computers*. 2022.
368. Wright, K., et al., *Benchmarking an 11-Qubit Quantum Computer*. Nature Communications, 2019. **10**(1).

369. Khoshaman, A.H., et al., *Quantum Variational Autoencoder.* Quantum Science and Technology, 2018. **4**(1): p. 014001.
370. Somma, R.D., et al., *Simulating Physical Phenomena by Quantum Networks.* Physical Review A, 2002. **65**(4).
371. Kurzyk, D., Ł. Pawela, and Z. Puchała, *Unconditional Security of a K-State Quantum Key Distribution Protocol.* Quantum Information Processing, 2018. **17**(9).
372. Laaji, E.H. and A. Azizi, *A Combination of BB84 Quantum Key Distribution and an Improved Scheme of NTRU Post-Quantum Cryptosystem.* Journal of Cyber Security and Mobility, 2022.
373. Anghel, C., *Research, Development and Simulation of Quantum Cryptographic Protocols.* Elektronika Ir Elektrotechnika, 2013. **19**(4).
374. Miyadera, T., *Relation Between Information and Disturbance in Quantum Key Distribution Protocol With Classical Alice.* International Journal of Quantum Information, 2011. **09**(06): p. 1427-1435.
375. Sun, S.H., et al., *Security Evaluation of Quantum Key Distribution With Weak Basis-Choice Flaws.* Scientific Reports, 2020. **10**(1).
376. Bocquet, A., R. Alléaume, and A. Leverrier, *Optimal Eavesdropping on Quantum Key Distribution Without Quantum Memory.* Journal of Physics a Mathematical and Theoretical, 2011. **45**(2): p. 025305.
377. Guan, J.-Y., et al., *Experimental Passive Round-Robin Differential Phase-Shift Quantum Key Distribution.* Physical Review Letters, 2015. **114**(18).
378. Jeong, Y.-C., Y.S. Kim, and Y.-H. Kim, *An Experimental Comparison of BB84 and SARG04 Quantum Key Distribution Protocols.* Laser Physics Letters, 2014. **11**(9): p. 095201.
379. Yang, H., et al., *Quantum Security Analysis of SM4 Algorithm.* 2024: p. 45.
380. Kundi, D.e.S., et al., *Resource-Shared Crypto-Coprocessor of AES Enc/Dec With SHA-3.* Ieee Transactions on Circuits and Systems I Regular Papers, 2020. **67**(12): p. 4869-4882.
381. Sodiya, E.O., et al., *Quantum Computing and Its Potential Impact on U.S. Cybersecurity: A Review: Scrutinizing the Challenges and Opportunities Presented by Quantum Technologies in Safeguarding Digital Assets.* Global Journal of Engineering and Technology Advances, 2024. **18**(2): p. 049-064.
382. Sedik, T.S., M. Gorbanyov, and M. Malaika, *Quantum Computing and the Financial System: Spooky Action at a Distance?* Imf Working Paper, 2021. **2021**(071): p. 1.
383. Azhari, R. and A.N. Salsabila, *Analyzing the Impact of Quantum Computing on Current Encryption Techniques.* Iaic Transactions on Sustainable Digital Innovation (Itsdi), 2024. **5**(2): p. 148-157.
384. Bova, F.J., A. Goldfarb, and R.G. Melko, *Commercial Applications of Quantum Computing.* Epj Quantum Technology, 2021. **8**(1).
385. Sun, Y., *Securing the Future: Shifting to Post-Quantum Cryptography Amidst Quantum Threats.* Applied and Computational Engineering, 2024. **110**(1): p. 154-160.
386. Wimmer, M. and T.G. Moraes, *Quantum Computing, Digital Constitutionalism, and the Right to Encryption: Perspectives From Brazil.* Digital Society, 2022. **1**(2).

Quantum Computing

387. Li, S., et al., *Post-Quantum Security: Opportunities and Challenges*. Sensors, 2023. **23**(21): p. 8744.
388. Raya, J.E., A.S. Yahya, and E.K. Ahmad, *Protection From a Quantum Computer Cyber-Attack*. Technium Romanian Journal of Applied Sciences and Technology, 2023. **5**: p. 1-12.
389. Mavroeidis, V., et al., *The Impact of Quantum Computing on Present Cryptography*. International Journal of Advanced Computer Science and Applications, 2018. **9**(3).
390. Wang, H. and J. Yu, *A Blockchain Consensus Protocol Based on Quantum Attack Algorithm*. Computational Intelligence and Neuroscience, 2022. **2022**: p. 1-6.
391. Liu, F., et al., *A Survey on Lattice-Based Digital Signature*. Cybersecurity, 2024. **7**(1).
392. Zhou, Z., et al., *A Software/Hardware Co-Design of Crystals-Dilithium Signature Scheme*. Acm Transactions on Reconfigurable Technology and Systems, 2021. **14**(2): p. 1-21.
393. Opiłka, F., et al., *Performance Analysis of Post-Quantum Cryptography Algorithms for Digital Signature*. Applied Sciences, 2024. **14**(12): p. 4994.
394. Borges, F., P.R. Reis, and D. Pereira, *A Comparison of Security and Its Performance for Key Agreements in Post-Quantum Cryptography*. Ieee Access, 2020. **8**: p. 142413-142422.
395. Barker, W., W.T. Polk, and M. Souppaya, *Getting Ready for Post-Quantum Cryptography: Exploring Challenges Associated With Adopting and Using Post-Quantum Cryptographic Algorithms*. 2021.
396. Farooq, M., R.S.A. Khan, and P.M.H. Khan, *QuantIoT Novel Quantum Resistant Cryptographic Algorithm for Securing IoT Devices: Challenges and Solution*. 2023.
397. Kim, S., Y. Lee, and K. Yoon, *Performance Evaluation of Isogeny-Based Digital Signature Algorithms : Introducing FIBS -- Fast Isogeny Based Digital Signature*. 2023.
398. Song, Y., et al., *High-Speed and Scalable FPGA Implementation of the Key Generation for the Leighton-Micali Signature Protocol*. 2021.
399. Bernstein, D.J., et al., *The SPHINCS ⁺ Signature Framework*. 2019: p. 2129-2146.
400. Kang, T.G., et al., *On the Performance Analysis of SPHINCS⁺ Verification*. Ieice Transactions on Information and Systems, 2019. **E102.D**(12): p. 2603-2606.
401. Desai, L.R., *Advanced Techniques in Post-Quantum Cryptography for Ensuring Data Security in the Quantum Era*. 2024. **35**(1s): p. 58-71.
402. N, S.K.A. and R.N. B, *VIHS With ROTR Technique for Enhanced Light-Weighted Cryptographic System*. International Journal of Advanced Computer Science and Applications, 2022. **13**(6).
403. Alagic, G., et al., *Status Report on the Third Round of the NIST Post-Quantum Cryptography Standardization Process*. 2022.
404. Singh, R., et al., *An End-to-End Analysis of EMFI on Bit-Sliced Post-Quantum Implementations*. 2022.
405. Snetkov, N., J. Vakarjuk, and P. Laud, *TOPCOAT: Towards Practical Two-Party Crystals-Dilithium*. 2024.

406. Aguilera, A.C., et al., *First End-to-end PQC Protected DPU-to-DPU Communications.* Electronics Letters, 2023. **59**(17).

407. Kwon, H.-Y., I. Bajuna, and M.K. Lee, *Compact Hybrid Signature for Secure Transition to Post-Quantum Era.* Ieee Access, 2024. **12**: p. 39417-39429.

408. Wang, T., et al., *Optimized Hardware-Software Co-Design for Kyber and Dilithium on RISC-V SoC FPGA.* Iacr Transactions on Cryptographic Hardware and Embedded Systems, 2024. **2024**(3): p. 99-135.

409. Zhou, R., et al., *A Survey on Post-Quantum Cryptography for 5g/6g Communications.* 2023.

410. Luc, N.-Q., et al., *Building Applications and Developing Digital Signature Devices Based on the Falcon Post-Quantum Digital Signature Scheme.* Engineering Technology & Applied Science Research, 2023. **13**(2): p. 10401-10406.

411. Marangon, D.G., et al., *Long-Term Test of a Fast and Compact Quantum Random Number Generator.* Journal of Lightwave Technology, 2018. **36**(17): p. 3778-3784.

412. Mannalath, V., S. Mishra, and A. Pathak, *A Comprehensive Review of Quantum Random Number Generators: Concepts, Classification and the Origin of Randomness.* 2022.

413. Ma, X., et al., *Quantum Random Number Generation.* NPJ Quantum Information, 2016. **2**(1).

414. Zhou, Q., et al., *Practical Quantum Random-number Generation Based on Sampling Vacuum Fluctuations.* Quantum Engineering, 2019. **1**(1): p. e8.

415. Zhang, J., et al., *Finite-Size Analysis of Continuous Variable Source-Independent Quantum Random Number Generation.* 2020.

416. Sanguinetti, B., et al., *Quantum Random Number Generation on a Mobile Phone.* Physical Review X, 2014. **4**(3).

417. Ma, X., et al., *Postprocessing for Quantum Random-Number Generators: Entropy Evaluation and Randomness Extraction.* Physical Review A, 2013. **87**(6).

418. Nie, Y.-Q., et al., *The Generation of 68 GBPS Quantum Random Number by Measuring Laser Phase Fluctuations.* Review of Scientific Instruments, 2015. **86**(6).

419. Haylock, B., et al., *Multiplexed Quantum Random Number Generation.* Quantum, 2019. **3**: p. 141.

420. Zhou, H., et al., *Quantum Random-Number Generator Based on Tunneling Effects in a Si Diode.* Physical Review Applied, 2019. **11**(3).

421. Raffaelli, F., et al., *Combining a Quantum Random Number Generator and Quantum-Resistant Algorithms Into the GnuGPG Open-Source Software.* Advanced Optical Technologies, 2019. **9**(5): p. 287-295.

422. Shakhovoy, R., et al., *Quantum Noise Extraction From the Interference of Laser Pulses in an Optical Quantum Random Number Generator.* Optics Express, 2020. **28**(5): p. 6209.

423. Ivanova, A.E., С.А. Чивилихин, and A.V. Gleim, *The Use of Beam- And Fiber Splitters in Quantum Random Number Generators Based on Vacuum Fluctuations.* Nanosystems Physics Chemistry Mathematics, 2016: p. 378-383.

424. Singh, S.K., M.S. Agarwal, and M.R. Gupta, *Quantum Computing: Fundamentals, Progress, and Implications*. International Journal for Research in Applied Science and Engineering Technology, 2023. **11**(9): p. 1106-1111.
425. Tan, T.G., et al., *Post-Quantum Adversarial Modeling: A User's Perspective*. Computer, 2023. **56**(8): p. 58-67.
426. Qi, F., et al., *Quantum Vulnerability Analysis to Guide Robust Quantum Computing System Design*. Ieee Transactions on Quantum Engineering, 2024. **5**: p. 1-11.
427. Ding, Y., et al., *Systematic Crosstalk Mitigation for Superconducting Qubits via Frequency-Aware Compilation*. 2020: p. 201-214.
428. Román, R., et al., *A Quantum-Resistant Face Template Protection Scheme Using Kyber and Saber Public Key Encryption Algorithms*. 2022.
429. Sreerangapuri, A., *Post-Quantum Cryptography for AI-Driven Cloud Security Solutions*. International Journal for Multidisciplinary Research, 2024. **6**(5).
430. Yunakovsky, S.E., et al., *Towards Security Recommendations for Public-Key Infrastructures for Production Environments in the Post-Quantum Era*. Epj Quantum Technology, 2021. **8**(1).
431. Krithika, S. and T. Kesavmurthy, *Securing IOT Network Through Quantum Key Distribution*. International Journal of Innovative Technology and Exploring Engineering, 2019. **8**(6S4): p. 693-696.
432. Guitouni, Z., et al., *Advanced Error Correction Method for Quantum Key Distribution in IoT Systems*. Physica Scripta, 2024. **99**(10): p. 105106.
433. Bounmy, S. and K. Sisavath, *Securing Internet of Things (IoT) Ecosystems: A Quantum Cryptography Approach*. 2023. **1**(1): p. 1-7.
434. Ramos-Calderer, S., *Efficient Quantum Interpolation of Natural Data*. 2022.
435. Kim, J.-H., et al., *Quantum Superposing Algorithm for Quantum Encoding*. Physica Scripta, 2024. **99**(11): p. 115110.
436. Schuld, M., et al., *Circuit-Centric Quantum Classifiers*. Physical Review A, 2020. **101**(3).
437. Wang, Y. and L. Cao, *Quantum Phase Transition Detection via Quantum Support Vector Machine*. Quantum Science and Technology, 2024. **10**(1): p. 015043.
438. Low, G.H. and I.L. Chuang, *Hamiltonian Simulation by Qubitization*. Quantum, 2019. **3**: p. 163.
439. Bokhan, D., et al., *Multiclass Classification Using Quantum Convolutional Neural Networks With Hybrid Quantum-Classical Learning*. 2022.
440. Havlíček, V., et al., *Supervised Learning With Quantum-Enhanced Feature Spaces*. Nature, 2019. **567**(7747): p. 209-212.
441. Nguyen, X.-B., et al., *Quantum Visual Feature Encoding Revisited*. 2024.
442. Schuld, M., *Supervised Quantum Machine Learning Models Are Kernel Methods*. 2021.
443. Schuld, M. and N. Killoran, *Quantum Machine Learning in Feature Hilbert Spaces*. Physical Review Letters, 2019. **122**(4).
444. Kordzanganeh, M., A. Utting, and A.M.M. Scaife, *Quantum Machine Learning for Radio Astronomy*. 2021.

445. Beer, K., et al., *Training Deep Quantum Neural Networks.* Nature Communications, 2020. **11**(1).
446. Jiang, W., J. Xiong, and Y. Shi, *A Co-Design Framework of Neural Networks and Quantum Circuits Towards Quantum Advantage.* Nature Communications, 2021. **12**(1).
447. Yang, Q., et al., *The Dilemma of Quantum Neural Networks.* 2021.
448. Küçükkara, M.Y., F. Atban, and C. Bayılmış, *Quantum-Neural Network Model for Platform Independent Ddos Attack Classification in Cyber Security.* Advanced Quantum Technologies, 2024.
449. Zhang, K., et al., *Toward Trainability of Quantum Neural Networks.* 2020.
450. Takahashi, K., Y. Shiotani, and M. Hashimoto, *Self-Tuning PID Controller Using Quantum Neural Network With Qubit-Inspired Neurons.* International Journal of Hybrid Intelligent Systems, 2015. **12**(1): p. 41-52.
451. Purushothaman, G. and N.B. Karayiannis, *Quantum Neural Networks (QNNs): Inherently Fuzzy Feedforward Neural Networks.* Ieee Transactions on Neural Networks, 1997. **8**(3): p. 679-693.
452. Wang, H., et al., *QOC: Quantum on-Chip Training With Parameter Shift and Gradient Pruning.* 2022.
453. Qi, J. and J. Tejedor, *Classical-to-Quantum Transfer Learning for Spoken Command Recognition Based on Quantum Neural Networks.* 2021.
454. Ezhov, A.A. and D. Ventura, *Quantum Neural Networks.* 2000: p. 213-235.
455. Park, S.-H., et al., *EQuaTE: Efficient Quantum Train Engine for Dynamic Analysis via HCI-based Visual Feedback.* 2023.
456. Markidis, S., *Programming Quantum Neural Networks on NISQ Systems: An Overview of Technologies and Methodologies.* Entropy, 2023. **25**(4): p. 694.
457. Simoes, R.D.M., et al., *Experimental Evaluation of Quantum Machine Learning Algorithms.* Ieee Access, 2023. **11**: p. 6197-6208.
458. Kattemölle, J. and G. Burkard, *Effects of Correlated Errors on the Quantum Approximate Optimization Algorithm.* 2022.
459. Chen, Q., et al., *Information Scrambling and Entanglement in Quantum Approximate Optimization Algorithm Circuits.* 2023.
460. McArdle, S., et al., *Variational Ansatz-Based Quantum Simulation of Imaginary Time Evolution.* NPJ Quantum Information, 2019. **5**(1).
461. LaRose, R., et al., *Variational Quantum State Diagonalization.* NPJ Quantum Information, 2019. **5**(1).
462. Araújo, L.M.M., et al., *Quantum Machine Learning for Drowsiness Detection With EEG Signals.* 2023: p. 940-941.
463. Wu, A., et al., *Towards Efficient Ansatz Architecture for Variational Quantum Algorithms.* 2021.
464. Santos, P.G.M.d., et al., *Quantum Enhanced Cross-Validation for Near-Optimal Neural Networks Architecture Selection.* International Journal of Quantum Information, 2018. **16**(08): p. 1840005.
465. Blank, C., et al., *Quantum Classifier With Tailored Quantum Kernel.* NPJ Quantum Information, 2020. **6**(1).

466. Jerbi, S., et al., *Quantum Machine Learning Beyond Kernel Methods.* Nature Communications, 2023. **14**(1).
467. Gil-Fuster, E., J. Eisert, and V. Dunjko, *On the Expressivity of Embedding Quantum Kernels.* Machine Learning Science and Technology, 2024. **5**(2): p. 025003.
468. Mengoni, R. and A.D. Pierro, *Kernel Methods in Quantum Machine Learning.* Quantum Machine Intelligence, 2019. **1**(3-4): p. 65-71.
469. Peters, E., et al., *Machine Learning of High Dimensional Data on a Noisy Quantum Processor.* NPJ Quantum Information, 2021. **7**(1).
470. Cao, Y., G.G. Guerreschi, and A. Aspuru-Guzik, *Quantum Neuron: An Elementary Building Block for Machine Learning on Quantum Computers.* 2017.
471. Zheng, J., Q. Gao, and Y. Lv, *Quantum Graph Convolutional Neural Networks.* 2021.
472. Suzuki, T., et al., *Quantum AI Simulator Using a Hybrid CPU–FPGA Approach.* Scientific Reports, 2023. **13**(1).
473. Niu, M.Y., et al., *Entangling Quantum Generative Adversarial Networks.* Physical Review Letters, 2022. **128**(22).
474. Zeng, J., et al., *Learning and Inference on Generative Adversarial Quantum Circuits.* Physical Review A, 2019. **99**(5).
475. Takahashi, K., *Remarks on Motion Control of Nonholonomic System (Falling Cat) by Using a Quantum Neural Controller.* 2012.
476. Beigi, S., *Quantum Kernel Method in the Presence of Noise.* 2022.
477. Mitrpanont, J.L. and A. Srisuphab, *The Realization of Quantum Complex-Valued Backpropagation Neural Network in Pattern Recognition Problem.* 2002. **1**: p. 462-466.
478. Huang, H.-Y., et al., *Power of Data in Quantum Machine Learning.* Nature Communications, 2021. **12**(1).
479. Saxena, A. and S. Saxena, *Pancreatic Cancer Data Classification With Quantum Machine Learning.* Journal of Quantum Computing, 2023. **5**(1): p. 1-13.
480. Zheng, S., et al., *Demonstration of Breast Cancer Detection Using QSVM on IBM Quantum Processors.* 2022.
481. Kumar, T., D. Kumar, and G. Singh, *Performance Analysis of Quantum Classifier on Benchmarking Datasets.* International Journal of Electrical and Electronics Research, 2022. **10**(2): p. 375-380.
482. Bartkiewicz, K., et al., *Experimental Kernel-Based Quantum Machine Learning in Finite Feature Space.* Scientific Reports, 2020. **10**(1).
483. Du, S. and Y. Li, *A Novel Deformation Forecasting Method Utilizing Comprehensive Observation Data.* Advances in Mechanical Engineering, 2018. **10**(9): p. 168781401879633.
484. Maouaki, W.E., T. Said, and M. Беннаи, *Quantum Support Vector Machine for Prostate Cancer Detection: A Performance Analysis.* 2024.
485. Saeedi, S., A. Panahi, and T. Arodz, *Quantum Semi-Supervised Kernel Learning.* Quantum Machine Intelligence, 2021. **3**(2).
486. Rastunkov, V., et al., *Boosting Method for Automated Feature Space Discovery in Supervised Quantum Machine Learning Models.* 2022.

487. Schuld, M. and F. Petruccione, *Quantum Ensembles of Quantum Classifiers.* Scientific Reports, 2018. **8**(1).
488. Li, Z., et al., *Resonant Quantum Principal Component Analysis.* Science Advances, 2021. **7**(34).
489. Wang, Y. and Y. Luo, *Resource-Efficient Quantum Principal Component Analysis.* Quantum Science and Technology, 2024. **9**(3): p. 035031.
490. Chen, Y., *Quantum Dilated Convolutional Neural Networks.* Ieee Access, 2022. **10**: p. 20240-20246.
491. Mari, A., et al., *Transfer Learning in Hybrid Classical-Quantum Neural Networks.* Quantum, 2020. **4**: p. 340.
492. Chen, T., et al., *Quantum Fuzzy Regression Model for Uncertain Environment.* Computers Materials & Continua, 2023. **75**(2): p. 2759-2773.
493. Liang, J.M., et al., *Quantum Anomaly Detection With Density Estimation and Multivariate Gaussian Distribution.* Physical Review A, 2019. **99**(5).
494. Park, G., J. Huh, and D.K. Park, *Variational Quantum One-Class Classifier.* Machine Learning Science and Technology, 2023. **4**(1): p. 015006.
495. Herr, D., B. Obert, and M. Rosenkranz, *Anomaly Detection With Variational Quantum Generative Adversarial Networks.* Quantum Science and Technology, 2021. **6**(4): p. 045004.
496. O'Malley, P., et al., *Scalable Quantum Simulation of Molecular Energies.* Physical Review X, 2016. **6**(3).
497. Kassal, I., et al., *Simulating Chemistry Using Quantum Computers.* Annual Review of Physical Chemistry, 2011. **62**(1): p. 185-207.
498. Cao, Y., et al., *Quantum Chemistry in the Age of Quantum Computing.* Chemical Reviews, 2019. **119**(19): p. 10856-10915.
499. Santoro, S., et al., *Elucidation of Mechanisms and Selectivities of Metal-Catalyzed Reactions Using Quantum Chemical Methodology.* Accounts of Chemical Research, 2016. **49**(5): p. 1006-1018.
500. Hariharan, S., S. Kinge, and L. Visscher, *Modeling Heterogeneous Catalysis Using Quantum Computers: An Academic and Industry Perspective.* Journal of Chemical Information and Modeling, 2024. **65**(2): p. 472-511.
501. Cuenya, B.R. and F. Behafarid, *Nanocatalysis: Size- And Shape-Dependent Chemisorption and Catalytic Reactivity.* Surface Science Reports, 2015. **70**(2): p. 135-187.
502. Motta, M. and J.E. Rice, *Emerging Quantum Computing Algorithms for Quantum Chemistry.* Wiley Interdisciplinary Reviews Computational Molecular Science, 2021. **12**(3).
503. McArdle, S., et al., *Quantum Computational Chemistry.* Reviews of Modern Physics, 2020. **92**(1).
504. Rice, J.E., et al., *Quantum Chemistry Simulations of Dominant Products in Lithium-Sulfur Batteries.* 2020.
505. Copenhaver, J., A. Wasserman, and B. Wehefritz–Kaufmann, *Using Quantum Annealers to Calculate Ground State Properties of Molecules.* The Journal of Chemical Physics, 2021. **154**(3).

506. Wang, J. and G. Li, *Full-scale Modeling of Chemical Experiments.* Smart Molecules, 2023. **2**(1).
507. Genin, S.N., et al., *Estimating Phosphorescent Emission Energies in Ir(III) Complexes Using Large-Scale Quantum Computing Simulations.* 2021.
508. Wang, Y., et al., *Quantum Simulation of Helium Hydride Cation in a Solid-State Spin Register.* Acs Nano, 2015. **9**(8): p. 7769-7774.
509. Babbush, R., et al., *Exponentially More Precise Quantum Simulation of Fermions in Second Quantization.* New Journal of Physics, 2016. **18**(3): p. 033032.
510. Gilyén, A., et al., *Quantum Singular Value Transformation and Beyond: Exponential Improvements for Quantum Matrix Arithmetics.* 2019: p. 193-204.
511. Arrazola, J.M., et al., *Differentiable Quantum Computational Chemistry With PennyLane.* 2021.
512. Xing, D.-X., et al., *An Exploration of Computational Software in Organic Chemistry Teaching.* Frontiers in Educational Research, 2021. **4**(1).
513. Su, Y., et al., *Fault-Tolerant Quantum Simulations of Chemistry in First Quantization.* 2021.
514. Nam, Y., et al., *Ground-State Energy Estimation of the Water Molecule on a Trapped-Ion Quantum Computer.* NPJ Quantum Information, 2020. **6**(1).
515. Barends, R., et al., *Digital Quantum Simulation of Fermionic Models With a Superconducting Circuit.* Nature Communications, 2015. **6**(1).
516. Bhattacharya, A., K. Dasgupta, and B. Paine, *Dynamics of a Free Particle Using Classical Computing and Quantum Computing: Introducing Quantum Computing to Chemistry Students.* Journal of Chemical Education, 2024. **101**(4): p. 1599-1609.
517. Shang, H., et al., *Large-Scale Simulation of Quantum Computational Chemistry on a New Sunway Supercomputer.* 2022.
518. Kowalski, P.M., S. Blouin, and P. Dufour, *Infrared Opacities in Dense Atmospheres of Cool White Dwarf Stars.* 2016.
519. Kim, J.M., et al., *Strain Engineering of Low-Dimensional Materials for Emerging Quantum Phenomena and Functionalities.* Advanced Materials, 2022. **35**(27).
520. Le, T.P., et al., *Spin-Chain Model of a Many-Body Quantum Battery.* Physical Review A, 2018. **97**(2).
521. Luo, T., et al., *Rational Designs of Crystal Solid-solution Materials for Lithium-ion Batteries.* Physica Status Solidi (B), 2011. **248**(9): p. 2027-2031.
522. Rodríguez, R.R., et al., *Optimal Quantum Control of Charging Quantum Batteries.* New Journal of Physics, 2024. **26**(4): p. 043004.
523. Ferrón, A., P. Serra, and O. Osenda, *Near-Threshold Properties of the Electronic Density of Layered Quantum Dots.* Physical Review B, 2012. **85**(16).
524. Paudel, H.P., et al., *Quantum Computing and Simulations for Energy Applications: Review and Perspective.* Acs Engineering Au, 2022. **2**(3): p. 151-196.
525. Fries, K.S. and S. Steinberg, *Fermi-Level Characteristics of Potential Chalcogenide Superconductors.* Chemistry of Materials, 2018. **30**(7): p. 2251-2261.
526. Fan, Y., et al., *Equation-of-Motion Theory to Calculate Accurate Band Structures With a Quantum Computer.* The Journal of Physical Chemistry Letters, 2021. **12**(36): p. 8833-8840.

527. Kandala, A., et al., *Hardware-Efficient Variational Quantum Eigensolver for Small Molecules and Quantum Magnets.* Nature, 2017. **549**(7671): p. 242-246.
528. Reiher, M., et al., *Elucidating Reaction Mechanisms on Quantum Computers.* Proceedings of the National Academy of Sciences, 2017. **114**(29): p. 7555-7560.
529. Polini, M., et al., *Materials and Devices for Fundamental Quantum Science and Quantum Technologies.* 2022.
530. Bauer, B., et al., *Hybrid Quantum-Classical Approach to Correlated Materials.* Physical Review X, 2016. **6**(3).
531. Gómez, A., et al., *A Survey on Quantum Computational Finance for Derivatives Pricing and VaR.* Archives of Computational Methods in Engineering, 2022. **29**(6): p. 4137-4163.
532. Stamatopoulos, N., et al., *Option Pricing Using Quantum Computers.* Quantum, 2020. **4**: p. 291.
533. Egger, D.J., et al., *Quantum Computing for Finance: State-of-the-Art and Future Prospects.* Ieee Transactions on Quantum Engineering, 2020. **1**: p. 1-24.
534. Chakrabarti, S., et al., *A Threshold for Quantum Advantage in Derivative Pricing.* Quantum, 2021. **5**: p. 463.
535. Atadoga, A., et al., *The Intersection of Ai and Quantum Computing in Financial Markets: A Critical Review.* Computer Science & It Research Journal, 2024. **5**(2): p. 461-472.
536. Herman, D., et al., *A Survey of Quantum Computing for Finance.* 2022.
537. Bouland, A., et al., *Prospects and Challenges of Quantum Finance.* 2020.
538. Valdez, F. and P. Melín, *A Review on Quantum Computing and Deep Learning Algorithms and Their Applications.* Soft Computing, 2022. **27**(18): p. 13217-13236.
539. Hung, S., *The Fundamental Theory of Quantum Computing, Applications in Practical Fields, and Future Challenges.* Applied and Computational Engineering, 2023. **18**(1): p. 117-123.
540. Qin, J., *Review of Ansatz Designing Techniques for Variational Quantum Algorithms.* Journal of Physics Conference Series, 2023. **2634**(1): p. 012043.
541. Otten, M., et al., *Localized Quantum Chemistry on Quantum Computers.* Journal of Chemical Theory and Computation, 2022. **18**(12): p. 7205-7217.
542. Izsák, R., et al., *Quantum Computing in Pharma: A Multilayer Embedding Approach for Near Future Applications.* Journal of Computational Chemistry, 2022. **44**(3): p. 406-421.
543. Lau, B., et al., *Insights From Incorporating Quantum Computing Into Drug Design workflows.* Bioinformatics, 2022. **39**(1).
544. Li, W., et al., *A Hybrid Quantum Computing Pipeline for Real World Drug Discovery.* Scientific Reports, 2024. **14**(1).
545. Zhou, Y., et al., *Quantum-Machine-Assisted Drug Discovery: Survey and Perspective.* 2024.
546. Chaudhari, Y., et al., *Computational Pharmaceutics: A Comprehensive Review of Past Achievements, Present Applications, and Future Challenges.* 2024.
547. Cavasotto, C.N. and M.G. Aucar, *High-Throughput Docking Using Quantum Mechanical Scoring.* Frontiers in Chemistry, 2020. **8**.

548. Choi, J. and J. Lee, *V-Dock: Fast Generation of Novel Drug-Like Molecules Using Machine-Learning-Based Docking Score and Molecular Optimization.* International Journal of Molecular Sciences, 2021. **22**(21): p. 11635.
549. Batra, K., et al., *Quantum Machine Learning Algorithms for Drug Discovery Applications.* Journal of Chemical Information and Modeling, 2021. **61**(6): p. 2641-2647.
550. Fu, Y., et al., *A New Approach for Flexible Molecular Docking Based on Swarm Intelligence.* Mathematical Problems in Engineering, 2015. **2015**: p. 1-10.
551. Goto, H., et al., *High-Performance Combinatorial Optimization Based on Classical Mechanics.* Science Advances, 2021. **7**(6).
552. Eyring, V., et al., *Overview of the Coupled Model Intercomparison Project Phase 6 (CMIP6) Experimental Design and Organization.* Geoscientific Model Development, 2016. **9**(5): p. 1937-1958.
553. Chakraborty, P., *Revolutionizing Climate Modeling With Quantum Computing and Machine Learning.* International Research Journal of Modernization in Engineering Technology and Science, 2024.
554. Berger, C., et al., *Quantum Technologies for Climate Change: Preliminary Assessment.* 2021.
555. Wang, X. and Q. Guo, *A Review on Quantum Computing for Analysing Cloud Droplet Dynamics.* International Journal of Computing Programming and Database Management, 2024. **5**(1): p. 43-45.
556. Choi, J., et al., *Deep Learning of Electrochemical CO₂ Conversion Literature Reveals Research Trends and Directions.* Journal of Materials Chemistry A, 2023. **11**(33): p. 17628-17643.
557. Anders, I., K. Peters, and H. Thiemann, *Canonical Workflows in Simulation-Based Climate Sciences.* Data Intelligence, 2022. **4**(2): p. 212-225.
558. Runge, J., et al., *Inferring Causation From Time Series in Earth System Sciences.* Nature Communications, 2019. **10**(1).
559. Hwang, M.-L.T., W.-L. Collin, and L.S. Wang-xu, *Quantum Computing and Supply Chain Optimization: Addressing Complexity and Efficiency Challenges.* International Journal of Enterprise Modelling, 2021. **15**(3): p. 148-162.
560. Chen, W.J., G.S. Marcus, and D.S. Leesburg, *Quantum Computing for Manufacturing and Supply Chain Optimization: Enhancing Efficiency, Reducing Costs, and Improving Product Quality.* International Journal of Enterprise Modelling, 2021. **15**(3): p. 130-147.
561. Yarkoni, S., et al., *Quantum Annealing for Industry Applications: Introduction and Review.* Reports on Progress in Physics, 2022. **85**(10): p. 104001.
562. Zhahir, A.A., et al., *Quantum Computing in the Cloud - A Systematic Literature Review.* International Journal of Electrical and Computer Engineering Systems, 2024. **15**(2): p. 185-200.
563. McClean, J.R., et al., *The Theory of Variational Hybrid Quantum-Classical Algorithms.* New Journal of Physics, 2016. **18**(2): p. 023023.
564. Karalekas, P.J., et al., *A Quantum-Classical Cloud Platform Optimized for Variational Hybrid Algorithms.* Quantum Science and Technology, 2020. **5**(2): p. 024003.

565. Anand, A., et al., *Noise Robustness and Experimental Demonstration of a Quantum Generative Adversarial Network for Continuous Distributions.* Advanced Quantum Technologies, 2021. **4**(5).
566. Li, R., et al., *Approximate Quantum Adders With Genetic Algorithms: An IBM Quantum Experience.* Quantum Measurements and Quantum Metrology, 2017. **4**(1): p. 1-7.
567. Ding, Y., et al., *Experimental Implementation of a Quantum Autoencoder via Quantum Adders.* Advanced Quantum Technologies, 2019. **2**(7-8).
568. Olivares-Sánchez, J., et al., *Measurement-Based Adaptation Protocol With Quantum Reinforcement Learning in a Rigetti Quantum Computer.* Quantum Reports, 2020. **2**(2): p. 293-304.
569. Ciliberto, C., et al., *Quantum Machine Learning: A Classical Perspective.* Proceedings of the Royal Society a Mathematical Physical and Engineering Sciences, 2018. **474**(2209): p. 20170551.
570. LaRose, R., *Overview and Comparison of Gate Level Quantum Software Platforms.* Quantum, 2019. **3**: p. 130.
571. Amazon, *Amazon Braket Pricing.* 2025.
572. Ding, Y., et al., *Toward Prediction of Financial Crashes With a D-Wave Quantum Annealer.* Entropy, 2023. **25**(2): p. 323.
573. Phillipson, F. and H. Bhatia, *Portfolio Optimisation Using the D-Wave Quantum Annealer.* 2020.
574. Mugel, S., et al., *Dynamic Portfolio Optimization With Real Datasets Using Quantum Processors and Quantum-Inspired Tensor Networks.* 2020.
575. Hu, F., et al., *Quantum Computing Cryptography: Finding Cryptographic Boolean Functions With Quantum Annealing by a 2000 Qubit D-Wave Quantum Computer.* Physics Letters A, 2020. **384**(10): p. 126214.
576. Subramanian, A.M., M. Lévesque, and V.v.d. Vrande, *"Pulling the Plug:" Time Allocation Between Drug Discovery and Development Projects.* Production and Operations Management, 2020. **29**(12): p. 2851-2876.
577. Bauer, J., et al., *How Can We Discover Developable Antibody-Based Biotherapeutics?* Frontiers in Molecular Biosciences, 2023. **10**.
578. Liu, G., et al., *Drug Research and Development Opportunities in Low- And Middle-Income Countries: Accelerating Traditional Medicine Through Systematic Utilization and Comprehensive Synergy.* Infectious Diseases of Poverty, 2022. **11**(1).
579. Das, P., A. Banerjee, and P. Karak, *Molecular Interplay Between Vitamin D and Immunity Can Aid Antitubercular Treatment Vitamin D in Immunomodulation of TB.* Defence Life Science Journal, 2022. **7**(3): p. 209-220.
580. Li, W. and G. Vottevor, *Towards a Truly General Intermolecular Binding Affinity Calculator for Drug Discovery &Amp; Design.* 2023.
581. Beale, S.J., et al., *Quantum Error Correction Decoheres Noise.* Physical Review Letters, 2018. **121**(19).
582. Nejabati, R., R. Wang, and D. Simeonidou, *Dynamic Quantum Network: From Quantum Data Centre to Quantum Cloud Computing.* 2022: p. Th3D.1.

Quantum Computing

583. Sarovar, M., et al., *Detecting Crosstalk Errors in Quantum Information Processors*. Quantum, 2020. **4**: p. 321.
584. Ghosh, D., et al., *Automated Error Correction in IBM Quantum Computer and Explicit Generalization*. Quantum Information Processing, 2018. **17**(6).
585. Moll, N., et al., *Quantum Optimization Using Variational Algorithms on Near-Term Quantum Devices*. Quantum Science and Technology, 2018. **3**(3): p. 030503.
586. Nguyen, H., M. Usman, and R. Buyya, *iQuantum: A Case for Modeling and Simulation of Quantum Computing Environments*. 2023.
587. Ma, L. and L. Ding, *Hybrid Quantum Edge Computing Network*. 2022.
588. Leymann, F., et al., *Quantum in the Cloud: Application Potentials and Research Opportunities*. 2020.
589. Zhang, F., et al., *Alibaba Cloud Quantum Development Platform: Large-Scale Classical Simulation of Quantum Circuits*. 2019.
590. Kaiiali, M., S. Sezer, and A. Khalid, *Cloud Computing in the Quantum Era*. 2019.
591. Takeda, K., et al., *Rapid Single-Shot Parity Spin Readout in a Silicon Double Quantum Dot With Fidelity Exceeding 99%*. NPJ Quantum Information, 2024. **10**(1).
592. Huang, C., et al., *Alibaba Cloud Quantum Development Platform: Surface Code Simulations With Crosstalk*. 2020.
593. Ouyang, Y. and P.P. Rohde, *A General Framework for the Composition of Quantum Homomorphic Encryption \&Amp; Quantum Error Correction*. 2022.
594. Nguyen, H., M. Usman, and R. Buyya, *iQuantum: A Toolkit for Modeling and Simulation of Quantum Computing Environments*. Software Practice and Experience, 2024. **54**(6): p. 1141-1171.
595. Dong, Y., et al., *Quantum Color Image Watermarking Scheme Based on Quantum Error Correction Coding [*]*. New Journal of Physics, 2023. **25**(11): p. 113016.
596. Wootton, J.R., *Benchmarking of Quantum Processors With Random Circuits*. 2018.
597. Arute, F., et al., *Quantum Supremacy Using a Programmable Superconducting Processor*. Nature, 2019. **574**(7779): p. 505-510.
598. Zeng, W., et al., *First Quantum Computers Need Smart Software*. Nature, 2017. **549**(7671): p. 149-151.
599. Steinbrecher, G.R., et al., *Quantum Optical Neural Networks*. NPJ Quantum Information, 2019. **5**(1).
600. Wootton, J.R. and D. Loss, *Repetition Code of 15 Qubits*. Physical Review A, 2018. **97**(5).
601. Bunyk, P., et al., *Architectural Considerations in the Design of a Superconducting Quantum Annealing Processor*. Ieee Transactions on Applied Superconductivity, 2014. **24**(4): p. 1-10.
602. Kortum, S. and J. Lerner, *Assessing the Contribution of Venture Capital to Innovation*. The Rand Journal of Economics, 2000. **31**(4): p. 674.
603. Katila, R., J.D. Rosenberger, and K.M. Eisenhardt, *Swimming With Sharks: Technology Ventures, Defense Mechanisms and Corporate Relationships*. Administrative Science Quarterly, 2008. **53**(2): p. 295-332.

604. Kupp, M., M. Marval, and P. Borchers, *Corporate Accelerators: Fostering Innovation While Bringing Together Startups and Large Firms.* Journal of Business Strategy, 2017. **38**(6): p. 47-53.
605. Marican, M.N.Y., et al., *Quantifying the Return of Security Investments for Technology Startups.* Baghdad Science Journal, 2024. **21**(7): p. 2449.
606. Weiblen, T. and H. Chesbrough, *Engaging With Startups to Enhance Corporate Innovation.* California Management Review, 2015. **57**(2): p. 66-90.
607. Köhler, T., *Corporate Accelerators: Building Bridges Between Corporations and Startups.* Business Horizons, 2016. **59**(3): p. 347-357.
608. Steiber, A. and S. Alänge, *Corporate-Startup Collaboration: Effects on Large Firms' Business Transformation.* European Journal of Innovation Management, 2020. **24**(2): p. 235-257.
609. Miyamoto, H., C. Mejía, and Y. Kajikawa, *A Study of Private Equity Rounds of Entrepreneurial Finance in EU: Are Buyout Funds Uninvited Guests for Startup Ecosystems?* Journal of Risk and Financial Management, 2022. **15**(6): p. 236.
610. Conti, A., M. Thursby, and F.T. Rothaermel, *Show Me the Right Stuff: Signals for High-Tech Startups.* Journal of Economics & Management Strategy, 2013. **22**(2): p. 341-364.
611. Schiffer, B.F., *Quantum Computers as an Amplifier for Existential Risk.* 2022.
612. Gill, S.S., et al., *Quantum Computing: A Taxonomy, Systematic Review and Future Directions.* Software Practice and Experience, 2021. **52**(1): p. 66-114.
613. Jenkins, J., N. Berente, and C.M. Angst, *The Quantum Computing Business Ecosystem and Firm Strategies.* 2022.
614. Umadevi, B. and D. Sundar, *Exploring Quantum Computing Software Platforms: A Comprehensive Study for Research and Development.* International Research Journal of Modernization in Engineering Technology and Science, 2024.
615. Wang, Y., *CsPbX3 Perovskite Quantum Dot Laser.* Highlights in Science Engineering and Technology, 2022. **27**: p. 334-342.
616. Huang, J.Y., et al., *High-Fidelity Spin Qubit Operation and Algorithmic Initialization Above 1 K.* Nature, 2024. **627**(8005): p. 772-777.
617. Eskandarpour, R., et al., *Quantum-Enhanced Grid of the Future: A Primer.* Ieee Access, 2020. **8**: p. 188993-189002.
618. Tychola, K.A., T. Kalampokas, and G.A. Papakostas, *Quantum Machine Learning—An Overview.* Electronics, 2023. **12**(11): p. 2379.
619. Raheman, F., *The Q-Day Dilemma and the Quantum Supremacy/Advantage Conjecture.* 2022.
620. Khan, K., *Quantum Data Science: Leveraging Data Analytics for Advancing Quantum Computing.* 2024.
621. Emmanni, P.S., *The Impact of Quantum Computing on Cybersecurity.* Journal of Mathematical & Computer Applications, 2023. **2**(2): p. 1-4.
622. Umar, D., *Cybersecurity Threats and Mitigation Strategies in the Age of Quantum Computing.* Journal of Technology and Systems, 2024. **6**(5): p. 1-14.

Quantum Computing

623. Daniel, S.A. and S.S. Victor, *Emerging Trends in Cybersecurity for Critical Infrastructure Protection: A Comprehensive Review.* Computer Science & It Research Journal, 2024. **5**(3): p. 576-593.
624. Ghaib, A.A., *Future Trends in Cybersecurity: Exploring Emerging Technologies and Strategies.* International Research Journal of Modernization in Engineering Technology and Science, 2024.
625. Li, Y., B. Li, and H. Sun, *Research on the Application of Quantum Communication in Intelligent and Connected Vehicle Cybersecurity.* 2023: p. 11.
626. Elendu, C., et al., *Legal Implications for Clinicians in Cybersecurity Incidents: A Review.* Medicine, 2024. **103**(39): p. e39887.
627. Jaipong, P., et al., *A Review of Metaverse and Cybersecurity in the Digital Era.* International Journal of Computing Sciences Research, 2023. **7**: p. 1125-1132.
628. Atadoga, A., et al., *Quantum Computing in Big Data Analytics: A Comprehensive Review: Assessing the Advancements, Challenges, and Potential Implications of Quantum Approaches in Handling Massive Data Sets.* Computer Science & It Research Journal, 2024. **5**(2): p. 498-517.
629. How, M.-L., *Advancing Multidisciplinary STEM Education With Mathematics for Future-Ready Quantum Algorithmic Literacy.* Mathematics, 2022. **10**(7): p. 1146.
630. Seskir, Z.C., R. Korkmaz, and A.U. Aydınoğlu, *The Landscape of the Quantum Start-Up Ecosystem.* Epj Quantum Technology, 2022. **9**(1).
631. Mallow, G.M., et al., *Quantum Computing: The Future of Big Data and Artificial Intelligence in Spine.* Spine Surgery and Related Research, 2022. **6**(2): p. 93-98.
632. Wilhelm, F.K., F. Struckmeier, and W. Maaß, *Introduction to the Minitrack on Quantum Computing Applications.* 2024.
633. Wang, Y., et al., *Quantum Computing in Community Detection for Anti-Fraud Applications.* Entropy, 2024. **26**(12): p. 1026.
634. Bayerstadler, A., et al., *Industry Quantum Computing Applications.* Epj Quantum Technology, 2021. **8**(1).
635. Baker, J.M., et al., *Time-Sliced Quantum Circuit Partitioning for Modular Architectures.* 2020: p. 98-107.
636. Ovide, A., et al., *Mapping Quantum Algorithms to Multi-Core Quantum Computing Architectures.* 2023.
637. Huang, H.-L., et al., *Superconducting Quantum Computing: A Review.* Science China Information Sciences, 2020. **63**(8).
638. Zhang, Z. and Q. Zhuang, *Distributed Quantum Sensing.* Quantum Science and Technology, 2021. **6**(4): p. 043001.
639. Gyöngyösi, L. and S. Imre, *Resource Prioritization and Balancing for the Quantum Internet.* Scientific Reports, 2020. **10**(1).
640. Gyöngyösi, L. and S. Imre, *Scalable Distributed Gate-Model Quantum Computers.* Scientific Reports, 2021. **11**(1).
641. Pirker, A. and W. Dür, *A Quantum Network Stack and Protocols for Reliable Entanglement-Based Networks.* New Journal of Physics, 2019. **21**(3): p. 033003.

642. Xu, M., et al., *Privacy-Preserving Intelligent Resource Allocation for Federated Edge Learning in Quantum Internet.* Ieee Journal of Selected Topics in Signal Processing, 2023. **17**(1): p. 142-157.
643. Rodrigo, S., et al., *On Double Full-Stack Communication-Enabled Architectures for Multicore Quantum Computers.* Ieee Micro, 2021. **41**(5): p. 48-56.
644. Jafarizadeh, S., *Optimizing the Convergence Rate of the Quantum Consensus: A Discrete-Time Model.* Automatica, 2016. **73**: p. 237-247.
645. Johnson, H., *Exploration of Optimizing FPGA-based Qubit Controller for Experiments on Superconducting Quantum Computing Hardware.* 2023.
646. Alghamdi, W., et al., *Quantum Computing: Algorithms,Architectures, and Applications.* E3s Web of Conferences, 2023. **399**: p. 04041.
647. Farhi, E. and H. Neven, *Classification With Quantum Neural Networks on Near Term Processors.* 2018.
648. Caves, C.M., I. Deutsch, and R. Blume-Kohout, *Physical-Resource Requirements and the Power of Quantum Computation.* Journal of Optics B Quantum and Semiclassical Optics, 2004. **6**(8): p. S801-S806.
649. Lloyd, S., *Ultimate Physical Limits to Computation.* Nature, 2000. **406**(6799): p. 1047-1054.
650. Jozsa, R. and N. Linden, *On the Role of Entanglement in Quantum-Computational Speed-Up.* Proceedings of the Royal Society a Mathematical Physical and Engineering Sciences, 2003. **459**(2036): p. 2011-2032.
651. Boixo, S., et al., *Characterizing Quantum Supremacy in Near-Term Devices.* Nature Physics, 2018. **14**(6): p. 595-600.
652. Wilson, E., S. Singh, and F. Mueller, *Just-in-Time Quantum Circuit Transpilation Reduces Noise.* 2020.
653. Blume-Kohout, R., C.M. Caves, and I. Deutsch, *Untitled.* Foundations of Physics, 2002. **32**(11): p. 1641-1670.
654. Kissinger, A. and A.M.-v.d. Griend, *CNOT Circuit Extraction for Topologically-Constrained Quantum Memories.* 2019.
655. Kukliansky, A., et al., *Network Anomaly Detection Using Quantum Neural Networks on Noisy Quantum Computers.* Ieee Transactions on Quantum Engineering, 2024. **5**: p. 1-11.
656. Reilly, D.J., *Challenges in Scaling-Up the Control Interface of a Quantum Computer.* 2019: p. 31.7.1-31.7.6.
657. Philathong, H., et al., *Computational Phase Transition Signature in Gibbs Sampling.* 2019.
658. Gheorghiu, V., et al., *Reducing the CNOT Count for Clifford+T Circuits on NISQ Architectures.* 2020.
659. Fenner, S., *A Physics-Free Introduction to the Quantum Computation Model.* 2004: p. 125-145.
660. Nam, Y., et al., *Automated Optimization of Large Quantum Circuits With Continuous Parameters.* NPJ Quantum Information, 2018. **4**(1).
661. Wehner, S., D. Elkouss, and R. Hanson, *Quantum Internet: A Vision for the Road Ahead.* Science, 2018. **362**(6412).

Quantum Computing

662. Dahlberg, A. and S. Wehner, *SimulaQron—a Simulator for Developing Quantum Internet Software.* Quantum Science and Technology, 2018. **4**(1): p. 015001.
663. Cacciapuoti, A.S., et al., *Quantum Internet: Networking Challenges in Distributed Quantum Computing.* Ieee Network, 2020. **34**(1): p. 137-143.
664. Cacciapuoti, A.S., et al., *When Entanglement Meets Classical Communications: Quantum Teleportation for the Quantum Internet.* Ieee Transactions on Communications, 2020. **68**(6): p. 3808-3833.
665. Cuomo, D., M. Caleffi, and A.S. Cacciapuoti, *Towards a Distributed Quantum Computing Ecosystem.* 2020.
666. Andrade, M.G.d., et al., *A Quantum Walk Control Plane for Distributed Quantum Computing in Quantum Networks.* 2021.
667. Bluvstein, D., et al., *Logical Quantum Processor Based on Reconfigurable Atom Arrays.* Nature, 2023. **626**(7997): p. 58-65.
668. Ghosal, A., et al., *Optimal Two-Qubit States for Quantum Teleportation Vis-À-Vis State Properties.* Physical Review A, 2020. **101**(1).
669. Lee, Y., et al., *A Quantum Router Architecture for High-Fidelity Entanglement Flows in Quantum Networks.* NPJ Quantum Information, 2022. **8**(1).
670. Laghari, A.A., et al., *A Review on Quantum Computing Trends &Amp; Future Perspectives.* Eai Endorsed Transactions on Cloud Systems, 2018: p. 173979.
671. Sonko, S., et al., *Quantum Cryptography and U.S. Digital Security: A Comprehensive Review: Investigating the Potential of Quantum Technologies in Creating Unbreakable Encryption and Their Future in National Security.* Computer Science & It Research Journal, 2024. **5**(2): p. 390-414.
672. Chen, S.Y.-C. and S. Yoo, *Federated Quantum Machine Learning.* Entropy, 2021. **23**(4): p. 460.
673. Majdoubi, C., S.E. Mendili, and Y. Gahi, *Quantum Cryptology in the Big Data Security Era.* International Journal of Advanced Computer Science and Applications, 2024. **15**(7).
674. Farooq, S., et al., *Resilience Optimization of Post-Quantum Cryptography Key Encapsulation Algorithms.* Sensors, 2023. **23**(12): p. 5379.
675. Bhowmik, D., et al., *Some New Classes of Homological Quantum Codes Associated With Surface Maps.* National Academy Science Letters, 2021. **45**(1): p. 83-86.
676. Chen, J., et al., *Lattice-Based Threshold Secret Sharing Scheme and Its Applications: A Survey.* Electronics, 2024. **13**(2): p. 287.
677. Mollick, E., *Establishing Moore's Law.* Ieee Annals of the History of Computing, 2006. **28**(3): p. 62-75.
678. Deng, L. and E. Williams, *Measures and Trends in Energy Use of Semiconductor Manufacturing.* 2008.
679. Tuomi, I., *The Lives and Death of Moore's Law.* First Monday, 2002.
680. Suzuki, A., *Advances in Optics and Exposure Devices Employed in Excimer Laser/Euv Lithography.* 2020: p. 1-42.
681. Wada, K. and Y. Ishikawa, *(Invited) Challenges of Si Photonics for Photonics Integration on Si LSIs.* Ecs Transactions, 2011. **41**(7): p. 265-271.
682. Mack, C.A., *The End of the Semiconductor Industry as We Know It.* 2003.

683. Wu, Z., *Microfluidic Devices Using Flexible Organic Electronic Materials.* 2015: p. 397-412.
684. Zhang, X., *Software System Research in Post-Moore's Law Era: A Historical Perspective for the Future.* Science China Information Sciences, 2019. **62**(9).
685. Childs, A.M. and W.v. Dam, *Quantum Algorithms for Algebraic Problems.* Reviews of Modern Physics, 2010. **82**(1): p. 1-52.
686. Guo, C., et al., *General-Purpose Quantum Circuit Simulator With Projected Entangled-Pair States and the Quantum Supremacy Frontier.* Physical Review Letters, 2019. **123**(19).
687. Kim, D. and G. Cho, *Machine Learning on Quantum Experimental Data Toward Solving Quantum Many-Body Problems.* 2023.
688. Li, J., et al., *Hybrid Quantum-Classical Approach to Quantum Optimal Control.* Physical Review Letters, 2017. **118**(15).
689. Colless, J., et al., *Computation of Molecular Spectra on a Quantum Processor With an Error-Resilient Algorithm.* Physical Review X, 2018. **8**(1).
690. Xiang, Q., et al., *Performance Analysis of Quantum Convolutional Layers for Image Classification.* Physica Scripta, 2024. **99**(12): p. 125120.
691. Li, Y., et al., *A Quantum Deep Convolutional Neural Network for Image Recognition.* Quantum Science and Technology, 2020. **5**(4): p. 044003.
692. Dam, T.J.v., et al., *Hybrid Helmholtz Machines: A Gate-Based Quantum Circuit Implementation.* Quantum Information Processing, 2020. **19**(6).
693. Chia, N.-H., K.-M. Chung, and C.Y. Lai, *On the Need for Large Quantum Depth.* 2019.
694. Liu, J., et al., *Hybrid Quantum-Classical Convolutional Neural Networks.* Science China Physics Mechanics and Astronomy, 2021. **64**(9).
695. Moll, M. and L. Kunczik, *Comparing Quantum Hybrid Reinforcement Learning to Classical Methods.* Human-Intelligent Systems Integration, 2021. **3**(1): p. 15-23.

Index

R

S

T

U

V

W

www.ingramcontent.com/pod-product-compliance
Lightning Source LLC
Chambersburg PA
CBHW081215220326
41598CB00037B/6783

* 9 7 8 1 7 6 3 8 8 1 1 7 4 *